N J A Sloane

THEORY AND APPLICATION OF SPECIAL FUNCTIONS

Publication No. 35
of the Mathematics Research Center
The University of Wisconsin

Theory and Application of Special Functions

Edited by Richard A. Askey

Proceedings of an Advanced Seminar
Sponsored by the Mathematics Research Center
The University of Wisconsin-Madison
March 31–April 2, 1975

Academic Press
New York · San Francisco · London   1975

A Subsidiary of Harcourt Brace Jovanovich, Publishers

ACADEMIC PRESS, INC.
111 Fifth Avenue, New York, New York 10003

United Kingdom Edition published by
ACADEMIC PRESS, INC. (LONDON) LTD.
24/28 Oval Road, London NW1

Library of Congress Cataloging in Publication Data

Advanced Seminar on Special Functions, Madison, Wis.,
    1975.
    Theory and application of special functions.

    (Publication of the Mathematics Research Center, the
University of Wisconsin ; no. 35)
        Bibliography: p.
        Includes index.
        1.   Functions, Special—Congresses.   I.   Askey,
Richard.   II.   Wisconsin.   University—Madison.   Mathe-
matics Research Center.   III.   Title.   IV.   Series:
Wisconsin.   University—Madison.   Mathematics Research
Center.   Publication ; 35.
QA3.U45   no. 35   [QA351] 510'.8s [515'.5]   75-30406
ISBN 0–12–064850–4

Contents

CONTENTS

Foreword

This volume contains the Proceedings of the Advanced Seminar on Special Functions held in Madison, Wisconsin, March 31-April 2, 1975, sponsored by the Mathematics Research Center, University of Wisconsin—Madison, with financial support from the National Science Foundation under Grant MP575-05571 and the United States Army under Contract No. DA-31-124-ARO-D-462. The five sessions were chaired by:

B. C. Carlson, *Iowa State University*
Yudell Luke, *University of Missouri, Kansas City*
A. Erdélyi, *University of Edinburgh*
Richard Askey, *University of Wisconsin, Madison*
W. J. Cody, Jr., *Argonne National Laboratory*

The program committee consisted of Professors L. Durand, J. Hirschfelder, and F. W. J. Olver, with the editor as chairman. Mrs. Gladys Moran handled all the organizational details in her usual efficient and pleasant way, yet even she could not make the beautiful weather of Monday last until Wednesday, when it was replaced by a spring snow storm which will be remembered by all those who spent an extra day in Madison or in a motel in the vicinity of Chicago. Mrs. Doris Whitmore did the typing of the manuscripts, and her skill will be evident to the reader.

Preface

Twenty years ago it was widely believed that the existence of large, fast, computing machines spelled the end of the study and use of special functions. Differential equations would be solved numerically, integrals would be evaluated numerically, and special functions would become a fossil. Many mathematicians acted on this belief, and special functions disappeared from their traditional place in the curriculum as part of a course in complex variables. They were often replaced in the mathematical physics course by a section on Hilbert space and functional analysis.

To paraphrase Mark Twain, reports of the death of special functions were premature. A sociologist of science would accept the following two facts as proof. Over 150,000 copies of the Handbook of Mathematical Functions have been sold. Two of the five most widely cited mathematics books (as measured by the citations in Science Citation Index) are this book and Higher Transcendental Functions by A. Erdélyi et al. Most of us are more concerned with quality than quantity, and so would not be convinced by these two facts. A more compelling proof is given in this book. Some interesting results, conjectures, and problems are given. In addition there are surprising connections with other fields.

J. J. Seidel told me about Delsarte's work on coding theory and Krawtchouk polynomials. My initial reaction was: What is coding theory? I still do not really understand it, but Sloane's article has given me some insight into a fascinating set of problems. These and related combinatorial problems will lead to important results on orthogonal polynomials in the next few years. Dunkl has used a finite group and the associated homogeneous space to find a deep addition formula for Krawtchouk polynomials, and in this formula there is a new set of orthogonal polynomials in two variables which is analogous to the disc polynomials described in Koornwinder's paper. Delsarte has made a start on the problem of finding an analogous space on which some of the q-orthogonal polynomials live. While some very deep results have been proved for q(or basic)-hypergeometric functions (see Andrews' paper) the results are not the type that come from considering orthogonal polynomials as spherical harmonics.

PREFACE

There are at least three ways to look at the classical orthogonal polynomials. One is to consider them as a special case of general orthogonal polynomials. From this point of view they can suggest interesting questions about general orthogonal polynomials. Case's paper can be looked at as showing that from the point of view of recurrence relations, general orthogonal polynomials look surprisingly like the classical polynomials. To be more precise, consider the Pollaczek polynomials which are described in the appendix of *Orthogonal Polynomials* by G. Szegö. If the Pollaczek polynomials which are orthogonal on (-1,1) are written in monic from (highest coefficient is one), then their recurrence relation takes the form $xp_n(x) = p_{n+1}(x) + a(n) p_n(x) + b(n) p_{n-1}(x)$, where $a(n)$ and $b(n)$ are certain rational functions satisfying $a(n) = A^{n-1} + O(n^{-2})$, $bn = \frac{1}{4} + Bn^{-1} + O(n^{-2})$. These Pollaczek polynomials are Jacobi polynomials if and only if $A = B = 0$, and it is exactly in this case when the weight function with respect to which they are orthogonal satisfies $\int \frac{|\log w(x)|}{(1-x^2)^{\frac{1}{2}}} dx < \infty$. Case shows that for an arbitrary $a(n) = O(n^{-2})$ and positive $b(n) = \frac{1}{4} + O(n^{-2})$, a similar conclusion is true.

A second way of looking at the classical orthogonal polynomials is to consider them as hypergeometric functions. Much of the work in Gasper's paper is of this sort. Many of the raising and lowering operators which played a prominent part in Miller's early work came from the differential-difference equations satisfied by hypergeometric functions. The limits of generality of this method have not been reached, especially when multiple series are considered, and the degree of complexity of some of the transformation formulas is almost overwhelming, yet some of these very complicated formulas are exactly what is needed to solve some interesting problems. Not only do these methods give results for the classical polynomials of Hermite, Laguerre, and Jacobi, which satisfy second order differential equations in x, but also for the classical discrete polynomials of Charlier, Krawtchouk, Meixner, and Hahn, which satisfy second order additive difference equations in x, and a class of q-polynomials studied by Hahn, which satisfy second order multiplicative difference equations in x. But the deeper formulas associated with the addition formlas do not arise in a natural way in this approach.

The third way to treat the classical orthogonal polynomials is to consider them as spherical functions on a homogeneous space. This is a very powerful method as is seen by the ease with which very deep addition formulas are obtained. However to get started one must find the right space on which the functions live, and this has not been done for all of these functions. There is much work to be done here, especially for the additive and multiplicative difference type orthogonal polynomials.

One striking fact about the talks at this meeting was the number of talks on orthogonal polynomials in several variables. When the Bateman Project was in operation twenty-five years ago there was very little known about orthogonal polynomials in several variables, and much of the chapter on orthogonal polynomials in several variables in *Higher Transcendental Functions, Vol. II*, deals

x

with biorthogonal sets of polynomials. It now looks like genuinely orthogonal sets of polynomials will be more interesting and useful than the classical biorthogonal sets, and it is clear that many decades will pass before the problems which have already come up are solved, and it is likely that only a start has been made in uncovering the right questions in this area.

The remaining papers are not as closely tied together as those mentioned above. Computational problems are still as improtant and interesting as when Jacob Bernoulli used the coefficients in a certain power series, which we now call Bernoulli numbers, to replace an elaborate set of tables of sums of the first six powers prepared by Ismail Bullialdus by a single page of work, or when Gauss calculated the zeros of low degree Lengendre polynomials to use them to approximate some integrals by a method we call Gaussian quadrature. Gautschi's survey gives an indication of the current interest in this field. It is extensive.

Berndt shows some of the consequences that can be obtained by use of Bernoulli numbers and some of their generalizations. This paper illustrates one surprising fact of special functions. They are often useful in areas disjoint from the one in which they were originally introduced. Bernoulli did not have number theory in mind when he introduced Bernoulli numbers, yet that is one of the places where they have been very useful. They have recently been used in algebraic topology, and the slight generalization to Bernoulli polynomials has been very useful in studying spline functions.

Olver's paper complements his recent book, *Asymptotics and Special Functions* in a very nice way. The book provides an up to date treatment of the methods used to find asymptotic expansions of special functions, and this paper contains many problems which need to be worked on.

There are many other uses of special functions which could not be surveyed in a meeting of three days, but a reader with an open eye will see them in most areas where mathematics is used. There was one other paper presented at this meeting which is not included here, since the author was unable to find time in his busy schedule to pull his notes together to make a paper he was willing to publish now. The talk was "Some relationships between commutative algebra and special functions" by Gian-Carlo Rota. I am particularly sorry about this omission, since the talk was very interesting and so are the notes which the author has sent me. I hope his paper will appear elsewhere in the not too distant future.

Computational Methods in Special Functions–A Survey
Walter Gautschi

Introduction

§4.   Computer software for special functions

   4.1 NATS software for special functions

   4.2 NAG software for special functions

   4.3 Other software for special functions

Introduction

   Scientific computing often requires special functions. In the past, the need for numerical values was partly satisfied by extensive mathematical tables. Today, with powerful digital computers available, such values are obtained almost invariably by direct computation. We wish to review here the principal methods used in computing special functions.

   We may group these methods into two large classes, namely those based on direct approximation, and those based on functional equations. Among the former, we consider only rational approximation methods (§1). We thus leave aside a multitude of possible expansions in terms of other special functions. These expansions, indeed, while often helpful, still leave us with the problem of evaluating the special functions involved. Among the functional equations most useful for computation are linear and nonlinear recurrence relations. These are discussed in §§2 and 3. We omit references to other functional equations, such as differential and integral equations, since we consider them of secondary importance in our context. In §4 we give a brief account of the current state of computer software development for special functions.

   Due to limitations in time and space, a number of important topics are omitted in this survey. Nothing is said, e.g., about elementary functions and special computational techniques related to them. Good accounts of this can be found in Lyusternik, Chervonenkis and Yanpol'skii [1965] and Fike [1968]. Other topics not covered include methods based on numerical quadrature and on Euler-Maclaurin and Poisson summation formulas, the computation of zeros of special functions and of inverse functions, and the computation of special constants to very high precision.

Few references are given to computer algorithms for special functions, as they can be retrieved from the indices in the journal "Communications of the ACM" and in "Collected Algorithms from CACM" (a looseleaf collection issued by ACM of all algorithms published in Comm. ACM since 1960). Another topic dealt with only superficially are asymptotic methods, as these are discussed more fully elsewhere in this volume.

There are not many general references on computational methods for special functions. The only book devoted entirely to this subject is Hart et al. [1968]. The two volumes of Luke [1969] also contain much relevant material, and informative survey articles have been written by Bulirsch and Stoer [1968] and Thacher [1969].

As to notations for special functions, we try to be consistent with Abramowitz and Stegun [1964]. With regard to bibliographic references, we give special emphasis to the literature of the past twenty years or so. Little attempt has been made to trace all results back to the original sources.

§1.   Methods based on preliminary approximation

Our concern in this paragraph is with the approximation of a given function of a real or complex variable by means of "simpler" functions. Most attractive among these simpler functions are polynomials and rational functions, since they can be evaluated by a finite number of rational operations. Hence we restrict ourselves to polynomial and rational approximation. One should keep in mind, however, that other means of approximation, e.g., expansions in special functions like Bessel functions, can be equally effective if one takes advantage of appropriate recursive schemes of computation. (cf., e.g., 1.5.3, 2.2.2.)

The selection of a particular rational approximation depends on a number of circumstances. If the region of interest is an interval on the real line and our objective is to produce an approximation of high efficiency, and if we are prepared to expend the necessary effort, then we may seek to obtain a best rational approximation, i.e., one whose maximum

4

error on the interval in question is as small as possible. This is often
the preferred choice in computer subroutines. If, on the other hand, we
are dealing with functions of a complex variable, or functions of several
variables, we are led to use analytic approximation methods, the constructive theory of best approximation in the multivariate case still being
in its infancy. (See, e.g., Collatz [1968], Williams [1972], Harris [1973],
Fletcher, Grant and Hebden [1974], Watson [1975].) Even if we decide to
construct a best approximation, in the process of doing so we still need
to be able to calculate the function to high accuracy. Here again, analytic methods can be useful.

   With regard to polynomial vs. rational approximation, folklore has
it that "in some overall sense, rational approximation is essentially no
better than polynomial approximation" (Newman [1964]). Precise theorems
to this effect (Walsh [1968b], Feinerman and Newman [1974, p. 71 ff]) add
further support to this contention. Experience, nevertheless, seems to
show that for the special functions encountered in everyday practice,
rational approximations are in fact somewhat superior.

   In designing a rational approximation, certain preliminary decisions
need to be made regarding the best form in which to approximate the function, the choice of auxiliary variables, and the best type of segmentation
of the independent variable. As there is little theory to go by, such decisions are usually made by trial and error. Taylor series, or asymptotic
expansions, usually suggest appropriate forms. For the problem of segmentation, see Lawson [1964], Collatz [1965], Meinardus [1966], [1964,
§11 of English translation], Hawkins [1972].

1.1. Best rational approximation

   Many computer subroutines for special functions employ rational
approximations in appropriate segments of the real line. If the subroutine operates in an environment in which every value of the independent variable is equally likely to occur, it is natural to design the approximation in such a way that the error on each segment is "uniformly

distributed", and about the same from segment to segment. In this way, no user is going to be punished if he happens to prefer one particular region over another. The logical conclusion of this philosophy is to employ the <u>principle of best uniform approximation</u> (Chebyshev approximation) on each segment and to arrange the maximum error to be about the same from segment to segment. The "uniform distribution" of the error is then guaranteed by the equi-oscillation property of the best approximation (cf. 1.1.1).

The theory of best uniform approximation is an important chapter of approximation theory, and is dealt with in a number of excellent books. We mention, e.g., Achieser [1956], Davis [1963], Meinardus [1964], Natanson [1964], Rice [1964b], [1969], Cheney [1966], Werner [1966], Rivlin [1969], Walsh [1969], Schönhage [1971], Feinerman and Newman [1974]. A treatise on numerical methods of Chebyshev approximation (not including, however, rational approximation) is Remez [1969]. Practical aspects of generating rational and polynomial approximations are reviewed by Cody [1970].

1.1.1. <u>Best uniform rational approximation.</u> We denote by $\mathbb{P}_n$ the class of polynomials of degree $\leq n$, and by $\mathbb{R}_{m,n}$ the family of rational functions

(1) $$r(x) = \frac{p(x)}{q(x)}, \quad p \in \mathbb{P}_n, \quad q \in \mathbb{P}_m, \quad q \not\equiv 0 .$$

Given a real-valued continuous function f on the compact interval $[a, b]$, there exists a unique element $r^*_{m,n} \in \mathbb{R}_{m,n}$ such that

(2) $$\|r^*_{m,n} - f\|_\infty \leq \|r - f\|_\infty \quad \text{for all } r \in \mathbb{R}_{m,n} .$$

Here the norm is $\|u\|_\infty = \max_{a \leq x \leq b} |u(x)|$ or, more generally, $\|u\|_\infty = \max_{a \leq x \leq b} w(x)|u(x)|$, where w is a positive weight function. One calls

$r_{m,n}^*$ the <u>rational function of best uniform approximation</u> to f from $\mathbb{R}_{m,n}$ (or briefly the rational <u>Chebyshev approximation</u> of f from $\mathbb{R}_{m,n}$). The associated error is denoted by

(3)
$$E_{m,n}(f) = \|r_{m,n}^* - f\|_\infty \ .$$

In particular, there is a unique polynomial $p_n^* \in \mathbb{P}_n$ of best uniform approximation, with associated error $E_n(f) = E_{0,n}(f)$. The array of rational functions

$$
\begin{array}{cccc}
r_{0,0}^* & r_{0,1}^* & r_{0,2}^* & \cdots \\[2mm]
r_{1,0}^* & r_{1,1}^* & r_{1,2}^* & \cdots \\[2mm]
r_{2,0}^* & r_{2,1}^* & r_{2,2}^* & \cdots \\[2mm]
\cdot\ \cdot\ \cdot\ \cdot\ \cdot\ \cdot\ \cdot\ \cdot\ \cdot\ \cdot\ \cdot
\end{array}
$$

is referred to as the L_∞ <u>Walsh array</u> of f on $[a,b]$.

The best approximation $r_{m,n}^*$ is characterized by the <u>equi-oscil-lation property</u>, which states (certain degenerate cases) that the error curve $w(r_{m,n}^* - f)$ assumes its extreme value (3) at $m+n+2$ consecutive points of $[a,b]$ with alternating signs (Achieser [1956, p. 55]). Moreover (barring again degeneracies), if $r \in \mathbb{R}_{m,n}$ is any rational function bounded on $[a,b]$ which has the <u>oscillation property</u>, i.e., an error curve $e = w(r-f)$ assuming values of alternating sign on $m+n+2$ consecutive points $x_i \in [a,b]$, say,

$$e(x_i) = (-1)^i \lambda_i, \quad \lambda_i > 0, \quad i = 1, 2, \ldots, m+n+2 \ ,$$

then (Achieser [1956, p. 52])

(4)
$$\min_i \lambda_i \le E_{m,n}(f) \le \|e\|_\infty \ .$$

7

Concerning the behavior of $E_{m,n}(f)$ as m and n both tend to infinity, little is known. If m, or n, remains fixed, there are asymptotic results for meromorphic functions, due to Walsh [1964b], [1965], [1968a], while in the polynomial case $m = 0$ one has the classical results of Jackson and Bernstein. The former states that $E_n(f) = o(n^{-p})$ if $f \in C^p[a, b]$, the latter that $\limsup [E_n(f)]^{1/n} < 1$ precisely if f is analytic on $[a, b]$, and $[E_n(f)]^{1/n} = o(1)$ precisely if f is entire (see, e.g., Natanson [1964, pp. 127, 183]).

1.1.2. <u>A list of available Chebyshev approximations</u>. Some entries of the Walsh array, often those along or near the diagonal $m = n$, have proven to yield remarkably efficient approximations for many of the special functions in current use. Table 1 lists those for which (numerically constructed) rational Chebyshev approximations are available. The first column shows the function being approximated, in the notation of Abramowitz and Stegun [1964]. The second column records the segmentation used, where $[a_0, a_1, \ldots, a_s]$ is written to indicate that the interval $[a_0, a_s]$ is broken up into segments $[a_{i-1}, a_i]$, $i = 1, 2, \ldots, s$. The exact form of the function which is being approximated, as well as the type (m, n) of rational function, usually changes from segment to segment in a manner not shown in the table. The third column tells whether the approximant is truly rational or polynomial. The fourth column indicates the approximate range of accuracy, where S is to be read as "significant decimal digits" and D as "decimal digits after the decimal point". The final column gives the source of the approximation. For an extensive bibliography of approximations see also Hart et al. [1968, pp. 161-179].

Table 1. Chebyshev approximations to special functions

f(x)	segmentation	type	accuracy	reference
$E_1(x)$	$[0, 1, 4, \infty]$	rat.	2-20S	Cody & Thacher [1968]
Ei(x)	$[0, 6, 12, 24, \infty]$	rat.	3-20S	Cody & Thacher [1969]

f(x)	segmentation	type	accuracy	reference
$\Gamma(x)$	$[2,3]$	pol.	7-18S	Werner & Collinge [1961]
$\ln\Gamma(x)$	$[.5, 1.5, 4, 12]$	rat.	2-17S	Cody & Hillstrom [1967]
$\Gamma(x)$	$[2,3]^{(1)}$	pol.&rat.	1-24D	Hart et al. [1968]
$\ln\Gamma(x)-(x-\frac{1}{2})\ln x$ $+ x - \ln\sqrt{2\pi}$	$[8, 1000]$	"	8-18D	"
"	$[12, 1000]$	"	9-23D	"
$\arg\Gamma(1+ix)$	$[0, 2, 4, \infty]$	rat.	4-20S	Cody & Hillstrom [1970]
$\psi(x)$	$[1, 2]$	pol.	6-8D	Moody [1967]
$\psi(x)$	$[.5, 3, \infty]$	rat.	2-20S	Cody, Strecok &Thacher [1973]
erfc x	$[0, 10]$	rat.	1-23D	Hart et al. [1968]
"	$[0, 20]$	"	4-6D	"
"	$[0, 4]$	"	1-9D	"
"	$[0, 8]$	"	1-16S	"
"	$[8, 100]$	"	3-17S	"
erf x	$[0, .5]$	rat.	5-19S	Cody [1969]
erfc x	$[.46875, 4, \infty]$	rat.	2-18S	"
$e^{-x^2}\int_0^x e^{t^2} dt$	$[0, 2.5, 3.5, 5, \infty]$	rat.	1-21S	Cody, Paciorek &Thacher [1970]
$C(x),\ S(x)$	$[0, 1.2, 1.6, 1.9, 2, 4, \infty]$	rat.	2-18S	Cody [1968]
$J_n(x), I_n(x), Y_n(x), K_n(x)$ $n=0,1$	$[0, 8]$	pol.	2-7D	Werner [1958/59]
$J_\nu(x), I_\nu(x)$ $\nu = -\frac{2}{3}, -\frac{1}{3}, \frac{1}{3}, \frac{2}{3}$	$[0, 4]$	pol.	10D	Bhagwandin [1962]
$K_\nu(x),\ \nu = \frac{1}{3}, \frac{2}{3}$	$[4, \infty]$	pol.	10D	"
$H_\nu^{(1)}(x),\ \nu = \frac{1}{3}, \frac{2}{3}$	$[4, \infty]$	rat.	10D	"
$I_0(x), I_1(x)$	$[0, 8, 70]$	rat.	8S	Gargantini [1966]
$K_0(x)$	$[0, .1, 8]$	rat.	8S	"

f(x)	segmentation	type	accuracy	reference
$K_1(x)$	$[0, 8]$	rat.	7S	Gargantini [1966]
$J_0(x), J_1(x), Y_0(x), Y_1(x)$	$[0, 8, \infty]$	rat.	3-25D	Hart et al. [1968]
$I_0(x), I_1(x)$	$[0, 1]$	rat.	2-23S	Russon & Blair [1969]
$K_0(x), K_1(x)$	$[0, 1, \infty]$	rat.	2-23S	"
$I_0(x), I_1(x)$	$[0, 15, \infty]$	rat.	8-23S	Blair [1974]
$I_0(x), I_1(x)$	$[0, 15, \infty]$	rat.	1-23S	Blair & Edwards [1974]
$Ki_r(x)$, r=1, 2, 3	$[0, \infty]$	rat.	2-7S	Gargantini & Pomentale [1964]
$\int_0^x I_0(t)dt$	$[0, 8, 30]$	rat.	8-9S	Gargantini [1966]
$\int_x^\infty K_0(t)dt$	$[0, .1, 8, 70]$	rat.	7S	"
$G_0(\eta, 2\eta), G_0'(\eta, 2\eta)$	$[1, 2, 3.5, 15]$	rat.	13-14S	Strecok & Gregory [1972]
$G_0(\eta, 1), G_0'(\eta, 1)$	$[0, 1]$	rat.	16S	"
$G_0(\eta, 30), G_0'(\eta, 30)$	$[15, 18.5, 22]$	rat.	13-14S	"
$\ell n(G_0(\eta, 30)),$ $\ell n(-G_0'(\eta, 30))$	$[22, 30]$	rat.	13S	"
$\int_0^{\pi/2} (1-x^2 \sin^2 t)^{\pm \frac{1}{2}} dt$	$[0, 1]$	pol.	4-17D	Cody [1965]
$\zeta(x)$	$[.5, 5, 11, 25, 55]$	rat.	8-22S	Cody, Hillstrom & Thacher [1971]
$x^{-1} \int_0^x t^k (e^t-1)^{-1} dt$ k = 1, 2, 3, 4	$[0, 10]$	rat.	2-5D	Thacher [1960]
$\int_0^\infty t^{\frac{1}{2}} (e^{t-x}+1)^{-1} dt$	$[-\infty, 1, \infty]$	pol.	3S	Werner & Raymann [1963]
$\int_0^\infty t^\nu (e^{t-x}+1)^{-1} dt$ $\nu = -\frac{1}{2}, \frac{1}{2}, \frac{3}{2}$	$[-\infty, 1, 4, \infty]$	rat.	2-10S	Cody & Thacher [1967]

(1) range incorrectly stated in Hart et al. [1968].

1.1.3. Computation of Chebyshev approximations. Most, if not all, of the approximations in Table 1 were generated by some version of Remez' second algorithm. This is a procedure, originally devised for polynomials (Remes [1934]) and later extended to rational functions, which attempts to achieve the equi-oscillation property in an iterative fashion. The object of the iteration, basically, is to move the two bounds in (4) ever closer together. There are many variants of the procedure, differing somewhat in the technical execution of each iteration step. Detailed descriptions of these can be found in some books on approximation theory, e.g., Meinardus [1964], Rice [1964b], Cheney [1966], Werner [1966], Remez [1969], Rivlin [1969], as well as in survey articles by Cheney and Southard [1963], Stiefel [1959], [1964], Fraser [1965], Ralston [1967], Krabs [1969], Cody [1970]. Computer algorithms are given in Stoer [1964], Werner [1966], Cody and Stoer [1966/67], Werner, Stoer and Bommas [1967], Cody, Fraser and Hart [1968], Huddleston [1972], Johnson and Blair [1973]. The construction of rational Chebyshev approximants, in spite of the many aids available, is still a tricky business due to the possibility of near-degeneracies. For a discussion of this, the reader is referred to Rice [1964a], Cody [1970], Huddleston [1972], Ralston [1973].

There are other methods of obtaining best rational approximations which rely more heavily on mathematical programming. Some of these are referenced in Lee and Roberts [1973] and compared there with Remez' algorithm. Others, more recently, are proposed by Har-El and Kaniel [1973] and Kaufman and Taylor [1974].

1.2. Truncated Chebyshev expansion

There is some effort involved in generating a best rational, or even polynomial, approximation to a given function f. Polynomials which approximate f "nearly best" can be obtained more easily by truncating the Chebyshev expansion of f.

Assuming that the interval of interest has been transformed to [-1, 1], we can formally expand f into a series of Chebyshev polynomials,

(1) $\qquad f(x) = \frac{1}{2}a_0 + \sum_{k=1}^{\infty} a_k T_k(x), \qquad -1 \leq x \leq 1$,

where

(2) $\qquad a_k = \frac{2}{\pi} \int_0^{\pi} f(\cos\theta)\cos k\theta\, d\theta, \qquad k = 0,1,2,\ldots$.

In effect, (1) is the Fourier cosine expansion of $f(\cos\theta)$. It converges uniformly and absolutely on $[-1,1]$ if $f \in C[-1,1]$ and $f' \in L_p[-1,1]$, $p > 1$ (Zygmund [1959, p. 242]). The polynomials referred to above are the partial sums of (1),

(3) $\qquad s_n(f;x) = \frac{1}{2}a_0 + \sum_{k=1}^{n} a_k T_k(x), \qquad n = 0,1,2,\ldots$.

The classical source on Chebyshev polynomials and their applications is Lanczos' introduction in National Bureau of Standards [1952]. More recent accounts can be found in the books of Fox and Parker [1968] and Rivlin [1974].

1.2.1. Convergence. The functions f encountered in practice are usually quite smooth, typically real-valued analytic on $[-1,1]$ and holomorphic in a domain of the complex plane enclosing the segment $[-1,1]$. If ϵ is the eccentricity of the largest ellipse, with foci at ± 1, in which f is holomorphic, then (1) converges like a geometric series with ratio $\epsilon/(1+\sqrt{1-\epsilon^2})$ (see, e.g., Werner [1966, §20], Rivlin [1974, p. 143]). For entire functions one has $\epsilon = 0$, and the convergence is supergeometric.

Scraton [1970] observes that convergence can be enhanced if one uses a suitable bilinear, rather than linear, transformation of variables to obtain the canonical interval $[-1,1]$. Experimental evidence of this has previously been presented by Thacher [1966].

Compared with expansions of f in other orthogonal polynomials, particularly ultraspherical polynomials $P_k^{(\alpha,\alpha)}$, Lanczos early recognized

(National Bureau of Standards [1952]) that convergence is most rapid when $\alpha = -1/2$, i.e., when the expansion is indeed in Chebyshev polynomials. Some firm results in this direction, for restricted classes of functions, are due to Rivlin and Wilson [1969] and Handscomb [1973].

Closely related to convergence is the asymptotic behavior of the expansion coefficients a_k as $k \to \infty$. This is studied in detail by Elliott [1964] for meromorphic functions, and also for functions with a branchpoint at an endpoint of the basic interval, and by Elliott and Szekeres [1965] for entire functions. The case of logarithmic and branch-point singularities on the real line, and combinations of such, is treated by Chawla [1966/67] and Piessens and Criegers [1974]. It is not uncommon to also find essential singularities at an endpoint or midpoint of $[-1,1]$. This occurs, e.g., if the original interval is infinite and f has an essential singularity at infinity. Mapping the interval onto $[-1,1]$ by a reciprocal transformation carries the singularity into a point of $[-1,1]$. The extent to which this slows down the convergence of (1) is studied by Miller [1966]. Asymptotic results for the expansion coefficients in the case of generalized hypergeometric functions are given by Németh [1974].

1.2.2. <u>Relation to best uniform approximation.</u> Letting

(4) $$S_n(f) = \max_{-1 \le x \le 1} |s_n(f;x) - f(x)|,$$

we clearly have $E_n(f) \le S_n(f)$, where $E_n(f)$ is the error of best uniform approximation of f by polynomials of degree n. The difference between $S_n(f)$ and $E_n(f)$ can be remarkably small if f is smooth. This can be seen from de La Vallée Poussin's inequality [1919, p. 107]

(5) $$\left| \sum_{r=0}^{\infty} a_{(2r+1)(n+1)} \right| \le E_n(f) \le S_n(f) \le \sum_{k=n+1}^{\infty} |a_k|,$$

and from other similar results (Hornecker [1958], [1960], Hewers and Zeller [1960/61], Blum and Curtis [1961], Cheney [1966, p. 131], Rivlin

13

[1974, p. 139ff]). If $a_{k+1} = o(a_k)$, for example, it follows from (5) that $S_n(f) \sim E_n(f)$ as $n \to \infty$. Even for larger classes of functions, e.g., the class $C^{n+1}_{M_n}$ of functions $f \epsilon C^{n+1}[-1, 1]$ with $\max_{-1 \le x \le 1} |f^{(n+1)}(x)| \le M_n$, the spread is still infinitesimal in the sense (Remez and Gavriljuk [1963])

$$(6) \qquad \sup_{f \epsilon C^{n+1}_{M_n}} S_n(f) = [1 + O(\tfrac{1}{n})] \sup_{f \epsilon C^{n+1}_{M_n}} E_n(f), \qquad n \to \infty .$$

Widening the class further to include all continuous functions $f \epsilon C[-1, 1]$ we have from the theory of orthogonal series (Alexits [1961, Theorem 4.5.1]) that

$$(7) \qquad 1 \le \frac{S_n(f)}{E_n(f)} \le 1 + \lambda_n ,$$

where λ_n is the Lebesgue constant for Fourier series (Zygmund [1959, p. 67]). Although these constants eventually grow logarithmically with n (Fejér [1910]), they are fairly small in the domain of common interest. It is known that λ_n is monotonically increasing, in fact totally monotone (Szegö [1921]), and $\lambda_1 = 1.436$, $\lambda_{1000} = 4.07$ (Powell [1967]). The error of the truncated Chebyshev expansion, in the range $1 \le n \le 1000$, is therefore never worse than five times the error of the corresponding best uniform approximation.

When f is a polynomial of degree $n + 1$, then in fact $S_n(f) = E_n(f)$. For polynomials of degree $> n+1$ the ratios in (7) are investigated by Clenshaw [1964], Lam and Elliott [1972] and Elliott and Lam [1973]. Some of this work, however, is based on conjectures. For related work, see also Riess and Johnson [1972].

It is possible to modify the truncated Chebyshev expansion so as to bring it closer to the best uniform approximation (Hornecker [1958], [1960], Korneičuk and Širikova [1968], Širikova [1970]). Other modifications can be made to fit interpolatory conditions at the end points (Cohen

14

[1971]). This may be useful in segmented approximation when continuity at the joints is desirable.

Using a method reminiscent of Lanczos' τ-method, Stolyarčuk [1974a, b] obtains explicit polynomial approximations to the sine integral, error function, and Bessel functions of integer order, which are valid on an arbitrary interval and are infinitesimally close to the best polynomial approximations on that interval as the degree tends to infinity.

1.2.3. Calculation of expansion coefficients. There are a number of methods available to calculate (or approximate) the expansion coefficients a_k. Some will now be considered.

(i) Fourier analysis. Since we are dealing with Fourier coefficients, we can enlist the techniques of harmonic analysis, and thus, for example, approximate a_k, $k \leq n$, by

(8) $$\alpha_k^{(n)} = \frac{2}{n} \sum_{j=0}^{n} {}'' f(x_j) T_k(x_j), \qquad x_j = \cos(j\pi/n) \ .$$

(The primes on the summation sign indicate that the first and last term is to be halved.) Since $T_k(x_j) = T_j(x_k)$, the sum in (8) can be evaluated effectively by Clenshaw's algorithm (cf. 1.5.1(ii)).

It is a relatively simple matter to increase the accuracy of (8), by doubling n, if one observes that about half of the terms in (8) can be reused, and only half of the $\alpha_k^{(n)}$ need to be computed, by virtue of

$$\alpha_k^{(n)} = \alpha_k^{(2n)} + \alpha_{2n-k}^{(2n)}$$

(Clenshaw [1964], Torii and Makinouchi [1968]).

(ii) Rearrangement of power series. The coefficients a_k of the Chebyshev expansion (1) are related to the coefficients c_k of the Maclaurin series, $f(x) = \sum_{k=0}^{\infty} c_k x^k$, by the linear transformation

15

$$(9) \quad \begin{bmatrix} a_0 \\ a_1 \\ a_2 \\ \vdots \end{bmatrix} = \begin{bmatrix} u_{00} & u_{01} & u_{02} & \cdots \\ 0 & u_{11} & u_{12} & \cdots \\ 0 & 0 & u_{22} & \cdots \\ \cdots\cdots\cdots\cdots\cdots \end{bmatrix} \begin{bmatrix} c_0 \\ c_1 \\ c_2 \\ \vdots \end{bmatrix} ,$$

where

$$u_{ij} = \begin{cases} 2^{1-j} \binom{j}{\frac{j-i}{2}} & \text{if } i+j \text{ is even}, \quad j \geq i \geq 0 , \\ \\ 0 & \text{otherwise} \end{cases}$$

(Minnick [1957], De Vogelaere [1959]). As some of the coefficients c_k may be quite large, and of different signs, the application of (9) is likely to require high-precision work. Another complication occurs if the power series converges very slowly (Clenshaw [1962]). The infinite series implied in (9) then also converge very slowly, although, sometimes, they respond well to nonlinear acceleration techniques (Thacher [1964]).

(iii) <u>Recurrence relations.</u> In many cases of practical interest it is possible to derive recurrence relations for the coefficients a_k, either directly from the integral representation (2), or indirectly via differential equations. In using these recursions, a certain amount of skill is required to maintain numerical stability (Clenshaw [1962], Luke and Wimp [1963], Németh [1965], [1974], Clenshaw and Picken [1966], Hangelbroek [1967], Wood [1967], Luke [1969 , Vol. II, §12. 5], [1971b, c], [1972a]).

(iv) <u>Numerical quadrature.</u> The integral in (2) can be approximated directly by numerical quadrature. Eq. (8), in fact, is an example. For others, see Rivlin [1974, p. 153ff] and Bjalkova [1963].

(v) <u>Explicit formulas.</u> Explicit formulas for a_k in terms of easily computed functions are known for a number of important special functions, e.g., Bessel functions J_ν, I_ν, Y_ν, K_ν (Wimp [1962], Cylkowski [1966/68]),

Dawson's integral (Hummer [1964]), $\psi(a+x)$, $\ln \Gamma(a+x)$, $Ci(x)$, $Si(x)$ (Wimp [1961]). Luke and Wimp [1963] express the expansion coefficients for confluent hypergeometric functions in terms of Meijer's G-function.

1.2.4. <u>Tables of Chebyshev expansions and computer programs.</u> The most extensive tables are those of Clenshaw [1962], Clenshaw and Picken [1966], and Luke [1969, Vol. II, Ch. XVII]. References to additional tables are given in Luke [1969, Vol. II, pp. 287-291]. Among the more recent specialized tables are those of Németh [1967] for Stirling's series, Strecok [1968] for the inverse error function, Wood [1968] for Clausen's integral, Ng, Devine and Tooper [1969] for Bose-Einstein functions, Wimp and Luke [1969] for modified Bessel functions and their incomplete Laplace transform, Kölbig, Mignaco and Remiddi [1970] for generalized polylogarithms, Németh [1971] for Airy functions, Németh [1972] for zeros of Bessel functions J_ν (considered as functions of ν), Németh [1974] for the integrals $\int_0^\infty t^{-\frac{1}{2}} \exp(-t-t^2/x^2)dt$, $\int_0^\infty (x+t)^{-1} \exp(-t^2)dt$, and Sheorey [1974] for Coulomb wave functions.

An interesting and potentially useful idea, advanced by Clenshaw and Picken [1966] and pursued further by Luke [1971b, c], [1972a], is to provide "miniaturized" tables for functions of several variables. These are tables of coefficients in multiple Chebyshev series. The idea is carried out for Bessel functions of real argument and real order.

A set of ALGOL procedures facilitating the use of Chebyshev expansions is given in Clenshaw, Miller and Woodger [1962/63]. FORTRAN programs for generating Chebyshev expansion coefficients can be found in Håvie [1968] and Amos and Daniel [1972].

1.3. <u>Taylor series and asymptotic expansion</u>

A special function f is often naturally represented in the form

(1) $$f(z) = \alpha(z)g(z), \qquad g(z) \sim \sum_{k=0}^\infty c_k(z-z_0)^k, \qquad c_0 = 1 ,$$

where the factor $\alpha(z)$ may vanish at z_0, be singular there, or represent

some other peculiar behavior. The expansion for g is a <u>Taylor series</u>
if it converges to $g(z)$ at some $z \neq z_0$, hence in some circle $|z-z_0| < \rho$,
$\rho > 0$. It is called an <u>asymptotic expansion</u> if it possibly diverges for
every $z \neq z_0$, but for each n $(n = 1, 2, 3, \ldots)$ obeys the law

$$(2) \qquad g(z) - \sum_{k=0}^{n-1} c_k (z-z_0)^k = O((z-z_0)^n) \text{ as } z \to z_0 .$$

It is customary, then, to write (2) in terms of descending powers of ζ ,
where $\zeta = (z-z_0)^{-1}$.

We will not give here a systematic account of Taylor's series and
of asymptotic expansions, but limit ourselves to a few remarks on the
computational uses of these expansions, and to an example. We refer
to Olver [1974] for a thorough treatment of asymptotic expansions and
their application to special functions.

1.3.1. <u>Computational uses.</u> As a computational tool, Taylor series
are most useful near the point of expansion, z_0, and then indeed may
be quite effective. Further away from z_0 one runs into several problems,
notably slow convergence, or absence of it, and severe cancellation of
terms, with the attendant loss of significant digits. Asymptotic expansions, likewise, may be quite useful sufficiently close to z_0 . The accuracy obtainable from a divergent asymptotic expansion, however, is
limited at any fixed $z \neq z_0$, in contrast to convergent expansions. Also,
error bounds are not always available, and the evaluation of higher order
terms may be laborious.

Both expansions may serve purposes other than direct evaluation
of functions. For one, they suggest an appropriate form in which to seek
best rational approximations. For another, they may be used as input to
some of the methods of 1.2, 1.4 for generating polynomial or rational
approximations (cf., in particular, 1.2.3(ii), 1.4.1, 1.4.2, 1.4.5).

Nontrivial problems arise in the expansion of functions of several
complex variables. Expanding in one variable leaves the coefficients to

be functions of the remaining variables. This creates challenging problems of effective computation, satisfactory rate of convergence, etc. An example in point is the Taylor expansion of the Bessel function $K_\nu(z)$ of complex order and complex argument, which is treated by Temme [1973]. Another example will be discussed below.

A further important problem is the computation of the Taylor expansion coefficients c_k, when z_0 is an arbitrary point in the complex plane. (In particular, this yields $g(z_0) = c_0$.) There are various approaches one can take: numerical quadrature on Cauchy's integral (Lyness and Sande [1971]), recursive computation of higher derivatives (as, e.g., in Gautschi [1966] and Gautschi and Klein [1970]), or more general backward recurrence techniques in cases where g satisfies a linear differential equation with polynomial coefficients (Thacher [1972], and work of Thacher in progress). The more obvious process of analytic continuation (Henrici [1966]), unfortunately, is inherently unstable.

1.3.2. An example (Van de Vel [1969]). Consider the incomplete elliptic integral of the first kind (cf. 3.1.1),

$$(3) \qquad F(\varphi, k) = \int_0^\varphi (1 - k^2 \sin^2 \theta)^{-\frac{1}{2}} d\theta, \quad 0 \le k \le 1, \quad 0 \le \varphi < \pi/2 ,$$

where φ is the amplitude and k the modulus of F . The developments to be made for (3) apply similarly to the integral of the second kind. The complementary modulus k' is defined by

$$(4) \qquad k' = \sqrt{1 - k^2} ,$$

and the complete integral by

$$(5) \qquad \mathbb{K}(k) = F(\frac{\pi}{2}, k) = \int_0^{\pi/2} (1 - k^2 \sin^2 \theta)^{-\frac{1}{2}} d\theta, \quad 0 \le k < 1 .$$

19

We are interested in Taylor's expansion of F with respect to the modulus k.

The most obvious attack is to expand the integrand in a binomial series and to integrate term by term. The result is

(6) $$F(\varphi, k) = \sum_{r=0}^{\infty} (-1)^r \begin{pmatrix} -\frac{1}{2} \\ r \end{pmatrix} \sigma_r(\varphi) k^{2r}, \qquad \sigma_r(\varphi) = \int_0^\varphi \sin^{2r} \theta \, d\theta .$$

For the σ_r one can find a simple recurrence formula. The series (6) converges geometrically, with an asymptotic quotient $k^2 \sin^2 \varphi$. We have rapid convergence, therefore, if k is small, but slow convergence, if k is near 1 and φ near $\pi/2$.

When k is near 1, then (4) suggests finding an expansion in k'. This can be achieved by writing (3) as

$$F(\varphi, k) = \int_0^\varphi \frac{d\theta}{\cos \theta [1 + k'^2 \tan^2 \theta]^{\frac{1}{2}}} ,$$

and again using the binomial expansion,

(7) $$F(\varphi, k) = \sum_{r=0}^{\infty} \begin{pmatrix} -\frac{1}{2} \\ r \end{pmatrix} \tau_r(\varphi) k'^{2r}, \qquad \tau_r(\varphi) = \int_0^\varphi \frac{\sin^{2r} \theta}{\cos^{2r+1} \theta} \, d\theta .$$

As before, the τ_r can be generated by a simple recursion. The asymptotic convergence quotient of the series (7) is now $k'^2 \tan^2 \varphi$, and thus satisfactory if k' is small and φ not too close to $\pi/2$.

It remains to deal with the last contingency, viz., φ near $\pi/2$. Here we write

$$\mathbb{K}(k) - F(\varphi, k) = \int_0^{\frac{\pi}{2} - \varphi} \frac{d\theta}{\cos \theta [k'^2 + \tan^2 \theta]^{\frac{1}{2}}} ,$$

and make the change of variables $\tan \theta = k' \tan \psi$. The result is

$$\mathbb{K}(k) - F(\varphi, k) = \int_0^u \frac{\cos\theta}{\cos\psi}\,d\psi = \int_0^u \frac{d\psi}{\cos\psi[1 + k'^2 \tan^2\psi]^{\frac{1}{2}}} \ ,$$

where

$$u = \cot^{-1}(k' \tan\varphi) \ .$$

Therefore, if $k'^2 \tan^2 u < 1$, i.e., $\varphi > \pi/4$, we can expand in a binomial series and find

(8) $$\mathbb{K}(k) - F(\varphi, k) = \sum_{r=0}^{\infty} \binom{-\frac{1}{2}}{r} \tau_r(u)k'^{2r} \ .$$

We now have a series whose convergence quotient is $k'^2 \tan^2 u = \cot^2\varphi$, thus independent of k, and which converges more rapidly, the closer φ is to $\pi/2$. Note, however, that (8) requires the computation of the complete elliptic integral. (For this, see 3.4.)

It is easily verified that for any k and φ in the region $0 \le k \le 1$, $0 \le \varphi < \pi/2$, at least one of the series (6), (7), (8) converges geometric-ally with an asymptotic quotient $\le 1/2$.

Other methods of computation, based on Gauss and Landen transformations, will be considered in 3.3. These are sometimes (but not always) more efficient than the expansions considered here.

1.4. Padé and continued fraction approximations

Given a formal power series about some point z_0 in the complex plane, one can associate with it certain rational functions having highest order contact with the power series at z_0. The rational functions in turn can be interpreted as convergents of continued fractions. These often converge faster, or in larger domains, than the original series, and may even converge when the series diverges. It is this property which makes them useful as a tool of approximation. Without loss of generality we shall assume the point of contact at the origin, $z_0 = 0$.

The basic references are Wall [1948], Perron [1957] and Khovanskii [1963]. On Padé approximation there are survey articles by Gragg [1972] and Chisholm [1973b], as well as a forthcoming book by Baker [1975]. Informative surveys on the use and application of Padé approximants and continued fractions can be found in the collection of articles edited by Baker and Gammel [1970], and in recent conference proceedings, e.g., Graves-Morris [1973a, b] and Jones and Thron [1974b]. We single out the extensive survey of Wynn [1974], containing many references, both to original sources and to newer developments. A good introduction into the numerical evaluation of continued fractions is Blanch [1964]. For a collection of computer algorithms see Wynn [1966b].

1.4.1. <u>Padé table</u>. Let

(1) $$f(z) \sim c_0 + c_1 z + c_2 z^2 + \ldots, \qquad c_0 \neq 0 ,$$

be a formal power series, and ν, μ two nonnegative integers. It is possible to determine polynomials $\hat{p}_{\nu,\mu} \in \mathbb{P}_\mu$, $\hat{q}_{\nu,\mu} \in \mathbb{P}_\nu$, with $\hat{q}_{\nu,\mu} \not\equiv 0$, such that

(2) $$\hat{q}_{\nu,\mu}(z)\, f(z) - \hat{p}_{\nu,\mu}(z) = (z^{\nu+\mu+1}) ,$$

where the symbol on the right stands for a formal power series beginning with a power z^k, $k \geq \nu + \mu + 1$. Although the polynomials $\hat{p}_{\nu,\mu}$ and $\hat{q}_{\nu,\mu}$ are not unique, they determine a unique rational function $\hat{p}_{\nu,\mu}(z)/\hat{q}_{\nu,\mu}(z)$, which may be expressed, in irreducible form, as

(3) $$[\nu,\mu]_f(z) = \frac{p_{\nu,\mu}(z)}{q_{\nu,\mu}(z)}, \quad p_{\nu,\mu} \in \mathbb{P}_\mu, \quad q_{\nu,\mu} \in \mathbb{P}_\nu, \quad q_{\nu,\mu}(0) = 1 .$$

One calls $[\nu,\mu]_f$ the <u>Padé approximant</u> of order ν, μ generated by $f(z)$ (Wall [1948, p. 377ff], Perron [1957, p. 235 ff]). We note from (2) and (3) that

(4)
$$[\nu, \mu]_f = \frac{1}{[\mu, \nu]_{\frac{1}{f}}} , \quad \nu \geq 0, \quad \mu \geq 0 .$$

The array of rational functions

(5)
$$[0, 0]_f \quad [0, 1]_f \quad [0, 2]_f \quad \cdots$$
$$[1, 0]_f \quad [1, 1]_f \quad [1, 2]_f \quad \cdots$$
$$[2, 0]_f \quad [2, 1]_f \quad [2, 2]_f \quad \cdots$$
$$\cdot \cdot \cdot \cdot \cdot \cdot \cdot \cdot \cdot \cdot \cdot \cdot \cdot \cdot \cdot \cdot \cdot \cdot \cdot \cdot \cdot \cdot \cdot$$

is called the Padé table of f .

If $f(z) - [\nu, \mu]_f(z) = (z^{r+1})$, and (z^{r+1}) cannot be replaced by (z^{s+1}) with $s > r$, we say that $[\nu, \mu]_f$ has contact of order r with f . A Padé table in which each approximant $[\nu, \mu]_f$ has contact of order $\nu + \mu$,

(6)
$$f(z) - [\nu, \mu]_f(z) = (z^{\nu+\mu+1}) ,$$

is called normal. A necessary and sufficient condition for this is (Wall [1948, p. 398])

(7)
$$\Delta_{m, n} = \det \begin{bmatrix} c_{n-m} & c_{n-m+1} & \cdots & c_n \\ c_{n-m+1} & c_{n-m+2} & \cdots & c_{n+1} \\ \cdots\cdots\cdots\cdots\cdots\cdots\cdots\cdots \\ c_n & c_{n+1} & \cdots & c_{n+m} \end{bmatrix} \neq 0, \quad n, m = 0, 1, 2, \ldots .$$

(The convention $c_k = 0$ for $k < 0$ is used here.) If this condition holds, $p_{\nu, \mu}$ and $q_{\nu, \mu}$ in (3) are of exact degrees μ and ν, respectively. In the abnormal case, identical approximants lie in square blocks of the Padé table of the form $[i+r, j+s]_f$ (r, s = 0, 1, ..., k), each approximant of this block having contact of order $i + j + k$.

23

The question of <u>convergence</u>, $[\nu, \mu]_f \to F$ as ν, μ, or both, tend to infinity, where F is a function associated in some way with f, is a difficult one, depending, as it does, on the behavior of the poles of $[\nu, \mu]_f$. We refer to Baker [1965], [1970], and Chisholm [1973c], for summaries of results and conjectures, and to Wynn [1972], Jones and Thron [1975], for more recent results.

    1.4.2 <u>Corresponding continued fractions.</u> If the series in (1) is such that

(8) $$\Delta_{m,m} \neq 0 \quad \text{for} \quad m = 0, 1, 2, \ldots \quad,$$

we can associate with it an infinite J-fraction,

(9) $$\sum_{k=0}^{\infty} c_k z^k \sim \frac{b_0}{1-a_0 z-} \frac{b_1 z^2}{1-a_1 z-} \frac{b_2 z^2}{1-a_2 z-} \cdots, \quad b_k \neq 0, \quad b_0 = c_0 .$$

If the series is such that

(10) $$\Delta_{m,m} \neq 0, \quad \Delta_{m,m+1} \neq 0 \quad \text{for} \quad m = 0, 1, 2, \ldots ,$$

we can also associate an infinite S-fraction,

(11) $$\sum_{k=0}^{\infty} c_k z^k \sim \frac{s_0}{1-} \frac{s_1 z}{1-} \frac{s_2 z}{1-} \frac{s_3 z}{1-} \cdots, \quad s_k \neq 0, \quad s_0 = c_0 .$$

Both continued fractions are completely characterized by their contact properties: the p-th convergent of the J-fraction (p = 1, 2, 3, ...) has contact of order 2p, that of the S-fraction contact of order p, with the series (1). The J-fraction, in fact, is a contraction of the S-fraction.

    The correspondences (9) and (11) are often written for series in descending powers of z (usually asymptotic series), in which case they assume the form (Wall [1948, pp. 197, 202])

(9') $$\sum_{k=0}^{\infty} \frac{c_k}{z^{k+1}} \sim \frac{b_0}{z-a_0-} \frac{b_1}{z-a_1-} \frac{b_2}{z-a_2-} \cdots ,$$

(11')
$$\sum_{k=0}^{\infty} \frac{c_k}{z^{k+1}} \sim \frac{s_0}{z-} \frac{s_1}{1-} \frac{s_2}{z-} \frac{s_3}{1-} \cdots .$$

An important special case of (8), namely $\Delta_{m,m} > 0$, occurs precisely when $\{c_k\}$ is a moment sequence (Wall [1948, p. 325]),

(12)
$$c_k = \int_{-\infty}^{\infty} t^k d\phi(t), \qquad k = 0, 1, 2, \cdots ,$$

with ϕ a bounded nondecreasing function having infinitely many points of increase. The series (1), called Stieltjes series, is then the formal expansion of a Stieltjes transform,

(13)
$$\int_{-\infty}^{\infty} \frac{d\phi(t)}{1-tz} \sim c_0 + c_1 z + c_2 z^2 + \cdots .$$

The continued fraction (9) associated with (13) has all a_k real, and all $b_k > 0$ (Perron [1957, p. 193]). Its convergents, as well as the convergents of (9'), are expressible in terms of the orthogonal polynomials $\{\pi_k(t)\}$ belonging to $d\phi(t)$, or in terms of Gaussian quadrature. For example, in the case of (9'),

(14)
$$\int_{-\infty}^{\infty} \frac{d\phi(t)}{z-t} \sim \frac{c_0}{z} + \frac{c_1}{z^2} + \cdots \sim \frac{b_0}{z-a_0-} \frac{b_1}{z-a_1-} \cdots ,$$

we have

(15)
$$\frac{b_0}{z-a_0-} \frac{b_1}{z-a_1-} \cdots \frac{b_{p-1}}{z-a_{p-1}} = \frac{1}{\pi_p(z)} \int_{-\infty}^{\infty} \frac{\pi_p(z) - \pi_p(t)}{z-t} d\phi(t)$$

(16)
$$= \sum_{k=1}^{p} \frac{\omega_k^{(p)}}{z-\tau_k^{(p)}} ,$$

where $\tau_k^{(p)}$ are the zeros of $\pi_p(t)$ and $\omega_k^{(p)}$ the associated Christoffel numbers. The polynomials $\pi_k(z)$ are thus the denominators of the continued fraction in (14), the <u>associated orthogonal polynomials</u>

$$\sigma_k(z) = \int_{-\infty}^{\infty} \frac{\pi_k(z) - \pi_k(t)}{z - t} \, d\phi(t)$$

the numerators. Both satisfy the same recurrence formula,

(17) $$y_{r+1} = (z - a_r)y_r - b_r y_{r-1}, \qquad r = 0, 1, 2, \ldots ,$$

where $y_0 = 1$, $y_{-1} = 0$ for $\{\pi_k\}$, and $y_0 = 0$, $y_{-1} = -1$ for $\{\sigma_k\}$. This is meaningful not only for Stieltjes series, but for any series which has an associated J-fraction, provided orthogonality is defined algebraically (Wall [1948, p. 192]). We also note that in terms of the continued fraction (11), we have

(18)
$$a_0 = s_1, \quad b_0 = s_0 ,$$
$$\left. \begin{array}{l} a_r = s_{2r} + s_{2r+1} \\[2mm] b_r = s_{2r-1} s_{2r} \end{array} \right\} \quad r = 1, 2, 3, \ldots .$$

A special case of (10), similarly, is $\Delta_{m,m} > 0$, $\Delta_{m,m+1} > 0$, and obtains precisely when (12) holds for some measure $d\phi(t)$ vanishing for $t < 0$ (Wall [1948, p. 327]). In this case, $s_k > 0$ for all $k \geq 0$ in (11), a source of useful inequalities when z is real and negative.

With regard to convergence of the continued fractions in (9) and (11), and their limits, we refer to Perron [1957, p. 145ff].

1.4.3. <u>Relation between Padé table and continued fractions.</u>
Assume that the series (1) is normal. The conditions (8) and (10) are then valid not only for the given series, but also for all delayed series

(1_m) $$f_m(z) \sim c_m + c_{m+1} z + c_{m+2} z^2 + \ldots, \qquad m = 0, 1, 2, \ldots .$$

26

Each of these, therefore, has an associated J-fraction

$$(9_m) \qquad f_m(z) \sim \frac{b_0^{(m)}}{1-a_0^{(m)}z-} \ \frac{b_1^{(m)}z^2}{1-a_1^{(m)}z-} \ \frac{b_2^{(m)}z^2}{1-a_2^{(m)}z-} \cdots, \qquad b_k^{(m)} \neq 0, \ b_0^{(m)} = c_m \ ,$$

and an associated S-fraction,

$$(11_m) \qquad f_m(z) \sim \frac{s_0^{(m)}}{1-} \ \frac{s_1^{(m)}z}{1-} \ \frac{s_2^{(m)}z}{1-} \ \frac{s_3^{(m)}z}{1-} \cdots, \qquad s_k^{(m)} \neq 0, \ s_0^{(m)} = c_m \ .$$

It turns out (Wall [1948, p. 380]) that the entries of the Padé table for $f = f_0$ in the stairlike sequence

$$[0, m-1] \quad [0, m]$$
$$[1, m] \quad [1, m+1]$$
$$[2, m+1] \quad [2, m+2]$$
$$\cdot \ \cdot$$

are identical with the successive convergents of the continued fraction

$$(19) \qquad c_0 + c_1 z + \ldots + c_{m-1}z^{m-1} + \frac{s_0^{(m)}z^m}{1-} \ \frac{s_1^{(m)}z}{1-} \ \frac{s_2^{(m)}z}{1-} \cdots \ ,$$

while those along the para-diagonal

$$[0, m-1]$$
$$[1, m]$$
$$[2, m+1]$$
$$\cdot \ \cdot$$

are the successive convergents of

$$(20) \quad c_0 + c_1 z + \ldots + c_{m-1}z^{m-1} + \frac{b_0^{(m)}z^m}{1-a_0^{(m)}z-} \ \frac{b_1^{(m)}z^2}{1-a_1^{(m)}z-} \ \frac{b_2^{(m)}z^2}{1-a_2^{(m)}z-} \cdots \ .$$

As in (15), the latter are expressible in terms of the orthogonal polynomials $\{\pi_k^{(m)}\}$ belonging to the measure $t^m \, d\phi(t)$. (See, in this connection, Allen, Chui, Madych, Narcowich and Smith [1974]). Similar statements can be obtained for the entries in the lower half of the Padé table by using (4).

We remark that in the case of convergence, the continued fraction

$$\frac{1}{1-} \; \frac{s_1^{(m)} z}{1-} \; \frac{s_2^{(m)} z}{1-} \; \cdots$$

in (19), and the analogous continued fraction in (20), serve as "<u>converging factor</u>", being the factor by which the last term $c_m z^m$ is to be multiplied in order to obtain the correct limit of the series (1).

1.4.4. <u>Algorithms</u>. The entries of the Padé table may be generated either in explicit form, as ratios of polynomials, or in their continued fraction form (19). For the former, there are a number of recursive schemes for generating the polynomials in question (Wynn [1960], Baker [1970], [1973], Longman [1971], Watson [1973]). For the latter, one has the <u>quotient-difference</u> (<u>qd-</u>) <u>algorithm</u> (Rutishauser [1954a, b], [1957], Henrici [1958], [1963], [1967]), which consists in generating the <u>qd-array</u>

$$
\begin{array}{ccccccc}
e_0^{(0)} & & & & & & \\
 & q_1^{(0)} & & & & & \\
e_0^{(1)} & & e_1^{(0)} & & & & \\
 & q_1^{(1)} & & q_2^{(0)} & & & \\
e_0^{(2)} & & e_1^{(1)} & & e_2^{(0)} & \cdot & \\
 & q_1^{(2)} & & q_2^{(1)} & & \cdot & \\
e_0^{(3)} & & e_1^{(2)} & & e_2^{(1)} & \cdot & \\
 & q_1^{(3)} & & q_2^{(2)} & & \cdot & \\
\vdots & & e_1^{(3)} & & e_2^{(2)} & \cdot & \\
 & & & q_2^{(3)} & & & \\
 & & & & e_2^{(3)} & \cdot & \\
\end{array}
$$

from left to right by means of

$$e_0^{(n)} = 0, \quad q_1^{(n)} = \frac{c_{n+1}}{c_n}, \qquad n = 0, 1, 2, \ldots \, ,$$

$$\left. \begin{aligned} e_k^{(n)} &= q_k^{(n+1)} - q_k^{(n)} + e_{k-1}^{(n+1)} \\ q_{k+1}^{(n)} &= \frac{e_k^{(n+1)}}{e_k^{(n)}} \, q_k^{(n+1)} \end{aligned} \right\} \quad k = 1, 2, 3, \ldots, \quad n = 0, 1, 2, \ldots \, .$$

The coefficients in the continued fraction (19) are then given by

$$s_0^{(m)} = c_m, \quad s_{2k-1}^{(m)} = q_k^{(m)}, \quad s_{2k}^{(m)} = e_k^{(m)}, \quad k = 1, 2, 3, \ldots, \quad m = 0, 1, 2, \ldots \, .$$

Unfortunately, the generation of the qd-array, as described, is unstable, and should be carried out in high precision, or with some other precautions (Gargantini and Henrici [1967]). Thacher [1971] notes, however, that inaccuracies in the higher order coefficients $s_k^{(m)}$ need not necessarily imply an inaccurate value of the continued fraction (19).

In some instances one has explicit expressions for the $e_k^{(n)}$, $q_k^{(n)}$, for example, in the case of the complex error function (Thacher [1967]), or for certain special hypergeometric and confluent hypergeometric functions (Wynn [1960], Henrici [1963]). For series (1), with $c_m = \prod_{\mu=0}^{m-1} \{ (a - q^{\alpha+\mu})(b-q^{\beta+\mu})^{-1} \}$, Wynn [1967] gives closed expressions for the numbers $e_k^{(n)}$, $q_k^{(n)}$, and also for the numerator and denominator polynomials of the approximants in the upper half of the Padé table. Limiting forms of these results (obtained, e.g., when $a = b = 1$, $q \to 1$) yield all cases in which these numbers and polynomials are known in closed form.

There are other algorithms, notably the ϵ-algorithm and related methods due to Wynn [1956], [1961], [1966a], which operate directly on the entries of the Padé table. Their most important use, probably, is in the calculation of numerical values for a sequence of Padé approximants,

e.g., the values at $z = 1$ in an attempt to speed the convergence of
$$\sum_{k=0}^{\infty} c_k \ .$$

1.4.5. <u>Applications to special functions</u>. The qd-algorithm, either applied to a Taylor series or to an asymptotic expansion, has been used by many authors to obtain the corresponding S-fraction explicitly or numerically. We mention the work of Gargantini and Henrici [1967] on the Bessel function $K_0(z)$ and more general confluent hypergeometric functions, the work of Thacher [1967] on the complex error function, of Cody and Thacher [1968] and Chipman [1972] on the exponential integral $E_1(z)$ and related integrals, of Strecok and Gregory [1972] on the irregular Coulomb wave function along the transition line, and the study of Shenton and Bowman [1971] on the polygamma functions $\psi^{(n)}(z)$. Jacobs and Lambert [1972] apply S-fractions to polylogarithms of a complex argument, while Barlow [1974] does the same to generalized polylogarithms.

Earlier, Fair [1964] uses Lanczos' τ-method for obtaining the J-fraction for functions defined by Riccati differential equations, and applies the technique to confluent hypergeometric functions and Bessel functions of the first and second kind. Fair and Luke [1967] further apply it to incomplete elliptic integrals (cf. also Luke [1969, Vol. II, p. 77ff]).

For large classes of functions, including Gauss hypergeometric functions and the incomplete gamma function, Luke [1969, Vol. II, Chs. XIII and XIV], [1970b], [1971a], [1975] gives explicit expressions for the Padé entries on the diagonal, and immediately above, as well as appraisals of the errors. Those for the incomplete gamma function also serve to approximate the gamma function in the complex plane. See, however, Ng [1975] for a comparison with other methods. Tables of Padé coefficients are given in Luke [1969, Vol. II, p. 402ff] for the exponential, sine, and cosine integrals and for the error function. Golden, McGuire and Nuttall [1973] give an experimental study of the diagonal Padé approximants in the case of Hankel functions of the first and second kind.

Gaussian quadrature, or the equivalent J-fraction in (15), have been used by Todd [1954] for evaluating the complex exponential integral, and by Gautschi [1970] for evaluating the complex error function. In the latter work, the continued fraction approach is combined with a Taylor series approach, there being a gradual transition from one to the other as the complex argument decreases in magnitude.

1.4.6. Error estimates. It is important to have reasonably good estimates of the error due to premature truncation of a continued fraction. One distinguishes between a priori estimates, which are expressed directly in terms of the elements of the continued fraction, and a posteriori estimates, which depend on the knowledge of a finite number (usually two or three) of convergents. Concerning the latter, we mention the elegant work of Henrici [1965] and Henrici and Pfluger [1966] on Stieltjes fractions, in which a sequence of nested lens-shaped regions is constructed the intersection of which contains the value of the continued fraction. For more recent extensions of this work, as well as for other types of estimates, we refer to the survey of Jones [1974].

For a large number of continued fraction expansions of special functions, Wynn [1962a, b] gives "efficiency profiles", i.e., tables from which the order of convergents can be determined as a function of the (real) argument and the accuracy desired.

1.4.7. Generalizations. In view of the contact properties of Padé and continued fraction approximations, one expects these approximations to be best near the point of contact, and to gradually worsen away from it. There is, in fact, a close relationship between the best uniform rational approximants on small discs $|z| \leq \epsilon$, or small intervals $0 \leq z \leq \epsilon$, and the Padé approximant, the former tending to the latter as $\epsilon \to 0$ (Walsh [1964a], [1974], Chui, Shisha and Smith [1974]). The reason for this behavior is largely due to the employment of powers in setting up the Padé table. To obtain a more balanced rational approximation on a given interval, it has been suggested to use systems of orthogonal

31

polynomials instead, and to proceed similarly as in 1.4.1, starting with the appropriate orthogonal expansion of f . It will be noted that the analogue of (2) is still a linear problem, but the analogue of (6) is not. The original work along this line is due to Maehly [1956], [1958] (see also Kogbetliantz [1960], Spielberg [1961b]), who uses Chebyshev polynomials, and is continued by Cheney [1966, p. 177ff], Holdeman [1969] and Fleischer [1972]. These authors use the linear approach. The nonlinear problem, which is closer in spirit to Padé approximation, has only recently been considered (Common [1969], Fleischer [1973a, b], Frankel and Gragg [1973], Clenshaw and Lord [1974], Gragg and Johnson [1974]). The use of Chebyshev polynomials often leads to nearly best rational approximations (Clenshaw [1974]).

In another direction, one might generalize Padé and continued fraction approximation by imposing contact conditions not only at one, but at several points (typically, at the origin and at infinity). See Baker, Rushbrooke and Gilbert [1964] and Baker [1970] for recent attempts in this direction, and McCabe [1974] for an interesting continued fraction approach. The potential of this approach remains largely to be explored.

Finally, we mention generalizations of Padé approximation to functions of two variables by Chisholm [1973a], Hughes Jones and Makinson [1974], Graves-Morris, Hughes Jones and Makinson [1974], Common and Graves-Morris [1974].

1.4.8. **Other rational approximations.** We already mentioned the τ-method (Lanczos [1956, pp. 464-507]) applied to linear and nonlinear differential equations as a source of rational approximations (Luke [1955], [1958], [1959/60], Guerra [1969], Verbeeck [1970]). Other sources are Maehly's economization of continued fractions and related techniques (Maehly [1960], Spielberg [1961a], Ralston [1963]), Hornecker's method of modifying the Chebyshev expansion (Hornecker [1959a, b], [1960]), the method of Luke and co-workers (Luke [1969, Vol. II, Ch. XI]) on generalized hypergeometric functions and functions representable as Laplace transforms, and the nonlinear sequence-to-sequence transformation of Levin applied to the partial sums of power series (Levin [1973], Longman

[1973]). Integrating Padé approximants for the square root, Luke [1968], [1970a] obtains rational approximations to the three normal forms of incomplete elliptic integrals, including asymptotic estimates of the error. We also mention the curious ad-hoc approximation to the gamma function $\Gamma(z)$ on $\text{Re } z \geq 1$ due to Lanczos [1964].

1.5. Representation and evaluation of approximations

Once an approximation to a special function has been constructed, it is often possible to represent this approximation in different mathematically equivalent forms. Each form in turn suggests one or several algorithms of evaluation. Although mathematically equivalent, these forms may behave quite differently under evaluation in finite precision. It is important to select a representation, and a corresponding evaluation algorithm, which to the maximum extent possible is invulnerable to the vagaries of finite precision arithmetic.

With regard to representation, what one aims for is well-conditioning. This means that the value of the particular functional form be insensitive to small perturbations in the parameters (coefficients) involved. With regard to algorithms, one strives for economy and stability, i.e., few arithmetic operations and maximum resistance to rounding errors. It is a rare instance where all three of these requirements are in complete harmony with each other.

We discuss some possible representations and algorithms for polynomial and rational approximations, and then consider an algorithm for evaluating approximations in the form of orthogonal sums.

1.5.1. Polynomials

(i) Power form. A polynomial of degree n is most frequently represented in the form

(1) $$p(x) = a_0 + a_1 x + a_2 x^2 + \ldots + a_n x^n ,$$

which can be evaluated rather economically by a scheme ascribed to Horner [1819] (but already known to Newton (Ostrowski [1954])),

$$(2) \quad \begin{cases} u_n = a_n \\ u_k = xu_{k+1} + a_k, \quad k = n-1, \ n-2, \ \ldots, \ 0, \\ p(x) = u_0. \end{cases}$$

The scheme requires n multiplications and n additions. With regard to addition, this is optimal (Ostrowski [1954]). The conditioning of the form (1) (at the point x) depends on the relative magnitudes of the quantities $\max_k |a_k x^k|$ and $|p(x)|$. If the former is much larger than the latter, then (1) is ill-conditioned at x. Horner's scheme is generally stable, but can be moderately, and in some cases severely, unstable (Wilkinson [1963, p. 36], Reimer and Zeller [1967], Reimer [1968]). The Chebyshev polynomials, of all, are particularly vulnerable (Reimer [1971]).

(ii) Chebyshev polynomial form. Every polynomial of degree n can be represented in terms of Chebyshev polynomials as (cf. 1.2.3(ii))

$$(3) \quad p(x) = \frac{1}{2} a_0 + \sum_{k=1}^{n} a_k T_k(x).$$

One of the attractive features of this form is the possibility of obtaining a sequence of approximations of varying accuracy by merely truncating (3) at consecutive terms. For the evaluation of $p(x)$ one has an algorithm due to Clenshaw [1955],

$$(4) \quad \begin{cases} u_n = a_n, \quad u_{n+1} = 0, \\ u_k = 2xu_{k+1} - u_{k+2} + a_k, \quad k = n-1, \ n-2, \ \ldots, \ 0, \\ p(x) = \frac{1}{2}(u_0 - u_2), \end{cases}$$

requiring $2n$ additions and n multiplications (cf. 1.5.3). Although more time-consuming than Horner's scheme, Clenshaw's algorithm is often preferred on account of its more favorable stability properties. See Newbery [1974] for a comparative study.

34

(iii) <u>Root product form</u>. This is the form obtained by factoring the polynomial into its linear and quadratic factors,

(5) $\quad p(x) = a_n \prod_{k=1}^{r} (x-x_k) \prod_{k=r+1}^{r+s} [(x-x_k)^2 + y_k], \quad y_k > 0, \quad r + 2s = n$.

Like Horner's scheme, this form requires n additions and n multiplications. For maximum stability, however, the differences $x - x_k$ must be evaluated with care: Assuming x machine representable (in floating-point arithmetic), and denoting by x_k^* the machine representable part of x_k, and by r_k the remainder,

$$x_k = x_k^* + r_k ,$$

one should evaluate $x - x_k$ in two steps as $(x - x_k^*) - r_k$, thereby preserving as much significance as possible when x is close to x_k. Note that this doubles the number of additions. The construction of the form (5) requires some effort, namely the calculation of all zeros of p, but this effort may be rewarded by a well-conditioned representation.

(iv) <u>Newton form</u>. In a sense intermediate between (1) and (5) is Newton's form

(6)
$$p(x) = a_0 + a_1(x-x_0) + a_2(x-x_0)(x-x_1) + \cdots$$
$$+ a_n(x-x_0)(x-x_1) \cdots (x-x_{n-1}) ,$$

which reduces to (1) if all $x_k = 0$, and to (5) (with $s = 0$), if $a_k = 0$ for $k < n$. We have the Horner-type evaluation scheme

(7)
$$\begin{cases} u_n = a_n , \\ u_k = (x-x_k)u_{k+1} + a_k, \qquad k = n-1, n-2, \ldots, 0 , \\ p(x) = u_0 , \end{cases}$$

35

which is quite stable if the differences $x - x_k$ are evaluated as above in (iii), and the parameters x_k, a_k are selected to make the two additive terms on the right of (7) of equal sign. This can always be done (Mesztenyi and Witzgall [1967]). A special form of (6) has proved useful, e. g., in approximating modified Bessel functions (Blair and Edwards [1974]).

(v) <u>Lagrange form</u>. Given any $n + 1$ distinct real numbers x_0, x_1, ..., x_n, we may represent a polynomial of degree n in its Lagrange form familiar from interpolation theory.

$$(8) \qquad p(x) = \sum_{k=0}^{n} a_k \ell_k(x), \quad \ell_k(x) = \prod_{\substack{r=0 \\ r \neq k}}^{n} \frac{x - x_r}{x_k - x_r}, \qquad a_k = p(x_k) ,$$

familiar from interpolation theory. It is evaluated most conveniently in the barycentric form (see, e. g., Bulirsch and Rutishauser [1968])

$$(9) \qquad p(x) = \frac{\displaystyle\sum_{k=0}^{n} a_k \frac{\lambda_k}{x - x_k}}{\displaystyle\sum_{k=0}^{n} \frac{\lambda_k}{x - x_k}} \qquad (x \neq x_i, \quad i = 0, 1, \ldots, n) ,$$

where $\lambda_k = \prod_{r \neq k} (x_k - x_r)^{-1}$ are precomputed constants.

(vi) <u>Ultraeconomic forms</u>. There are a number of representations, due to Motzkin, Belaga, Pan, and others, which require only of the order $n/2$ multiplications and n additions. While these forms are highly interesting from the standpoint of complexity theory, their practical merits are not entirely clear. For one thing, they tend to be poorly conditioned (Rice [1965], Fike [1967]), although this matter deserves further analysis. For another, the time saving gained by fewer multiplications may well be lost on some computers by the need for more memory transactions (Cody [1967]).

1. 5. 2. <u>Rational functions</u>

(i) <u>Polynomial ratio form</u>. This is the collective name given to all the forms that can be obtained by representing the polynomials p and q in

$$(10) \qquad r(x) = \frac{p(x)}{q(x)}$$

in any one of the forms discussed in 1.5.1. Since division is a stable operation, the conditioning and stability properties of r depend entirely on those of p and q. Occasionally it is preferable (see, e.g., Cody and Hillstrom [1970, p. 676]) to write the two polynomials in descending powers of x.

(ii) Continued fraction forms. Intrinsically different are representations of r in terms of continued fractions. There are many different types of continued fractions that can be used in this connection. We mention only the J-fractions (cf. 1.4.2), which are of the form

$$(11) \qquad r(x) = \frac{r_1}{x+s_1+} \ \frac{r_2}{x+s_2+} \cdots \frac{r_n}{x+s_n}, \quad r_k \neq 0 \ \text{all} \ k \ ,$$

and refer to Hart et al. [1968, p. 73ff] for others. The continued fraction (11) represents a rational function in $\mathbb{R}_{n,n-1}$. Conversely, a rational function in $\mathbb{R}_{n,n-1}$ can be represented in the form (11), unless certain determinants in the coefficients of p and q happen to vanish (Wall [1948, p. 165]). Conversion algorithms are given in Hart et al. [1968, pp. 155-160].

For the evaluation of (11) one proceeds most easily "from tail to head", according to

$$(12) \qquad \begin{cases} u_{n+1} = 0 \ , \\[2ex] u_k = \dfrac{r_k}{x+s_k+u_{k+1}} \ , \qquad k = n, \ n-1, \ \ldots, \ 1 \ , \\[2ex] r(x) = u_1 \ . \end{cases}$$

This requires 2n-1 additions and n divisions, which, unless division is very slow, compares favorably with the 2n-1 additions, 2n-1 multiplications, and 1 division, required with Horner's scheme in (10), and even more favorably with the evaluation of the continued fraction by means of the fundamental three-term recurrence relation. The algorithm (12) is not only more economical than Horner's scheme, but also more stable, in general. There are, however, exceptions (Cody and Hillstrom [1967, p. 203]). The stability of evaluation schemes for continued fractions is discussed by Macon and Baskervill [1956], Blanch [1964] and Jones and Thron [1974a, c].

1.5.3. <u>Orthogonal sums.</u> The Chebyshev polynomials T_k in (3) are a special case of orthogonal polynomials, $\{\pi_k\}$, which are known to satisfy a recurrence relation of the form (cf. 1.4.2 (17))

$$(13) \qquad \pi_{r+1} = \alpha_r(x)\,\pi_r + \beta_r(x)\,\pi_{r-1}, \qquad r = 1, 2, 3, \ldots \ .$$

Other (nonpolynomial) systems of special functions also satisfy relations of this type. When expanding a given function in terms of π_k, it is useful to have an efficient algorithm for evaluating a partial sum,

$$(14) \qquad s(x) = \sum_{k=0}^{n} a_k\,\pi_k(x) \ .$$

One such algorithm is <u>Clenshaw's algorithm</u> (Clenshaw [1955]), a generalization of the algorithm in (4),

$$(15) \qquad
\begin{cases}
u_n = a_n, \quad u_{n+1} = 0 \ , \\[2mm]
u_k = \alpha_k(x)\,u_{k+1} + \beta_{k+1}(x)\,u_{k+2} + a_k \\[2mm]
\qquad\qquad k = n-1, n-2, \ldots, 0 \ , \\[2mm]
s(x) = u_0\pi_0(x) + u_1[\pi_1(x) - \alpha_0(x)\,\pi_0(x)] \ .
\end{cases}$$

38

The validity of (15) is best seen by writing (13) in matrix form (Deuflhard [1974]) as

$$
\begin{bmatrix}
1 & & & & \\
-\alpha_0 & 1 & & \text{\large O} & \\
-\beta_1 & -\alpha_1 & 1 & & \\
& \cdot & \cdot & \cdot & \cdot \\
\text{\large O} & & \cdot & \cdot & \cdot \\
& & -\beta_{n-1} & -\alpha_{n-1} & 1
\end{bmatrix}
\begin{bmatrix}
\pi_0 \\ \pi_1 \\ \pi_2 \\ \vdots \\ \\ \pi_n
\end{bmatrix}
=
\begin{bmatrix}
\pi_0 \\ \pi_1 - \alpha_0 \pi_0 \\ 0 \\ \vdots \\ \\ 0
\end{bmatrix},
$$

or briefly as $L\pi = \rho$, and noting that the recurrence in (15) is simply $L^T u = a$. Thus,

$$ s(x) = a^T \pi = a^T L^{-1} \rho = ((L^T)^{-1} a)^T \rho = u^T \rho \ , $$

which is the last line of (15). The argument clearly extends to functions π_k satisfying recurrence relations of order larger than two. Other possible extensions are considered in Saffren and Ng [1971].

For orthogonal polynomials $\{\pi_k\}$ one has $\pi_1 - \alpha_0(x)\pi_0 = 0$, and the last line in (15) simplifies to $s(x) = u_0 \pi_0$. The derivative $s'(x)$ can be computed by a similar algorithm (Smith [1965], Cooper [1967]).

Applying (15) to the Chebyshev polynomials T_k, U_k of the first and second kind, and noting that $T_k(\cos \theta) = \cos k\theta$, $(\sin \theta) U_k(\cos \theta) = \sin(k+1)\theta$, one is led to an algorithm for evaluating trigonometric sums, known as <u>Goertzel's algorithm</u> (Goertzel [1958], [1960]), or <u>Watt's algorithm</u> (Watt [1958/59]).

Floating-point error analyses of Clenshaw's and Goertzel's algorithms are given by Gentleman [1969/70], Newbery [1973] and Cox [unpubl.]. Although usually quite stable, these algorithms are not without pitfalls. Clenshaw's algorithm (15), e.g., should be avoided if $\{\pi_k\}$ is a minimal solution of (13) (cf. 2.2.1). The computation of $s(x)$ in the final step of (15) then likely leads to large cancellation errors (Elliott [1968]). Goertzel's algorithm, in turn, suffers from substantial accumulation of rounding errors if θ is small modulo π (Gentleman [1969/70]). It can be stabilized either by incorporating phase shifts (Newbery [1973]), or by reformulating the recurrence in a manner proposed by Reinsch (Stoer [1972, p. 64]).

For computational experiments with Clenshaw's algorithm see Ng [1968/69].

§2. Methods based on linear recurrence relations

It is often necessary to compute not just one particular function, but a whole sequence of special functions. The task is considerably simplified if the members of the sequence satisfy a recurrence relation. It is then possible to compute each member recursively in terms of those already computed. The process is not only fast, but also well adapted to modern computing machinery, and may be useful even if only one member of the sequence is desired.

Most recurrences of interest in special functions are linear difference equations. The particular solution desired is often rapidly decaying, but embedded in a family of growing solutions. The question of numerical stability then becomes a central issue. In order to keep the dominant solutions in check, special precautions need to be adopted. The nature of these precautions is the subject of this paragraph.

Computational aspects of recurrence relations have been reviewed by several writers, notably Fox [1965], Gautschi [1967], [1972], Wimp [1970], and Amos [1970].

2.1. First-order recurrence relations

The simplest linear recurrence is

(1) $$y_{n+1} = a_n y_n, \qquad n = 0, 1, 2, \ldots ,$$

where y_0 and $a_n \neq 0$ are given numbers. Multiplication being a stable operation, errors due to rounding will essentially accumulate linearly

with n, making (1) a stable computational process. A classic example
is the recurrence relation for the gamma function.

As we proceed to inhomogeneous recurrences,

$$(2) \qquad y_{n+1} = a_n y_n + b_n, \qquad n = 0, 1, 2, \dots ,$$

the stability characteristics may change significantly. The relation (2)
indeed involves repeated additions, thus potentially unstable operations.
It suffices that the two terms on the right be nearly equal in magnitude
and opposite in sign to cause significant loss of accuracy, due to "cancellation". If this happens repeatedly, the computation may quickly de-
teriorate, giving rise to numerical instability.

2.1.1. A simple analysis of numerical stability. Suppose $f_n \neq 0$
is a solution of (2) that we wish to compute. It is instructive to examine
how a relative error ϵ in f_n, committed at $n = s$ (s for "starting"),
affects the value of f_n at $n = t$ (t for "terminal"), where $t \lessgtr s$,
assuming that no further errors are being introduced. If we denote the
perturbed solution by f_n^*, so that $f_s^* = (1 + \epsilon)f_s$, we find by a simple
computation that

$$(3) \qquad f_t^* = (1 + \frac{\rho_t}{\rho_s} \epsilon)f_t ,$$

where

$$(4) \qquad \rho_n = \frac{f_0 h_n}{f_n} ,$$

and h_n is the solution of the homogeneous recurrence (1), with $h_0 = 1$.
Going from s to t, the relative error is thus amplified if $|\rho_t| > |\rho_s|$,
and damped if $|\rho_t| < |\rho_s|$. In an effort to maintain optimal numerical
stability, the recurrence (2), therefore, should be applied in the direc-
tion of decreasing $|\rho_n|$, whenever practicable.

An important special case is

(5)
$$\lim_{n \to \infty} |\rho_n| = \infty ,$$

where $|\rho_n|$ diverges monotonically. The recurrence (2) is then <u>unstable</u> <u>in the forward direction</u>, the ratio $|\rho_t/\rho_s|$ being unbounded for $t > s$, but <u>stable in the backward direction</u>, the same ratio now being bounded by 1 . More than that, we can start the recursion arbitrarily with $f_\nu^* = 0$, for some ν sufficiently large, and recur downward to some fixed n, thereby obtaining f_n to arbitrarily high accuracy. This is because the initial error, $\epsilon = -1$, according to (3), will be damped by a factor of $|\rho_n/\rho_\nu|$, which can be made arbitrarily small by choosing ν large enough. All intermediate rounding errors, moreover, are being consistently damped.

We can interprete (5) by saying that the particular solution of (2) desired is <u>dominated</u> by the "complementary solution" of (2), i.e., the solution of the corresponding homogeneous recurrence (1). It should be clear on intuitive grounds that forward recurrence cannot be stable under these circumstances.

We remark that similar stability considerations apply to general systems of linear difference equations (Gautschi [1972]).

2.1.2. <u>Applications to special functions</u>. Although not many special functions obey relations of the type (2), there are some which do, e.g., certain integrals in the theory of molecular structure (Gautschi [1961]), the incomplete gamma function (Kohútová [1970], Amos and Burgmeier [1973]), in particular the exponential integrals $E_n(z)$ (Gautschi [1973]), and successive derivatives of $f(z)/z$ (Gautschi [1966], [1972], Gautschi and Klein [1970]). The techniques indicated above provide effective schemes of computation in all these cases.

2.2. <u>Homogeneous second-order recurrence relations</u>

We assume now, more importantly, that f_n satisfies a three-term recurrence relation

(1) $y_{n+1} + a_n y_n + b_n y_{n-1} = 0,$ $n = 1, 2, 3, \ldots ,$

where, for simplicity,

(2) $f_n \neq 0, \quad b_n \neq 0 \quad \text{for all} \quad n .$

Given f_0 and f_1, we can use (1) in turn for $n = 1, 2, \ldots$ to successively calculate $f_2, f_3, \ldots$. This is quite effective if f_0 and f_1 are
easily calculated and the recurrence (1) is numerically stable. We expect the latter to be the case if no solution of (1) grows faster than f_n .
An important example of such a recursion is the one for orthogonal polynomials, $f_n = \pi_n$, where the second solution is the sequence of associated orthogonal polynomials, $g_n = \sigma_n$ (cf. 1.4.2), and where by a
theorem of Markov the ratios σ_n/π_n converge to a finite limit, the corresponding Stieltjes integral, at least outside the interval of orthogonality
(Perron [1957, p. 198ff]). The recurrence relation is reputed to be stable
even on the interval of orthogonality, except possibly in the vicinities of
the endpoints.

If there are solutions which grow much faster than f_n, then forward recursion on (1), as in 2.1(5), is bound to fail. Such is the case
if the solution f_n is minimal.

2.2.1. Minimal solutions. We call a solution f_n of (1) minimal,
if for every other, linearly independent, solution g_n we have

(3) $\lim_{n \to \infty} f_n/g_n = 0 .$

All solutions g_n, for which (3) holds, are called dominant. A minimal
solution, if one exists, is unique apart from a constant factor. It can be
specified by imposing a single condition, e.g.,

(4) $f_0 = s ,$

or more generally,

$$(5) \qquad \sum_{m=0}^{\infty} \lambda_m f_m = s ,$$

where s and λ_m are given numbers.

Defining

$$(6) \qquad r_n = f_{n+1}/f_n , \qquad n = 0, 1, 2, \ldots ,$$

we have by a result of Pincherle (see, e.g., Perron [1957, Satz 2.46C], Gautschi [1967]) that

$$(7) \qquad r_{n-1} = \frac{f_n}{f_{n-1}} = \frac{-b_n}{a_n -} \frac{b_{n+1}}{a_{n+1} -} \frac{b_{n+2}}{a_{n+2} -} \cdots , \qquad n = 1, 2, 3, \ldots ,$$

where the continued fractions converge precisely if (1) has a (nonvanishing) minimal solution, f_n. In principle, therefore, all ratios r_{n-1} are known, and (5) gives

$$(8) \qquad f_0 = \frac{s}{\displaystyle\sum_{m=0}^{\infty} \lambda_m r_0 r_1 \cdots r_{m-1}} ,$$

from which

$$(9) \qquad f_n = r_{n-1} f_{n-1} , \qquad n = 1, 2, 3, \ldots ,$$

by virtue of (6).

2.2.2. <u>Algorithms for minimal solutions.</u> Any implementation of the approach just described will involve, explicitly or implicitly, the truncated continued fractions

$$(10) \qquad r_{n-1}^{(\nu)} = \frac{-b_n}{a_n -} \frac{b_{n+1}}{a_{n+1} -} \cdots \frac{b_\nu}{a_\nu} , \qquad n = 1, 2, \ldots, \nu .$$

We assume they all exist. (They do for ν sufficiently large.) For simplicity of exposition, we consider the case of prescribed f_0, see (4).

(i) <u>Nonlinear backward recursion.</u> Evaluating (10) recursively
from behind, and then using an approximate version of (9), we get the
algorithm (Gautschi [1967])

(11)
$$\begin{cases} r_\nu^{(\nu)} = 0, \quad r_{n-1}^{(\nu)} = -\dfrac{b_n}{a_n + r_n^{(\nu)}}, & n = \nu, \nu-1, \ldots, 1, \\[4mm] f_0^{(\nu)} = f_0, \quad f_n^{(\nu)} = r_{n-1}^{(\nu)} f_{n-1}^{(\nu)}, & n = 1, 2, \ldots, \nu. \end{cases}$$

From Pincherle's result it follows that

(12)
$$\lim_{\nu \to \infty} f_n^{(\nu)} = f_n$$

for any fixed n.

The major inconvenience with (11) is the fact that we do not always
know an appropriate value of ν ahead of time, and may have to repeat
(11) several times, with increasing ν, until the $f_n^{(\nu)}$ converge to the
desired accuracy.

Replacing $r_\nu^{(\nu)} = 0$ in (11) by $r_\nu^{(\nu)} = \rho_\nu$, a suitable approximation
of $r_\nu = f_{\nu+1}/f_\nu$, often leads to improved convergence (Gautschi [1967,
pp. 38, 40], Scraton [1972]).

(ii) <u>Linear algebraic system.</u> The approximations $f_n^{(\nu)}$ in (11) can
be identified with the solution of the tridiagonal system

(13)
$$\begin{bmatrix} a_1 & 1 & & & O \\ b_2 & a_2 & 1 & & \\ & \cdot & \cdot & \cdot & \\ & & \cdot & \cdot & \cdot \\ O & & & b_\nu & a_\nu \end{bmatrix} \begin{bmatrix} f_1^{(\nu)} \\ f_2^{(\nu)} \\ \vdots \\ \vdots \\ f_\nu^{(\nu)} \end{bmatrix} = \begin{bmatrix} -b_1 f_0 \\ 0 \\ \vdots \\ \vdots \\ 0 \end{bmatrix},$$

which is formally obtained from (1) by setting $y_{\nu+1} = 0$.

45

(iii) <u>Miller's backward recurrence algorithm</u>. We may start the recurrence (1) with

(14) $$\eta_\nu = 1, \quad \eta_{\nu+1} = 0 \ ,$$

and use it in the backward direction to obtain $\eta_n = \eta_n^{(\nu)}$, $n = \nu-1, \nu-2,$ $\ldots, 1$. In effect, we produce a solution of the linear system (13), where f_0 on the right is replaced by $\eta_0^{(\nu)}$. Consequently,

(15) $$f_n^{(\nu)} = \frac{f_0}{\eta_0^{(\nu)}} \eta_n^{(\nu)}, \quad n = 0, 1, \ldots, \nu \ .$$

Generating $\eta_n^{(\nu)}$ as described, and then $f_n^{(\nu)}$ by (15), is known as <u>Miller's algorithm</u> (British Association for the Advancement of Science [1952, p. xvii]). It has the same disadvantage as noted in (i). In addition, the quantities $\eta_n^{(\nu)}$ may become large enough to cause overflow on a computer.

(iv) <u>Olver's algorithm</u>. Miller's algorithm can be thought of as solving the system (13) by a form of Gauss elimination, in which the elimination is performed backwards, from the last equation to the first, and the solution then obtained by forward substitution. The algorithm proposed by Olver [1967a] uses the more conventional forward elimination followed by back substitution. To describe it, let

$$p_0 = 0, \quad p_1 = 1, \quad e_0 = f_0 \ ,$$

(16) $$\left.\begin{array}{l} p_{n+1} = -a_n p_n - b_n p_{n-1} \\[2mm] e_n = b_n e_{n-1} \end{array}\right\} \quad n = 1, 2, \ldots, \nu \ .$$

Then

(17) $$f_{\nu+1}^{(\nu)} = 0, \quad p_{n+1} f_n^{(\nu)} - p_n f_{n+1}^{(\nu)} = e_n, \quad n = \nu, \nu-1, \ldots, 1 \ ,$$

which yields $f_\nu^{(\nu)}, f_{\nu-1}^{(\nu)}, \ldots, f_1^{(\nu)}$ in this order, provided none of the p_n vanishes.

We note from (17) that

$$\frac{f_n^{(\nu)}}{p_n} - \frac{f_{n+1}^{(\nu)}}{p_{n+1}} = \frac{e_n}{p_n p_{n+1}} \quad ,$$

so that

(18) $$f_n^{(\nu)} = p_n \sum_{k=n}^{\nu} \frac{e_k}{p_k p_{k+1}} \quad , \qquad n = 1, 2, \ldots, \nu \quad .$$

In particular, by (12),

(19) $$f_n = p_n \sum_{k=n}^{\infty} \frac{e_k}{p_k p_{k+1}} \quad .$$

It follows that $f_n^{(\nu)}$ has relative error

(20) $$\frac{f_n - f_n^{(\nu)}}{f_n} = \frac{\sum_{k=\nu+1}^{\infty} e_k / p_k p_{k+1}}{\sum_{k=n}^{\infty} e_k / p_k p_{k+1}} \doteq \frac{e_{\nu+1}}{p_{\nu+1} p_{\nu+2}} \Big/ \frac{e_n}{p_n p_{n+1}} \quad ,$$

the approximation on the far right being valid if the series in (19) converges rapidly. (Using the techniques in Olver [1967b] one could estimate the series more carefully and thus obtain a rigorous error bound).
If we wish to obtain f_n to within a relative error ϵ, we may thus iter-
ate with (16) until a value of ν is reached for which

(21) $$\left| \frac{e_{\nu+1}}{p_{\nu+1} p_{\nu+2}} \right| \leq \epsilon \min_n \left| \frac{e_n}{p_n p_{n+1}} \right| \quad ,$$

the minimum being taken over all values n of interest. With ν so determined, the $f_n^{(\nu)}$ are then obtained as described in (17). It is this
feature of automatically determining ν, which makes Olver's algorithm
attractive.

47

(v) <u>Olver's and Miller's algorithm combined</u>. In some applications, the recursion in (16) for p_n is mildly unstable, initially, although ultimately it is always stable. Olver and Sookne [1972] therefore suggest applying the procedure (16), which serves mainly to determine the cutoff-index ν, only in a region $n \geq n_0$ of perfect stability for the p-recursion, starting with $p_{n_0} = 0$, $p_{n_0+1} = 1$ as before, but with $e_{n_0} = 1$. Once ν is determined, the desired approximations are then obtained by recurring backward, as in Miller's procedure, starting with $f_{\nu+1}^{(\nu)} = 0$, $f_{\nu}^{(\nu)} = e_{\nu}/p_{\nu+1}$, and by a final normalization, as in (15).

We remark that all algorithms described can be extended to accommodate the more general "normalization condition" (5). This is an important point, inasmuch as the algorithms so extended do not require the calculation of any particular value of f_n (such as f_0 above). For details, we refer to the cited references.

2.2.3. <u>Applications to special functions</u>. The algorithms of 2.2.2 have been applied to a large number of special functions. The first major applications involved Bessel functions and Coulomb wave functions, whose recurrence relations are similar in nature. Further applications soon followed, e.g., to Legendre functions, incomplete beta and gamma functions, repeated integrals of the error function, and others. Detailed references, up to about 1965, can be found in Gautschi [1967]. More recently, in connection with Bessel functions, Mechel [1968] and Cylkowski [1971] discuss appropriate choices of the starting index ν in Miller's algorithm, while Amos [1974] proposes accurate starting values from uniform asymptotic expansions. The latter approach, combined with Taylor expansion where appropriate, is carefully implemented in Amos and Daniel [1973] and Amos, Daniel and Weston [unpubl.]. Ratios of successive Bessel functions (and of other functions, e.g., the repeated integrals of the error function) can also be computed by an iterative algorithm based on certain inequalities satisfied by these ratios (Amos [1973], [1974]). For Bessel functions, this is implemented in Amos and Daniel [1973]. Still on Bessel functions, we mention the work of Luke

48

[1972b], which relates Miller's algorithm to certain rational approximations in the theory of hypergeometric functions, and the computer implementation and certification of Olver's algorithm by Sookne [1973a, b, c, d].
Sidonskiĭ [1967] has a related algorithm for Bessel functions of integer
order and real argument, furnishing upper and lower bounds. Hitotumatu
[1967/68] recommends a nonlinear normalization condition in place of the
linear condition (5). On Coulomb wave functions we note a recent improvement by Gautschi [1969] on the recurrence algorithm (i), and refer
to Wills [1971] for a procedure very similar to Olver's. Kölbig [1972]
gives a survey of computational methods for Coulomb wave functions.
Legendre functions are discussed by Fettis [1967] and more recently by
Amos and Bulgren [1969] in connection with series expansions for the bivariate t-distribution in statistics. Bardo and Ruedenberg [1971] revisit
the repeated integrals of the error function. Temme [1972] applies algorithm (i) to certain Laplace integrals connected with van Wijngaarden's
transformation of formal series.

The stability of forward recurrence is analyzed by Wimp [1971/72],
and in the case of orthogonal polynomials of the Laguerre and Hermite
type, by Baburin and Lebedev [1967].

2.3. Inhomogeneous second-order and higher-order recurrence
relations

Some of the more esoteric functions are solutions of inhomogeneous
second-order recurrence relations,

(1) $$y_{n+1} + a_n y_n + b_n y_{n-1} = c_n, \qquad n = 1, 2, 3, \ldots .$$

Others satisfy recurrences of even higher order. The latter are also encountered in the computation of expansion coefficients, e.g., the coefficients in a Taylor series or a series in Chebyshev polynomials. Frequently, the solutions of interest are of the recessive type, in which
case some of the algorithms described in 2.2.2, suitably extended, are
again effective.

49

2.3.1. <u>Subdominant solutions of inhomogeneous second-order recurrence relations</u>. Assume that the homogeneous recurrence associated with (1) has a pair of linearly independent solutions g_n and h_n, of which g_n is minimal (with $g_0 \neq 0$), hence h_n dominant. We then call a solution f_n of (1) <u>subdominant</u> if

(2)
$$\lim_{n \to \infty} f_n / h_n = 0 .$$

A subdominant solution may or may not dominate the minimal solution g_n. If it does, neither forward nor backward recurrence is entirely satisfactory.

In analogy to 2.2(13) we consider the linear algebraic system

(3)
$$
\begin{bmatrix}
a_1 & 1 & & & O \\
b_2 & a_2 & 1 & & \\
& \ddots & \ddots & \ddots & \\
O & & & b_\nu & a_\nu
\end{bmatrix}
\begin{bmatrix}
f_1^{(\nu)} \\
f_2^{(\nu)} \\
\vdots \\
f_\nu^{(\nu)}
\end{bmatrix}
=
\begin{bmatrix}
c_1 - b_1 f_0 \\
c_2 \\
\vdots \\
c_\nu
\end{bmatrix} .
$$

If this system has a solution for all ν sufficiently large, and if f_n is a subdominant solution of (1), then by a result of Olver [1967a],

(4)
$$\lim_{\nu \to \infty} f_n^{(\nu)} = f_n .$$

The algorithms of Olver and Olver and Sookne (cf. 2.2.2(iv), (v)) thus extend readily to the case of subdominant solutions. So does in particular Olver's device for determining the appropriate ν and estimating the error (Olver [1967a, b]). Related algorithms are also discussed in Amos and Burgmeier [1973].

Olver applies his algorithm to Anger-Weber and Struve functions, while Amos and Burgmeier apply theirs to numerous other special functions, including incomplete Laplace transforms, and moments, of Bessel and Struve functions, the incomplete gamma function and Lommel functions. Sadowski and Lozier [1972] give an interesting application of Olver's algorithm to certain definite integrals in plasma physics, involving Chebyshev polynomials. Similar integrals are also treated by Piessens and Branders [1973].

2.3.2. Higher-order recurrence relations. Miller's algorithm is applicable to recurrence relations of arbitrary order, but, unless substantially modified, is effective only for solutions which are "sufficiently minimal". For a penetrating study of this we refer to Wimp [1969]. There are applications to hypergeometric and confluent hypergeometric functions in Wimp [1969], as well as in Wimp [1974], and another application in Wimp and Luke [1969]. Thacher [1972] discusses Miller's algorithm in connection with the solution in power series of linear differential equations with polynomial coefficients and relates minimality of the expansion coefficients to the singularities of the differential equation.

Given enough information about the growth pattern of fundamental solutions, approaches via boundary value problems appear to be more widely applicable. By imposing the right boundary conditions, it is sometimes possible to filter out a desired solution which is neither minimal nor dominant. The principal references in this direction are Oliver [1966/67], [1968 a, b].

§3. Nonlinear recurrence algorithms for elliptic integrals and elliptic functions

Some functions of several variables, notably elliptic integrals, have the remarkable property that their values remain unchanged as the variables undergo certain nonlinear transformations. Repeated application of these transformations, moreover, causes the variables to converge rapidly to certain limiting values, for which the functions can be evaluated by elementary means. These invariance properties thus give rise to interesting and powerful recursive algorithms for computing the functions in question.

3.1. Elliptic integrals and Jacobian elliptic functions

3.1.1. <u>Definitions and special values.</u> The best known functions enjoying invariance properties of the type indicated are the <u>elliptic integrals</u> of the first, second, and third kind. In Legendre's canonical form, they are, respectively,

$$
(1) \qquad F(\varphi, k) = \int_0^\varphi \frac{d\theta}{\sqrt{1-k^2 \sin^2 \theta}} ,
$$

$$
(2) \qquad E(\varphi, k) = \int_0^\varphi \sqrt{1-k^2 \sin^2 \theta} \, d\theta ,
$$

$$
(3) \qquad \Pi(\varphi, n, k) = \int_0^\varphi \frac{d\theta}{(1+n \sin^2 \theta)\sqrt{1-k^2 \sin^2 \theta}} .
$$

The variable k is known as the <u>modulus</u>; we assume it in the interval $0 \le k \le 1$. The <u>complementary modulus</u> k' is defined by

$$
(4) \qquad k' = \sqrt{1 - k^2} .
$$

The variable φ is called the <u>amplitude</u>, and we assume that $0 \le \varphi \le \pi/2$. The variable n in (3) may take on arbitrary values, provided the integral is interpreted in the sense of a Cauchy principal value, should n be negative and $1 + n \sin^2 \varphi < 0$.

The integrals (1)-(3) are called <u>complete</u>, or <u>incomplete</u>, depending on whether $\varphi = \pi/2$, or $\varphi < \pi/2$. The complete elliptic integrals are usually denoted by

$$
(5) \qquad \mathbb{K}(k) = F(\tfrac{\pi}{2}, k), \quad \mathbb{E}(k) = E(\tfrac{\pi}{2}, k), \quad \Pi(n, k) = \Pi(\tfrac{\pi}{2}, n, k) .
$$

As $k \downarrow 0$, or $k \uparrow 1$, we have the limiting values

(6)
$$\lim_{k \downarrow 0} F(\varphi, k) = \lim_{k \downarrow 0} E(\varphi, k) = \varphi ,$$

(7)
$$\lim_{k \uparrow 1} F(\varphi, k) = \tanh^{-1}(\sin \varphi) \ (0 \le \varphi < \pi/2), \ \lim_{k \uparrow 1} E(\varphi, k) = \sin \varphi .$$

Similar, but more complicated formulas hold for $\Pi(\varphi, n, k)$ (see, e.g., Byrd and Friedman [1971, p. 10]). We also note

(8)
$$F(\varphi, k) \sim \ell n \ \frac{4}{\cos \varphi + \sqrt{1-k^2 \sin^2 \varphi}}, \ E(\varphi, k) \sim 1 \ \text{as} \ k \uparrow 1, \ \varphi \uparrow \pi/2 ,$$

where the first relation is given by Carlson [1965, p. 39]; see also Nellis and Carlson [1966, p. 228].

Considering k fixed, the function $u = F(\varphi, k)$ is monotone in φ, and thus possesses an inverse function,

(9)
$$\varphi = \text{am} \, u ,$$

the amplitude function. In terms of it one defines Jacobian elliptic functions by

(10)
$$\text{sn} \, u = \sin \varphi, \ \text{cn} \, u = \cos \varphi, \ \text{dn} \, u = \sqrt{1-k^2 \sin^2 \varphi} .$$

3.1.2. Gauss transformations vs. Landen transformations. One distinguishes between Gauss transformations and Landen transformations, and for each between descending and ascending transformations. (Terminology, however, varies). In a descending transformation, the modulus k always decreases; in an ascending transformation, it always increases.

53

In a Gauss transformation, the amplitude φ varies in parallel with k (i. e., φ and k both increase or both decrease), while in a Landen transformation they vary in opposite directions. Repeated application of a descending transformation causes k to converge down to zero, while φ converges down to some limiting value φ_∞ in a Gauss transformation and up to ∞ in a Landen transformation. The former, therefore, eventually invokes the equations in (6). Repeated application of an ascending transformation, instead, causes k to converge upward to 1, while φ converges upward to $\pi/2$ in a Gauss transformation and down to some limiting value φ_∞ in a Landen transformation. The former, therefore, eventually invokes the relations in (8), the latter those in (7).

In describing these transformations, we limit ourselves to elliptic integrals of the first kind, and must refer to the literature for the others. An early treatment of computational algorithms for elliptic functions and integrals is King [1924]. We follow more closely the work of Carlson [1965], who develops the algorithms in a unified way, at least for integrals of the first two kinds. Hofsommer and van de Riet [1963] have ALGOL programs for integrals of the first and second kind, using Landen transformations, as well as programs for elliptic functions, based on ascending Landen and descending Gauss transformations. See also Neuman [1969/70a, b] and Kami, Kiyoto and Arakawa [1971a, b]. Descending transformations for integrals of the third kind are discussed by Ward [1960] in the case of complete integrals, and by Fettis [1965] in the case of incomplete integrals. A thorough treatment of descending Gauss and Landen transformations for integrals of all three kinds, complete with ALGOL procedures, is given in Bulirsch [1965a, b], and more definitively, especially as regards integrals of the third kind, in Bulirsch [1969a, b]. In the latter work, more general transformations, ascribed to Bartky, and extensions thereof, are used effectively. A good introduction into these developments is Bulirsch and Stoer [1968]. For the theory of elliptic integrals and elliptic functions we refer to the books of Neville [1944], [1971] and Tricomi [1948], [1951].

We begin with Gauss' process of the arithmetic-geometric mean, which underlies all algorithms for elliptic functions.

3.2. Gauss' algorithm of the arithmetic-geometric mean

Starting with $a_0 > b_0 > 0$, Gauss' algorithm generates two sequences $\{a_n\}$, $\{b_n\}$ by compounding the arithmetic and the geometric mean in the following manner,

(1)
$$\begin{cases} a_{n+1} = \frac{1}{2}(a_n + b_n) , \\[2mm] b_{n+1} = \sqrt{a_n b_n} , \end{cases} \qquad n = 0, 1, 2, \ldots .$$

Since the iteration functions are homogeneous of degree 1, only the ratio b_0/a_0 matters.

The arithmetic mean being larger than the geometric mean, we have $a_n > b_n$ for all n, and therefore $b_0 < a_{n+1} < a_n$, $b_n < b_{n+1} < a_0$. It follows that $\{a_n\}$ and $\{b_n\}$ both converge monotonically to certain limits, which, by letting $n \to \infty$ in (1), are readily found to be equal. The common limit is denoted by $M = M(a_0, b_0)$, and called the arithmetic-geometric mean of a_0 and b_0. Clearly, $b_0 < M(a_0, b_0) < a_0$.

In order to discuss the rate of convergence, it is convenient to introduce

(2)
$$\epsilon_n = \frac{a_n - b_n}{a_n + b_n} .$$

One finds by a simple computation that

(3)
$$\epsilon_{n+1} = \left(\frac{\epsilon_n}{1 + \sqrt{1 - \epsilon_n^2}} \right)^2 , \qquad n = 0, 1, 2, \ldots .$$

The sequence $\{\epsilon_n\}$, therefore, converges monotonically and quadratically to zero. Since

55

(4)
$$0 < a_n - M < (a_0 + M)\epsilon_n, \quad 0 < M - b_n < 2M\epsilon_n ,$$

we see that also $\{a_n\}$ and $\{b_n\}$ converge quadratically. We note from (2) that

(5)
$$\frac{b_n}{a_n} = 1 - 2\epsilon_n + O(\epsilon_n^2), \quad n \to \infty .$$

Quadratic convergence is a common feature of more general processes of compounding means (Lehmer [1971]). For variants of Gauss' algorithm (none of which quadratically convergent, however), and for many historical notes, see also Carlson [1971]. For complex variables, the algorithm is discussed by Fettis and Caslin [1969] and Morita and Horiguchi [1972/73].

In applications to elliptic integrals, the ratio b_0/a_0 will be identified with either the modulus k, or the complementary modulus k'. The algorithm (1) then generates a sequence of <u>transformed moduli</u> $k_n = b_n/a_n$, or $k_n' = b_n/a_n$, respectively, where in the former case

(6)
$$k_{n+1} = \frac{2\sqrt{k_n}}{1 + k_n}, \quad n = 0, 1, 2, \ldots, \quad k_0 = k ,$$

and in the latter,

(7')
$$k_{n+1}' = \frac{2\sqrt{k_n'}}{1 + k_n'}, \quad n = 0, 1, 2, \ldots, \quad k_0' = k' .$$

An equivalent form of (7') is

(7)
$$k_{n+1} = \frac{1 - k_n'}{1 + k_n'}, \quad n = 0, 1, 2, \ldots, \quad k_0 = k .$$

Since the modulus increases in (6), and decreases in (7), we call (6) an <u>ascending</u> and (7) a <u>descending transformation</u>. The convergence is to 1 and 0, respectively, and quadratic in both cases.

The choice of the transformation is dictated by the speed of convergence, which depends on the magnitude of $\epsilon_0 = (1-b_0/a_0)/(1+b_0/a_0)$. Since we want ϵ_0 small, we choose an ascending transformation if $k^2 > \frac{1}{2}$, and a descending transformation otherwise, so that in either case $1 > (b_0/a_0)^2 \geq \frac{1}{2}$, and thus

$$\epsilon_0 \leq \frac{1-2^{-\frac{1}{2}}}{1+2^{-\frac{1}{2}}} < .172 \ .$$

From (3) we then find that

$$(8) \qquad \epsilon_{n+1} = \left(\frac{\epsilon_n}{1+\sqrt{1-\epsilon_n^2}} \right)^2 < \frac{\epsilon_n^2}{(1+\sqrt{1-\epsilon_0^2})^2} < \frac{\epsilon_n^2}{3.94} \ ,$$

and so,

$$(9) \quad \epsilon_1 < .00751, \quad \epsilon_2 < 1.44 \times 10^{-5}, \quad \epsilon_3 < 5.27 \times 10^{-11}, \quad \epsilon_4 < 7.05 \times 10^{-22}, \ldots ,$$

illustrating the quadratic nature of convergence.

3.3. Computational algorithms based on Gauss and Landen transformations

3.3.1. Descending Gauss transformation.   We define

$$(1) \qquad \begin{cases} a_0 = 1, \quad b_0 = k', \quad t_0 = \csc \varphi \ , \\[2mm] a_{n+1} = \frac{1}{2}(a_n + b_n) \ , \\[2mm] b_{n+1} = \sqrt{a_n b_n} \ , \qquad\qquad n = 0, 1, 2, \ldots \ . \\[2mm] t_{n+1} = \frac{1}{2}(t_n + \sqrt{t_n^2 - a_n^2 + b_n^2}) \ , \end{cases}$$

One verifies without difficulty that t_n and a_n/t_n both decrease. Hence, $a_n/t_n \leq 1$, and t_n must converge,

$$t_n \downarrow T, \quad n \to \infty \; ,$$

where $T \geq M$. The speed of convergence is comparable to that of ϵ_n , in the sense

(2) $$t_n - T \sim \frac{M^2}{T} \epsilon_n, \quad n \to \infty \; .$$

To see this, observe from the last relation in (1), and from 3.2(5), that

$$t_{n+1} = \frac{1}{2} t_n \left\{ 1 + \sqrt{ 1 - \left(\frac{a_n}{t_n}\right)^2 \left[1 - \left(\frac{b_n}{a_n}\right)^2 \right] } \right\} = \frac{1}{2} t_n \left\{ 1 + \sqrt{ 1 - \left(\frac{a_n}{t_n}\right)^2 \left[4\epsilon_n + O(\epsilon_n^2) \right] } \right\}$$

$$= \frac{1}{2} t_n \left\{ 1 + 1 - 2\left(\frac{M}{T}\right)^2 \epsilon_n + o(\epsilon_n) \right\} = t_n \left\{ 1 - \left(\frac{M}{T}\right)^2 \epsilon_n + o(\epsilon_n) \right\} \; ,$$

from which

$$t_n - t_{n+1} = \frac{M^2}{T} \epsilon_n + o(\epsilon_n) \; .$$

Since ϵ_n converges quadratically, in particular $\epsilon_{n+1} < \epsilon_n^2$, we easily obtain, for any $p \geq 0$,

$$t_n - t_{n+p+1} = \frac{M^2}{T} \epsilon_n + o(\epsilon_n) \; ,$$

from which (2) follows by letting $p \to \infty$.

We now set

(3) $$\frac{a_n}{t_n} = \sin \varphi_n \; (0 < \varphi_n < \frac{\pi}{2}), \quad \frac{b_n}{a_n} = k_n', \quad n = 0, 1, 2, \ldots \; ,$$

which for $n = 0$ is consistent with the first relations in (1) (if $\varphi_0 = \varphi$, $k_0' = k'$) . The last relation in (1) can then be written in trigonometric form as

$$\sin \varphi_{n+1} = \frac{(1+k_n') \sin \varphi_n}{1 + \sqrt{1 - k_n^2 \sin^2 \varphi_n}} \quad , \qquad n = 0, 1, 2, \ldots \quad .$$

If in the integral $F(\varphi_n, k_n) = \int_0^{\varphi_n} (1 - k_n^2 \sin^2 \theta)^{-\frac{1}{2}} d\theta$ we make the change of variables

$$\sin \lambda = \frac{(1+k_n') \sin \theta}{1 + \sqrt{1 - k_n^2 \sin^2 \theta}} \quad , \qquad 0 < \theta \le \varphi_n \quad ,$$

we find, after a little computation, that

(4) $\qquad \dfrac{1}{a_n} F(\varphi_n, k_n) = \dfrac{1}{a_{n+1}} F(\varphi_{n+1}, k_{n+1}), \qquad n = 0, 1, 2, \ldots \quad .$

This is the <u>descending Gauss transformation</u> for elliptic integrals $F(\varphi, k)$. Since $k_n \downarrow 0$, and recalling 3.1(6), we conclude

(5) $\qquad F(\varphi, k) = \lim_{n \to \infty} \dfrac{1}{a_n} F(\varphi_n, k_n) = \dfrac{1}{M} \sin^{-1} \dfrac{M}{T} \quad .$

Thus, $F(\varphi, k)$ may be approximated by evaluating $a_n^{-1} \sin^{-1}(a_n/t_n)$ for some n sufficiently large. Observing that

$$0 < \frac{a_n}{t_n} - \frac{M}{T} = \frac{(a_n - M)T + M(T - t_n)}{t_n T} < \frac{a_n - M}{T} \quad ,$$

and using Taylor's theorem, and 3.2(4), we find

$$\left| \frac{1}{M} \sin^{-1} \frac{M}{T} - \frac{1}{a_n} \sin^{-1} \frac{a_n}{t_n} \right| \le \frac{M+1}{M^2} \left(\frac{\pi}{2} + \sec \varphi \right) \epsilon_n \quad .$$

For $a_n^{-1} \sin^{-1}(a_n/t_n)$ to be an acceptable approximation to $F(\varphi, k)$, it

suffices, therefore, that ϵ_n be sufficiently small [which for most purposes will be the case when $n = 3$ or $n = 4$; cf. 3.2(9)].

3.3.2. <u>Ascending Landen transformation.</u>   We define

$$
(6) \quad
\begin{cases}
a_0 = 1, \quad b_0 = k, \quad s_0 = \cot \varphi , \\[2mm]
a_{n+1} = \frac{1}{2}(a_n + b_n) , \\[2mm]
b_{n+1} = \sqrt{a_n b_n} , \qquad\qquad n = 0, 1, 2, \ldots . \\[2mm]
s_{n+1} = \frac{1}{2}\left(s_n + \sqrt{s_n^2 + a_n^2 - b_n^2} \right) ,
\end{cases}
$$

Clearly, s_n increases, while a_n/s_n decreases. An argument similar
to the one surrounding (2) shows that $s_n \uparrow S$, where $S < \infty$, and in
fact,

$$
(7) \qquad S - s_n \sim \frac{M^2}{S} \epsilon_n, \qquad n \to \infty .
$$

Letting

$$
(8) \qquad \frac{a_n}{s_n} = \tan \varphi_n \ (0 < \varphi_n < \frac{\pi}{2}), \quad \frac{b_n}{a_n} = k_n, \qquad n = 0, 1, 2, \ldots ,
$$

we can recast the last relation in (6) in the trigonometric form

$$
\tan \varphi_{n+1} = \frac{(1+k_n)\tan \varphi_n}{1 + \sqrt{1 + k_n'^2 \tan^2 \varphi_n}} , \qquad n = 0, 1, 2, \ldots .
$$

Similarly as in 3.3.1, it follows that

$$
(9) \qquad \frac{1}{a_n} F(\varphi_n, k_n) = \frac{1}{a_{n+1}} F(\varphi_{n+1}, k_{n+1}), \qquad n = 0, 1, 2, \ldots ,
$$

which is known as the <u>ascending Landen transformation.</u>   Making use of
3.1(7), we now obtain

(10) $\qquad F(\varphi, k) = \lim_{n \to \infty} \frac{1}{a_n} F(\varphi_n, k_n) = \frac{1}{M} \sinh^{-1} \frac{M}{S}$.

When S is small, there is some loss of significant figures in computing $s_n^2 + a_n^2 - b_n^2$. We can avoid this by introducing

(11) $\qquad\qquad d_n = \sqrt{a_n^2 - b_n^2}$,

and computing $s_n^2 + a_n^2 - b_n^2 = s_n^2 + d_n^2$, where the d_n are generated recursively by

(12) $\qquad\qquad d_{n+1} = \frac{d_n^2}{4a_{n+1}}$, $\qquad n = 0, 1, 2, \ldots$.

3.3.3. Ascending Gauss transformation. We define

(13) $\qquad\begin{cases} a_0 = 1, \quad b_0 = k, \quad q_0 = \csc\varphi , \\[2mm] a_{n+1} = \frac{1}{2}(a_n + b_n) , \\[2mm] b_{n+1} = \sqrt{a_n b_n} , \qquad\qquad n = 0, 1, 2, \ldots . \\[2mm] q_{n+1} = \frac{1}{2}\left(q_n + \frac{a_n b_n}{q_n}\right) , \end{cases}$

One verifies without difficulty that $q_n \geq a_n$ for all n . Consequently, $q_{n+1} < q_n$, and the sequence $\{q_n\}$, being monotone decreasing and bounded from below by M, converges to some limit. It is easily seen that the limit is M ,

$$q_n \downarrow M, \quad n \to \infty .$$

We set

(14) $\quad \dfrac{a_n}{q_n} = \sin \varphi_n \ \left(0 < \varphi_n < \dfrac{\pi}{2}\right), \quad \dfrac{b_n}{a_n} = k_n, \qquad n = 0, 1, 2, \ldots \ ,$

and rewrite the last relation in (13) as

$$\sin \varphi_{n+1} = \dfrac{(1+k_n)\sin \varphi_n}{1+k_n \sin^2 \varphi_n} \ , \qquad n = 0, 1, 2, \ldots \ ,$$

which shows that φ_n indeed increases. The <u>ascending Gauss transformation</u> takes the form

(15) $\quad \dfrac{1}{2^n a_n} F(\varphi_n, k_n) = \dfrac{1}{2^{n+1} a_{n+1}} F(\varphi_{n+1}, k_{n+1}), \qquad n = 0, 1, 2, \ldots \ .$

In contrast to the previous transformations, $F(\varphi_n, k_n)$ no longer remains bounded as $n \to \infty$. Indeed, simultaneously $\varphi_n \uparrow \pi/2$ and $k_n \uparrow 1$, and so, by 3.1(8) ,

$$F(\varphi_n, k_n) \sim \ln \dfrac{4}{\cos \varphi_n + \sqrt{1 - k_n^2 \sin^2 \varphi_n}} \ , \qquad n \to \infty \ .$$

From (15) we obtain

(16) $\quad F(\varphi, k) = \dfrac{1}{M} \lim_{n \to \infty} \left\{ 2^{-n} \ln \dfrac{2q_n + 2a_n}{\sqrt{q_n^2 - a_n^2} + \sqrt{q_n^2 - b_n^2}} \right\} \ .$

It suffices to evaluate the expression on the right for n large enough so that ϵ_n is negligible compared to 1 .

The denominator

$$e_n = \sqrt{q_n^2 - a_n^2} + \sqrt{q_n^2 - b_n^2}$$

can be computed without loss of significance by means of

$$4q_n e_{n+1} = e_n^2 + (a_n - b_n)^2 \; ,$$

where the second term on the right is substantially smaller than the first, $(a_n - b_n)^2 \le \epsilon_n e_n^2$. The cancellation error incurred in computing $a_n - b_n$, therefore, is of no consequence.

3.3.4. <u>Descending Landen transformation.</u> We define

(17)
$$\begin{cases} a_0 = 1, \quad b_0 = k_0', \quad p_0 = \cot \varphi \; , \\[2mm] a_{n+1} = \frac{1}{2}(a_n + b_n) \; , \\[2mm] b_{n+1} = \sqrt{a_n b_n} \; , \qquad\qquad n = 0, 1, 2, \ldots \; . \\[2mm] p_{n+1} = \frac{1}{2}\left(p_n - \frac{a_n b_n}{p_n} \right) \; , \end{cases}$$

This time, p_n cannot possibly tend to a finite limit P, as this would imply $P^2 = -M^2$, which is absurd. Neither need p_n preserve its sign.

Letting

(18)
$$\frac{a_n}{p_n} = \tan \varphi_n, \quad \frac{b_n}{a_n} = k_n', \qquad n = 0, 1, 2, \ldots \; ,$$

and writing the last relation of (17) in trigonometric form,

$$\tan \varphi_{n+1} = \frac{(1 + k_n') \tan \varphi_n}{1 - k_n' \tan^2 \varphi_n}, \qquad n = 0, 1, 2, \ldots \; ,$$

we find however that φ_n increases, if we take (Carlson [1965])

(19)
$$i_n \frac{\pi}{2} < \varphi_n \le (i_n + 1) \frac{\pi}{2} \; ,$$

where

63

$$(20) \qquad i_0 = 0, \qquad i_n = \begin{cases} 2i_{n-1} & \text{if } p_n \geq 0 \;, \\[2mm] 2i_{n-1}+1 & \text{if } p_n < 0 \text{ or } p_n = \infty \;. \end{cases}$$

The descending Landen transformation states that

$$(21) \qquad \frac{1}{2^n a_n} F(\varphi_n, k_n) = \frac{1}{2^{n+1} a_{n+1}} F(\varphi_{n+1}, k_{n+1}), \qquad n = 0, 1, 2, \dots \;,$$

and consequently, since $k_n \downarrow 0$, that

$$(22) \qquad F(\varphi, k) = \frac{1}{M} \lim_{n \to \infty} 2^{-n} \varphi_n, \qquad \varphi_n = \tan^{-1} \frac{a_n}{p_n} \;.$$

The branch of the inverse tangent is to be taken in conformity with (19) and (20).

3.4. Complete elliptic integrals

All four transformations discussed in 3.3 apply equally for complete integrals. Some of them, however, simplify.

Thus, in the descending Gauss transformation, we find that $t_n = a_n$ for all n, which reduces the algorithm 3.3(1), and 3.3(5), to

$$(1) \qquad \begin{cases} a_0 = 1, \quad b_0 = k' \;, \\[2mm] a_{n+1} = \dfrac{1}{2}(a_n + b_n) \;, \\[2mm] b_{n+1} = \sqrt{a_n b_n} \;, \\[2mm] \mathbb{K}(k) = \dfrac{\pi}{2M} \;. \end{cases} \qquad n = 0, 1, 2, \dots \;,$$

The arithmetic-geometric mean $M = M(1, k')$ is thus seen to be related to the complete elliptic integral of the first kind, $\mathbb{K}(k)$.

Similarly, in the descending Landen transformation, we have $p_0 = 0$, and thus $p_n = \infty$ for all $n \geq 1$, which, by 3.3(20) has the consequence that $i_n = 2^n - 1$. By 3.3(19), therefore, $\varphi_n = 2^{n-1} \pi$, and 3.3(22) then reestablishes (1). The descending Gauss and Landen transformations thus become identical.

Not so for the ascending transformations. In the <u>ascending Gauss transformation</u>, we find $q_n = a_n$ for all n, and 3.3(13), together with 3.3(16), where n is conveniently replaced by $n + 1$, simplify to

(2)
$$
\begin{cases}
a_0 = 1, \quad b_0 = k, \\
a_{n+1} = \dfrac{1}{2}(a_n + b_n), \\
b_{n+1} = \sqrt{a_n b_n}, \\
\mathbb{K}(k) = \dfrac{1}{2M} \lim_{n \to \infty} \left(2^{-n} \ln \dfrac{4}{\epsilon_n}\right).
\end{cases}
\qquad n = 0, 1, 2, \ldots ,
$$

The <u>ascending Landen transformation</u>, finally, neither simplifies, nor does it preserve the completeness of the integral.

3.5. Jacobian elliptic functions

All four algorithms of 3.3, suitably reversed, yield algorithms for computing Jacobian elliptic functions. We recall that, by definition,

(1)
$$\operatorname{sn} u = \sin \varphi, \quad \text{where} \quad u = F(\varphi, k), \quad 0 \leq u \leq \mathbb{K}(k).$$

In the case of the <u>descending Gauss transformation</u>, e.g., we need to compute $\operatorname{sn} u = 1/t_0$ in 3.3(1), knowing that $T = \lim_{n \to \infty} t_n = M/\sin(Mu)$ by virtue of 3.3(5). We accomplish this by using the Gauss arithmetic-geometric mean process to compute M, hence T, and then reversing the recursion for t_n in 3.3(1) to compute t_0. Thus (Salzer [1962], Hofsommer and van de Riet [1963], Carlson [1965]),

65

$$(2) \quad \begin{cases} a_0 = 1, \quad b_0 = k' , \\[4pt] \left. \begin{array}{l} a_{n+1} = \dfrac{1}{2}(a_n + b_n) \\[8pt] b_{n+1} = \sqrt{a_n b_n} \end{array} \right\} \quad n = 0, 1, \ldots, \nu-1 , \\[20pt] t_\nu^{(\nu)} = a_\nu / \sin(a_\nu u) , \\[12pt] t_{n-1}^{(\nu)} = t_n^{(\nu)} + \dfrac{a_{n-1}^2 - b_{n-1}^2}{4 t_n^{(\nu)}} , \quad n = \nu, \nu-1, \ldots, 1 , \\[16pt] sn\, u = \dfrac{1}{t_0^{(\nu)}} , \quad cn\, u = \sqrt{[t_0^{(\nu)}]^2 - 1}\, sn\, u, \quad dn\, u = (2 t_1^{(\nu)} - t_0^{(\nu)})\, sn\, u , \end{cases}$$

where ν is chosen large enough for $a_\nu - M$ to be negligible. (By 3.2(4), this will be the case if ϵ_ν is negligible compared to 1/2). A simpler form of the t-recursion results from using 3.3.2(11) and (12),

$$(2') \quad t_{n-1}^{(\nu)} = t_n^{(\nu)} + \frac{a_n d_n}{t_n^{(\nu)}} , \quad n = \nu, \nu-1, \ldots, 1 .$$

From the <u>ascending Landen transformation</u> we obtain (Southard [1963], Hofsommer and van de Riet [1963], Carlson [1965]), similarly,

$$(3) \quad \begin{cases} a_0 = 1, \quad b_0 = k , \\[4pt] \left. \begin{array}{l} a_{n+1} = \dfrac{1}{2}(a_n + b_n) \\[8pt] b_{n+1} = \sqrt{a_n b_n} \end{array} \right\} \quad n = 0, 1, \ldots, \nu-1 , \\[20pt] s_\nu^{(\nu)} = a_\nu / \sinh(a_\nu u) , \\[12pt] s_{n-1}^{(\nu)} = s_n^{(\nu)} - \dfrac{a_n d_n}{s_n^{(\nu)}} , \quad n = \nu, \nu-1, \ldots, 1 , \\[16pt] sn\, u = \dfrac{1}{\sqrt{1+[s_0^{(\nu)}]^2}} , \quad cn\, u = s_0^{(\nu)}\, sn\, u, \quad dn\, u = (2 s_1^{(\nu)} - s_0^{(\nu)})\, sn\, u . \end{cases}$$

66

According to the discussion at the end of 3.2, the ascending algorithm
(3) is faster than the descending algorithm (2) when $k^2 > \frac{1}{2}$.

§4. Computer software for special functions

Good numerical methods need to be made easily accessible to the
interested user. One way of doing this is by providing computer programs written in one of the higher-level languages such as FORTRAN or
ALGOL. For special functions, as well as for many other mathematical
problem areas, a great number of such programs are in fact available,
and have been so for some time. There are published algorithms in
specialized journals (e.g., Comm. ACM, Numer. Math., BIT, Computer
Physics Comm., Applied Statistics, and ACM Trans. Mathematical Software), and many others in user's group libraries, commercial libraries,
local subroutine libraries, etc. Unfortunately, the quality of these algorithms and programs varies enormously. It has been felt, therefore,
in recent years, that libraries should be established by selecting a few
algorithms, known for their outstanding quality, implementing them carefully into reliable and thoroughly tested pieces of computer software,
integrating the pieces into larger, well-streamlined, and easy-to-use
collections of subroutines, and finally releasing these collections to the
computing public, with provisions for updating them at regular intervals.

This is not the place to enter into a discussion of the many design
objectives and desirable attributes of such packages, nor of explaining
the considerable difficulties in trying to attain them; we refer for this to
Rice [1971] and Cody [1974]. We would like to draw attention, however,
to two current efforts in this direction, one in the United States known
as the NATS project (National Activity to Test Software), the other in
England, known as the NAG project (Numerical Algorithms Group, former-
ly Nottingham Algorithms Group). The former's original objective is to
produce high-quality software for two restricted problem areas, namely
matrix eigensystem problems, and special functions, for which initial
packages have been released in 1972 and 1973 under the names EISPACK

67

and FUNPACK, respectively. The latter's objectives are similar, but embrace a wider problem area – essentially all the major numerical analysis problems. The most recent version ("mark 4") was completed in 1973. For a general description of the NATS project we refer to Boyle et al. [1972] and Smith, Boyle and Cody [1974], and for a discussion of the NAG project to Ford and Hague [1974] and Ford and Sayers [1974]. We briefly compare the two efforts, as far as they concern special functions.

4.1. NATS software for special functions

The special function package of the NATS project – FUNPACK – is developed and maintained under the direction of Cody at Argonne National Laboratory (Cody [1975]). His principal decisions in designing FUNPACK are, first of all, to adopt FORTRAN as the exclusive language of the package, and, secondly, to limit the programs to three different lines of computers, namely the IBM 360-370 series, the CDC6000-7000 series, and Univac 1108. Accordingly, only three accuracy requirements have to be dealt with, roughly 14 significant decimal digits on CDC equipment, and 16, respectively 18, decimal digits for the hardware double-precision arithmetic on IBM and Univac equipment. The package, therefore, is designed to perform well on these particular machines, and is not expected, nor intended, to be easily transportable to other machines.

The limitation to three different precisions has a major influence in the selection of approximation methods. Most attractive, under the circumstances, are rational Chebyshev approximations, both by virtue of their efficiency and uniform accuracy. This, in fact, is the choice made in FUNPACK. The current version I includes subroutines for six special functions – the exponential integral, the complete elliptic integrals of the first and second kind, Dawson's integral, and the Bessel functions K_0 and K_1. All of them are computed from appropriate best rational approximations. Plans are underway to extend the package to include sequences of functions and multivariate functions.

68

All the programming of the package, as well as the initial testing, was done at Argonne National Laboratory, which has IBM equipment. Similar tests were run on CDC equipment at the University of Texas, and on Univac equipment at the University of Wisconsin. After this initial testing and "tuning" of the routines, they were subjected to additional field tests on the same type of computers, running, however, with different FORTRAN compilers and under a variety of operating systems, some in batch mode, others in time sharing mode. Only after successful completion of all field tests, in September of 1973, was the first version of the package released.

4.2. NAG software for special functions

The special function chapter of the NAG library is being developed by Schonfelder at the University of Birmingham (Schonfelder [1974a, b]). While the basic objectives, and methods of testing, are similar to those of the NATS project, there are some significant differences. For one, all programs in the NAG library are written separately in two languages, FORTRAN and ALGOL. For another, the library is designed to be highly portable, i.e., to run, with a minimal amount of changes, on a wide variety of different machines. Finally, coverage is hoped to eventually include all the major functions in Abramowitz and Stegun [1964] — roughly fifty separate functions. At the moment (Schonfelder [1975]), the list of functions for the forthcoming edition ("mark 5") is to include the exponential, sine, and cosine integral, the gamma function, the error function and Fresnel integrals, and the Bessel functions J_0, J_1, Y_0, Y_1, I_0, I_1, K_0, K_1. Plans exist to cover functions of several variables, and of complex variables, but implementation appears to be several years in the future (Schonfelder [1975]).

The choices made for the methods of computation reflect the multi-machine character of the NAG library. Preference, in fact, is given to expansions in Chebyshev polynomials, which can be truncated easily to fit various machine precisions, although they may be somewhat inferior in efficiency compared to rational approximations.

69

4. 3. Other software for special functions

Good subroutines for special functions can be found in other mathematical subroutine libraries, e. g. , the Boeing library and handbook
(Newbery [1971]), containing programs in FORTRAN, and the NUMAL library
(Numerical procedures in ALGOL 60) developed at the Mathematical Centre
in Amsterdam (den Heijer et al. [1974]). The latter has appeared in seven
volumes, volume 6 being devoted to special functions. In addition, there
are a number of commercial subroutine packages. IBM offers SSP
(Scientific Subroutine Package), currently in its 5th edition, and SLMATH
(Subroutine Library Mathematics) and its PL/1 version, PLMATH, while
IMSL (International Mathematical Statistical Libraries) regularly issues
revised editions of its library.

We listed only those library projects, relevant to special functions,
which are most familiar to us, realizing that there are undoubtedly many
others.

Acknowledgments. The author is indebted to Profs. F. W. J. Olver
and H. Thacher, Jr. , who read the entire manuscript and suggested several improvements. He also gratefully acknowledges helpful comments
by Prof. P. Wynn on section 1.4 of the manuscript.

REFERENCES

Abramowitz, M. , and Stegun, I. A. (1964): Handbook of mathematical
functions, Nat. Bur. Standards Appl. Math. Ser. 55.
Achieser, N. I. (1956): Theory of Approximation, Frederick Ungar Publ.
Co. , New York.
Alexits, G. (1961): Convergence problems of orthogonal series, Pergamon
Press, New York-Oxford-Paris.
Allen, G. D. , Chui, C. K. , Madych, W. R. , Narcowich, F. J. , and Smith,
P. W. (1974): Padé approximation and orthogonal polynomials, Bull.
Austral. Math. Soc. 10, 263-270.

70

Amos, D. E. (1970): Significant digit computation of certain distribution functions, Proc. Sympos. on Empirical Bayes Estimation and Computing in Statistics, pp. 165-180. Texas Tech. Press, Lubbock, Tex.

_____ (1973): Bounds on iterated coerror functions and their ratios, Math. Comp. 27, 413-427.

_____ (1974): Computation of modified Bessel functions and their ratios, Math. Comp. 28, 239-251.

_____ and Bulgren, W. G. (1969): On the computation of a bivariate t-distribution, Math. Comp. 23, 319-333.

_____ and Burgmeier, J. W. (1973): Computation with three-term, linear, nonhomogeneous recursion relations, SIAM Rev. 15, 335-351.

_____ and Daniel, S. L. (1972): CDC 6600 utility routines for Chebyshev approximation and function inversion, Report SC-DR-720917, Sandia Laboratories, Albuquerque, New Mexico.

_____ and _____ (1973): CDC 6600 codes for Bessel functions $I_\nu(x)$, $ber_\nu(x)$, $bei_\nu(x)$ and ratios $I_{\nu+1}(x)/I_\nu(x)$, Report SLA-73-0072, Sandia Laboratories, Albuquerque, New Mexico.

_____, _____, and Weston, M, K. (unpubl.): CDC 6600 subroutines IBESS and JBESS for Bessel functions $I_\nu(x)$ and $J_\nu(x)$, $x \geq 0$, $\nu \geq 0$.

Baburin, O. V., and Lebedev, V. I. (1967): The computation of tables of roots and weights of Hermite and Laguerre polynomials for $n = 1(1)101$ (Russian), Ž. Vyčisl. Mat. i Mat. Fiz. 7, 1021-1030.

Baker, G. A., Jr. (1965): The theory and application of the Padé approximant method, in "Advances in Theoretical Physics" (K. A. Brueckner, ed.), Vol. 1, pp. 1-58. Academic Press, New York-London.

_____ (1970): The Padé approximant method and some related generalizations, in "The Padé Approximant in Theoretical Physics" (G. A. Baker, Jr. and J. L. Gammel, eds.), pp. 1-39. Academic Press, New York-London.

71

Baker, G. A., Jr. (1973): Recursive calculation of Padé approximants, in "Padé Approximants and their Applications" (P. R. Graves-Morris, ed.), pp. 83-91. Academic Press, London-New York.

_____ (1975): Essentials of Padé Approximants, Academic Press, New York -London.

_____ and Gammel, J. L., eds. (1970): The Padé Approximant in Theoretical Physics, Academic Press, New York -London.

_____, Rushbrooke, G. S., and Gilbert, H. E. (1964): High-temperature series expansions for the spin-$\frac{1}{2}$ Heisenberg model by the method of irreducible representations of the symmetric group, Phys. Rev. 135, A1272-A1277.

Bardo, R. D., and Ruedenberg, K. (1971): Numerical analysis and evaluation of normalized repeated integrals of the error function and related functions, J. Computational Phys. 8, 167-174.

Barlow, R. H. (1974): Convergent continued fraction approximants to generalised polylogarithms, BIT 14, 112-116.

Bhagwandin, K. (1962): L'approximation uniforme des fonctions d'Airy-Stokes et fonctions de Bessel d'indices fractionnaires, 2^e Congr. Assoc. Française Calcul Traitement Information, Paris, 1961, pp. 137-145. Gauthier-Villars, Paris.

Bjalkova, A. I. (1963): Computation of Fourier-Chebyshev coefficients (Russian), Metody Vyčisl. 1, 27-29.

Blair, J. M. (1974): Rational Chebyshev approximations for the modified Bessel functions $I_0(x)$ and $I_1(x)$, Math. Comp. 28, 581-583.

_____ and Edwards, C. A. (1974): Stable rational minimax approximations to the modified Bessel functions $I_0(x)$ and $I_1(x)$, Report AECL - 4928, Atomic Energy of Canada Limited, Chalk River Nuclear Laboratories, Chalk River, Ontario.

Blanch, G. (1964): Numerical evaluation of continued fractions, SIAM Rev. 6, 383-421.

Blum, E. K., and Curtis, P. C., Jr. (1961): Asymptotic behavior of the best polynomial approximation, J. Assoc. Comput. Mach. 8, 645-647.

Boyle, J. M., Cody, W. J., Cowell, W. R., Garbow, B. S., Ikebe, Y., Moler, C. B., and Smith, B. T. (1972): NATS - a collaborative effort to certify and disseminate mathematical software, Proc. ACM Annual Conference, 1972, vol. II, pp. 630-635. Assoc. Comput. Mach., New York.

British Association for the Advancement of Science (1952): Mathematical Tables, vol. X, Bessel functions, Part II, Functions of Positive Integer Order, Cambridge University Press.

Bulirsch, R. (1965a): Numerical calculation of elliptic integrals and elliptic functions, Numer. Math. 7, 78-90.

_____ (1965b): Numerical calculation of elliptic integrals and elliptic functions II, Numer. Math. 7, 353-354.

_____ (1969a): An extension of the Bartky-transformation to incomplete elliptic integrals of the third kind, Numer. Math. 13, 266-284.

_____ (1969b): Numerical calculation of elliptic integrals and elliptic functions III, Numer. Math. 13, 305-315.

_____ and Rutishauser, H. (1968): Interpolation und genäherte Quadratur, in "Mathematische Hilfsmittel des Ingenieurs" (R. Sauer and I. Szabó, eds.), Teil III, pp. 232-319. Springer-Verlag, Berlin-Heidelberg-New York.

_____ and Stoer, J. (1968): Darstellung von Funktionen in Rechenautomaten, in "Mathematische Hilfsmittel des Ingenieurs" (R. Sauer and I. Szabó, eds.), Teil III, pp. 352-446. Springer-Verlag, Berlin-Heidelberg-New York.

Byrd, P. F., and Friedman, M. D. (1971): Handbook of elliptic integrals for engineers and scientists, 2nd ed., Springer-Verlag, New York-Heidelberg-Berlin.

Carlson, B. C. (1965): On computing elliptic integrals and functions, J. Math. and Phys. 44, 36-51.

_____ (1971): Algorithms involving arithmetic and geometric means, Amer. Math. Monthly 78, 496-505.

73

Chawla, M. M. (1966/67): A note on the estimation of the coefficients in the Chebyshev series expansion of a function having a logarithmic singularity, Comput. J. 9, 413.

Cheney, E. W. (1966): Introduction to Approximation Theory, McGraw-Hill, New York-Toronto-London.

_____ and Southard, T. H. (1963): A survey of methods for rational approximation, with particular reference to a new method based on a formula of Darboux, SIAM Rev. 5, 219-231.

Chipman, D. M. (1972): The numerical computation of two transcendental functions related to the exponential integral, Math. Comp. 26, 241-249.

Chisholm, J. S. R. (1973a): Rational approximants defined from double power series, Math. Comp. 27, 841-848.

_____ (1973b): Mathematical theory of Padé approximants, in "Padé Approximants" (P. R. Graves-Morris, ed.), Lectures delivered at a summer school held at the University of Kent, July 1972, pp. 1-18. The Institute of Physics, London-Bristol.

_____ (1973c): Convergence properties of Padé approximants, in "Padé Approximants and their Applications" (P. R. Graves-Morris, ed.), pp. 11-21. Academic Press, London-New York.

Chui, C.K., Shisha, O. and Smith, P. W. (1974): Padé approximants as limits of best rational approximants, J. Approximation Theory 12, 201-204.

Clenshaw, C. W. (1955): A note on the summation of Chebyshev series, Math. Tables Aids Comput. 9, 118-120.

_____ (1962): Chebyshev series for mathematical functions, National Physical Laboratory Mathematical Tables, Vol. 5. Her Majesty's Stationery Office, London.

_____ (1964): A comparison of "best" polynomial approximations with truncated Chebyshev series expansions, J. Soc. Indust. Appl. Math. Ser. B Numer. Anal. 1, 26-37.

Clenshaw, C. W. (1974): Rational approximations for special functions, in "Software for Numerical Mathematics" (D. J. Evans, ed.), pp. 275-284. Academic Press, London-New York.

_____ and Picken, S. M. (1966): Chebyshev series for Bessel functions of fractional order, National Physical Laboratory Mathematical Tables, Vol. 8. Her Majesty's Stationery Office, London.

_____ and Lord, K. (1974): Rational approximations from Chebyshev series, in "Studies in Numerical Analysis" (B. K. P. Scaife, ed.), pp. 95-113. Academic Press, London-New York.

_____, Miller, G. F., and Woodger, M. (1962/63): Algorithms for special functions I, Numer. Math. 4, 403-419.

Cody, W. J. (1965): Chebyshev approximations for the complete elliptic integrals K and E, Math. Comp. 19, 105-112. {Corrigenda: ibid. 20 (1966), 207}.

_____ (1967): Another aspect of economical polynomials, Letters to the Editor, Comm. ACM 10, 531.

_____ (1968): Chebyshev approximations for the Fresnel integrals, Math. Comp. 22, 450-453.

_____ (1969): Rational Chebyshev approximations for the error function, Math. Comp. 23, 631-637.

_____ (1970): A survey of practical rational and polynomial approximation of functions, SIAM Rev. 12, 400-423. {Reprinted in: SIAM Studies in Appl. Math. 6 (1970), 86-109}.

_____ (1974): The construction of numerical subroutine libraries, SIAM Rev. 16, 36-46.

_____ (1975): The FUNPACK package of special function subroutines, ACM Trans. Mathematical Software 1, 13-25.

_____ and Hillstrom, K. E. (1967): Chebyshev approximations for the natural logarithm of the gamma function, Math. Comp. 21, 198-203.

_____ and _____ (1970): Chebyshev approximations for the Coulomb phase shift, Math. Comp. 24, 671-677. {Corrigendum: ibid. 26 (1972), 1031}.

75

Cody, W.J., and Stoer, J. (1966/67): Rational Chebyshev approximation using interpolation, Numer. Math. 9, 177-188.

_____ and Thacher, H. C., Jr. (1967): Rational Chebyshev approximations for Fermi-Dirac integrals of orders - 1/2, 1/2 and 3/2, Math. Comp. 21, 30-40.

_____ and _____ (1968): Rational Chebyshev approximations for the exponential integral $E_1(x)$, Math. Comp. 22, 641-649.

_____ and _____ (1969): Chebyshev approximations for the exponential integral Ei(x), Math. Comp. 23, 289-303.

_____, Fraser, W., and Hart, J. F. (1968): Rational Chebyshev approximation using linear equations, Numer. Math. 12, 242-251.

_____, Hillstrom, K. E., and Thacher, H. C., Jr. (1971): Chebyshev approximations for the Riemann zeta function, Math. Comp. 25, 537-547.

_____, Paciorek, K. A., and Thacher, H. C., Jr. (1970): Chebyshev approximations for Dawson's integral, Math. Comp. 24, 171-178.

_____, Strecok, A. J., and Thacher, H. C., Jr. (1973): Chebyshev approximations for the psi function, Math. Comp. 27, 123-127.

Cohen, E. A., Jr. (1971): Note on a truncated Chebyshev series modified to match function values at interval endpoints, SIAM J. Numer. Anal. 8, 754-756.

Collatz, L. (1965): Einschliessungssatz für die Minimalabweichung bei der Segmentapproximation, Simpos. Internaz. Appl. Analisi Fis. Mat., Cagliari - Sassari 1964, pp. 11-21. Edizioni Cremonese, Roma.

_____ (1968): Zur numerischen Behandlung der rationalen Tschebyscheff-Approximation bei mehreren unabhängigen Veränderlichen, Apl. Mat. 13, 137-146.

Common, A. K. (1969): Properties of Legendre expansions related to series of Stieltjes and applications to π - π scattering, Nuovo Cimento 63A, 863-891.

Common, A. K., and Graves-Morris, P. R. (1974): Some properties of Chisholm approximants, J. Inst. Math. Appl. 13, 229-232.

Cooper, G. J. (1967): The evaluation of the coefficients in a Chebyshev expansion, Comput. J. 10, 94-100.

Cox, M. G. (unpubl.): Numerical computations associated with Chebyshev polynomials.

Cylkowski, Z. (1966/68): Chebyshev series expansions of the functions $J_\nu(kx)/(kx)^\nu$ and $I_\nu(kx)/(kx)^\nu$, Zastos. Mat. 9, 413-415.

_____ (1971): Remarks on the evaluation of the Bessel functions from the recurrent formula, Zastos. Mat. 12, 217-220.

Davis, P. J. (1963): Interpolation and Approximation, Blaisdell Publ. Co., New York-Toronto-London.

Deuflhard, P. (1974): On algorithms for the summation of certain special functions, Bericht Nr. 7407, Techn. Univ. München, Abteilung Mathematik.

De Vogelaere, R. (1959): Remarks on the paper "Tchebysheff approximations for power series", J. Assoc. Comput. Mach. 6, 111-114.

Elliott, D. (1964): The evaluation and estimation of the coefficients in the Chebyshev series expansion of a function, Math. Comp. 18, 274-284.

_____ (1968): Error analysis of an algorithm for summing certain finite series, J. Austral. Math. Soc. 8, 213-221.

_____ and Szekeres, G. (1965): Some estimates of the coefficients in the Chebyshev series expansion of a function, Math. Comp. 19, 25-32.

_____ and Lam, B. (1973): An estimate of $E_n(f)$ for large n, SIAM J. Numer. Anal. 10, 1091-1102.

Fair, W. (1964): Padé approximation to the solution of the Ricatti equation, Math. Comp. 18, 627-634.

_____ and Luke, Y. L. (1967): Rational approximations to the incomplete elliptic integrals of the first and second kinds, Math. Comp. 21, 418-422.

Feinerman, R. P. , and Newman, D. J. (1974): Polynomial approximation, The Williams and Wilkins Co. , Baltimore.

Fejér, L. (1910): Lebesguesche Konstanten und divergente Fourierreihen, J. Reine Angew. Math. 138, 22-53.

Fettis, H. E. (1965): Calculation of elliptic integrals of the third kind by means of Gauss' transformation, Math. Comp. 19, 97-104.

_____ (1967): Calculation of toroidal harmonics without recourse to elliptic integrals, in: "Blanch Anniversary Volume" (B. Mond, ed.), pp. 21-34. Aerospace Research Lab. , U. S. Air Force, Washington, D. C

_____ and Caslin, J. C. (1969): A table of the complete elliptic integral of the first kind for complex values of the modulus I, II, Reports ARL 69-0172 and 69-0173, Wright-Patterson Air Force Base, Ohio.

Fike, C. T. (1967): Methods of evaluating polynomial approximations in function evaluation routines, Comm. ACM 10, 175-178.

_____ (1968): Computer Evaluation of Mathematical Functions, Prentice-Hall, Englewood Cliffs, N. J.

Fleischer, J. (1972): Analytic continuation of scattering amplitudes and Padé approximants, Nucl. Phys. B37 , 59-76. {Erratum: ibid. B44 (1972), 641}.

_____ (1973a): Nonlinear Padé approximants for Legendre series, J. Mathematical Phys. 14, 246-248.

_____ (1973b): Generalizations of Padé approximants, in: "Padé Approximants" (P. R. Graves-Morris, ed.), Lectures delivered at a summer school held at the University of Kent, July 1972, pp. 126-131. The Institute of Physics, London-Bristol.

Fletcher, R. , Grant, J. A. , and Hebden, M. D. (1974): Linear minimax approximation as the limit of best L_p-approximation, SIAM J. Numer. Anal. 11, 123-136.

Ford, B. , and Hague, S. J. (1974): The organisation of numerical algorithms libraries, in: "Software for Numerical Mathematics" (D. J. Evans, Ed.), pp. 357-372. Academic Press, London-New York.

Ford, B., and Sayers, D. (1974): Developing a single numerical algorithms library for different machine ranges, in: "Mathematical Software II", pp. 234-237, Informal Proceedings of a Conference, Purdue Univ., May 29-31, 1974.

Fox, L. (1965): The proper use of recurrence relations, Math. Gaz. $\underline{49}$, 371-387.

_____, and Parker, I. B. (1968): Chebyshev Polynomials in Numerical Analysis, Oxford University Press, London-New York-Toronto.

Frankel, A. P., and Gragg, W. B. (1973): Algorithmic almost best uniform rational approximation with error bounds (abstract), SIAM Rev. $\underline{15}$, 418-419.

Fraser, W. (1965): A survey of methods of computing minimax and near-minimax polynomial approximations for functions of a single independent variable, J. Assoc. Comput. Mach. $\underline{12}$, 295-314.

Gargantini, I. (1966): On the application of the process of equalization of maxima to obtain rational approximation to certain modified Bessel functions, Comm. ACM $\underline{9}$, 859-863.

_____ and Henrici, P. (1967): A continued fraction algorithm for the computation of higher transcendental functions in the complex plane, Math. Comp. $\underline{21}$, 18-29.

_____ and Pomentale, T. (1964): Rational Chebyshev approximations to the Bessel function integrals $Ki_s(x)$, Comm. ACM $\underline{7}$, 727-730.

Gautschi, W. (1961): Recursive computation of certain integrals, J. Assoc. Comput. Mach. $\underline{8}$, 21-40.

_____ (1966): Computation of successive derivatives of $f(z)/z$, Math. Comp. $\underline{20}$, 209-214.

_____ (1967): Computational aspects of three-term recurrence relations, SIAM Rev. $\underline{9}$, 24-82.

_____ (1969): An application of minimal solutions of three-term recurrences to Coulomb wave functions, Aequationes Math. $\underline{2}$, 171-176.

Gautschi, W. (1970): Efficient computation of the complex error function, SIAM J. Numer. Anal. 7, 187-198.

_____ (1972): Zur Numerik rekurrenter Relationen, Computing 9, 107-126. {English translation in: Report ARL 73-0005, Aerospace Research Laboratories, Wright-Patterson Air Force Base, Ohio, 1973}.

_____ (1973): Algorithm 471-Exponential integrals, Comm. ACM 16, 761-763.

_____ and Klein, B. J. (1970): Recursive computation of certain derivatives - A study of error propagation, Comm. ACM 13, 7-9.

Gentleman, W. M. (1969/70): An error analysis of Goertzel's (Watt's) method for computing Fourier coefficients, Comput. J. 12, 160-165.

Goertzel, G. (1958): An algorithm for the evaluation of finite trigonometric series, Amer. Math. Monthly 65, 34-35.

_____ (1960): Fourier analysis, in "Mathematical Methods for Digital Computers" (A. Ralston and H. S. Wilf, eds.), pp. 258-262. Wiley, New York-London.

Golden, J. E., McGuire, J. H., and Nuttall, J. (1973): Calculating Bessel functions with Padé approximants, J. Math. Anal. Appl. 43, 754-767.

Gragg, W. B. (1972): The Padé table and its relation to certain algorithms of numerical analysis, SIAM Rev. 14, 1-62.

_____ and Johnson, G. D. (1974): The Laurent-Padé table, Proc. IFIP Congress 74, pp. 632-637. North-Holland Publ. Co.

Graves-Morris, P. R., ed. (1973a): Proceedings of the Canterbury Summer School on Padé Approximants and their Applications, Institute of Physics.

_____ (1973b): Proceedings of the 1972 Canterbury International Conference on Padé Approximants and their Applications. Academic Press, London-New York.

_____ , Hughes Jones, R., and Makinson, G. J. (1974): The calculation of some rational approximants in two variables, J. Inst. Math. Appl. 13, 311-320.

Guerra, S. (1969): Sul calcolo approssimato di particolari funzioni iper-
geometriche confluenti con la tecnica del "τ-method", Calcolo
6, 213-223.

Handscomb, D. C. (1973): The relative sizes of the terms in Chebyshev
and other ultraspherical expansions, J. Inst. Math. Appl. 11, 241-
246.

Hangelbroek, R. J. (1967): Numerical approximation of Fresnel integrals
by means of Chebyshev polynomials, J. Engrg. Math. 1, 37-50.

Har-El, J., and Kaniel, S. (1973): Linear programming method for rational approximation, Israel J. Math. 16, 343-349.

Harris, R. M. (1973): Uniform approximation of functions through parti-
tioning, J. Approximation Theory 7, 239-255.

Hart, J. F., Cheney, E. W., Lawson, C. L., Maehly, H. J., Mesztenyi,
C. K., Rice, J. R., Thacher, H. C., Jr., and Witzgall, C. (1968):
Computer Approximations, Wiley, New York-London-Sydney.

Havie, T. (1968): CHECOF - double precision calculation of the coefficients in a Chebyshev expansion, CERN 6600 Series Program
Library C320.

Hawkins, D. M. (1972): On the choice of segments in piecewise approximation, J. Inst. Math. Appl. 9, 250-256.

den Heijer, C., Hemker, P. W., van der Houwen, P. J., Temme, N. M.,
and Winter, D. T., eds. (1974): NUMAL - A library of numerical
procedures in ALGOL 60, vols. 0-7. Mathematisch Centrum,
Amsterdam.

Henrici, P. (1958): The quotient-difference algorithm, Nat. Bur.
Standards Appl. Math. Ser. 49, 23-46.

_____ (1963): Some applications of the quotient-difference algorithm, in: "High Speed Computing and Experimental Arithmetic",
pp. 159-183. Proc. Sympos. Appl. Math. 15, Amer. Math. Soc.,
Providence, R. I.

81

Henrici, P. (1965): Error bounds for computations with continued fractions, in: "Error in Digital Computation", Vol. 2, pp. 39-53. Proc. Sympos. Math. Res. Center, U. S. Army, Univ. Wisconsin, Madison, Wis., 1965. Wiley, New York.

_____ (1966): An algorithm for analytic continuation, SIAM J. Numer. Anal. 3, 67-78.

_____ (1967): Quotient-difference algorithms, in: "Mathematical Methods for Digital Computers", Vol. II (A. Ralston and H. S. Wilf, eds.), pp. 37-62. Wiley, New York-London-Sydney.

_____ and Pfluger, P. (1966): Truncation error estimates for Stieltjes fractions, Numer. Math. 9, 120-138.

Hewers, W., and Zeller, K. (1960/61): Tschebyscheff-Approximation und Tschebyscheff-Entwicklung, Ann. Univ. Sci. Budapest. Eötvös. Sect. Math. 3-4, 91-93.

Hitotumatu, S. (1967/68): On the numerical computation of Bessel functions through continued fraction, Comment. Math. Univ. St. Paul. 16, 89-113.

Hofsommer, D. J., and van de Riet, R. P. (1963): On the numerical calculation of elliptic integrals of the first and second kind and the elliptic functions of Jacobi, Numer. Math. 5, 291-302.

Holdeman, J. T., Jr. (1969): A method for the approximation of functions defined by formal series expansions in orthogonal polynomials, Math. Comp. 23, 275-287.

Hornecker, G. (1958): Evaluation approchée de la meilleure approximation polynomiale d'ordre n de f(x) sur un segment fini [a, b], Chiffres 1, 157-169.

_____ (1959a): Approximations rationnelles voisines de la meilleure approximation au sens de Tchebycheff, C. R. Acad. Sci. Paris 249, 939-941.

_____ (1959b): Détermination des meilleures approximations rationnelles (au sens de Tchebichef) de fonctions réelles d'une variable sur un segment fini et des bornes d'erreur correspondantes, C. R. Acad. Sci. Paris 249, 2265-2267.

Hornecker, G. (1960): Méthodes pratiques pour la détermination approchée de la meilleure approximation polynômiale ou rationnelle, Chiffres 3, 193-228.

Horner, W. G. (1819): A new method of solving numerical equations of all orders by continuous approximation, Philos. Trans. Roy. Soc. London, part I, 308-335.

Huddleston, R. E. (1972): REHRAT - A program for best min-max rational approximation, Report SCL-DR-720370, Sandia Laboratories, Livermore, California.

Hughes Jones, R., and Makinson, G. J. (1974): The generation of Chisholm rational polynomial approximants to power series in two variables, J. Inst. Math. Appl. 13, 299-310.

Hummer, D. G. (1964): Expansion of Dawson's function in a series of Chebyshev polynomials, Math. Comp. 18, 317-319.

Jacobs, D., and Lambert, F. (1972): On the numerical calculation of polylogarithms, BIT 12, 581-585.

Johnson, J. H., and Blair, J. M. (1973): REMES2: A Fortran program to calculate rational minimax approximations to a given function, Report AECL-4210, Atomic Energy of Canada Limited, Chalk River Nuclear Laboratories, Chalk River, Ontario.

Jones, W. B. (1974): Analysis of truncation error of approximations based on the Padé table and continued fractions, Rocky Mountain J. Math. 4, 241-250.

_____ and Thron, W. J. (1974a): Numerical stability in evaluating continued fractions, Math. Comp. 28, 795-810.

_____ and _____, eds. (1974b): Proceedings of the international conference on Padé approximants, continued fractions and related topics, Rocky Mountain J. Math. 4, 135-397.

_____ and _____ (1974c): Rounding error in evaluating continued fraction expansions, Proceedings ACM Annual Conference, November 1974, San Diego, pp. 11-18. Association for Computing Machinery, New York.

Jones, W. B. and Thron, W. J. (1975): On convergence of Padé approximants, SIAM J. Math. Anal. 6, 9-16.

Kami, Y., Kiyoto, S., and Arakawa, T. (1971a): Method for numerical calculation of the standard elliptic integrals of the first and second kind (Japanese), Rep. Univ. Electro-Commun. 22, 99-108.

_____, _____, and _____ (1971b): Programming method on accurate values of the elliptic integral of the third kind (Japanese), Rep. Univ. Electro-Commun. 22, 109-118.

Kaufman, E. H., Jr., and Taylor, G. D. (1974): An application of linear programming to rational approximation, Rocky Mountain J. Math. 4, 371-373.

King, L. V. (1924): On the Direct Numerical Calculation of Elliptic Functions and Integrals, Cambridge Univ. Press, London.

Khovanskii, A. N. (1963): The Application of Continued Fractions and their Generalizations to Problems in Approximation Theory, translated from Russian by Peter Wynn. P. Noordhoff N. V., Groningen.

Kogbetliantz, E. G. (1960): Generation of elementary functions, in: "Mathematical Methods for Digital Computers" (A. Ralston and H.S. Wilf, eds.), pp. 7-35. Wiley, New York-London.

Kohútová, E. (1970): Stabilitätsbedingungen von rekurrenten Relationen und deren Anwendung, Apl. Mat. 15, 207-212.

Kölbig, K. S. (1972): Remarks on the computation of Coulomb wavefunctions, Computer Physics Comm. 4, 214-220.

_____, Mignaco, J. A., and Remiddi, E. (1970): On Nielsen's generalized polylogarithms and their numerical calculation, BIT 10, 38-73.

Korneĭčuk, A. A., and Širikova, N. Ju. (1968): An iterational method of determining the polynomial of best approximation (Russian), Ž.Vyčisl. Mat. i Mat. Fiz. 8, 670-674.

Krabs, W. (1969): Gleichmässige Approximation von Funktionen, B. I - Hochschultaschenbücher 247/247a, Überblicke Math. 3, 39-69.

Lam, B., and Elliott, D. (1972): On a conjecture of C. W. Clenshaw, SIAM J. Numer. Anal. 9, 44-52.

Lanczos, C. (1956): Applied Analysis, Prentice Hall, Englewood Cliffs, N. J.

_____ (1964): A precision approximation to the gamma function, J. Soc. Indust. Appl. Math. Ser. B Numer. Anal. 1, 86-96.

de La Vallée Poussin, C. (1919): Leçons sur l'approximation des fonctions d'une variable réelle, Gauthier-Villars, Paris.

Lawson, C. L. (1964): Characteristic properties of the segmented rational minimax approximation problem, Numer. Math. 6, 293-301.

Lee, C. M., and Roberts, F. D. K. (1973): A comparison of algorithms for rational ℓ_∞ approximation, Math. Comp. 27, 111-121.

Lehmer, D. H. (1971): On the compounding of certain means, J. Math. Anal. Appl. 36, 183-200.

Levin, D. (1973): Development of non-linear transformations for improving convergence of sequences, Intern. J. Comput. Math. 3, 371-388.

Longman, I. M. (1971): Computation of the Padé table, Intern. J. Comput. Math. 3, 53-64.

_____ (1973): On the generation of rational function approximations for Laplace transform inversion with an application to viscoelasticity, SIAM J. Appl. Math. 24, 429-440.

Luke, Y. L. (1955): Remarks on the τ-method for the solution of linear differential equations with rational coefficients, J. Soc. Indust. Appl. Math. 3, 179-191.

_____ (1958): The Padé table and the τ-method, J. Math. and Phys. 37, 110-127.

_____ (1959/60): On economic representations of transcendental functions, J. Math. and Phys. 38, 279-294.

_____ (1968): Approximations for elliptic integrals, Math. Comp. 22, 627-634.

Luke, Y. L. (1969): The Special Functions and their Approximations, Vols. I, II, Academic Press, New York-London.

_____ (1970a): Further approximations for elliptic integrals, Math. Comp. 24, 191-198.

_____ (1970b): Evaluation of the gamma function by means of Padé approximations, SIAM J. Math. Anal. 1, 266-281.

_____ (1971a): Rational approximations for the logarithmic derivative of the gamma function, Applicable Anal. 1, 65-73.

_____ (1971b): Miniaturized tables of Bessel functions, Math. Comp. 25, 323-330.

_____ (1971c): Miniaturized tables of Bessel functions II, Math. Comp. 25, 789-795.

_____ (1972a): Miniaturized tables of Bessel functions III, Math. Comp. 26, 237-240. {Corrigendum: ibid 26 (1972), no.120, loose microfiche supplement Al-A7 }.

_____ (1972b): On generating Bessel functions by use of the backward recurrence formula, Report ARL 72-0030, Aerospace Research Laboratories, Wright-Patterson Air Force Base, Ohio.

_____ (1975): On the error in the Padé approximants for a form of the incomplete gamma function including the exponential function, to appear in: SIAM J. Math. Anal.

_____ and Wimp, J. (1963): Jacobi polynomial expansions of a generalized hypergeometric function over a semi-infinite ray, Math. Comp. 17, 395-404.

Lyness, J. N., and Sande, G. (1971): Algorithm 413-ENTCAF and ENTCRE: Evaluation of normalized Taylor coefficients of an analytic function, Comm. ACM 14, 669-675.

Lyusternik, L. A., Chervonenkis, O. A., and Yanpol'skii, A. R. (1965): Handbook for Computing Elementary Functions, Pergamon Press, New York.

Macon, N., and Baskervill, M. (1956): On the generation of errors in the digital evaluation of continued fractions, J. Assoc. Comput. Mach. 3, 199-202.

Maehly, H. J. (1956): Monthly Progress Report, Oct. 1956, Institute for Advanced Study.

_____ (1958): First Interim Progress Report on Rational Approximations, June 23, 1958, Project NR 044-196, Princeton University.

_____ (1960): Methods for fitting rational approximations I. Telescoping procedures for continued fractions, J. Assoc. Comput. Mach. 7, 150-162.

Mechel, Fr. (1968): Improvement in recurrence techniques for the computation of Bessel functions of integral order, Math. Comp. 22, 202-205.

Meinardus, G. (1964): Approximation von Funktionen und ihre numerische Behandlung, Springer, Berlin-New York. {Expanded English translation in: Springer Tracts in Natural Philosophy, Vol. 13, 1967}.

_____ (1966): Zur Segmentapproximation mit Polynomen, Z. Angew. Math. Mech. 46, 239-246.

Mesztenyi, C., and Witzgall, C. (1967): Stable evaluation of polynomials, J. Res. Nat. Bur. Standards 71B, 11-17.

Miller, G. F. (1966): On the convergence of the Chebyshev series for functions possessing a singularity in the range of representation, SIAM J. Numer. Anal. 3, 390-409.

Minnick, R. C. (1957): Tshebysheff approximations for power series, J. Assoc. Comput. Mach. 4, 487-504.

Moody, W. T. (1967): Approximations for the psi (digamma) function, Math. Comp. 21, 112.

Morita, T., and Horiguchi, T. (1972/73): Convergence of the arithmetic-geometric mean procedure for the complex variables and the calculation of the complete elliptic integrals with complex modulus, Numer. Math. 20, 425-430.

McCabe, J. H. (1974): A continued fraction expansion, with a truncation error estimate, for Dawson's integral, Math. Comp. 28, 811-816.

Natanson, I. P. (1964): Constructive Function Theory, vol. I, Uniform Approximation. Frederick Ungar Publ. Co., New York.

National Bureau of Standards (1952): Tables of Chebyshev polynomials $S_n(x)$ and $C_n(x)$, Appl. Math. Ser. 9 .

Nellis, W. J., and Carlson, B. C. (1966): Reduction and evaluation of elliptic integrals, Math. Comp. 20, 223-231.

Németh, G. (1965): Chebyshev expansions for Fresnel integrals, Numer. Math. 7, 310-312.

_____ (1967): Chebyshev expansion of the Stirling series (Hungarian), Mat. Lapok 18, 329-333.

_____ (1971): Chebyshev polynomial expansions of Airy functions, their zeros, derivatives, first and second integrals (Hungarian), Magyar Tud. Akad. Mat. Fiz. Oszt. Közl. 20, 13-33.

_____ (1972): Tables of the expansions of the first 10 zeros of Bessel functions (Russian), Comm. Joint Inst. Nuclear Res., Dubna, Report 5-6336, March 1972.

_____ (1974): Expansion of generalized hypergeometric functions in Chebyshev polynomials, Collection of Scientific Papers in Collaboration of Joint Institute for Nuclear Research, Dubna, USSR and Central Research Institute for Physics. Algorithms and Programs for Solution of Some Problems in Physics, pp. 57-91. Central Research Institute, Budapest.

Neuman, E. (1969/70a): On the calculation of elliptic integrals of the second and third kinds, Zastos. Mat. 11, 91-94.

_____ (1969/70b): Elliptic integrals of the second and third kinds, Zastos. Mat. 11, 99-102.

Neville, E. H. (1944): Jacobian Elliptic Functions, Oxford University Press, London.

_____ (1971): Elliptic Functions: a Primer, prepared for publication by W. J. Langford. Pergamon Press, Oxford-New York-Toronto-Sydney-Braunschweig.

Newbery, A. C. R. (1971): The Boeing library and handbook of mathematical routines, in "Mathematical Software" (J. R. Rice, ed.), pp. 153-169. Academic Press, New York-London.

Newbery, A. C. R. (1973): Error analysis for Fourier series evaluation, Math. Comp. 27, 639-644.

_____ (1974): Error analysis for polynomial evaluation, Math. Comp. 28, 789-793.

Newman, D. J. (1964): Rational approximation to $|x|$, Michigan Math. J. 11, 11-14.

Ng, E. W. (1968/69): On the direct summation of series involving higher transcendental functions, J. Computational Phys. 3, 334-338.

_____ (1975): A comparison of computational methods and algorithms for the complex gamma function, ACM Trans. Mathematical Software 1, 56-70.

_____, Devine, C. J., and Tooper, R. F. (1969): Chebyshev polynomial expansion of Bose-Einstein functions of orders 1 to 10, Math. Comp. 23, 639-643.

Oliver, J. (1966/67): Relative error propagation in the recursive solution of linear recurrence relations, Numer. Math. 9, 323-340.

_____ (1968a): The numerical solution of linear recurrence relations, Numer. Math. 11, 349-360.

_____ (1968b): An extension of Olver's error estimation technique for linear recurrence relations, Numer. Math. 12, 459-467.

Olver, F. W. J. (1967a): Numerical solution of second-order linear difference equations, J. Res. Nat. Bur. Standards 71B, 111-129.

_____ (1967b): Bounds for the solutions of second-order linear difference equations, J. Res. Nat. Bur. Standards 71B, 161-166.

_____ (1974): Asymptotics and Special Functions, Academic Press, New York-London.

_____ and Sookne, D. J. (1972): Note on backward recurrence algorithms, Math. Comp. 26, 941-947.

Ostrowski, A. (1954): On two problems in abstract algebra connected with Horner's rule, in: "Studies in Mathematics and Mechanics Presented to Richard von Mises", pp. 40-48. Academic Press, New York.

Perron, O. (1957): Die Lehre von den Kettenbrüchen, Vol. II, 3rd ed.,
Teubner Verlagsges., Stuttgart.

Piessens, R., and Branders, M. (1973): The evaluation and application
of some modified moments, BIT 13, 443-450.

_____ and Criegers, R. (1974): Estimation asymptotique des coefficients du développement en série de polynômes de Chebyshev
d'une fonction ayant certaines singularités, C. R. Acad. Sci.
Paris A278, 405-407.

Powell, M. J. D. (1967): On the maximum errors of polynomial approximations defined by interpolation and by least squares criteria,
Comput. J. 9, 404-407.

Ralston, A. (1963): On economization of rational functions, J. Assoc.
Comput. Mach. 10, 278-282.

_____ (1967): Rational Chebyshev approximation, in "Mathematical
Methods for Digital Computers", Vol. II (A. Ralston and H. S. Wilf,
eds.), pp. 264-284. Wiley, New York-London-Sydney.

_____ (1973): Some aspects of degeneracy in rational approximations, J. Inst. Math. Appl. 11, 157-170.

Reimer, M. (1968): Bounds for the Horner sums, SIAM J. Numer. Anal.
5, 461-469.

_____ (1971): Numerische Stabilität beim Horner-Schema, Z. Angew.
Math. Mech. 51, T71-T72.

_____ and Zeller, K. (1967): Abschätzung der Teilsummen reeller
Polynome, Math. Z. 99, 101-104.

Remes, E. [Remez, E. Ja.] (1934): Sur le calcul effectif des polynomes
d'approximation de Tchebichef, C. R. Acad. Sci. Paris 199, 337-
340.

_____ (1969): Fundamentals of Numerical Methods of Chebyshev
Approximation (Russian), "Naukova Dumka", Kiev.

_____ and Gavriljuk, V. T. (1963): Some remarks on polynomial
Chebyshev approximations of functions compared to the intervals
of expansions in Chebyshev polynomials (Russian), Ukrain. Mat.
Ž. 15, 46-57.

Rice, J. R. (1964a): On the L_∞ Walsh arrays for $\Gamma(x)$ and $\mathrm{Erfc}(x)$, Math. Comp. 18, 617-626.

_____ (1964b): The Approximation of Functions, Vol. I: Linear Theory. Addison-Wesley Publ. Co., Reading, Mass. -London.

_____ (1965): On the conditioning of polynomial and rational forms, Numer. Math. 7, 426-435.

_____ (1969): The Approximation of Functions, Vol. II: Nonlinear and Multivariate Theory. Addison-Wesley Publ. Co., Reading, Mass. -London-Don Mills, Ont.

_____ (1971): The challenge for mathematical software, in: "Mathematical Software" (J. R. Rice, ed.), pp. 27-41. Academic Press, New York-London.

Riess, R. D., and Johnson, L. W. (1972): Estimates for $E_n(x^{n+2m})$, Aequationes Math. 8, 258-262.

Rivlin, T. J. (1969): An Introduction to the Approximation of Functions, Blaisdell Publ. Co., Waltham, Mass. -Toronto-London.

_____ (1974): The Chebyshev Polynomials, Wiley, New York-London-Sydney-Toronto.

_____ and Wilson, M. W. (1969): An optimal property of Chebyshev expansions, J. Approximation Theory 2, 312-317.

Russon, A. E., and Blair, J. M. (1969): Rational function minimax approximations for the Bessel functions $K_0(x)$ and $K_1(x)$, Report AECL-3461, Atomic Energy of Canada Limited, Chalk River, Ontario.

Rutishauser, H. (1954a): Der Quotienten-Differenzen-Algorithmus, Z. Angew. Math. Phys. 5, 233-251.

_____ (1954b): Anwendungen des Quotienten-Differenzen-Algorithmus, Z. Angew. Math. Phys. 5, 496-508.

_____ (1957): Der Quotienten-Differenzen-Algorithmus, Birkhäuser Verlag, Basel.

Sadowski, W. L., and Lozier, D. W. (1972): Use of Olver's algorithm to evaluate certain definite integrals of plasma physics involving Chebyshev polynomials, J. Computational Phys. 10, 607-613.

91

Saffren, M. M. , and Ng, E. W. (1971): Recursive algorithms for the summation of certain series, SIAM J. Math. Anal. 2, 31-36.

Salzer, H. E. (1962): Quick calculation of Jacobian elliptic functions, Comm. ACM 5, 399.

Schonfelder, J. L. (1974a): The NAG library and its special function chapter, in: "International Computing Symposium 1973" (A. Günther et al. , eds.), pp. 109-116. North-Holland Publ. Co.

_____ (1974b): Special functions in the NAG library, in: "Software for Numerical Mathematics" (D. J. Evans, ed.), pp. 285-300. Academic Press, London-New York.

_____ (1975): private communication.

Schönhage, A. (1971): Approximationstheorie, Walter de Gruyter, Berlin-New York.

Scraton, R. E. (1970): A method for improving the convergence of Chebyshev series, Comput. J. 13, 202-203.

_____ (1972): A modification of Miller's recurrence algorithm, BIT 12, 242-251.

Shenton, L. R. , and Bowman, K. O. (1971): Continued fractions for the psi function and its derivatives, SIAM J. Appl. Math. 20, 547-554.

Sheorey, V. B. (1974): Chebyshev expansions for wave functions, Computer Physics Comm. 7, 1-12.

Sidonskiǐ, O. B. (1967): Computation of Bessel functions from the recurrence relation by the double-sweep method (Russian), Izv. Sibirsk. Otdel. Akad. Nauk SSSR 1967, 3-7.

Širikova, N. Ju. (1970): A formula for the polynomial of best approximation, obtained by means of a computer (Russian), Ž. Vyčisl. Mat. i Mat. Fiz. 10, 181-183.

Smith, F. J. (1965): An algorithm for summing orthogonal polynomial series and their derivatives with applications to curve-fitting and interpolation, Math. Comp. 19, 33-36.

Smith, B. T., Boyle, J. M., and Cody, W. J. (1974): The NATS approach to quality software, in "Software for Numerical Mathematics" (D. J. Evans, ed.), pp. 393-405. Academic Press, London-New York.

Sookne, D. J. (1973a): Bessel functions I and J of complex argument and integer order, J. Res. Nat. Bur. Standards 77B, 111-114.

_____ (1973b): Certification of an algorithm for Bessel functions of real argument, J. Res. Nat. Bur. Standards 77B, 115-124.

_____ (1973c): Bessel functions of real argument and integer order, J. Res. Nat. Bur. Standards 77B, 125-132.

_____ (1973d): Certification of an algorithm for Bessel functions of complex argument, J. Res. Nat. Bur. Standards 77B, 133-136.

Southard, T. H. (1963): On the evaluation of the Jacobian elliptic and related functions, Mathematical Note No. 329, Boeing Scientific Research Laboratory.

Spielberg, K. (1961a): Representation of power series in terms of polynomials, rational approximations and continued fractions, J. Assoc. Comput. Mach. 8, 613-627.

_____ (1961b): Efficient continued fraction approximations to elementary functions, Math. Comp. 15, 409-417.

Stiefel, E. L. (1959): Numerical methods of Tchebycheff approximation, in: "On Numerical Approximation" (R. E. Langer, ed.), pp. 217-232. University of Wisconsin Press, Madison.

_____ (1964): Methods - old and new - for solving the Tchebycheff approximation problem, J. Soc. Indust. Appl. Math. Ser. B Numer. Anal. 1, 164-176.

Stoer, J. (1964): A direct method for Chebyshev approximation by rational functions, J. Assoc. Comput. Mach. 11, 59-69.

_____ (1972): Einführung in die numerische Mathematik I, Springer Verlag, Berlin-Heidelberg-New York.

Stoljarčuk, V. K. (1974a): The construction, for functions si x and $\phi(x) = 2\pi^{-\frac{1}{2}} \int_0^x e^{-t^2} dt$, of polynomials that realize a close-to-best approximation (Russian), Ukrain. Mat. Ž. 26, 216-226.

Stoljarčuk, V. K. (1974b): Uniform approximation, by polynomials on a
segment, of Bessel functions with integer index (Russian), Ukrain.
Mat. Ž. 26, 683-686.

Strecok, A. J. (1968): On the calculation of the inverse of the error function, Math. Comp. 22, 144-158.

_____ and Gregory, J. A. (1972): High precision evaluation of the
irregular Coulomb wave functions, Math. Comp. 26, 955-961.

Szegö, G. (1921): Über die Lebesgueschen Konstanten bei den Fourier-
schen Reihen, Math. Z. 9, 163-166.

Temme, N. M. (1972): Numerical evaluation of functions arising from
transformations of formal series, Report TW 134/72, Mathematisch
Centrum Amsterdam.

_____ (1973): On the numerical evaluation of the modified Bessel
function of the third kind, Report TN 72/73, Mathematisch Centrum
Amsterdam.

Thacher, H. C., Jr. (1960): Rational approximations for the Debye functions, J. Chem. Phys. 32, 638.

_____ (1964): Conversion of a power to a series of Chebyshev
polynomials, Comm. ACM 7, 181-182.

_____ (1966): Independent variable transformations in approximation, Proc. IFIP Congress 65, Vol. 2, pp. 576-577. Spartan Books,
Washington, D. C.

_____ (1967): Computation of the complex error function by continued fractions, in: "Blanch Anniversary Volume" (B. Mond, ed.), pp.
315-337. Aerospace Research Lab., U. S. Air Force, Washington,
D. C.

_____ (1969): Computational methods for mathematical functions,
Report 32-1324, Jet Propulsion Laboratory, Pasadena, California.

_____ (1971): Making special arithmetics available, in: "Mathematical Software" (J. R. Rice, ed.), pp. 113-119. Academic Press, New
York-London.

94

Thacher, H. C., Jr. (1972): Series solutions to differential equations by backward recurrence, Proc. IFIP Congress 71, Vol. 2, pp. 1287-1291. North-Holland Publ. Co., Amsterdam-London.

Todd, J. (1954): Evaluation of the exponential integral for large complex arguments, J. Res. Nat. Bur. Standards 52, 313-317.

Torii, T., and Makinouchi, S. (1968): An efficient algorithm for Chebyshev expansion, Information Processing in Japan 8, 89-92.

Tricomi, F. (1948): Elliptische Funktionen, translated and edited by Maximilian Krafft. Akad. Verlagsges., Leipzig.

_____ (1951): Funzioni ellittiche, 2nd ed., Nicola Zanichelli Editore, Bologna.

Van de Vel, H. (1969): On the series expansion method for computing incomplete elliptic integrals of the first and second kinds, Math. Comp. 23, 61-69.

Verbeeck, P. (1970): Rational approximations for exponential integrals $E_n(x)$, Acad. Roy. Belg. Bull. Cl. Sci. (5) 56, 1064-1072.

Wall, H. S. (1948): Analytic Theory of Continued Fractions, D. Van Nostrand Co., New York.

Walsh, J. L. (1964a): Padé approximants as limits of rational functions of best approximation, J. Math. Mech. 13, 305-312.

_____ (1964b): The convergence of sequences of rational functions of best approximation, Math. Ann. 155, 252-264.

_____ (1965): The convergence of sequences of rational functions of best approximation II, Trans. Amer. Math. Soc. 116, 227-237.

_____ (1968a): The convergence of sequences of rational functions of best approximation III, Trans. Amer. Math. Soc. 130, 167-183.

_____ (1968b): Degree of approximation by rational functions and polynomials, Michigan Math. J. 15, 109-110.

_____ (1969): Interpolation and Approximation by Rational Functions in the Complex Domain, Amer. Math. Soc. Colloq. Publ., Vol. 20, 5th ed., Amer. Math. Soc., Providence, R. I.

Walsh, J. L. (1974): Padé approximants as limits of rational functions of best approximation, real domain, J. Approximation Theory 11, 225-230.

Ward, M. (1960): The calculation of the complete elliptic integral of the third kind, Amer. Math. Monthly 67, 205-213.

Watson, G. A. (1975): A multiple exchange algorithm for multivariate Chebyshev approximation, SIAM J. Numer. Anal. 12, 46-52.

Watson, P. J. S. (1973): Algorithms for differentiation and integration, in: "Padé Approximants and their Applications" (P. R. Graves-Morris, ed.), pp. 93-97. Academic Press, London-New York.

Watt, J. M. (1958/59): A note on the evaluation of trigonometric series, Comput. J. 1, 162.

Werner, H. (1958/59): Tschebyscheffsche Approximationen für Bessel-Funktionen, Nukleonik 1, 60-63.

_____ (1966): Vorlesung über Approximationstheorie, Lecture Notes in Mathematics 14, Springer-Verlag, Berlin-New York.

_____ and Collinge, R. (1961): Chebyshev approximations to the gamma function, Math. Comp. 15, 195-197.

_____ and Raymann, G. (1963): An approximation to the Fermi integral $F_{\frac{1}{2}}(x)$, Math. Comp. 17, 193-194.

_____, Stoer, J., and Bommas, W. (1967): Rational Chebyshev approximation, Numer. Math. 10, 289-306.

Wilkinson, J. H. (1963): Rounding Errors in Algebraic Processes, Prentice-Hall, Englewood Cliffs, N. J.

Williams, J. (1972): Numerical Chebyshev approximation in the complex plane, SIAM J. Numer. Anal. 9, 638-649.

Wills, J. G. (1971): On the use of recursion relations in the numerical evaluation of spherical Bessel functions and Coulomb functions, J. Computational Phys. 8, 162-166.

Wimp, J. (1961): Polynomial approximations to integral transforms, Math. Comp. 15, 174-178.

Wimp, J. (1962): Polynomial expansions of Bessel functions and some associated functions, Math. Comp. 16, 446-458.

_____ (1969): On recursive computation, Report ARL 69-0186, Wright-Patterson Air Force Base, Ohio.

_____ (1970): Recent developments in recursive computation, SIAM Studies in Appl. Math. 6, 110-123.

_____ (1971/72): Forward computation in second order difference equations, Applicable Anal. 1, 325-329.

_____ (1974): On the computation of Tricomi's ψ function, Computing 13, 195-203.

_____ and Luke, Y. L. (1969): An algorithm for generating sequences defined by nonhomogeneous difference equations, Rend. Circ. Mat. Palermo (2) 18, 251-275.

Wood, V. E. (1967): Chebyshev expansions for integrals of the error function, Math. Comp. 21, 494-496.

_____ (1968): Efficient calculation of Clausen's integral, Math. Comp. 22, 883-884.

Wynn, P. (1956): On a device for computing the $e_m(S_n)$ transformation, Math. Tables Aids Comput. 10, 91-96.

_____ (1960): The rational approximation of functions which are formally defined by a power series expansion, Math. Comp. 14, 147-186.

_____ (1961): L'ϵ-algoritmo e la tavola di Padé, Rend. Mat. e Appl. (5) 20, 403-408.

_____ (1962a): Numerical efficiency profile functions, Nederl. Akad. Wetensch. Proc. Ser. A 65 = Indag. Math. 24, 118-126.

_____ (1962b): The numerical efficiency of certain continued fraction expansions IA, Nederl. Akad. Wetensch. Proc. Ser. A65 = Indag. Math. 24, 127-137; IB, ibid., 138-148.

_____ (1966a): Upon systems of recursions which obtain among the quotients of the Padé table, Numer. Math. 8, 264-269.

Wynn, P.   (1966b): An arsenal of Algol procedures for the evaluation of continued fractions and for effecting the epsilon algorithm, Rev. Française Traitement Information Chiffres 9, 327-362.

_____ (1967): A general system of orthogonal polynomials, Quart. J. Math. Oxford Ser. (2) 18, 81-96.

_____ (1972): Upon a convergence result in the theory of the Padé table, Trans. Amer. Math. Soc. 165, 239-249.

_____ (1974): Some recent developments in the theories of continued fractions and the Padé table, Rocky Mountain J. Math. 4, 297-323.

Zygmund, A. (1959): Trigonometric Series, Vol. I, Cambridge University Press.

Unsolved Problems in the Asymptotic Estimation of Special Functions

F. W. J. Olver

Abstract

The first part of this paper surveys the tools of asymptotic analysis that are presently available and those that are needed for the next stages of development of the asymptotic theory of special functions. Methods are grouped according to the number of free variables to which they achieve a uniform reduction. Topics include the approximation of functions defined parametrically by a definite integral or infinite sum, and approximate solutions of linear ordinary differential equations. The main areas in which work appears to be needed are (i) problems of confluence, that is, coalescing saddle points and singularities of integrals, or coalescing turning points and singularities of differential equations; (ii) rigorous error analysis.

The second part of the paper discusses the more important special functions in increasing order of the total number of variables and parameters involved. Almost all asymptotic problems concerning special functions of one variable have been solved. For functions of two variables the problems are solved, or can be solved by use of existing asymptotic tools. It is in the case of functions depending on three variables (including parameters) that the most significant work can be expected in the immediate future. Two new asymptotic tools have been developed very recently, one for integrals with three coalescing saddle points, the other for second-order differential equations having two coalescing turning points. Applications of these tools are needed, as well as the development of similar tools for other three-variable problems involving confluence. As the total number of variables increases beyond three, our knowledge of the asymptotic behavior of the special functions becomes more fragmentary. It will undoubtedly be many years before we have complete knowledge of the asymptotic behavior of any of the functions of four variables, including for example, the hypergeometric function $F(a,b;c;z)$.

CONTENTS

1. INTRODUCTION

1.1. <u>Classification of problems</u>. The problems referred to in the title of this paper are of two types: problems for which the tools of asymptotic analysis are already available, and problems for which the asymptotic tools need to be developed. For the most part, problems of the first type are the easier to solve. I hasten to add, however, that in most instances it would be wrong to replace the words "the easier" by "easy". Quite apart from the heavy algebraic computations that plague applications of asymptotic analysis, incorrect choice of tool or poor application of a

good tool may lead to results that are meaningless, crude, or at best un-
necessarily restricted. Judgment and experience are of great importance.

Problems in the asymptotic estimation of special functions for
which the asymptotic tools are not on the shelf vary enormously in difficulty. For some, the method is known in principle and merely needs
development and refinement. For others, the difficulties are profound
and await new inspirations. An index of the difficulty of a given problem
is the number of free variables (including parameters) that are involved.
The simplest type is the determination of an asymptotic approximation,
in terms of elementary functions, for a function $f(x)$ of a single real or
complex variable x, as x tends to infinity or a finite singularity of
$f(x)$.

For a function $f(x, y)$ of two variables, we consider its behavior
as one of the variables x, say, tends to infinity or a singularity, and
seek asymptotic approximations that are uniform with respect to the
second variable y when $y \in Y$. Here Y is a given interval, complex
domain, or other infinite point set which may depend on x, and may
contain singularities of $f(x, y)$ regarded as a function of y. The concept of uniformity is of utmost importance; without it we would be considering a finite set of fixed values of y in effect, and the problem re-
verts to the type mentioned in the preceding paragraph. With the formu-
lation just given, x is called the asymptotic variable and y the nonasymptotic variable. The roles are interchangeable, that is, we may require asymptotic approximations as y tends to infinity or a singularity
that are uniform with respect to x. As approximants we naturally try to
use elementary functions wherever possible. Sometimes, however, the
neighborhoods of certain points in the plane of the nonasymptotic variable can be included only by employing transcendental functions as approximants. Furthermore, these exceptional points need not be singularities of $f(x, y)$. [†]

[†] An example is furnished by the Bessel functions $J_n(x)$ and $Y_n(x)$.
Asymptotic expansions for large n, which are uniform with respect to
unrestricted positive real values of x, can be found in terms of Airy

For a function $f(x, y, z)$ of three variables, the problem is to construct asymptotic approximations as one of the variables x, say, tends to infinity or a singularity, which are uniform with respect to both y and z . Thus there is one asymptotic variable and two nonasymptotic variables; again, the roles are interchangeable. In some regions of the $Y \times Z$ space it will be possible to employ elementary functions as approximants. In others, transcendental functions that depend on a single variable will suffice. To cover the whole of the $Y \times Z$ space, however, it will be necessary in some regions to resort to transcendental functions of two variables as approximants.

For a function of r free variables, approximants that depend on as many as $r - 1$ variables will be needed, as a rule, in order to have a complete description of the function when any one of its variables tends to infinity or a singularity. To stress the importance of the concept of uniformity with respect to the nonasymptotic variables, we introduce the following terminology. An asymptotic formula that approximates a function of r variables is said to achieve a <u>uniform reduction of free variables</u> if it satisfies both of the following conditions: (i) the approximation is uniform with respect to all $r - 1$ nonasymptotic variables as each ranges independently over an infinite set of values; (ii) the approximation involves only elementary functions or transcendental functions that depend on fewer than r free variables, and the total number of such functions is finite.

As an illustration[†] if we approximate Legendre functions $P_n^m(z)$ and $Q_n^m(z)$ for large n by an asymptotic formula which involves the parabolic cylinder function $U(a, x)$ and is uniform with respect to both m and z, then we achieve a uniform reduction from three free variables to two. But if this problem were modified by fixing m, or fixing m/n ,

---

functions, but not in terms of elementary functions. The troublesome point here is $x = n$, which is not a singularity of $J_n(x)$ or $Y_n(x)$.

† This example is discussed more fully in §6.1 below.

then it would not be permissible to use a parabolic cylinder function to achieve a uniform reduction of variables, unless for example, the variable a in the approximant were fixed.

For the purpose of providing analytical insight into the nature of a given function, or the construction of algorithms for computing numerical values, asymptotic approximations that achieve a uniform reduction of free variables are much superior to other kinds. For if the number of free variables in an approximant equals or exceeds the number in the original function, then it is just as easy, as a rule, to evaluate the original function by direct numerical methods (such as numerical integration of an ordinary differential equation) as it would be to compute the approximant.

1.2. <u>Arrangement of the paper</u>. The first part of the present paper describes the tools of asymptotic analysis that are presently available, and indicates extensions and new tools that are needed. Although any method of asymptotic analysis may eventually prove to have applications to the special functions, the account is restricted to methods that strike the author as having significant applications in the next stages of development. The topics treated are the approximation of functions defined parametrically by a definite integral or infinite sum, and approximate solutions of ordinary linear differential equations. For each topic, methods are grouped according to the number of free variables to which they achieve uniform reduction.

In the second part of the paper, the commoner special functions are discussed in increasing order of the number of variables on which they depend. Perhaps it should be clarified that by "special functions" we mean throughout the functions defined and discussed in the Bateman Manuscript Project [1] and the N.B.S. Handbook [35]. Among the more important omissions in the present paper are the Mathieu and Lamé functions. Other omissions that have been made in order to keep the paper to a reasonable length are associated problems, such as asymptotic

103

approximations of integrals of the special functions, asymptotic approximations of zeros, and the construction of converging factors. [†]

Some special functions, for example the generalized hypergeometric function, have an arbitrarily large number of variables and it is possible that a complete description of their asymptotic behavior will have to proceed through a chain of lower functions, successive functions in the chain having fewer and fewer variables. As a corollary, from the standpoint of asymptotic analysis attempts to unify the theory of special functions within the framework of a single function of many variables, for example Meijer's G-function, are usually unfruitful because we know so little about the general asymptotic behavior of such functions. [‡] On the whole it is the particular specializations of the parameters in the more general functions – in other words the special properties of the special functions – that lead to solvable problems concerning their asymptotic behavior.

1.3. References. In the rest of this paper references are given in the usual manner, except that to avoid frequent repetition almost all references to the Bateman Manuscript Project [1] and the N. B. S. Handbook [35], and some references to the author's recent book [42] are omitted. The first two supply all standard properties of the special functions that are needed, and the third may be consulted for further details concerning standard asymptotic methods.

PART I.   THE TOOLS OF ASYMPTOTIC ANALYSIS
2.   INTEGRALS

2.1. Standard methods. The approach that has been used most frequently for deriving asymptotic approximations for the special functions is via

---

[†] An extensive (albeit formal) treatment of converging factors for asymptotic expansions of the special functions is included in Dingle's recent book [10]. Dingle also introduces the name "terminant" in preference to "converging factor".

[‡] Compare §7.3 below.

integral representations of the form

$$(2.1) \qquad\qquad I(z) = \int_P K(z, t)dt \ .$$

Here P is an interval of the real axis or a contour in the complex plane, often infinite in length, and z is the asymptotic variable under consideration. Sometimes repeated integration by parts is an effective procedure: each integration yields a new term in the asymptotic expansion of I(z), and the error term is given explicitly as an integral. More often than not, however, other methods are needed.

For real variables, standard procedures are the method of Laplace, which is applicable to integrands of the form

$$(2.2) \qquad\qquad K(z, t) = e^{-zp(t)} q(t) \ ,$$

where p(t) and q(t) are independent of z, and the method of stationary phase, applicable when

$$(2.3) \qquad\qquad K(z, t) = e^{izp(t)} q(t) \ .$$

Both methods apply to the case $z \to \infty$ and yield asymptotic approximations for I(z) in terms of elementary functions. Often the approximations are extendable to asymptotic expansions in descending fractional powers of z .

For contour integrals, there is the complex-variable analogue of Laplace's method, which is often called the method of steepest descents, though this name is not entirely appropriate [41]. Probably the most frequently used method of all is Watson's lemma, which may be regarded as the special case of Laplace's method in which p(t) = t and P is the positive real axis.

All of the above methods date back at least half a century, and although several papers have been published showing how to estimate or otherwise control the error terms associated with the resulting approximations, for example [21], [37], and [28], it is only quite recently that theories have been devised to enable the error terms to be bounded in a systematic and rigorous manner; see [40], [41], [43], and [22]. Mention may also be made here of the work of Stenger [52] in adapting quadrature formulas to the derivation of asymptotic approximations, complete with strict error bounds. All of these error analyses need extensive applications in order to test their effectiveness.

2.2. <u>Movable saddle points and singularities</u>. Of decisive importance in the construction of asymptotic approximations for functions I(z) represented by integrals of the form (2.1) are the points at which $\partial K(z,t)/\partial t$ vanishes. These are the so-called <u>stationary points</u> (real variables) or <u>saddle points</u> (complex variables). Their importance stems from the fact that for large $|z|$ the major contribution to the integral comes from the neighborhoods of these points or the endpoints. In the case of a contour integral it may be necessary to deform the contour to pass through an appropriate saddle point before Laplace's method can be applied successfully. In the next class of problems the integral depends on a parameter α; thus

$$(2.4) \qquad I(\alpha, z) = \int_P K(\alpha, z, t)dt \ .$$

Asymptotic approximations are needed for large $|z|$ that are uniform with respect to α when α ranges over a given infinite point set. In general, the saddle points (or stationary points) depend on α. If, as α varies continuously, the saddle points neither coalesce with each other nor coincide with an endpoint, then the modifications to the methods of §2.1 are straightforward and the same approximations are obtained. When these conditions are not fulfilled the problem is more profound, and it is no longer possible to construct asymptotic approximations that are uniform with respect to α in terms of elementary functions. Surveys of methods for treating these problems have been made by Erdélyi [12] and Jones [22]. We describe these methods briefly and mention some more recent developments.

Suppose first that the variables are real, P is a fixed finite or infinite interval $(0, T)$, and the integrand in (2.4) has the form

$$(2.5) \qquad K(\alpha, z, t) = e^{-zp(\alpha, t)} q(\alpha, t) \ ,$$

where $p(\alpha, t)$ and $q(\alpha, t)$ are independent of z. Furthermore, assume that $\partial p(\alpha, t)/\partial t$ has exactly one zero in $[0, T)$ which varies continuously with α, and coincides with the endpoint $t = 0$ for a certain critical value $\alpha = \alpha_0$, say. Following earlier writers, especially van der Waerden [56], Bleistein [3] has shown that by making a quadratic change of integration variable it is possible to obtain an asymptotic approximation for large $|z|$ that is uniform with respect to α in a closed interval containing α_0. The approximant employed is the simplest integral possessing the essential features of the problem, given by

$$(2.6) \qquad \int_0^\infty \exp\{-z(\tfrac{1}{2}\tau^2 - a\tau)\}d\tau \ .$$

Here a is a function of α that vanishes when $\alpha = \alpha_0$. The integral (2.6) is transformable into the error function, accordingly the method achieves a uniform reduction from two free variables to one.

The foregoing discussion is predicated on the assumption that $p(\alpha, t)$ and $q(\alpha, t)$ are free from singularity when $t \in [0, T)$. If, for example, $q(\alpha, t)$ behaves as a multiple of $t^{\lambda-1}$ when $t \to 0+$, where λ denotes a positive constant, then the uniform approximant (2.6) is replaced by

$$\int_0^\infty \exp\{-z(\tfrac{1}{2}\tau^2 - a\tau)\}\tau^{\lambda-1}d\tau \ ,$$

which is expressible in terms of the parabolic cylinder function

107

$U(\lambda - \frac{1}{2}, -az^{\frac{1}{2}})$. Since λ is fixed, this is a function of the single variable $az^{\frac{1}{2}}$.

Next, consider the case in which P is a real interval or contour in the complex plane, $K(\alpha, z, t)$ again has the form (2.5), and $\partial p(\alpha, t)/\partial t$ has exactly two zeros which coalesce into a double zero lying on P for a certain critical value of α . To handle this case a cubic change of integration variable is needed, and the uniform approximant is either

$$(2.7) \qquad \int_{\infty e^{-\pi i/3}}^{\infty e^{\pi i/3}} \exp\{z(\frac{1}{3} \tau^3 - a\tau)\} d\tau ,$$

or

$$(2.8) \qquad \int_0^\infty \exp\{z(-\frac{1}{3} \tau^3 + a\tau)\} d\tau ,$$

depending whether the limiting double zero is located at an interior point of P, or happens to be an endpoint. The integral (2.7) is transformable into the Airy function, and (2.8) is transformable into Scorer's function; furthermore, each of these functions depends on a single variable. This is the method of Chester, Friedman, and Ursell [7].

Other methods of similar type that achieve a uniform reduction from two free variables to one have been described by Wong [64] and Erdélyi [13]. In Wong's problem there are no stationary points on the path of integration, except for a critical value of the nonasymptotic variable α when an exterior saddle point coincides with an endpoint. Furthermore, this endpoint is permitted to be an algebraic singularity of the function $q(\alpha, t)$. In Erdélyi's problem there are no stationary points or singularities on the path, except for a critical value of α when an exterior algebraic singularity coincides with an endpoint.

To summarize the present subsection, the main problems associated with Laplace's method concerning confluence of saddle points, algebraic singularities, and endpoints have been solved, at least in

108

principle, for integrals representing functions of two free variables. Subsidiary problems of varying difficulty remain, however, including extensions to logarithmic singularities, determination of maximal regions of validity when the variables are complex, and the construction of strict and realistic error bounds.

2. 3. <u>Several movable saddle points and singularities</u>. The next level of difficulty is presented by functions of three variables. It is now possible for any three saddle points or singularities to move independently and coincide with each other, or for any two of these points to move independently and coincide with each other and also with an endpoint of the path of integration.

The only problem of this type that has been solved successfully to date is Ursell's treatment [55] of the case in which three saddle points are allowed to coalesce into a triple saddle point at an interior point of the path of integration. By introducing a quartic change of integration variable Ursell is able to construct uniform asymptotic approximations in terms of functions of the form

$$(2.9) \qquad \int_{\infty e^{-\pi i/4}}^{\infty e^{\pi i/4}} \tau^j \exp\{z(\frac{1}{4}\tau^4 - a\tau^2 - b\tau)\}d\tau, \qquad j = 0, 1, 2 \ ,$$

where a and b are parameters that vanish in the critical situation. Since each of these integrals is effectively a function of two variables, the method achieves a uniform reduction from three free variables to two. Moreover, in the same paper Ursell notes that his method can be extended in a straightforward manner to the case of a function of r variables expressed as an integral with s coalescing saddle points. Provided that $s \leq r$ this extension, also, achieves a uniform reduction of free variables.

Some of the unsolved problems involving three or more free variables would have immediate applications to the special functions. An example is provided by the case of two saddle points coalescing into a double saddle point which, in turn, may coincide with a singularity at an endpoint for an appropriate combination of variables. This problem arises with Whittaker functions with large parameters; see §6.2 below.

The general problem of an arbitrary number of saddle points coinciding with an arbitrary number of algebraic singularities has been studied by Bleistein [4]. His analysis is incomplete and Erdélyi and Ursell have suggested that there are severe difficulties in rendering it complete [12], [55]. It may therefore be worthwhile to pursue special cases, such as that mentioned in the preceding paragraph, both for their own applications and for any light that may be shed on the general problem. Subsidiary problems of the kind mentioned in the closing paragraph of §2.2 also arise in connection with the methods of the present subsection.

2.4. <u>Other kernels</u>. Another direction for the generalization of Laplace's method and the method of stationary phase is to replace the exponential function in (2.2), (2.3), or (2.5), by a transcendental function, for example, an Airy or Bessel function. Some of these problems are amenable to fairly straightforward extensions of the standard methods. Other cases, particularly real integrals with oscillatory integrands, require more sophisticated techniques such as the use of the Parseval formula for the Mellin transform. For introductions to these generalizations the reader should consult the recent books by Bleistein and Handelsman [5] and Riekstiņš [47]. All the extensions mentioned in §§2.2 and 2.3 also present themselves for integrals of the more general type.

Strict error bounds are lacking for most approximations obtained by these methods. An exception is the Hankel transform, which has recently been studied by Wong [65].

3. SUMS AND SEQUENCES

3.1. <u>Sums</u>. Most methods for the asymptotic estimation of a function defined as the sum of an infinite power series depend upon the replacement, or approximate replacement, of the sum by an integral, which is then estimated by the residue calculus or the methods mentioned in Section 2.

The Euler-Maclaurin formula estimates the sum

$$f(0) + f(1) + \ldots + f(n)$$

by the integral

$$\int_0^n f(t)dt \ ,$$

with a correction expressed as a finite series involving the values of $f(t)$ and its odd derivatives at $t = 0$ and n, together with an explicit integral representation for the error term.

Methods of Barnes' type depend on the use of the residue theorem to express an infinite sum depending on a variable z

$$S(z) = \sum_{n=0}^{\infty} f(z, n)$$

as an integral of the type

$$S(z) = \frac{1}{2i} \int_P \cot(\pi t) \ f(z, t)dt \ ,$$

in which P is a loop contour enclosing the points $t = 0, 1, 2, \cdots$, but not enclosing $t = -1, -2, -3, \cdots$, or the singularities of $f(z, t)$.

3.2. Sequences. The commonest way in which special functions, especially the orthogonal polynomials, appear as members of an infinite sequence is in the expansion of a generating function, thus

$$G(z,t) = \sum_{n=0}^{\infty} \phi_n(z) t^n \ .$$

The problem of determining the asymptotic behavior of $\phi_n(z)$ as $n \to \infty$ can be attacked by use of Cauchy's integral formula

$$(3.1) \qquad \phi_n(z) = \frac{1}{2\pi i} \int_P \frac{G(z,t)}{t^{n+1}} dt \quad ,$$

where P is a simple closed contour encircling $t = 0$, but not any singularity of $G(z,t)$. Thus the methods mentioned in Section 2 again become relevant.

A different procedure for approximating integrals of the form (3.1) for large n, which is applicable when $G(z,t)$ is not an entire function of t, is Darboux's method, as follows. The generating function $G(z,t)$ is suitably matched by another function $H(z,t)$ for which the coefficients $\psi_n(z)$ in its Maclaurin expansion

$$H(z,t) = \sum_{n=0}^{\infty} \psi_n(z) t^n$$

are known explicitly. Then

$$\phi_n(z) \sim \psi_n(z), \quad n \to \infty \ .$$

The matching process consists of identifying the singularities of $G(z,t)$ on its circle of convergence in the t-plane, and arranging for $H(z,t)$ to have the same kind of singularity at exactly the same points.

Darboux's method dates back almost a century, and although often used it does not appear to have been extended until quite recently. Wong and Wyman [66] have shown how to treat the case in which the singularities of $G(z,t)$ on its circle of convergence are of logarithmic type. Fields [16] has examined the confluent case in which two algebraic singularities on the circle of convergence coalesce. Fields' asymptotic expansions are in terms of certain transcendental functions of one variable, and are uniform with respect to the distance separating the singularities, including zero distance. Accordingly, the expansions achieve a uniform reduction from two free variables to one.

112

Fields' work has pointed the way to further problems of confluence, for example, three coalescing singularities on the circle of convergence. First, however, it may be desirable to investigate whether any simplifications are feasible, since the results in [16] are rather complicated to apply in their present form.

Other unsolved problems in the asymptotic approximation of sequences are discussed in a recent survey by Bender [2], written from a broader standpoint than the estimation of the special functions.

4. LINEAR ORDINARY DIFFERENTIAL EQUATIONS

4.1. _Introduction._ Many of the special functions originated as solutions of ordinary differential equations after separating variables in Laplace's equation. These equations are therefore linear and of the second order, and as we know now general methods for constructing asymptotic solutions can be devised in a variety of circumstances. The development of these methods lagged well behind that of methods for the asymptotic approximation of integrals, however. Consequently, in the early history of special functions asymptotic approximations were almost invariably derived from integral representations rather than directly from the defining differential equation. An impression of the meager state of the art in the 1920's can be formed by perusal of Ince's classic text [20]. Well before then, the powerful methods for integrals mentioned in §2.1 were established and applied frequently.

The situation has changed in recent years, and certainly as far as differential equations of the second order are concerned, a considerable range of powerful tools is available. Furthermore, compared with methods for integrals the differential-equation approach often has the advantages of easing the determination of maximal regions of validity, easing the calculation of higher terms in asymptotic expansions, and yielding simpler and sharper error bounds.

4.2. _Irregular singularities._ The first asymptotic problem associated

113

with an ordinary differential equation is the nature of its solutions in the
neighborhood of a singularity, particularly an irregular singularity. Here
the theory of error bounds has paid a remarkable dividend. Bounds that
were constructed for the errors associated with the Liouville-Green approximation (or JWKB approximation in the terminology of theoretical
physicists) have established that this approximation furnishes an asymptotic representation of solutions of second-order differential equations
in the neighborhood of virtually all kinds of irregular singularity, not
merely those of finite rank. Accordingly, when a new type of irregular
singularity is encountered we no longer need follow the approach of most
standard texts and begin by trying to guess the type of asymptotic approximation that is appropriate. A further consequence is that the concept of
rank is of diminished importance, because there are now numerous treat-
able types of singularity that do not have classifiable rank.

4.3. <u>Two-variable problems</u>. In the next class of problems the differential equation contains a real or complex parameter u, say; thus

$$d^2w/dz^2 = f(u, z)w \ . \tag{4.1}$$

We seek asymptotic approximations for the solutions when $u \to \infty$, or
some other distinguished point, that are uniform with respect to $z \in Z$,
where Z is a given real interval or complex domain. The form of the
asymptotic solutions is governed by the number and nature of the singularities and zeros of $f(u, z)$ in Z.

In the case of equations satisfied by the special functions, the coefficient $f(u, z)$ often can be decomposed in the form

$$f(u, z) = u^2 f(z) + g(u, z) \ , \tag{4.2}$$

where $f(z)$ is independent of u, and $|g(u, z)|$ is uniformly small com-
pared with $|u^2 f(z)|$ when $|u|$ is large, except possibly in fixed
neighborhoods of the singularities and zeros of $f(z)$. If the only singularities of the differential equation in Z are irregular and Z contains
no zeros of $f(z)$, then the Liouville-Green functions

114

(4.3)
$$f^{-\frac{1}{4}}(z) \exp\{\pm u \int f^{\frac{1}{2}}(z)dz\}$$

furnish uniform asymptotic approximations to the solutions as $|u| \to \infty$.
Furthermore, when $g(u, z)$ can be expanded uniformly in descending
powers of u, the asymptotic approximations to the solutions can be extended to asymptotic expansions; in particular, this includes the case
in which $g(u, z)$ is independent of u.

Zeros of $f(z)$ are called turning points or transition points of the
differential equation. At these points the Liouville-Green functions (4.3)
become infinite and no longer represent the solutions. In order to approximate the solutions uniformly in a region containing a simple turning
point we employ, as approximants, solutions of the simplest kind of differential equation possessing the essential features, given by

(4.4)
$$d^2w/dz^2 = u^2 zw .$$

Solutions of (4.4) are $Ai(u^{2/3}z)$ and $Bi(u^{2/3}z)$, where Ai and Bi denote the Airy functions. Since these approximants depend on the single
variable $u^{2/3}z$, a uniform reduction from two free variables to one is
achieved by this procedure. Again, the asymptotic approximations to the
solutions can be extended to asymptotic expansions in descending powers
of u when $g(u, z)$ depends on u in an appropriate manner.

In a similar way, in a region containing a turning point of multiplicity ℓ, uniform asymptotic approximations to the solutions can be
constructed in terms of solutions of the equation

$$d^2w/dz^2 = u^2 z^\ell w .$$

Again, these approximants are functions of a single variable, namely
$u^{2/(\ell+2)}z$; they can be expressed as Bessel functions of order $1/(\ell+2)$.
When $\ell > 1$, however, there is no straightforward extension from asymptotic approximations to asymptotic expansions in descending powers of
u [46].

Returning from turning points to singularities, we have already noted that the Liouville-Green functions furnish uniform asymptotic approximations for large $|u|$ in regions containing irregular singularities. This is also the case when the region contains a regular singularity which is a double pole of $f(z)$, provided that the parameter u is suitably modified. In the case of a simple pole of $f(z)$, however, it is no longer possible to use elementary functions as approximants. Instead, we employ Bessel functions, the order of which depends on the behavior of the term $g(u, z)$ at the pole of $f(z)$. Provided that the order is independent of u, and this is the usual case, we do not violate the principle of reduction of free variables.

For the most part, the methods mentioned in the present subsection are well developed, both for real and complex variables, including the construction of error bounds. Among the problems that need more detailed treatment is the case in which $f(z)$ has a simple pole and the order of the approximating Bessel functions is purely imaginary.

4.4. <u>Three or more variables</u>. With the inclusion of an extra parameter α, say, in (4.2), the differential equation (4.1) takes the form

$$d^2w/dz^2 = \{u^2 f(\alpha, z) + g(u, \alpha, z)\} w .$$

The turning points and singularities now depend on α, and may coalesce as α assumes certain critical values. The situation is analogous to that for integrals with movable saddle points and singularities discussed in §2.2, and the problem is to obtain asymptotic solutions for large $|u|$ that are uniform with respect to α, including the critical values, as well as with respect to z.

The only problem of this kind that has been solved successfully so far is the case of two coalescing simple turning points, treated by the present writer in [44]. This reference gives asymptotic approximations, complete with error bounds, these approximations being uniform with respect to α in closed intervals containing the critical value, and also

116

uniform with respect to z in intervals that extend to infinity or to the nearest singularities of the differential equation. The approximants that are used are the solutions of the simplest form of differential equation having the essential features, given by

(4.5) $$d^2 w/dz^2 = u^2 (z^2 - a^2) w \ .$$

Here a is a parameter that depends on α, and vanishes at the critical value. Solutions of (4.5) are the parabolic cylinder functions $U\{-\frac{1}{2}ua^2 , \pm(2u)^{\frac{1}{2}} z\}$, consequently the analysis achieves a uniform reduction from three free variables to two. The investigation in [44] is restricted to real values of the variables; a significant open problem is to extend the results to complex variables.

Similar problems arise when a turning point and pole coalesce. This occurs, for example, in the case of the associated Legendre equation, confluent hypergeometric equation, and hypergeometric equation; see §§6.1, 6.2, and 7.2 below. If the pole is double, then the natural choice of approximating differential equation is given by

(4.6) $$\frac{d^2 w}{dz^2} = \left(u^2 \frac{a-z}{z^2} + \frac{\rho}{z^2} \right) w \ ,$$

where a is a parameter that depends on α and vanishes at the critical value, and ρ is a constant. Solutions of (4.6) are the two-variable functions $z^{\frac{1}{2}} C_\nu (2uz^{\frac{1}{2}})$, where

$$\nu^2 = 1 + 4\rho + 4au^2 \ ,$$

and C_ν denotes any cylinder function of order ν . Alternatively, if the pole that coalesces with the turning point is simple, then the appropriate approximating differential equation would appear to be

(4.7) $$\frac{d^2 w}{dz^2} = \left(u^2 \frac{z-a}{z} + \frac{\rho}{z^2} \right) w \ ,$$

117

where a and ρ have the same significance as in (4.6). Solutions of (4.7) are the Whittaker functions $M_{au/2,\mu}(2uz)$ and $W_{au/2,\mu}(2uz)$, where

$$\mu^2 = \rho + \frac{1}{4} \ .$$

Again, with the given conditions they are functions of two variables. As far as the present writer is aware, neither of these problems has been attempted. For real variables they may be solvable by methods similar to those of [44].

As the number of free variables in the original differential equation advances from three to four and higher, the number of possible problems involving coalescing turning points and singularities proliferates. None has been solved, or is likely to be solved in the near future.

4.5. <u>Inhomogeneous equations</u>. For each problem mentioned in the preceding subsections there are corresponding problems for inhomogeneous differential equations, the solution of which often requires considerably more effort than straightforward application of the method of variation of parameters. Naturally the nature of the inhomogeneous term affects the problem. In the case of irregular singularities, mentioned in §4.2, and also in the simplest case involving a large parameter, mentioned in the second paragraph of §4.3, some fairly general results are included in [42] for the types of problem that arise with the special functions. The case of a simple turning point is also treated briefly in [42], and more completely in [9]. Inhomogeneous equations with double turning points are the subject of [29], but the asymptotic expansions obtained in this reference do not achieve a uniform reduction of free variables.

4.6. <u>Differential equations of higher order</u>. Another direction for generalization is to differential equations of order higher than two, or matrix systems of first-order differential equations. There is an extensive body of literature on this topic; see Wasow's book [57], and also [26], [19], [30], and [36]. The present writer is not aware of any significant applications to the special functions so far, however.

Other open problems associated with higher-order equations, or systems, include the development of error analyses. Stenger [50], [51] has given bounds for the error terms in the asymptotic expansions of solutions in the neighborhood of an irregular singularity of finite rank. The next problem ripe for error analysis would appear to be the generalization of the problem mentioned in the second paragraph of §4.3, that is, the case of a large parameter with elementary functions used as approximants.

PART II. ASYMPTOTIC ESTIMATES OF THE SPECIAL FUNCTIONS

5. FUNCTIONS OF ONE OR TWO VARIABLES

5.1. <u>Functions of one variable.</u> Included in this category are the Gamma function, exponential integral, sine and cosine integrals, error function, Fresnel integrals, Airy functions, and related functions. Each has an essential singularity, which is located at infinity. In all cases asymptotic expansions are available, these expansions being in series of descending powers of the argument and involving only elementary functions. Furthermore, in almost all cases satisfactory error bounds have been found, both for real and complex arguments.

5.2. <u>Incomplete Gamma function.</u> The asymptotic behavior of the incomplete Gamma function $\Gamma(\alpha, x)$ for large values of the variable x and fixed (or bounded) values of the parameter α is well known, as is the behavior for large α and fixed (or bounded) x. Asymptotic expansions for large values of the parameter that are uniform with respect to unbounded values of the variable (or vice versa) can be found as follows.

Suppose first we are considering $\Gamma(1-\alpha, x)$ for positive values of α and x. Then we have

$$x^{\alpha-1} e^x \Gamma(1-\alpha, x) = \int_0^\infty e^{-\alpha\{yt + \ell n(1+t)\}} dt ,$$

where $y = x/\alpha$. Since $y > 0$ the function $yt + \ell n(1+t)$ attains its

119

minimum in the interval of integration at the left endpoint $t = 0$. Accordingly, Laplace's method is applicable, and yields an expansion in descending powers of α. This expansion begins

$$x^{\alpha-1} e^x \, \Gamma(1-\alpha, x) \sim \frac{1}{1+y} \frac{1}{\alpha} + \frac{1}{(1+y)^3} \frac{1}{\alpha^2} + \ldots, \quad \alpha \to \infty ,$$

and is uniformly valid for $y \in [0, \infty)$, or, correspondingly, $x \in [0, \infty)$. The result is not new; it is in fact Blanch's uniform expansion for the generalized exponential integral $E_n(x)$ for large n; see [34], p. 61. This integral is related to the incomplete Gamma function by the formula

$$E_n(x) = x^{n-1} \, \Gamma(1-n, x) .$$

Next, consider $\Gamma(1+\alpha, x)$, where α and x again are positive. In this case

$$x^{-1-\alpha} e^x \Gamma(1+\alpha, x) = \int_0^\infty e^{-\alpha \{yt - \ln(1+t)\}} \, dt ,$$

where, as before, $y = x/\alpha$. The function $yt - \ln(1+t)$ has a minimum at $t = (1/y) - 1$. This lies inside the interval of integration when $0 < y < 1$, coincides with the left endpoint when $y = 1$, and is outside the interval of integration when $y > 1$. Since y is a free variable we have exactly the situation described in §2.2 in which a stationary point coalesces with an endpoint. In consequence it is possible to construct uniform asymptotic expansions in terms of the error function. These expansions have been supplied recently by Wong [64], §4 for the interval $y \in [\delta, 1]$, or equivalently $x \in [\delta\alpha, \alpha]$, where δ is a positive constant, and by Temme [53] for the more extensive range $y \in [0, \infty)$, or equivalently $x \in [0, \infty)$. Extensions to complex variables and the construction of error bounds are still needed.

5.3. Bessel functions and associated functions. Among all functions of two variables those having the most thoroughly studied asymptotic behavior are undoubtedly the Bessel functions. All of the methods sketched in §§2 and 4 that can be applied to these functions have been applied; see, for example, [42], Chapters 4 and 7 to 11.

For large argument, asymptotic expansions in terms of elementary functions are available, complete with error bounds. These expansions are uniformly valid with respect to bounded values of the order. For large order, asymptotic expansions are available in terms of elementary functions or in terms of Airy functions. These expansions are uniformly valid in restricted, though unbounded, regions of the plane of the argument; the regions overlap and their union covers the whole complex plane. The only essential problem that remains is the evaluation of bounds for the error terms in the uniform expansions for large order in the case when the order is complex.

A similar, though less complete, situation exists for the Struve functions, Anger functions, and functions of H. F. Weber, each of which satisfies an inhomogeneous form of Bessel's equation. Asymptotic expansions for large argument are available in terms of elementary functions, and uniform expansions for large order are available in terms of elementary functions or transcendental functions of a single variable; see [60], Chapter 10 and [42], Chapters 7 and 9. In most cases, however, error bounds have been calculated only for real variables.

5.4. <u>Parabolic cylinder functions</u>. These functions are entire in the argument and entire in the parameter. The situation is much the same as with the Bessel functions. Asymptotic expansions for large argument and fixed (or bounded) values of the parameter are available in terms of elementary functions. Asymptotic expansions for large values of the parameter are available in terms of elementary functions or Airy functions, these expansions being uniformly valid in overlapping regions whose union covers the entire complex plane of the argument. These results are given in [14], [33], [38], and [25]. Some error bounds have been calculated for the approximations in terms of elementary functions when the variables are real; see [42], pp. 206-208. These need extending to complex variables, and similar bounds also need to be calculated for the approximations in terms of Airy functions.

An important specialization is to the Hermite polynomials, via the formula

$$H_n(x) = 2^{n/2} \exp(\tfrac{1}{2} x^2) \, U(-n-\tfrac{1}{2}, \; x\sqrt{2}) \; .$$

In consequence, uniform asymptotic expansions of $H_n(x)$ for large n are available in terms of Airy functions, complete with error bounds when x is real; see [42], pp. 403 and 412.

6. FUNCTIONS OF THREE VARIABLES

6.1. <u>Legendre functions</u>. The singularities of the associated Legendre equation

$$(1-z^2) \frac{d^2 L}{dz^2} - 2z \frac{dL}{dz} + \left\{ n(n+1) - \frac{m^2}{1-z^2} \right\} L = 0$$

are located at $z = \pm 1$ and ∞. Since each is regular, no problems of an asymptotic nature arise when z lies in their neighborhoods.

To discuss cases in which the parameters m and n are large we begin by removing the term in the first derivative by the substitution $w = (z^2 - 1)^{\frac{1}{2}} L$; thus

(6.1) $$\frac{d^2 w}{dz^2} = \left\{ \frac{n(n+1)}{z^2-1} + \frac{m^2-1}{(z^2-1)^2} \right\} w \; .$$

For simplicity in the main discussion we shall assume that the symbols m and n denote real numbers such that $m \geq 0$ and $n \geq -\tfrac{1}{2}$; neither need be an integer, however. The variable z may be real or complex.

When n is large and m is fixed (or bounded) the coefficient of the large parameter in (6.1) has simple poles at $z = \pm 1$ and no zeros. Hence from §4.3 uniform asymptotic expansions can be constructed in terms of Bessel functions of fixed order. These expansions are given in

[42], Chapter 12, complete with error bounds; the order of the Bessel functions is, in fact, m, and the expansions descend in powers of $n + \frac{1}{2}$. Each expansion has a domain of validity that includes one or other of the singularities $z = \pm 1$, and extends to $z = \infty$.

In the complementary situation in which m is large and n is fixed (or bounded) the coefficient of the large parameter in (6.1) has double poles at $z = \pm 1$, and again no zeros. Hence from §4.3 asymptotic expansions can be constructed in terms of elementary functions that are uniform in a domain containing one or other of the singularities $z = \pm 1$. These expansions are none other than the well-known representations in hypergeometric series, typified by

$$
P_n^{-m}(z) = \frac{(z-1)^{m/2}}{\Gamma(m+1)(z+1)^{m/2}} \sum_{s=0}^{\infty} \frac{(n+1)_s (-n)_s}{(m+1)_s \, s!} \left(\frac{1-z}{2}\right)^s ,
$$

where, for example, $(n+1)_s$ is Pochhammer's symbol for $(n+1)(n+2)\ldots$ $(n+s)$. This series converges when $|z-1| < 2$, and as $m \to \infty$ it furnishes an asymptotic expansion that is uniformly valid in any compact z-domain lying within the sector $|\mathrm{ph}(z+1)| < \pi$.

Asymptotic expansions for large m that are uniform in unbounded z-domains can be found as follows. We introduce a new variable ζ related to z by

$$
\frac{\dot{z}^2}{(z^2-1)^2} = \frac{1}{\zeta} ,
$$

where the dot signifies differentiation with respect to ζ. Integrating and making $\zeta = 0$ correspond to $z = \infty$, we obtain

$$
z = \coth(2\sqrt{\zeta}) .
$$

The Liouville transformation is completed by introducing the new dependent variable

$$W = \dot{z}^{-\frac{1}{2}} w \quad .$$

With these substitutions equation (6.1) becomes

$$\frac{d^2 W}{d\zeta^2} = \left[\frac{m^2}{\zeta} + \frac{(n+\frac{1}{2})^2 - 1}{4\zeta^2} + \frac{n(n+1)}{\zeta} \left\{ \operatorname{cosech}^2(2\sqrt{\zeta}) - \frac{1}{4\zeta} \right\} \right] W \quad .$$

In this form the differential equation is ready for direct application of the approximation theorems given in [42], Chapter 12. Identifying the solutions in the usual manner, we arrive at

$$(6.2) \quad \begin{cases} P_n^{-m}(z) = \left(\frac{m}{\pi}\right)^{\frac{1}{2}} \frac{2\zeta^{\frac{1}{4}}}{\Gamma(m+1)} K_{n+\frac{1}{2}}(2m\zeta^{\frac{1}{2}})\{1+\varepsilon_1(\zeta)\} \quad , \\[2ex] Q_n^m(z) = \pi^{\frac{1}{2}} e^{m\pi i} \frac{\Gamma(m+n+1)}{m^{n+\frac{1}{2}}} \zeta^{\frac{1}{4}} I_{n+\frac{1}{2}}(2m\zeta^{\frac{1}{2}})\{1+\varepsilon_2(\zeta)\} \quad , \end{cases}$$

where the error terms have the properties

$$\varepsilon_1(\zeta) \to 0, \quad \zeta \to \infty; \quad \varepsilon_2(\zeta) \to 0, \quad \zeta \to 0 \quad .$$

Furthermore, as $m \to \infty$

$$(6.3) \qquad \varepsilon_1(\zeta) = O(m^{-1}), \quad \varepsilon_2(\zeta) = O(m^{-1}) \quad ,$$

uniformly in unbounded domains containing $\zeta = 0$, or, correspondingly, in unbounded z-domains containing $z = 1$. Relations (6.3) are the desired asymptotic properties, and the results (6.2) appear to be new. Further details need working out, such as the maximal domains of validity, explicit expressions for higher terms, and error bounds.

The next problem to be considered is the construction of asymptotic approximations for large n that are uniform with respect to unbounded values of m, as well as being uniform with respect to z. To this end we set

$$u = n + \frac{1}{2}, \quad \alpha = \frac{m}{u} \quad .$$

Then equation (6.1) becomes

$$\frac{d^2w}{dz^2} = \left\{ u^2 \frac{z^2-1+\alpha^2}{(z^2-1)^2} - \frac{z^2+3}{4(z^2-1)^2} \right\} w \ ,$$

in which u is a large positive parameter and α is a nonnegative real parameter. In this form the characterizing features are the double poles at $z = \pm 1$ and turning points at $z = \pm(1-\alpha^2)^{\frac{1}{2}}$. Four cases arise, which we discuss in turn.

(i) If the turning points are bounded away from each other, and also bounded away from the singularities at $z = \pm 1$ and ∞, that is, if

$$\delta \leq \alpha \leq 1-\delta \quad \text{or} \quad 1+\delta \leq \alpha \leq 1+\Delta \ ,$$

where δ and Δ are positive constants such that $\delta < \frac{1}{2}$ and $\delta < \Delta$, then asymptotic expansions that are uniform with respect to α, as well as z, can be constructed in terms of Airy functions (§4.3). Moreover, the regions of validity of these expansions will extend to the singularities at ± 1 and ∞. For the case $\delta \leq \alpha \leq 1-\delta$ these expansions have been investigated by Thorne [54].

(ii) When $1-\delta \leq \alpha \leq 1+\delta$, we have the problem of coalescing turning points. Accordingly, asymptotic approximations that are uniform with respect to α and z can be found in terms of parabolic cylinder functions (§4.4). For real z these approximations have been given recently by the present writer, complete with error bounds [45].

(iii) When $0 \leq \alpha \leq \delta$, each turning point may coalesce with one of the double poles at $z = \pm 1$. As we noted in §4.4, a general theory for this case is not yet available. When the theory has been supplied, we should be able to construct uniform asymptotic approximations in terms of Bessel functions of variable order. Some of the results of Thorne given in [54], §5 may prove to be pertinent to this problem, even though Thorne considered only fixed values of α in his analysis.

(iv) The remaining interval to be covered is $\alpha \in [1+\Delta, \infty)$. In this case both turning points may coalesce with the singularity at infinity.

125

To treat this problem, we interchange the roles of m and $n + \frac{1}{2}$, regarding m now as the large parameter, and $n + \frac{1}{2}$ as a free variable such that

$$0 \le n + \frac{1}{2} \le (1+\Delta)^{-1} m \quad .$$

Writing

(6.4)
$$\beta = \frac{1}{\alpha} = \frac{n + \frac{1}{2}}{m} \quad ,$$

and applying the Liouville transformation

$$\xi = \frac{1}{z^2} \; , \quad \hat{w} = \left(\frac{dz}{d\xi} \right)^{-\frac{1}{2}} w \; ,$$

to (6.1), we obtain

$$\frac{d^2 \hat{w}}{d\xi^2} = \left\{ m^2 \frac{(1-\beta^2)\xi + \beta^2}{4\xi^2 (1-\xi)^2} - \frac{3\xi^2 - 3\xi + 4}{16\xi^2 (1-\xi)^2} \right\} \hat{w} \quad .$$

This equation has a single turning point, located at $\xi = -\beta^2/(1-\beta^2)$. As $\beta \to 0$ this turning point coalesces with the double pole at $\xi = 0$. Hence from §4.4, again, we expect to be able to construct approximations in terms of Bessel functions of variable order[†] that are uniform with respect to $\beta \in [0, (1+\Delta)^{-1}]$ and ξ lying in a domain that includes $\xi = 0$. This domain may possibly extend to $\xi = 1$ and $\xi = \infty$.

Once the problems (i) to (iv) have all been solved we shall have a complete picture of the behavior of the Legendre functions not only for large n, but also for large m. This can be seen as follows. Let us again define β by (6.4), so that $\beta \in [0, \infty)$. The case $\beta \in [0, (1+\Delta)^{-1}]$ has been treated directly in (iv). Alternatively, when $\beta \ge (1+\Delta)^{-1}$ it

---

[†] The order is in fact $n + \frac{1}{2}$.

follows that $n + \frac{1}{2} \to \infty$ as $m \to \infty$, consequently the results of (i), (ii), and (iii) supply the required information.

Each of the problems discussed in this subsection is capable of generalization to complex values of m and n. The effort required to solve such problems completely would be very substantial, however, and would seem to be justifiable at present only in cases for which physical applications are known. This includes purely imaginary values of m and $n + \frac{1}{2}$, for which some results have already been obtained; see [8], [42], pp. 473-474, and [45].

6.2. Confluent hypergeometric functions. Whittaker's form of the confluent hypergeometric equation is given by

$$(6.5) \qquad \frac{d^2 w}{dz^2} = \left(\frac{1}{4} - \frac{k}{z} + \frac{m^2 - \frac{1}{4}}{z^2} \right) w \ .$$

This differential equation has a regular singularity at $z = 0$, and an irregular singularity at infinity. Asymptotic expansions of the solutions for large $|z|$ and fixed (or bounded) values of the parameters k and m are well known, and error bounds for real or complex values of k, m, and z are supplied in [39].

When $|k|$ is large and m is fixed (or bounded) we replace z by $4k\xi$; thus

$$\frac{d^2 w}{d\xi^2} = \left(4k^2 \frac{\xi - 1}{\xi} + \frac{m^2 - \frac{1}{4}}{\xi^2} \right) w \ .$$

The coefficient of k^2 has a zero at $\xi = 1$ and a simple pole at $\xi = 0$. In consequence (§4.3) asymptotic expansions can be found in terms of Airy functions that are uniform in an unbounded domain containing $\xi = 1$ but not $\xi = 0$, or in terms of Bessel functions of order $2m$ that are uniform in an unbounded domain containing $\xi = 0$ but not $\xi = 1$. For real

127

or complex values of k, m, and z these expansions have been supplied by Erdélyi and Swanson [15] and, more generally, by Skovgaard [48]. Error bounds, mainly for real variables, are given in [42], pp. 412 and 446-447.

We turn now to asymptotic approximations for large m that are uniform with respect to k, as well as z . For simplicity in the main discussion we assume that k is real, and m and z are real and positive. We write

(6.6) $$\alpha = k/m, \quad \zeta = z/m ,$$

so that $\alpha \in (-\infty, \infty)$ and $\zeta \in (0, \infty)$. Equation (6.5) becomes

$$\frac{d^2 w}{d\zeta^2} = \left(m^2 \frac{\zeta^2 - 4\alpha\zeta + 4}{4\zeta^2} - \frac{1}{4\zeta^2} \right) w .$$

In this form the characterizing features are the singularities at $\zeta = 0$ and ∞ and the turning points at

$$\zeta = 2\alpha \pm 2(\alpha^2 - 1)^{\frac{1}{2}} .$$

Four cases arise, which we discuss in turn.

(i) Assume that $-1 - \Delta \leq \alpha \leq 1 - \delta$, where δ and Δ are positive constants such that $\delta < 1$ and $\delta < \Delta$. Both turning points are then bounded away from the positive real axis. Consequently asymptotic approximations can be found in terms of elementary functions that are uniform with respect to $\alpha \in [-1-\Delta, 1-\delta]$ as well as $\zeta \in (0, \infty)$. These approximations are given in [42], pp. 260-261.[†] It should be observed that the problem in which m is large and k is fixed (or bounded) is included as a special case of these results.

[†] This reference discusses explicitly only values of α in the interval $[0, 1-\delta]$, but the results do in fact hold in the larger interval $[-1-\Delta, 1-\delta]$.

(ii)   When $1+\delta \leq \alpha \leq 1+\Delta$ both turning points lie in the interval $(0,\infty)$ and are bounded away from each other and also away from the singularities at $\zeta = 0$ and ∞ . Consequently two sets of asymptotic expansions can be constructed in terms of Airy functions.   The first set is valid in an interval that includes $\zeta = 2\alpha - 2(\alpha^2 - 1)^{\frac{1}{2}}$ and extends to $\zeta = 0$; the second set is uniformly valid in an interval that includes $\zeta = 2\alpha + 2(\alpha^2 - 1)^{\frac{1}{2}}$ and extends to infinity.   The intervals overlap and each expansion is uniformly valid with respect to α as well as ζ .   At present, these expansions do not appear to be in the literature, but some related results are given in [42], pp. 401-403.

(iii)   When $1-\delta \leq \alpha \leq 1+\delta$ we have the problem of coalescing turning points.   The theory of [44] is relevant, and has been applied by the present writer to derive uniform approximations in terms of parabolic cylinder functions similar to those mentioned in §6.1(ii) for Legendre functions.   This work is unpublished.[†]

(iv)   Lastly, when $\alpha \in (-\infty, -1-\Delta]$ or $\alpha \in [1+\Delta, \infty)$, one turning point may coincide with the irregular singularity at infinity whilst the other simultaneously coincides with the regular singularity at the origin. To treat these cases we regard $|k|$ as the large parameter, set

$$z = 4k\xi, \quad m = k\beta \quad ,$$

and restrict

$$-(1+\Delta)^{-1} \leq \beta \leq (1+\Delta)^{-1} \quad .$$

Equation (6.5) becomes

---

† Kazarinoff [24] has given asymptotic approximations in terms of parabolic cylinder functions for large m that are uniform with respect to bounded values of $(m^2 - k^2)/k$.   These approximations are less powerful than those derived by the present writer for the following reason.   If $(m^2 - k^2)/k$ is bounded, then $(1-\alpha^2)/\alpha = O(1/m)$, implying that the α-interval of uniform validity is not fixed; indeed it shrinks on $\alpha = 1$ as $m \to \infty$.   We may also note that some formal results pertinent to the problem have been given by Jorna [23].

$$\frac{d^2 w}{d\xi^2} = \left(k^2 \frac{4\xi^2 - 4\xi + \beta^2}{\xi^2} - \frac{1}{4\xi^2} \right) w \ .$$

This has turning points at

$$\xi = \frac{1}{2} \pm \frac{1}{2}(1-\beta^2)^{\frac{1}{2}} \ .$$

Since

$$\beta^2 \le (1+\Delta)^{-2} < 1 \ ,$$

the turning points are bounded away from each other, but the smaller one may coalesce with the double pole at $\xi = 0$. Therefore from §4.4 we expect to be able to construct asymptotic approximations in terms of Bessel functions of variable order [†] that are uniform with respect to β, and also with respect to ξ lying in a compact interval that includes the origin, but not the point $\frac{1}{2} + \frac{1}{2}(1-\beta^2)^{\frac{1}{2}}$. [‡]

When the problems (i), (ii), (iii), and (iv) have been solved, we shall also have a complete picture of the asymptotic behavior of the Whittaker functions for large positive or negative values of k; compare the penultimate paragraph of §6.1. Again, as in the case of the Legendre functions, each of the problems generalizes to complex values of k, m, and z, but first efforts in this direction should be aimed at the physical-ly important combinations of purely imaginary values of k, m, and z, for which Whittaker's equation assumes a real form; compare [42], pp. 401-403 and 433.

It is instructive to look briefly at the approach to Cases (i), (ii), and (iii) via integral representations. We have

---

[†] The order is in fact 2m .

[‡] If we were considering complex values of ξ, then the domain of val-idity may well extend to $\xi = \infty$, but the second turning point would still be excluded.

$$\Gamma(m-k+\tfrac{1}{2})\, e^{z/2}\, z^{-m-\frac{1}{2}}\, W_{k,\,m}(z) = \int_0^\infty e^{-zt}\, t^{m-k-\frac{1}{2}}(1+t)^{m+k-\frac{1}{2}}\, dt \quad .$$

With the substitutions (6.6), the integral on the right-hand side becomes

$$\int_0^\infty e^{-m\zeta t}\, t^{m(1-\alpha)-\frac{1}{2}}(1+t)^{m(1+\alpha)-\frac{1}{2}}\, dt \quad .$$

Since m is the large parameter, the stationary points are given by the equation

$$-\zeta + \frac{1-\alpha}{t} + \frac{1+\alpha}{1+t} = 0 \quad ,$$

the roots of which are

$$t = \frac{2-\zeta \pm (\zeta^2 - 4\alpha\zeta + 4)^{\frac{1}{2}}}{2\zeta} \quad .$$

It is easily verified that in Case (i) one stationary point lies within the interval of integration and the other stationary point lies outside; compare §2.1. In Case (ii) either both stationary points lie within the interval $(0, \infty)$ or both lie outside this interval, furthermore they coalesce at an interior point of the interval when

$$\zeta = 2\alpha - 2(\alpha^2 - 1)^{\frac{1}{2}} \quad ;$$

compare §2.2. In Case (iii) the stationary points coalesce with each other and also with the endpoint $t = 0$ when $\alpha = 1$ and $\zeta = 2$; as we saw in §2.3 this type of problem is unsolved.

The results for Whittaker's equation (6.5) can be reformulated for the solutions of the standard confluent hypergeometric equation

(6.7)
$$z\frac{d^2 W}{dz^2} + (b-z)\frac{dW}{dz} - aW = 0 \quad ,$$

by means of the substitutions

131

$$k = \frac{1}{2}b - a, \quad m = \frac{1}{2}b - \frac{1}{2}, \quad w = e^{-z/2} z^{b/2} W \; .$$

Many kinds of asymptotic expansions for the solutions of (6.7) have been given explicitly by Slater [49], pp. 65-88 and Luke [27], pp. 129-134. On the whole, however, because of their more restricted regions of validity these expansions are less powerful than the transformed versions of the expansions and approximations described above.

Another important renormalization is to the Laguerre polynomials $L_n^{(\sigma)}(z)$, by means of the substitutions $k = n + \frac{1}{2}\sigma + \frac{1}{2}, \quad m = \frac{1}{2}\sigma$.

6.3. <u>Unnamed special functions</u>. A somewhat different type of problem is posed by the differential equation

(6.8)
$$\frac{d^2 w}{dz^2} = z(z^2 + az + b)w \; ,$$

in which a and b are parameters. This is the simplest type of second-order differential equation having three free turning points, just as Airy's equation is the simplest equation with one turning point, and Weber's equation is the simplest equation having two turning points. The solutions of (6.8) do not appear to be expressible in closed form in terms of named special functions. In order to attack the problem of constructing asymptotic solutions of a general second-order differential equation having three turning points that may coalesce in any manner, a thorough study of equation (6.8) would be needed. This includes the definition of standard solutions that are numerically satisfactory in the sense of Miller [32], the determination of connection formulas, and the determination of the uniform asymptotic behavior of the solutions for large a, b, or z .

In a similar way, a study of the solutions of the equation

$$\frac{d^2 w}{dz^2} = \frac{z^2 + az + b}{z} w$$

appears to be a necessary preliminary to an attack on the general problem of second-order differential equations having two turning points which may coalesce with a simple pole. On the other hand, the prototype equation for the case of two simple turning points coalescing with a double pole, given by

$$\frac{d^2 w}{dz^2} = \frac{z^2 + az + b}{z^2} w \quad ,$$

is solvable in terms of Whittaker functions.

Analogous problems also arise in connection with methods for approximating integrals. For example, we know very little about the functions (2.9) that are used as uniform approximants in Ursell's method for integrals with three coalescing saddle points.

7. FUNCTIONS OF FOUR OR MORE VARIABLES

7.1. <u>General remarks</u> . The dividing line between functions of three variables and functions of four variables can be regarded as a watershed in our present stage of knowledge of asymptotic analysis. Although a great deal of work remains to be completed, most of the necessary tools are available to describe completely the asymptotic behavior, as the variables change in any manner, of several special functions of three variables, including the Legendre functions (§6.1) and the Whittaker functions (§6.2). The remaining tools that are needed probably can be constructed by methods similar to those used to develop existing tools; compare §2.3 and §4.4. The prospect of achieving a complete description of the asymptotic behavior of any special function of four variables, however, appears quite remote at present. Available methods will not suffice and fundamentally new work is needed.

7.2. <u>Hypergeometric function</u>. Consider, for example, one of the simplest special functions of four variables, the hypergeometric function. Using the differential-equation approach, we start with the hypergeometric equation

133

$$z(1-z)\frac{d^2w}{dz^2} + \{c-(a+b+1)z\}\frac{dw}{dz} - abw = 0 .$$

Removing the term in the first derivative by setting

$$W = z^{c/2}(1-z)^{(a+b-c+1)/2} w ,$$

we arrive at

(7.1) $$\frac{d^2W}{dz^2} = \frac{\{(a-b)^2-1\}z^2 + \{4ab+2c(1-a-b)\}z + c(c-2)}{4z^2(1-z)^2} W .$$

This equation has regular singularities at $z = 0, 1,$ and $\infty,$ and turning points at the zeros of the numerator on the right-hand side.

If, as the parameters a, b, and c vary, the turning points are bounded away from the singularities and also away from each other, then uniform asymptotic expansions can be found in terms of elementary functions or Airy functions. Moreover, the regions of validity of these expansions will overlap and cover the whole z-plane, including the singularities. Secondly, if the turning points coalesce, then the methods described in §4. 4 will yield asymptotic approximations in terms of parabolic cylinder functions that are uniform in intervals or domains containing the singularities. Thirdly, if one of the turning points coalesces with a singularity, then, as we also noted in §4.4, uniform asymptotic approximations probably can be constructed in terms of Bessel functions of variable order. Lastly, however, it may happen that both turning points coalesce with one of the singularities. In this event, as we suggested in §6. 3, it is probably necessary to employ functions of three variables (in fact, the Whittaker functions) in order to obtain uniform approximations, but we are unable to cope with this complicated problem with our present knowledge.

Results that have been found so far for the hypergeometric function permit at most three degrees of freedom in the four variables. For example, Watson [59], followed by Luke [27], Chapter 7, used the integral representation

$$F(a, b; c; z) = \frac{\Gamma(c)}{\Gamma(b)\Gamma(c-b)} \int_0^1 t^{b-1}(1-t)^{c-b-1}(1-zt)^{-a} dt$$

and Laplace's method to obtain asymptotic expansions of $F(a+\theta_1 u,$ $b+\theta_2 u; c+\theta_3 u; z)$ as $u \to +\infty$, where $\theta_1, \theta_2, \theta_3 = 0$ or ± 1. These expansions are in terms of elementary functions and are uniform with respect to z in specified complex domains.

Another example permitting two degrees of freedom is provided by Elliott's expansions for Jacobi polynomials of large degree [11]. These polynomials are related to the hypergeometric function by

$$P_n^{(\alpha, \beta)}(x) = \binom{n+\alpha}{n} F(a, b; c; z) ,$$

where

$$a = -n, \quad b = \alpha+\beta+n+1, \quad c = \alpha+1, \quad z = \frac{1}{2} - \frac{1}{2} x .$$

In Elliott's analysis n is large, and α and β are fixed (or bounded). Thus on the right-hand side of (7.1) the only terms in the numerator in which the large parameter appears are those in z^2 and z. In consequence, the singularity at $z = 0$ of the term in the differential equation involving the large parameter reduces to a simple pole. Similarly, this term has a simple pole at $z = 1$. Hence from §4.3, we can construct expansions that are uniform with respect to z in unbounded domains containing one or other of the singularities by using Bessel functions of fixed order (in fact, α or β). These expansions have been found by Elliott. Similar results for the same problem are sketched by Fields in [16], the approach being via Fields' extension of Darboux's method (§3.2).

Examples of uniform asymptotic approximations for the hypergeometric function that permit three degrees of freedom among the four variables are supplied by the results mentioned in §6.1 for Legendre functions. For example, we have

$$P_n^{-m}(z) = \frac{(z-1)^{m/2}}{\Gamma(m+1)(z+1)^{m/2}} F(n+1,\ -n;\ m+1;\ \frac{1}{2} - \frac{1}{2}z)\ .$$

7.3. <u>Generalizations of the hypergeometric function.</u> The situation is similar for generalized hypergeometric functions, basic hypergeometric functions, and the G-function. Apart from the specializations to the confluent hypergeometric function and the ordinary hypergeometric function which we discussed in §6.2 and the present section, almost all existing asymptotic approximations and expansions either pertain to the case in which the variable z tends to infinity and the parameters are kept fixed or bounded (for example, [58], [63], [31], [6], [27], Chapter 5, and [17]), or permit some or all of the parameters to depend linearly on a single variable which tends to infinity ([62], [27], Chapter 7, [61], and [18]). The asymptotic expansions are in terms of elementary functions, and in most cases the derivations proceed via integral representations using Laplace's method or the residue calculus.

ACKNOWLEDGMENTS

This research was supported in part by the U. S. Army Research Office, Durham under Contract DA ARO D 31 124 73 G204, and the National Science Foundation under Grant GP 32841X. The writer is indebted to Dr. L. Maximon for suggestions for improving the presentation of the results.

REFERENCES

[1]   Bateman Manuscript Project, <u>Higher transcendental functions</u> (A. Erdélyi, ed.), Vols. 1, 2, 3, McGraw-Hill, New York (1953, 1953, 1955).

[2]   E. A. Bender, <u>Asymptotic methods in enumeration</u>, SIAM Rev., <u>16</u> (1974), 485-515.

[3]   N. Bleistein, Uniform asymptotic expansions of integrals with stationary point near algebraic singularity, Comm. Pure Appl. Math., 19 (1966), 353-370.

[4]   N. Bleistein, Uniform asymptotic expansions of integrals with many nearby stationary points and algebraic singularities, J. Math. Mech., 17 (1967), 533-559.

[5]   N. Bleistein and R. A. Handelsman, Asymptotic expansions of integrals, Holt, Rinehart, and Winston, New York (1975).

[6]   B. L. J. Braaksma, Asymptotic expansions and analytic continuations for a class of Barnes-integrals, Noordhoff, Groningen (1963).

[7]   C. Chester, B. Friedman, and F. Ursell, An extension of the method of steepest descents, Proc. Cambridge Philos. Soc., 53 (1957), 599-611.

[8]   N. K. Chuhrukidze, Asymptotic expansions for spherical Legendre functions of imaginary argument (Russian), Ž. Vyčisl. Mat. i Mat. Fiz., 8 (1968), 3-12. [English translation: U. S. S. R. Computational Math. and Math. Phys., 8 (1968), No. 1, 1-13.]

[9]   R. A. Clark, Asymptotic solutions of a nonhomogeneous differential equation with a turning point, Arch. Rational Mech. Anal., 12 (1963), 34-51.

[10] R. B. Dingle, Asymptotic expansions: Their derivation and interpretation, Academic Press, New York (1973).

[11] D. Elliott, Uniform asymptotic expansions of the Jacobi polynomials and an associated function, Math. Comp., 25 (1971), 309-315.

[12] A. Erdélyi, Uniform asymptotic expansion of integrals, Analytic methods in mathematical physics (R. P. Gilbert and R. G. Newton, eds.), 149-168, Gordon and Breach, New York (1970).

[13] A. Erdélyi, Asymptotic evaluation of integrals involving a fractional derivative, SIAM J. Math. Anal., 5 (1974), 159-171.

[14] A. Erdélyi, M. Kennedy, and J. L. McGregor, Parabolic cylinder functions of large order, J. Rational Mech. Anal., 3 (1954), 459-485.

[15] A. Erdélyi and C. A. Swanson, Asymptotic forms of Whittaker's confluent hypergeometric functions, Mem. Amer. Math. Soc., No. 25 (1957).

[16] J. L. Fields, A uniform treatment of Darboux's method, Arch. Rational Mech. Anal., $\underline{27}$ (1968), 289-305.

[17] J. L. Fields, The asymptotic expansion of the Meijer G-function, Math. Comp., $\underline{26}$ (1972), 757-765.

[18] J. L. Fields, Uniform asymptotic expansions of certain classes of Meijer G-functions for a large parameter, SIAM J. Math. Anal., $\underline{4}$ (1973), 482-507.

[19] R. J. Hanson, Reduction theorems for systems of ordinary differential equations with a turning point, J. Math. Anal. Appl., $\underline{16}$ (1966), 280-301.

[20] E. L. Ince, Ordinary differential equations, Longmans, Green, London (1927). Reprinted by Dover, New York (1956).

[21] H. Jeffreys, The remainder in Watson's lemma, Proc. Roy. Soc. London Ser. A, $\underline{248}$ (1958), 88-92.

[22] D. S. Jones, Asymptotic behavior of integrals, SIAM Rev., $\underline{14}$ (1972), 286-317.

[23] S. Jorna, Derivation of Green-type, transitional, and uniform asymptotic expansions from differential equations III. The confluent hypergeometric function $U(a, c, z)$ for large $|c|$. Proc. Roy. Soc. London Ser. A, $\underline{284}$ (1965), 531-539.

[24] N. D. Kazarinoff, Asymptotic forms for the Whittaker functions with both parameters large, J. Math. Mech., $\underline{6}$ (1957), 341-360.

[25] M. K. Kerimov, Certain new results of the theory of Weber functions (Russian), Biblioteka Mat. Tablits, No. 45, Vyčisl. Tsentr SSSR, Moscow (1968), 88-116. [Translated into English by A. Werner and issued as Report No. TR 71-86 of the Bell Laboratories.]

[26] K. H. Kiyek, Zur Theorie der linearen Differentialgleichungssysteme mit einem grossen Parameter, Doctoral thesis, Univ. of Würzburg (1963).

[27] Y. L. Luke, The special functions and their approximations, Vol. 1,
      Academic Press, New York (1969).

[28] J. J. Mahony, Error estimates for asymptotic representations of
      integrals, J. Austral. Math. Soc., 13 (1972), 395-410.

[29] D. J. McGuinness, Nonhomogeneous differential equation with a
      second-order turning point, J. Mathematical Phys., 7 (1966), 1030-
      1037.

[30] J. A. M. McHugh, An historical survey of ordinary linear differen-
      tial equations with a large parameter and turning points, Arch.
      History Exact Sci., 7 (1971), 277-324.

[31] C. S. Meijer, On the G-function. Parts I, VI, VII, and VIII,
      Nederl. Akad. Wetensch. Proc. Ser. A, 49 (1946), 227-237, 936-
      943, 1063-1072, and 1165-1175.

[32] J. C. P. Miller, On the choice of standard solutions for a homogen-
      ous linear differential equation of the second order, Quart. J.
      Mech. Appl. Math., 3 (1950), 225-235.

[33] J. C. P. Miller, Tables of Weber parabolic cylinder functions,
      H. M. Stationery Office, London (1955).

[34] National Bureau of Standards, Tables of functions and of zeros of
      functions, Appl. Math. Ser. No. 37, U. S. Govt. Printing Office,
      Washington, D. C. (1954).

[35] National Bureau of Standards, Handbook of mathematical functions,
      Appl. Math. Ser. No. 55 (M. Abramowitz and I. A. Stegun, eds.),
      U. S. Govt. Printing Office, Washington, D. C. (1964).

[36] A. H. Nayfeh, Perturbation methods, Wiley (Interscience), New
      York (1973).

[37] F. Oberhettinger, On a modification of Watson's lemma, J. Res.
      Nat. Bur. Standards Sect. B, 63 (1959), 15-17.

[38] F. W. J. Olver, Uniform asymptotic expansions for Weber parabolic
      cylinder functions of large orders, J. Res. Nat. Bur. Standards
      Sect. B, 63 (1959), 131-169.

[39]   F. W. J. Olver, On the asymptotic solutions of second-order dif-
        ferential equations having an irregular singularity of rank one,
        with an application to Whittaker functions, SIAM J. Numer. Anal.
        Ser. B, 2 (1965), 225-243.

[40]   F. W. J. Olver, Error bounds for the Laplace approximation for
        definite integrals, J. Approximation Theory, 1 (1968), 293-313.

[41]   F. W. J. Olver, Why steepest descents?, Studies in applied math-
        ematics, No. 6 (D. Ludwig and F. W. J. Olver, eds.), 44-63, Soc.
        Indust. and Appl. Math., Philadelphia (1970). Reprinted in SIAM
        Rev., 12 (1970), 228-247.

[42]   F. W. J. Olver, Asymptotics and special functions, Academic
        Press, New York (1974).

[43]   F. W. J. Olver, Error bounds for stationary phase approximations,
        SIAM J. Math. Anal., 5 (1974), 19-29.

[44]   F. W. J. Olver, Second-order linear differential equations with two
        turning points, Philos. Trans. Roy. Soc. London Ser. A, 278
        (1975), 137-174.

[45]   F. W. J. Olver, Legendre functions with both parameters large,
        Philos. Trans. Roy. Soc. London Ser. A, 278 (1975), 175-185.

[46]   F. W. J. Olver, Second-order differential equations with fractional
        transition points. [To be published.]

[47]   E. Ya. Riekstiņš, Asymptotic expansions of integrals, Vol. 1
        (Russian), Zinātne, Riga (1974).

[48]   H. Skovgaard, Uniform asymptotic expansions of confluent hyper-
        geometric functions and Whittaker functions, Gjellerups Publ.,
        Copenhagen (1966).

[49]   L. J. Slater, Confluent hypergeometric functions, Cambridge Univ.
        Press, London and New York (1960).

[50]   F. Stenger, Error bounds for asymptotic solutions of differential
        equations I. The distinct eigenvalue case, J. Res. Nat. Bur.
        Standards Sect. B, 70 (1966), 167-186.

[51]   F. Stenger,   Error bounds for asymptotic solutions of differential equations II. The general case,   J. Res. Nat. Bur. Standards Sect. B, $\underline{70}$ (1966), 187-210.

[52]   F. Stenger,   The asymptotic approximation of certain integrals, SIAM J. Math. Anal., $\underline{1}$ (1970), 392-404.

[53]   N. M. Temme,   Uniform asymptotic expansions of the incomplete Gamma functions and the incomplete Beta function, Math. Comp. [In press.]

[54]   R. C. Thorne,   The asymptotic expansion of Legendre functions of large degree and order, Philos. Trans. Roy. Soc. London Ser. A, $\underline{249}$ (1957), 597-620.

[55]   F. Ursell,   Integrals with a large parameter. Several nearly coincident saddle-points, Proc. Cambridge Philos. Soc., $\underline{72}$ (1972), 49-65.

[56]   B. L. van der Waerden,   On the method of saddle points, Appl. Sci. Res. Ser. B, $\underline{2}$ (1951), 33-45.

[57]   W. Wasow,   Asymptotic expansions for ordinary differential equations, Wiley (Interscience),   New York (1965).

[58]   G. N. Watson,   The continuations of functions defined by generalised hypergeometric series, Trans. Cambridge Philos. Soc., $\underline{21}$ (1910), 281-299.

[59]   G. N. Watson,   Asymptotic expansions of hypergeometric functions, Trans. Cambridge Philos. Soc., $\underline{22}$ (1918), 277-308.

[60]   G. N. Watson,   A treatise on the theory of Bessel functions, 2nd ed., Cambridge Univ. Press, London and New York (1944).

[61]   M. W. Wilson,   On the Hahn polynomials, SIAM J. Math. Anal., $\underline{1}$ (1970), 131-139.

[62]   J. Wimp,   The asymptotic representation of a class of G-functions for large parameter, Math. Comp., $\underline{21}$ (1967), 639-646.

[63]   E. M. Wright,   The asymptotic expansion of the generalized hypergeometric function, Proc. London Math. Soc., $\underline{46}$ (1940), 389-408.

[64]   R. Wong,   On uniform asymptotic expansion of definite integrals,
       J. Approximation Theory, 7 (1973), 76-86.

[65]   R. Wong,   Error bounds for asymptotic expansions of Hankel trans-
       forms.   [To be published.]

[66]   R. Wong and M. Wyman,   The method of Darboux, J. Approximation
       Theory, 10 (1974), 159-171.

University of Maryland and
National Bureau of Standards

Periodic Bernoulli Numbers, Summation Formulas and Applications

Bruce C. Berndt

1. <u>Introduction</u>. Let $A = \{a_j\}$, $-\infty < j < \infty$, be a sequence of complex numbers with period $k > 0$. In [19], L. Schoenfeld and the author introduced periodic Bernoulli numbers $B_n(A)$ and periodic Bernoulli polynomials $B_n(x, A)$, $0 \leq n < \infty$. If $A = I \equiv \{1\}$ and $k = 1$, then $B_n(I)$ and $B_n(x, I)$ are the ordinary Bernoulli numbers B_n and polynomials $B_n(x)$, respectively.

Periodic Bernoulli numbers appear in periodic analogues of the Euler-Maclaurin summation formula. These periodic Euler-Maclaurin formulas yield expressions for $\sum_{c < n < d} a_n f(n)$, where f is sufficiently smooth on $[c, d]$. The periodic Euler-Maclaurin formula may be derived from the periodic Poisson summation formula, both of which were first established by Schoenfeld and the author [19].

In Section 2, we give alternative definitions of the periodic Bernoulli numbers and polynomials and indicate a few of their properties. The periodic Poisson and Euler-Maclaurin summation formulas are then stated in Section 3. These formulas have proven to be enormously useful, especially in applications to number theory. Therefore we devote the main portion of this paper to the exposition of several applications to number theory, to the theory of Bessel functions, and to classical analysis. Most of the applications yield new results.

2. <u>Periodic Bernoulli numbers and polynomials</u>. Let $A = \{a_j\}$, $-\infty < j < \infty$, have period k. In analogy with Gaussian sums, let

$$(2.1) \qquad G(z, A) = \sum_{n=0}^{k-1} a_n e^{2\pi i n z / k} \quad ,$$

where z is complex.

Definition 2.1. The periodic Bernoulli numbers $B_n(A)$, $0 \le n < \infty$, are defined by the generating function

$$(2.2) \qquad \frac{xG(kx/2\pi i, A)}{e^{kx} - 1} = \sum_{n=0}^{\infty} \frac{B_n(A)}{n!} x^n \qquad (|x| < 2\pi/k) \ .$$

In particular, we note that

$$(2.3) \qquad B_0(A) = \frac{1}{k} \sum_{j=0}^{k-1} a_j \ .$$

It is clear from (2.2) that if $A = I \equiv \{1\}$, then $B_n(I) = B_n$, $n \ge 0$, where B_n denotes the nth ordinary Bernoulli number. If $A = \{(-1)^j\}$ and $k = 2$, then an elementary calculation shows that $B_n(A) = (2^n - 1)B_n$, $0 \le n < \infty$. Let $k = 4$ and let χ_4 be the sequence defined by the values $a_{2j} = 0$ and $a_{2j+1} = (-1)^j$, $-\infty < j < \infty$. Note that χ_4 is the nonprincipal quadratic character modulo 4 . Then,

$$(2.4) \qquad \frac{xG(kx/2\pi i, A)}{e^{kx} - 1} = \frac{x(e^x - e^{3x})}{e^{4x} - 1} = -\frac{x \sinh x}{\sinh 2x}$$

$$= -\frac{x}{2 \cosh x} = -\frac{x}{2} \sum_{n=0}^{\infty} \frac{E_n}{n!} x^n \quad (|x| < \pi/2) \ ,$$

where we have used a familiar generating function for the Euler numbers E_n, $0 \le n < \infty$ [39, p. 35]. Thus, by comparing (2.2) with (2.4), we deduce that $B_n(A) = -\frac{1}{2} n E_{n-1}$, $1 \le n < \infty$.

Let $A = \chi \equiv \{\chi(j)\}$, where $\chi(j)$ is a character with modulus k . If χ is primitive, then $B_n(\chi)$, $0 \le n < \infty$, denotes the nth generalized

Bernoulli number. For real χ, the generalized Bernoulli numbers were first defined by Ankeny, Artin and Chowla [2], and for complex χ, they were first defined by Leopoldt [51]. Generalized Bernoulli numbers have recently appeared very prominently in applications to algebraic number theory.

Definition 2.2. Let $A^* = \{a_{-j}\}$. The periodic Bernoulli polynomials $B_n(t, A)$, $0 \le n < \infty$, are defined by the generating function

(2.5)
$$\frac{xe^{xt}G(kx/2\pi i, A^*)}{e^{kx} - 1} = \sum_{n=0}^{\infty} \frac{B_n(t, A)}{n!} x^n \quad (|x| < 2\pi/k) \ .$$

From (2.5), it is easily seen that if $A = I$, then $B_n(t, A) = B_n(t)$, $0 \le n < \infty$, where $B_n(t)$ denotes the nth ordinary Bernoulli polynomial. If $A = \{(-1)^j\}$, then $B_n(t, A) = 2^n B_n(t/2) - B_n(t)$, $0 \le n < \infty$. If $A = \chi_4$, then by (2.4),

(2.6)
$$\frac{xe^{xt}G(kx/2\pi i, A^*)}{e^{kx} - 1} = \frac{xe^{xt}}{2\cosh x} = \frac{x\,e^{2x\{(t+1)/2\}}}{e^{2x} + 1}$$

$$= \frac{x}{2} \sum_{n=0}^{\infty} E_n(\frac{t+1}{2}) \frac{(2x)^n}{n!} \quad (|x| < \pi/2) \ ,$$

where we have used a well-known generating function for the Euler polynomials $E_n(x)$, $0 \le n < \infty$ [1, p. 804]. By comparing (2.5) and (2.6), we deduce that $B_n(t, A) = n2^{n-2} E_{n-1}(\frac{t+1}{2})$, $1 \le n < \infty$. Let $A = \chi$, where χ is an arbitrary primitive character of modulus k . Then (2.5) yields the generalized Bernoulli polynomials $B_n(t, \chi)$, $0 \le n < \infty$, as originally defined by Leopoldt [51].

We now state a few of the most important properties of periodic Bernoulli numbers and periodic Bernoulli polynomials. The proofs may be found in [19].

145

The connection between the periodic Bernoulli numbers and periodic Bernoulli polynomials is given by the relations $B_n(0, A) = (-1)^n B_n(A)$, $n \neq 1$, and $B_1(0, A) = -B_1(A) - a_0$. The periodic Bernoulli numbers and polynomials are related to the ordinary Bernoulli polynomials by the very important formula

$$B_n(x, A) = k^{n-1} \sum_{j=0}^{k-1} a_{-j} B_n(\frac{x+j}{k}) \qquad (n \geq 0) .$$

This fact can be easily deduced from the Fourier expansion of $B_n(x, A)$.

To enunciate this Fourier series, we must first define the complementary sequence $B = \{b_j\}$, $-\infty < j < \infty$, of period k by

$$(2.7) \qquad b_j = \frac{1}{k} \sum_{m=0}^{k-1} a_m e^{-2\pi i j m/k} .$$

These numbers $\{b_j\}$ are the finite Fourier series coefficients of $\{a_j\}$. From (2.1), we note that $b_j = G(-j, A)/k$, and from (2.3), we observe that $b_0 = B_0(A)$. By the inversion of (2.7), we deduce that

$$(2.8) \qquad a_j = \sum_{m=0}^{k-1} b_m e^{2\pi i j m/k} = G(j, B) \quad (-\infty < j < \infty) .$$

Unfortunately, (2.7) and (2.8) are asymmetric. If $a_n = \chi(n)$, where χ is a primitive character of modulus k, we find from (2.7) and the factorization theorem for Gaussian sums [4, p. 312] that

$$(2.9) \qquad b_j = \frac{1}{k} \overline{\chi}(-j) G(\chi) ,$$

where $G(\chi) = G(1, \chi)$.

The Fourier expansion of $B_n(x, A)$, $n \geq 1$, is given by

$$(2.10) \quad B_n(x, A) = -n! \sum_{m=1}^{\infty} (k/2\pi i m)^n \left\{ b_m e^{2\pi i m x/k} + (-1)^n b_{-m} e^{-2\pi i m x/k} \right\},$$

where $0 \leq x < 1$ if $n \geq 2$ and $0 < x < 1$ if $n = 1$. If N is an integer, then on the interval $N < x < N + 1$, the series on the right side of (2.10) converges to a polynomial of degree at most n, but, in general, if $k \geq 2$, the series converges to different polynomials on adjacent intervals with integral endpoints. If $A = I$, (2.10) reduces to the well-known Fourier series of $B_n(x)$ [48, p. 522]. If $A = \chi$, where χ is primitive, (2.9) has been previously given by Hasse [46], Schmidt [65], and the author [16].

Definition 2.3. We say that the sequence A is even if $a_j = a_{-j}$, $-\infty < j < \infty$, and that A is odd if $a_j = -a_{-j}$, $-\infty < j < \infty$. Let $\gamma = 1$ if A is even, and let $\gamma = -1$ if A is odd.

In particular, if $(-1)^n \gamma = -1$, $n \geq 2$, we see from (2.10) that $B_n(A) = 0$.

Definition 2.4. Let x and a be real. For $\mathrm{Re}(s) > 1$, define

$$\varphi(x, a, s; A) = \sum_{n=0}^{\infty}{}' a_n \, e^{2\pi i n x/k} \, (n+a)^{-s} \, ,$$

where the prime ' on the summation sign indicates that if a is a non-positive integer, then the index $n = -a$ is to be omitted from the summation. Furthermore, put $\zeta(s, a; A) = \varphi(0, a, s; A)$ and $\zeta(s; A) = \zeta(s, 0, A)$.

Observe that $\zeta(s; I) = \zeta(s)$, the Riemann zeta-function; $\zeta(s; \chi) = L(s, \chi)$, Dirichlet's L-function; and, if $0 < a \leq 1$, $\zeta(s, a; I) = \zeta(s, a)$, the Hurwitz zeta-function.

If A is even or odd, certain values of $\zeta(s; A)$ and $\zeta(s, a; A)$ may be given in terms of values of periodic Bernoulli numbers and polynomials. Thus, if $(-1)^n \gamma = 1$ and $n \geq 2$,

(2.11) $$\zeta(n; A) = -\frac{1}{2} k (2\pi i/k)^n B_n(B)/n!$$

and

(2.12) $$\zeta(n; B) = -\frac{1}{2} (-2\pi i/k)^n B_n(A)/n! \, .$$

147

If $A = I$ and $n = 2N$, then both (2.11) and (2.12) reduce to Euler's famous formula for $\zeta(2N)$. In the case that $a_n = (\frac{n}{p})$, where p is an odd prime, and where $(\frac{n}{p})$ denotes the Legendre symbol, formula (2.11) appears to have been first given by L. Carlitz [21], although the terminology of generalized Bernoulli numbers was not used by him. When $A = \chi$, where χ is a general primitive character, (2.11) gives Leopoldt's formulas for the values of Dirichlet L-functions [51]. For real primitive characters, (2.11) was also shown by S. Chowla [27].

If $n \geq 2$ is an integer, and if A is <u>any</u> periodic sequence, then $\zeta(1-n;A) = -B_n(A)/n$. In addition, if A is even or odd, and if $0 \leq a < 1$, then $\zeta(1-n, a;A) = -\gamma B_n(a, A)/n$. These facts generalize well-known facts about $\zeta(s)$ and $\zeta(s, a)$ [74, pp. 267-268].

3. <u>The periodic Poisson and periodic Euler-Maclaurin summation formulas.</u> The proofs of the results in this section may be found in [19]. We first state the periodic Poisson summation formula.

<u>Theorem 3.1.</u> If f is of bounded variation on $[c, d]$, then

$$(3.1) \quad \frac{1}{2} \sideset{}{'}\sum_{c \leq n \leq d} a_n \{f(n+0) + f(n-0)\} = \lim_{N \to \infty} \sum_{n=-N}^{N} b_n \int_c^d e^{2\pi i n x/k} f(x) dx ,$$

where the prime ' on the summation sign indicates that if c is an integer, then the first term of the sum at the left is $a_c f(c+0)$, and if d is an integer, then the last term of that sum is $a_d f(d-0)$.

<u>Corollary 3.2.</u> Let f be of bounded variation on $[c, d]$. If A is even, then

$$(3.2) \quad \frac{1}{2} \sideset{}{'}\sum_{c \leq n \leq d} a_n \{f(n+0) + f(n-0)\} = b_0 \int_c^d f(x) dx + 2 \sum_{n=1}^{\infty} b_n \int_c^d f(x) \cos(2\pi n x/k) dx ;$$

if A is odd, then

$$(3.3) \quad \frac{1}{2} \sum_{c \leq n \leq d}{}' a_n \{f(n+0) + f(n-0)\} = 2i \sum_{n=1}^{\infty} b_n \int_c^d f(x) \sin(2\pi nx/k)dx \ .$$

If $A = I$, (3.1) and (3.2), of course, reduce to the ordinary Poisson summation formula. A less general version of Theorem 3.1 was first given by L. J. Mordell [56]. In the case that a_n is the Legendre-Jacobi symbol, a version of Theorem 3.1 was first established by Lerch [52, pp. 418, 421]. When $A = \chi$, where $\chi(n)$ is a primitive character, versions of Theorem 3.1 were established by A. P. Guinand [40], [41] and the author [10], [16].

We now state a version of the periodic Euler-Maclaurin formula.

Theorem 3.3. Let $f \in C^{(r)}[Mk, Nk]$, where $r \geq 1$ and M and N are integers. Then

$$\sum_{Mk \leq n < Nk} a_n f(n) = B_0(A) \int_{Mk}^{Nk} f(x)dx$$

$$+ \sum_{j=1}^{r} \frac{B_j(A)}{j!} \left\{ f^{(j-1)}(Nk) - f^{(j-1)}(Mk) \right\} + (-1)^{r+1} \int_{Mk}^{Nk} P_r(x, A) \, f^{(r)}(x) \, dx \ ,$$

where $n! \, P_n(x, A)$ is defined by the right side of (2.10), except for when $n = 1$ and x is an integer.

More general versions of the periodic Euler-Maclaurin formula are found in the papers of Schoenfeld and the author [19] and Rosser and Schoenfeld [62]. If $A = I$, Theorem 3.3, of course, gives the ordinary Euler-Maclaurin formula. If $A = \{(-1)^j\}$, Theorem 3.3 reduces to Boole's summation formula [20, p. 102]. In the case $A = \chi$, a form of the periodic Euler-Maclaurin formula was established by Davies and Haselgrove [31] and the author [16]. For $A = \chi_4$, a special case of Theorem 3.3 was given by Chowla [28].

The remainder of the paper is devoted to several applications of the theorems of this section. The applications in Sections 4, 5, 8, 9, 10 and 11 are completely new.

4. The distribution of quadratic residues. Let f be continuous and of bounded variation on $[c, d]$. If χ is primitive and even, then we have from (3.2) and (2.9)

$$(4.1) \quad \sum_{c \leq n \leq d}' \chi(n)\, f(n) = \frac{2G(\chi)}{k} \sum_{n=1}^{\infty} \bar{\chi}(n) \int_c^d f(x) \cos(2\pi n x/k)\,dx \quad;$$

if χ is primitive and odd, then we have from (3.3) and (2.9)

$$(4.2) \quad \sum_{c \leq n \leq d}' \chi(n)\, f(n) = -\frac{2iG(\chi)}{k} \sum_{n=1}^{\infty} \bar{\chi}(n) \int_c^d f(x) \sin(2\pi n x/k)\,dx \quad.$$

The primes on the summation signs on the left sides of (4.1) and (4.2) indicate that if c or d is integral, the associated summands must be halved.

Formula (4.2) may be used to give an amazingly short proof of a classical theorem of Dirichlet [34] on quadratic residues.

Theorem 4.1. Let p be a prime with $p \equiv 3 \pmod 4$. Then in the interval $(0, p/2)$ the number of quadratic residues modulo p exceeds the number of quadratic nonresidues.

Proof. In (4.2), let $\chi(n)$ be the Legendre symbol $(\frac{n}{p})$, let $f(x) \equiv 1$, and let $c = 0$ and $d = p/2$. Since here $G(\chi) = ip^{\frac{1}{2}}$, we have

$$\sum_{0 < n < p/2} (\frac{n}{p}) = 2p^{-\frac{1}{2}} \sum_{n=1}^{\infty} (\frac{n}{p}) \int_0^{p/2} \sin(2\pi n x/p)\,dx$$

$$= \frac{2p^{\frac{1}{2}}}{\pi} \sum_{n=0}^{\infty} \left(\frac{2n+1}{p}\right) \frac{1}{2n+1} \quad,$$

which may be shown to be positive by a well-known argument [58]. Hence, Dirichlet's theorem follows immediately.

We remark that if we replace $(\frac{n}{p})$ in the argument above by $\chi(n)$, where $\chi(n)$ denotes an arbitrary odd, real, primitive character of modulus k, then we may determine the sign of

$$\sum_{0 < n < k/2} \chi(n) \ .$$

Dirichlet's proof of Theorem 4.1 is a consequence of one of his class number formulas. Proofs, independent of class number considerations have been given by K. Chung [29], Chowla [26], A. L. Whiteman [73], L. Moser [58], and Carlitz [21]. Like the proof above, all make essential use of Fourier series. The author [18] has recently found two new proofs of Theorem 4.1 that do not use Fourier series, but unfortunately neither is elementary. A truly elementary proof of Theorem 4.1 has never been found.

The following theorem also follows from one of Dirichlet's class number formulas [34]. There do not seem to be any proofs of this theorem in the literature which are independent of class number formulas. Another proof is given in a forthcoming paper by the author [18].

Theorem 4.2. Let p be a prime with $p \equiv 1 \pmod 4$. Then in the interval $(0, p/4)$ the number of quadratic residues modulo p exceeds the number of quadratic nonresidues.

Proof. In (4.1), let $\chi(n) = (\frac{n}{p})$, $f(x) \equiv 1$, $c = 0$, and $d = p/4$. Since $G(\chi) = p^{\frac{1}{2}}$, we get

$$\sum_{0 < n < p/4} (\frac{n}{p}) = 2 p^{-\frac{1}{2}} \sum_{n=1}^{\infty} (\frac{n}{p}) \int_0^{p/4} \cos(2\pi n x/p)dx$$

$$= \frac{p^{\frac{1}{2}}}{\pi} \sum_{n=0}^{\infty} \left(\frac{2n+1}{p}\right) \frac{(-1)^n}{2n+1} \ ,$$

which may be shown to be positive by a standard argument [4, pp. 27-28], and hence Theorem 4.2 follows.

151

The previous two theorems are but two of a large class of theorems that can be obtained by similar applications of (4.1) and (4.2). The author has recently learned that Y. Yamamoto [75] has used essentially the same method to obtain results of this type. We state just two more such theorems and refer the reader to [18] for additional results.

Theorem 4.3. Let p be any odd prime. Then in the interval $(0, p/3)$ the number of quadratic residues modulo p exceeds the number of quadratic nonresidues.

Theorem 4.4. Let p be an odd prime, and let

$$S_{6j} = \sum_{(j-1)p/6 < n < jp/6} \left(\frac{n}{p}\right) \qquad (j = 1, 2, 3) \ .$$

Then

1) $S_{61} > 0, \ S_{62} = S_{63} < 0$ $(p \equiv 1 \ (\text{mod } 8))$;

2) $S_{62} \geq 0, \ S_{63} > 0$ $(p \equiv 3 \ (\text{mod } 8))$;

3) $S_{61} = 0, \ S_{62} = -S_{63} > 0$ $(p \equiv 5 \ (\text{mod } 8))$;

4) $S_{61} > 0, \ S_{62} = -S_{63} \geq 0$ $(p \equiv 7 \ (\text{mod } 8))$.

In connection with the remark made after the proof of Theorem 4.1, Theorems 4.2 - 4.4 have obvious generalizations.

The next theorem is also due to Dirichlet [34]. By an elementary argument [73], it is easily shown to be equivalent to Theorem 4.1. However, a very simple, direct proof may be given by the use of (4.2).

Theorem 4.5. Let p be a prime with $p \equiv 3(\text{mod } 4)$. Then in the interval $(0, p)$, the sum of the quadratic nonresidues modulo p exceeds the sum of the quadratic residues.

Proof. In (4.2), let $\chi(n) = \left(\frac{n}{p}\right)$, $f(x) = x$, $c = 0$, and $d = p$. We find that

152

$$\sum_{n=1}^{p-1} (\frac{n}{p})n = 2 p^{-\frac{1}{2}} \sum_{n=1}^{\infty} (\frac{n}{p}) \int_0^p x \sin(2\pi nx/p)dx$$

$$= -\frac{p^{3/2}}{\pi} \sum_{n=1}^{\infty} (\frac{n}{p}) \frac{1}{n} < 0 \ ,$$

after an integration by parts and the use of a standard argument [73].
Theorem 4.5 now follows.

The following theorems may be readily proved by arguments similar to that above.

Theorem 4.6. Let p denote an odd prime. Then in the interval
(0, p/2) ,

1) the sum of the quadratic residues modulo p is equal to the sum of the nonresidues if p ≡ 7 (mod 8) ;

2) the sum of the quadratic residues modulo p exceeds the sum of the nonresidues if p ≡ 3(mod 8) ;

3) the sum of the quadratic nonresidues modulo p exceeds the sum of the residues if p ≡ 1(mod 4) .

Theorem 4.7. Let p be a prime with p ≡ 3 (mod 4). Then in the interval (0, p/3) the sum of the quadratic residues modulo p exceeds the sum of the nonresidues.

Theorem 4.8. Let p be an odd prime. Then,

1) in the interval (0, p/4), the sum of the quadratic residues modulo p exceeds the sum of the nonresidues if p ≡ 7(mod 8) ;

2) in the interval (p/4, p/2), the sum of the quadratic nonresidues modulo p exceeds the sum of the residues if p ≡ 1 (mod 4) ;

3) in the interval (p/2, 3p/4), the sum of the quadratic nonresidues modulo p exceeds the sum of the residues if p ≡ 3 (mod 4) .

153

Theorem 4.9. Let p denote an odd prime. Then in the interval (0, p) ,

1) the sum of the squares of the quadratic residues modulo p exceeds the sum of the squares of the nonresidues if $p \equiv 1 \pmod 4$;

2) the sum of the squares of the quadratic nonresidues modulo p exceeds the sum of the squares of the residues if $p \equiv 3 \pmod 4$.

Theorem 4.10. Let p be a prime with either $p \equiv 1 \pmod 4$ or $p \equiv 7 \pmod 8$. Then in the interval (0, p/2), the sum of the squares of the quadratic nonresidues modulo p exceeds the sum of the squares of the residues.

The second part of Theorem 4.9 was observed by Ayoub, Chowla and Walum [5]. However, this result seems to have been first noticed by Cauchy in 1840 [32, pp. 102, 103].

For the most part, in the six previous theorems, we have refrained from stating results that follow trivially from other results that are given. Other than these few additional easy deductions, Theorems 4.5 - 4.10 appear to be exhaustive. In this connection, one should consult a paper of N. J. Fine [37].

5. Power sums and cotangent sums. Let $A = \{a_j\}$, $-\infty < j < \infty$, have period $k \geq 2$. (The case $k = 1$ yields vacuous results in this application.) Let m be any integer, and let r be a positive integer. Define

$$M_r(m, A) = \sum_{n=1}^{k-1} a_n n^r e^{2\pi i m n/k} .$$

Theorem 5.1. For each positive integer r, $M_r(m, A)$ can be written as a linear combination of the cotangent sums

$$\sum_{n=1}^{k-1} b_{n-m} \cot^j(\pi n/k) \qquad (0 \leq j \leq r)$$

plus a constant multiple of b_{-m} .

154

Proof. We will give a complete proof for the case $r = 1$. For $r > 1$, the procedure is of the same nature.

First, in general, apply Theorem 3.1 with $f(x) = x^r e^{2\pi i m x/k}$, $c = 0$, and $d = k$ to get

$$(5.1) \qquad M_r(m, A) + \frac{1}{2} a_0 k^r = \sum_{n=-\infty}^{\infty *} b_n \int_0^k x^r e^{2\pi i (m+n)x/k} dx$$

$$= \sum_{j=-\infty}^{\infty *} b_{j-m} \int_0^k x^r e^{2\pi i j x/k} dx \quad ,$$

where, in this proof, the symbol $\sum_{n=-\infty}^{\infty *}$ will mean $\lim_{N \to \infty} \sum_{n=-N}^{N}$.

We now suppose that $r = 1$. Using the values

$$\int_0^k x \, e^{2\pi i j x/k} dx = \begin{cases} k^2/2, & j = 0 \ , \\ k^2/(2\pi i j), & j \neq 0 \ , \end{cases}$$

and putting $j = \mu k + \nu$, $-\infty < \mu < \infty$, $0 \leq \nu \leq k-1$, in (5.1), we find that

$$(5.2) \qquad M_1(m, A) + \frac{1}{2} a_0 k = \frac{1}{2} b_{-m} k^2 + \frac{k}{2\pi i} \sum_{\nu=1}^{k-1} b_{\nu-m} \sum_{\mu=-\infty}^{\infty *} \frac{1}{\mu + \nu/k}$$

$$= \frac{1}{2} b_{-m} k^2 + \frac{k}{2i} \sum_{\nu=1}^{k-1} b_{\nu-m} \cot(\pi\nu/k) \quad ,$$

where we also used the trivial fact $\sum_{\substack{\mu=-\infty \\ \mu \neq 0}}^{\infty *} 1/\mu = 0$. Using the representation of a_0 from (2.8), we deduce from (5.2) that

$$(5.3) \qquad M_1(m, A) = \frac{1}{2} k(k-1) b_{-m} - \frac{1}{2} k \sum_{\nu=1}^{k-1} b_{\nu-m} \{i \cot(\pi\nu/k) + 1\} \quad ,$$

which completes the proof of the theorem for $r = 1$.

155

If $m = 0$ and $A = \chi$, where $\chi(n)$ is primitive, our results have an obvious connection with results in the preceding section. In the case that χ is odd, (5.3) yields, with the aid of (2.9),

$$(5.4) \qquad M_1(0, \chi) = \frac{1}{2} iG(\chi) \sum_{\nu=1}^{k-1} \bar{\chi}(\nu) \cot(\pi\nu/k) \ .$$

(If χ is even, a triviality occurs.) In the instance that $\chi(n)$ is the Legendre-Jacobi symbol, formula (5.4) has a long history, especially in its connection with class number formulas. The right side of (5.4) appears in a class number formula stated without proof by Gauss [32, p. 97]. The first proof of (5.4) appears to have been given by V. A. Lebesgue [32, p. 103]. A very simple proof has been given by Whiteman [73]. For further connections with class number formulas, consult Dickson's history [32, Chapter 6].

In the case that $m = 0$, a_n is the Legendre-Jacobi symbol, and $1 \le r \le 3$, the results of Theorem 5.1 can be found in a paper of Chowla [25]. For $m = 0$ and an arbitrary primitive character χ, Theorem 5.1 was first established by Apostol [3]. Another proof of Apostol's result was given by the author [13]. In the case that $b_{-m} = 0$, a more explicit version of Theorem 5.1 was proved by the author [14].

6. <u>Gauss sums</u>. In [19], Schoenfeld and the author proved the following reciprocity theorem.

<u>Theorem 6.1</u>. Let a, b and c be integers with $ac \ne 0$ and $ack + b$ even. Then

$$(6.1) \qquad \sum_{n=0}^{|c|k-1} a_n e^{\pi i a n^2/ck + \pi i bn/ck}$$

$$= (|c|k/|a|)^{\frac{1}{2}} e^{\frac{1}{4}\pi i \operatorname{sgn}(ac) - \pi i b^2/(4ack)} \sum_{n=0}^{|a|k-1} b_n e^{-\pi i c n^2/ak - \pi i bn/ak} \ .$$

The proof is in the same spirit as that of Dirichlet [33] who used the ordinary Poisson formula to evaluate the ordinary Gaussian sums. To obtain these values, in (6.1), put $A = I$, $a = 2$ and $b = 0$. Dirichlet's proof may be found in Davenport's book [30, pp. 16-17].

Another result contained in (6.1) is the reciprocity theorem for ordinary Gaussian sums, often called Schaar's identity [53, p. 75]. This is the relation obtained by letting $A = I$ and $b = 0$ above. The reciprocity theorem for generalized Gaussian sums is obtained by setting $A = I$ above. The result for $a_n = \chi(n)$, where χ is primitive and even, $b = 0$, and $a, c > 0$ was first established by Guinand [41].

In fact, the values of certain biquadratic Gaussian sums may be determined from (6.1).

Corollary 6.2. Let p be an odd prime, and let b be any odd integer. Then

$$\sum_{n=0}^{p-1} e^{\pi i n^4/p + \pi i b n^2/p} = p^{\frac{1}{2}} e^{\frac{\pi i}{4p}(p-b^2)} \quad ,$$

and, in particular,

$$\sum_{n=0}^{p-1} (-1)^n e^{\pi i n^4/p} = p^{\frac{1}{2}} e^{\frac{\pi i}{4}(1-p)} \quad .$$

We will not give the proof of Corollary 6.2 but will just mention that the result arises from the instance $a_n = (\frac{n}{p})$ and may be derived easily from [14, Corollary 4].

7. Functional equations. The periodic Euler-Maclaurin formula has been used by Schoenfeld and the author [19] to derive the functional equation of $\varphi(x, a, s; A)$ given in Definition 2.4.

Theorem 7.1. For x and a real, $\varphi(x, a, s; A)$ has an analytic continuation into the full complex s-plane with at most a simple pole at

157

$s = 1$. Furthermore, if $B_1^* = \{b_{-n-1}\}$, $0 \leq a \leq 1$, and $x \leq 1$, we have, for all s,

$$(7.1) \quad \varphi(x, a, 1-s; A) = (k/2\pi)^s \Gamma(s) \left\{ e^{\pi i s/2 - 2\pi i a x/k} \varphi(-a, x, s; B) \right.$$

$$\left. + e^{-\pi i s/2 + 2\pi i a(1-x)/k} \varphi(a, 1-x, s; B_1^*) \right\}.$$

The functional equations of $\zeta(s)$, $\zeta(s, a)$, $L(s, \chi)$, and Lerch's function $\varphi(x, a, s; I)$ are special cases of (7.1). Proofs of the functional equations of $\zeta(s)$, $\zeta(s, a)$, and $\varphi(x, a, s; I)$ by the ordinary Euler-Maclaurin formula are given in [70, pp. 14-15], [11], and [12], respectively.

The functional equation of $\varphi(x, a, s; A)$ may also be achieved by applying the periodic Poisson summation formula. Mordell [56] and Guinand [40] have used the ordinary Poisson formula to derive the functional equation of $\zeta(s)$. Mordell [57] has also derived Hurwitz's formula for $\zeta(s, a)$ by the same method, and he [56] has used a character version of the Poisson summation formula to derive the functional equation of Dirichlet's L-functions associated with primitive characters.

8. **A trigonometric series of Hardy and Littlewood.** In their study of Lambert series and Tauberian theorems, Hardy and Littlewood [45] examined the trigonometric sum

$$\sum_{n \leq x} \frac{1}{n} \sin(\frac{x}{n}) \ ,$$

which differs from

$$(8.1) \quad \sum_{n=1}^{\infty} \frac{1}{n} \sin(\frac{x}{n})$$

by a function which is just $O(1)$ as x tends to ∞. S. L. Segal [66] considered the series resulting from the integration of (8.1) and showed that for $x > 0$,

$$(8.2) \qquad 2 \sum_{n=1}^{\infty} \sin^2(\frac{x}{n}) = \pi x - \frac{1}{2} + \sum_{n=1}^{\infty} (\frac{\pi x}{n})^{\frac{1}{2}} J_1(4\sqrt{\pi x n}) \ ,$$

where J_ν denotes the ordinary Bessel function of order ν . On the other hand, T. Kano [47] demonstrated that the formal, termwise differentiation of both sides of (8.2) is not valid. Segal gave two proofs of (8.2), but, in fact, a simpler proof of (8.2) may be obtained by the use of the ordinary Poisson summation formula. We will use below the periodic Poisson formula to give a considerable generalization of (8.2).

Theorem 8.1. Let A have period k, and let B be the complementary sequence to A . Let $x > 0$. If A is even, then

$$(8.3) \qquad 2 \sum_{n=1}^{\infty} b_n \sin^2 (\frac{x}{n}) = a_0 \frac{\pi x}{k} - \frac{1}{2} b_0 + \sum_{n=1}^{\infty} a_n (\frac{\pi x}{kn})^{\frac{1}{2}} J_1(4\sqrt{\pi x n/k}) \ ;$$

if A is odd, then

$$(8.4) \qquad 2i \sum_{n=1}^{\infty} b_n \sin (\frac{x}{n}) = \sum_{n=1}^{\infty} a_n (\frac{2\pi x}{kn})^{\frac{1}{2}} J_1(2\sqrt{2\pi x n/k}) \quad .$$

Proof. First, suppose that A is even. We shall employ the periodic Poisson formula (3.2) with $f(t) = t^{-\frac{1}{2}} J_1(4\sqrt{\pi x t/k})$. As $J_0'(u) = -J_1(u)$ [72, p. 45],

$$(8.5) \qquad \int_0^{\infty} f(t)dt = \frac{1}{2} (\frac{k}{\pi x})^{\frac{1}{2}} \int_0^{\infty} J_1(u)du = \frac{1}{2} (\frac{k}{\pi x})^{\frac{1}{2}} \ ,$$

since $J_0(0) = 1$, and since as x tends to ∞ [72, p. 199]

159

$$(8.6) \quad J_\nu(x) = \left(\frac{2}{\pi x}\right)^{\frac{1}{2}} \left\{ \cos(x - \frac{1}{2}\nu\pi - \frac{1}{4}\pi) + (\nu^2 - \frac{1}{4})\sin(x - \frac{1}{2}\nu\pi - \frac{1}{4}\pi)\frac{1}{2x} \right.$$
$$\left. + O(1/x^2) \right\} \ .$$

Next, we have [39, p. 742]

$$(8.7) \quad \int_0^\infty J_1(4\sqrt{\pi x t/k})\cos(2\pi nt/k)dt = \left(\frac{k}{\pi x}\right)^{\frac{1}{2}}\sin^2\left(\frac{x}{n}\right) \ .$$

In (3.2), let $c = 0$ and $d = Nk$, where N is a positive integer. Letting N tend to ∞, employing (8.5) and (8.7), and using the value $f(0) = 2(\pi x/k)^{\frac{1}{2}}$, we arrive at

$$(8.8) \quad a_0\left(\frac{\pi x}{k}\right)^{\frac{1}{2}} + \sum_{n=1}^\infty a_n n^{-\frac{1}{2}} J_1(4\sqrt{\pi xn/k}) = \frac{1}{2} b_0\left(\frac{k}{\pi x}\right)^{\frac{1}{2}}$$

$$+ 2 \lim_{N\to\infty} \sum_{n=1}^\infty b_n \left\{ \left(\frac{k}{\pi x}\right)^{\frac{1}{2}}\sin^2\left(\frac{x}{n}\right) - \int_{Nk}^\infty t^{-\frac{1}{2}} J_1(4\sqrt{\pi xt/k})\cos(2\pi nt/k)dt \right\} \ .$$

If we can show that

$$(8.9) \quad \lim_{N\to\infty} \sum_{n=1}^\infty b_n \int_{Nk}^\infty t^{-\frac{1}{2}} J_1(4\sqrt{\pi xt/k})\cos(2\pi nt/k)dt = 0 \ ,$$

then we may conclude (8.3) immediately from (8.8). The proof of (8.9) is easily accomplished by two integrations by parts with the help of the differentiation formula [72, p. 45]

$$(8.10) \quad \frac{d}{dz}\left\{z^{-\nu} J_\nu(z)\right\} = -z^{-\nu} J_{\nu+1}(z)$$

and also with the help of (8.6). This completes the proof of (8.3).

Secondly, let A be odd. To prove (8.4), we proceed in the same fashion as above. This time we use (3.3) with $c = 0$, $d = Nk$, and

$f(t) = t^{-\frac{1}{2}} J_1(2\sqrt{2\pi xt/k})$. Since A is odd, we see from (2.7) and (2.8) that $a_0 = b_0 = 0$. Now [39, p. 742],

(8.11) $\qquad \int_0^\infty t^{-\frac{1}{2}} J_1(2\sqrt{2\pi xt/k}\,) \sin(2\pi nt/k)dt = (\frac{k}{2\pi x})^{\frac{1}{2}} \sin(\frac{x}{n})$.

Letting N tend to ∞ in (3.3) and using (8.11), we get

(8.12) $\qquad \sum_{n=1}^\infty a_n n^{-\frac{1}{2}} J_1(2\sqrt{2\pi xn/k}\,)$

$= 2i \lim_{N\to\infty} \sum_{n=1}^\infty \left\{ (\frac{k}{2\pi x})^{\frac{1}{2}} \sin(\frac{x}{n}) - \int_{Nk}^\infty t^{-\frac{1}{2}} J_1(2\sqrt{2\pi xt/k}) \sin(2\pi nt/k)dt \right\}$.

To show that

$$\sum_{n=1}^\infty b_n \int_{Nk}^\infty t^{-\frac{1}{2}} J_1(2\sqrt{2\pi xt/k}\,) \sin(2\pi nt/k)dt$$

tends to 0 as N tends to ∞, we proceed as before, but this time we must also use the fact that $\sum_{n=1}^\infty b_n/n$ converges, which, since $a_0 = 0$, may be shown by a routine application of partial summation. Thus, (8.4) now readily follows from (8.12).

By letting A = I in (8.3), we get (8.2) at once. The result for primitive characters is stated in the following corollary.

Corollary 8.2. Let χ be a primitive character modulo k . If χ is even, then

(8.13) $\qquad 2 \sum_{n=1}^\infty \chi(n) \sin^2(\frac{x}{n}) = G(\chi) \sum_{n=1}^\infty \overline{\chi}(n) (\frac{\pi x}{kn})^{\frac{1}{2}} J_1(4\sqrt{\pi xn/k})$;

if χ is odd, then

(8.14) $\qquad 2i \sum_{n=1}^\infty \chi(n) \sin(\frac{x}{n}) = G(\chi) \sum_{n=1}^\infty \overline{\chi}(n) (\frac{2\pi x}{kn})^{\frac{1}{2}} J_1(2\sqrt{2\pi xn/k}\,)$.

161

BRUCE C. BERNDT

Proof. Let $a_n = \overline{\chi}(n)$ in Theorem 8.1. Using (2.9) and the fundamental property [4, p. 313]

$$(8.15) \qquad G(\chi)\, G(\overline{\chi}) = \chi(-1)k \ ,$$

we easily arrive at (8.13) and (8.14).

The series of Bessel functions on the right sides of (8.3) and (8.4) are exactly the same type that are found in the representations of summatory functions for arithmetical functions generated by Dirichlet series satisfying a functional equation involving $\Gamma(s)$ [6], [9], [22]. According as to whether A is even or odd, let $E(x, A)$ denote the infinite series on the right side of (8.3) or (8.4). $E(x, A)$ may be thought of as the "error term." Much of the theory that has been developed for the error terms of the aforementioned summatory functions can be applied here to $E(x, A)$. Hence, by applying a method originally due to W. Sierpiński [50, pp. 207-208], we obtain the following theorem.

Theorem 8.3. Let A be even or odd. Then as x tends to ∞,

$$E(x, A) = O(x^{1/3}) \ .$$

Theorem 8.4. Let A be even or odd. Suppose that $\mathrm{Re}\{a_n\} \neq 0$ for at least one value of n. Then,

$$\overline{\lim_{x \to \infty}} \ x^{-\frac{1}{4}} \, \mathrm{Re}\,\{E(x, A)\} = \pm\infty \ .$$

If $\mathrm{Im}\{a_n\} \neq 0$ for at least 1 value of n, then

$$\overline{\lim_{x \to \infty}} \ x^{-\frac{1}{4}} \, \mathrm{Im}\,\{E(x, A)\} = \pm\infty \ .$$

The proof of Theorem 8.4 uses a technique of A. E. Ingham considerably generalized by K. Chandrasekharan and R. Narasimhan [23,

162

Theorem 3.1]. If $a_n \geq 0$, $1 \leq n \leq k$, then a further improvement is possible. See [66, p. 388] or [50, p. 248].

By using an argument of J. Steinig [68], we can deduce the next theorem.

Theorem 8.5.   Let A be even or odd. Suppose that $\mathrm{Re}\{a_n\} \neq 0$ for at least one value of n. Then on the interval $0 \leq x \leq t$, $\mathrm{Re}\{E(x, A)\}$ changes sign at least $c_1 t^{\frac{1}{2}} - c_2$ times, where c_1 and c_2 are positive constants dependent upon A but independent of t. If $\mathrm{Im}\{a_n\} \neq 0$ for at least one value of n, the same conclusion holds with $\mathrm{Re}\{E(x, A)\}$ replaced by $\mathrm{Im}\{E(x, A)\}$.

9.   Infinite series of ordinary Bessel functions.   Two classes of results are derived in this section. In each case, the infinite series of Bessel functions is reminiscent of the series of Bessel functions in the preceding section.

Theorem 9.1.   Let $x > 0$. Then for even A and $\mathrm{Re}\,\nu > -1/2$,

$$(9.1) \quad \sum_{0 < n \leq x}' b_n (x^2 - n^2)^{\nu - \frac{1}{2}} = a_0 \frac{\pi^{\frac{1}{2}} \Gamma(\nu + \frac{1}{2})}{2k\,\Gamma(\nu+1)} x^{2\nu} - \frac{1}{2} b_0 x^{2\nu - 1}$$

$$+ \frac{\pi^{\frac{1}{2}} \Gamma(\nu + \frac{1}{2})}{k} \sum_{n=1}^{\infty} a_n \left(\frac{kx}{\pi n}\right)^{\nu} J_\nu(2\pi nx/k) \;;$$

for odd A and $\mathrm{Re}\,\nu > 1/2$,

$$(9.2) \quad i \sum_{0 < n \leq x}' b_n n (x^2 - n^2)^{\nu - 3/2} = \frac{\pi^{\frac{1}{2}} \Gamma(\nu - \frac{1}{2})x}{k} \sum_{n=1}^{\infty} a_n \left(\frac{kx}{\pi n}\right)^{\nu - 1} J_\nu(2\pi nx/k) .$$

The prime ' on the summation signs at the left sides of (9.1) and (9.2) indicates that if $\nu = 1/2$ and $\nu = 3/2$, respectively, and if x is a positive integer, then the summand for $n = x$ is to be halved.

163

Proof.   Suppose first that A is even. In (3.2), put $c = 0$, $d = Nk$, where N is a positive integer, and $f(t) = t^{-\nu} J_\nu(2\pi tx/k)$, where $x > 0$ and $\text{Re } \nu > -1/2$. Eventually, we shall let N tend to ∞. First [39, p. 684],

$$(9.3) \qquad \int_0^\infty t^{-\nu} J_\nu(2\pi tx/k) dt = \frac{1}{2} \left(\frac{\pi x}{k}\right)^{\nu-1} \frac{\pi^{\frac{1}{2}}}{\Gamma(\nu+\frac{1}{2})} \; .$$

Secondly [69, p. 178], [72, p. 404],

$$(9.4) \qquad \int_0^\infty t^{-\nu} J_\nu(2\pi tx/k) \cos(2\pi tn/k) dt$$

$$= \frac{1}{2} \left(\frac{k}{\pi x}\right)^{-\nu+1} \frac{\pi^{\frac{1}{2}}}{\Gamma(\nu+\frac{1}{2})} \; g(n, x) \; ,$$

where

$$g(n, x) = \begin{cases} (1 - n^2/x^2)^{\nu-\frac{1}{2}}, & \text{if } n < x \; , \\ 1/2, & \text{if } n = x \text{ and } \nu = 1/2 \; , \\ 0, & \text{if } n \geq x \text{ and } \text{Re } \nu \geq 1/2, \text{ except when } n = x \\ & \underline{\text{and }} \nu = 1/2; \text{ or if } n > x \text{ and } |\text{Re } \nu| < 1/2 \; . \end{cases}$$

If we let N tend to ∞ in (3.2), employ (9.3) and (9.4), and use the value $f(0) = (\pi x/k)^\nu/\Gamma(\nu+1)$, we get

$$(9.5) \quad \frac{1}{2} a_0 \left(\frac{\pi x}{k}\right)^\nu \frac{1}{\Gamma(\nu+1)} + \sum_{n=1}^\infty n^{-\nu} J_\nu(2\pi nx/k)$$

$$= \frac{1}{2} b_0 \left(\frac{\pi x}{k}\right)^{\nu-1} \frac{\pi^{\frac{1}{2}}}{\Gamma(\nu+\frac{1}{2})} + \left(\frac{\pi x}{k}\right)^{\nu-1} \frac{\pi^{\frac{1}{2}}}{\Gamma(\nu+\frac{1}{2})} \sum_{0 < n \leq x}' b_n (1 - n^2/x^2)^{\nu-1/2}$$

$$+ \lim_{N \to \infty} 2 \sum_{n=1}^\infty b_n \int_{Nk}^\infty t^{-\nu} J_\nu(2\pi tx/k) \cos(2\pi tn/k) dt \; .$$

164

We see from (9. 5) that the proof of (9.1) will be complete if we can show that the last expression on the right side of (9. 5) is zero. If we integrate by parts twice, each with the use of (8.10), and then make full use of (8. 6), the desired limit is readily shown to be zero.

Suppose next that A is odd. In (3. 3), we let $c = 0$, $d = Nk$, where N is a positive integer, and $f(t) = t^{-\nu+1} J_\nu(2\pi tx/k)$, where $\mathrm{Re}\ \nu > 1/2$ and $x > 0$. Subsequently, we shall let N tend to ∞ . We use the value of the integral [69, p. 179], [72, p. 404],

$$\int_0^\infty t^{1-\nu} J_\nu(2\pi tx/k)\ \sin(2\pi nt/k)dt = \frac{1}{2}(\frac{\pi x}{k})^{\nu-2}\ \frac{\pi^{\frac{1}{2}}}{\Gamma(\nu-\frac{1}{2})}\ h(n, x)\ ,$$

where

$$h(n, x) = \begin{cases} (n/x)\ (1 - n^2/x^2)^{\nu-3/2}, & \text{if } n < x\ , \\ 1/2, & \text{if } n = x \text{ and } \nu = 3/2\ , \\ 0, & \text{if } n \geq x \text{ and } \mathrm{Re}\ \nu \geq 3/2, \text{ except when } n = x \text{ and} \\ & \nu = 3/2;\text{ or if } n > x \text{ and } 1/2 < \mathrm{Re}\ \nu < 3/2\ . \end{cases}$$

The remainder of the proof of (9. 2) follows along the same lines as the proof of (9. 1).

We now examine several interesting consequences of Theorem 9.1. If $A = I$, (9.1) is very old. In particular, the special case with $\nu = 0$ may be found in Nielsen's book [59, p. 336]. For $A = I$, a generalization of (9. 1) in another sense may be found in papers of G. Doetsch [35], A. Erdélyi [36], and H. Kober [49]. For real ν, the case $A = I$ in (9.1) is also covered by the general theorems in [6], [9], and [22]. The special case for $A = \{(-1)^j\}$ and $\nu = 0$ in Theorem 9.1 can also be found in Nielsen's book [59, p. 336].

Corollary 9.2.   Let $x > 0$, and let χ be a primitive character modulo k . If χ is even and $\mathrm{Re}\ \nu > -1/2$, then

165

$$G(\overline{\chi}) \sum_{n \le x}' \chi(n)(x^2 - n^2)^{\nu - \frac{1}{2}} = \pi^{\frac{1}{2}} \Gamma(\nu + \tfrac{1}{2}) \sum_{n=1}^{\infty} \overline{\chi}(n)\left(\frac{kx}{\pi n}\right)^{\nu} J_{\nu}(2\pi nx/k) \ ;$$

if χ is odd and $\mathrm{Re}\ \nu > 1/2$, then

$$G(\overline{\chi}) \sum_{n \le x}' \chi(n)\, n(x^2 - n^2)^{\nu - 3/2} = i\pi^{\frac{1}{2}} \Gamma(\nu - \tfrac{1}{2})x \sum_{n=1}^{\infty} \overline{\chi}(n)\left(\frac{kx}{\pi n}\right)^{\nu - 1} J_{\nu}(2\pi nx/k) \ .$$

For ν real and χ even, Corollary 9.2 is a special case of the results in [6], [9], and [22]. For ν real and χ odd, a result similar to Corollary 9.2 may be found in [7, Example 11.1].

Corollary 9.3.   For $0 < y < \pi$,

$$\sum_{n=1}^{\infty} (-1)^{n+1} J_0(ny) = \frac{1}{2} \ .$$

Proof.   In Theorem 9.1, let $A = \{(-1)^j\}$, $k = 2$, $\nu = 0$, and $0 < x < 1$. Recalling that $b_0 = 0$ and letting $x = y/\pi$, we obtain the desired result forthwith.

Corollary 9.4.   Let $0 < y < 2\pi/k$, and let χ be a primitive character of modulus k. If $\mathrm{Re}\ \nu > -1/2$ and χ is even, then

$$\sum_{n=1}^{\infty} \chi(n)\, n^{-\nu} J_{\nu}(ny) = 0 \ ;$$

if $\mathrm{Re}\ \nu > 1/2$ and χ is odd, then

$$\sum_{n=1}^{\infty} \chi(n)\, n^{-\nu+1} J_{\nu}(ny) = 0 \ .$$

Proof.   The results follow from Corollary 9.2 with $x = ky/(2\pi)$.

The last two corollaries are but two of several results of a similar type that can be deduced from Theorem 9.1. Corollary 9.3 may be found in [39, p. 976].

166

If we set $\nu = 1/2$ in (9.1) and use the fact that $J_{1/2}(x) = (2/\pi x)^{\frac{1}{2}}$
$\cdot \sin x$, then (9.1) reduces to the Fourier series of $B_1(x, B)$ given by (2.10).
If we set $\nu = 3/2$ in (9.2) and use the fact that $J_{3/2}(x) = (2/\pi x)^{\frac{1}{2}}$
$\cdot \{(1/x)\sin x - \cos x\}$, then (9.2) reduces to the Fourier series of $B_1(x, B)$
as well.

The series of Bessel functions on the right sides of (9.1) and (9.2)
are like those for summatory functions of arithmetical functions generated by Dirichlet series satisfying a functional equation involving $\Gamma(s/2)$
and $\Gamma(\{s+1\}/2)$, respectively. Thus comments analogous to those made
in the latter part of Section 8 can be made here. We shall not pursue
the matter further, however.

Theorem 9.5. Let $0 < x < 2\pi/k$. If A is even and $\mathrm{Re}\,\nu > 1/2$,
then

$$(9.6) \quad \sum_{n=1}^{\infty} a_n\, n^{-\nu+1} J_\nu(nx) = \frac{1}{2} \sum_{n=0}^{\infty} \frac{(-1)^n B_{2n}(A)(x/2)^{\nu+2n-2}}{n!\, \Gamma(\nu+n)} \quad ;$$

if A is odd and $\mathrm{Re}\,\nu > -1/2$, then

$$(9.7) \quad \sum_{n=1}^{\infty} a_n\, n^{-\nu} J_\nu(nx) = \sum_{n=0}^{\infty} \frac{(-1)^{n+1} B_{2n+1}(A)(x/2)^{\nu+2n}}{n!\, \Gamma(\nu+n+1)(2n+1)} \quad .$$

Proof. First, suppose that A is even. In (3.2), let $c = 0$,
$d = Nk$, where N is a positive integer, and $f(t) = t^{-\nu+1} J_\nu(tx)$, with
$\mathrm{Re}\,\nu > 1/2$ and $0 < x < 2\pi/k$. Again, we shall subsequently let N
tend to ∞. This is justified in exactly the same manner as before, i.e.,
by two integrations by parts with the aid of (8.10), and by the use of
(8.6). Using (8.10), we have

$$(9.8) \quad \int_0^{\infty} t^{-\nu+1} J_\nu(tx)dt = -x^{\nu-2} \int_0^{\infty} \frac{d}{du}\left\{ u^{-(\nu-1)} J_{\nu-1}(u) \right\} du$$

$$= \frac{x^{\nu-2}}{2^{\nu-1}\,\Gamma(\nu)} ,$$

167

provided that $\operatorname{Re} \nu > 1/2$. Next [39, p. 747],

$$(9.9) \qquad \int_0^\infty t^{-\nu+1} J_\nu(tx) \cos(2\pi nt/k)dt$$

$$= -\frac{k^2 x^\nu}{2^{\nu+2} \pi^2 n^2 \Gamma(\nu+1)} F(1, 3/2; \nu+1; (\frac{kx}{2\pi n})^2) \ ,$$

provided that $\operatorname{Re} \nu > 1/2$ and $0 < x < 2\pi n/k$, where $F(a, b; c; z)$ denotes the hypergeometric function. Thus, letting N tend to ∞ in (3.2) and using (9.8) and (9.9), we arrive at

$$(9.10) \qquad \sum_{n=1}^\infty a_n n^{-\nu+1} J_\nu(nx) - \frac{x^{\nu-2} B_0(A)}{2^{\nu-1} \Gamma(\nu)}$$

$$= -\frac{k^2 x^\nu}{2^{\nu+1} \pi^2 \Gamma(\nu+1)} \sum_{n=1}^\infty \frac{b_n}{n^2} F(1, 3/2; \nu+1; (\frac{kx}{2\pi n})^2)$$

$$= -\frac{k^2 x^\nu}{2^{\nu+1} \pi^2 \Gamma(\nu+1)} \sum_{n=1}^\infty \frac{b_n}{n^2} \sum_{j=0}^\infty \frac{\Gamma(3/2+j) \Gamma(\nu+1)}{\Gamma(3/2) \Gamma(\nu+1+j)} (\frac{kx}{2\pi n})^{2j}$$

$$= -\frac{k^2 x^\nu}{2^\nu \pi^{5/2}} \sum_{j=0}^\infty \frac{\Gamma(3/2+j)}{\Gamma(\nu+1+j)} (\frac{kx}{2\pi})^{2j} \zeta(2j+2; B)$$

$$= \sum_{j=0}^\infty \frac{(-1)^{j+1} B_{2j+2}(A) (x/2)^{\nu+2j}}{j! \ \Gamma(\nu+j+1) (2j+2)} \ .$$

In the foregoing calculation, we first used the series representation for the hypergeometric function and then inverted the order of summation, which is justified by absolute convergence. In the last step, we used (2.12) and the duplication formula for the gamma function. It is now easily seen that (9.6) follows from (9.10).

Now let A be odd. In (3.3), put $c = 0$, $d = Nk$, and $f(t) = t^{-\nu} J_\nu(tx)$, where $\text{Re } \nu > -1/2$ and $x > 0$. We again let N tend to ∞ with the same justification as before. We have [39, p. 747]

(9.11)
$$\int_0^\infty t^{-\nu} J_\nu(tx) \sin(2\pi nt/k)dt$$

$$= \frac{kx^\nu}{2^{\nu+1} \pi n \, \Gamma(\nu+1)} F(1, 1/2; \nu+1; \left(\frac{kx}{2\pi n}\right)^2) \; ,$$

provided that $\text{Re } \nu > -1/2$ and $0 < x < 2\pi n/k$. From (3.3) and (9.13), we find that

$$\sum_{n=1}^\infty a_n n^{-\nu} J_\nu(nx) = \frac{ikx^\nu}{2^\nu \pi \, \Gamma(\nu+1)} \sum_{n=1}^\infty \frac{b_n}{n} F(1, 1/2; \nu+1; \left(\frac{kx}{2\pi n}\right)^2)$$

$$= \frac{ikx^\nu}{2^\nu \pi^{3/2}} \sum_{j=0}^\infty \frac{\Gamma(\frac{1}{2}+j)}{\Gamma(\nu+j+1)} \left(\frac{kx}{2\pi}\right)^{2j} \zeta(2j+1; B)$$

$$= \sum_{j=0}^\infty \frac{(-1)^{j+1} B_{2j+1}(A)(x/2)^{\nu+2j}}{j! \, \Gamma(\nu+j+1)(2j+1)} \; ,$$

where we used the series representation for the hypergeometric function, inverted the order of summation, employed the duplication formula for the gamma function, and applied (2.12). Due to the presence of the term corresponding to $j = 0$, we cannot justify the inversion in order of summation by absolute convergence, but a direct verification is quite easy. This completes the proof.

We cite two examples to illustrate Theorem 9.5. If $A = \{(-1)^j\}$, then $B_{2n}(A) = (2^{2n} - 1)B_{2n}$, and so we get from (9.6),

$$\sum_{n=1}^\infty (-1)^n n^{-\nu+1} J_\nu(nx) = \frac{1}{2} \sum_{n=0}^\infty \frac{(-1)^n (2^{2n}-1)B_{2n}(x/2)^{\nu+2n-2}}{n! \, \Gamma(\nu+n)} \; ,$$

where $\mathrm{Re}\,\nu > 1/2$ and $0 < x < \pi$. If $A = \chi_4$, then $B_{2n+1}(A) = -\frac{1}{2}(2n+1)\,E_{2n}$. Thus, for $\mathrm{Re}\,\nu > -1/2$ and $0 < x < \pi/2$, we conclude from (9.7) that

$$(9.12) \qquad \sum_{n=0}^{\infty} (-1)^n (2n+1)^{-\nu} J_\nu((2n+1)x)$$

$$= \frac{1}{2} \sum_{n=0}^{\infty} \frac{(-1)^n E_{2n}(x/2)^{\nu+2n}}{n!\,\Gamma(\nu+n+1)}\ .$$

In the case $\nu = 1$, (9.12) was formally established by G. G. Macfarlane [54]. Macfarlane uses Mellin transforms and makes no attempt to justify his shifting of the vertical line of integration "to $-\infty$." Indeed, this justification seems to be very difficult. Moreover, Macfarlane derives his result under the apparent assumption $0 < x < \infty$. This is quite definitely false, for if we assumed that $x > 2\pi/k$ is fixed, then some of the integrals calculated in the proof of Theorem 9.5 would have different evaluations, and so we would no longer arrive at (9.6) or (9.7), as the case may be. Moreover, the radius of convergence of the series on the right side of (9.12), is $\pi/2$. Macfarlane's result is reproduced in Sneddon's book [67, p. 286] with the same inattention to rigor. We remark that a very short, formal proof of Theorem 9.5 may be achieved by resorting to the periodic Euler-Maclaurin formula, given in Theorem 3.3, and letting r tend to ∞. However, it appears very difficult to make this use rigorous.

The contrast between Theorems 9.1 and 9.5 should be noted. Theorem 9.1 may be further generalized along the lines of the aforementioned papers of Doetsch and Erdélyi. Theorem 9.5 may be further generalized, since more general integrals than those calculated in the proof may be evaluated in terms of hypergeometric functions [39, p. 747]. Furthermore, formulas similar to those in Theorem 9.5 may be derived for other ranges of x.

10. Infinite series of modified Bessel functions. Throughout this section, $K_\nu(z)$ denotes the modified Bessel function of order ν [72, p. 78].

Theorem 10.1. Let $x > 0$. If A is even and $\operatorname{Re}\nu > 0$, then

(10.1)
$$\frac{1}{4} a_0 (\frac{k}{\pi x})^\nu \Gamma(\nu) + \sum_{n=1}^{\infty} a_n n^\nu K_\nu(2\pi nx/k)$$

$$= \frac{1}{4} b_0 \pi^{\frac{1}{2}} (\frac{k}{\pi x})^{\nu+1} \Gamma(\nu+\tfrac{1}{2}) + \frac{\pi^{\frac{1}{2}}}{2x} (\frac{kx}{\pi})^{\nu+1} \Gamma(\nu+\tfrac{1}{2}) \sum_{n=1}^{\infty} b_n (n^2 + x^2)^{-\nu-\frac{1}{2}} \ ;$$

if A is odd and $\operatorname{Re}\nu > -1$, then

(10.2)
$$\sum_{n=1}^{\infty} a_n n^{\nu+1} K_\nu(2\pi nx/k)$$

$$= \frac{i\pi^{\frac{1}{2}}}{2x^2} (\frac{kx}{\pi})^{\nu+2} \Gamma(\nu+3/2) \sum_{n=1}^{\infty} b_n n(n^2 + x^2)^{-\nu-3/2} \ .$$

Proof. Suppose first that A is even. In (3.2), let $f(t) = t^\nu K_\nu(2\pi tx/k)$ with $x, \operatorname{Re}\nu > 0$, let $c = 0$, and let d tend to ∞, which is easily justified by two integrations by parts. Now for $\operatorname{Re}\nu > 0$ [72, p. 388],

$$\int_0^\infty t^\nu K_\nu(2\pi tx/k)dt = \frac{1}{4} \pi^{\frac{1}{2}} (\frac{k}{\pi x})^{\nu+1} \Gamma(\nu+1/2)$$

and [39, p. 749]

$$\int_0^\infty t^\nu K_\nu(2\pi tx/k) \cos(2\pi nt/k)dt$$

$$= \frac{\pi^{\frac{1}{2}}}{4x} (\frac{kx}{\pi})^{\nu+1} \frac{\Gamma(\nu+\tfrac{1}{2})}{(n^2+x^2)^{\nu+\frac{1}{2}}} \ .$$

Also $f(0) = \frac{1}{2}(k/\pi x)^{\nu} \Gamma(\nu)$ [1, p. 375]. If we use all of the foregoing information in (3.2), we deduce (10.1) at once.

Now suppose that A is odd. Let $f(t) = t^{\nu+1} K_{\nu}(2\pi tx/k)$, where $x > 0$ and $\operatorname{Re}\nu > -1/2$. Using (3.3), we proceed as above, except that now we use [39, p. 749]

$$\int_0^{\infty} t^{\nu+1} K_{\nu}(2\pi tx/k) \sin(2\pi nt/k)dt$$

$$= \frac{\pi^{\frac{1}{2}}}{4x^2} (\frac{kx}{\pi})^{\nu+2} \frac{n \Gamma(\nu+3/2)}{(n^2+x^2)^{\nu+3/2}} .$$

Using the above in (3.3), we deduce (10.2) at once for $\operatorname{Re}\nu > -1/2$. We now observe that the Dirichlet series on the right side of (10.2) converges uniformly in the half plane $\operatorname{Re}\nu \geq -1+\epsilon$, $\epsilon > 0$, since $a_0 = 0$. The series on the left side of (10.2) certainly converges uniformly in that same half plane. Thus, by analytic continuation, (10.2) is valid for $\operatorname{Re}\nu > -1$. This completes the proof.

In the case A = I, Theorem 10.1 was first established by Watson [71], who also achieved a result for $\nu = 0$. A generalization of Watson's result in another direction was done by Kober [49]. For other results of the nature of Theorem 10.1, see the papers of the author [7, Section 8], [8, Example 5] and the paper of Chandrasekharan and Narasimhan [22, Lemma 6], as well as the several references cited in these papers.

We shall examine two special cases of Theorem 10.1. Suppose that A is even and that $\nu = 1/2$. Since $K_{\frac{1}{2}}(z) = (\pi/2z)^{\frac{1}{2}} e^{-z}$, we find that (10.1) yields immediately

(10.3) $\frac{1}{2} a_0 + \sum_{n=1}^{\infty} a_n e^{-2\pi nx/k} = \frac{b_0 k}{2\pi x} + \frac{kx}{\pi} \sum_{n=1}^{\infty} \frac{b_n}{n^2+x^2}$.

Equation (10.3) was derived by Hamburger [42] under seemingly different assumptions, but it can be shown that his hypotheses are essentially equivalent to ours.

Now assume that A is odd and that $\nu = -1/2$. Since $K_{-\frac{1}{2}}(z) =$
$K_{\frac{1}{2}}(z) = (\pi/2z)^{\frac{1}{2}} e^{-z}$, we find that (10.2) gives forthwith

$$(10.4) \qquad \sum_{n=1}^{\infty} a_n e^{-2\pi nx/k} = \frac{ik}{\pi} \sum_{n=1}^{\infty} \frac{n b_n}{n^2 + x^2} \ .$$

In fact, the series on the left sides of (10.3) and (10.4) are easily
summed by letting $n = \mu k + \nu$, $0 \le \mu < \infty$, $1 \le \nu \le k$, and then evaluating
the resulting geometric series. Thus, the values for the series on the
right sides of (10.3) and (10.4) may be given in finite terms. See [15]
for another proof of these evaluations.

Interesting formulas arise from integrating (10.3) and (10.4) termwise over the interval $[0, y]$, $y > 0$. In the cases of real primitive
characters, the resulting formulas are due originally to Lerch [52,
pp. 419, 422]. Lerch then used his results to easily derive new formulas
for class numbers. In this connection, see also a paper of Mordell [56].

11. <u>Entries from Ramanujan's Notebooks and kindred formulae.</u> In
this section, we establish many interesting identities for and evaluations
of infinite series. Several of these formulae may be found in Ramanujan's
Notebooks [61]. Most, in fact, are new. Some may be found in a paper
by the author [17], where it is shown that a large mass of such results
scattered in the literature may be deduced quite simply from a couple of
general theorems. Other special cases of results found in this section
are due to J. W. L. Glaisher [38], O. Schlömilch [63], [64], Chowla [24],
and others. Periodic Bernoulli numbers arise in several of the results
which follow.

<u>Theorem 11.1.</u>   Let A be even, and let α and β be positive
numbers such that $\alpha\beta = \pi^2$. If n is a positive integer, then

(11.1) $\quad \alpha^n \sum\limits_{m=1}^{\infty} a_m \dfrac{m^{2n-1}}{e^{2\alpha m}-1} - (-\beta)^n \sum\limits_{m=1}^{\infty} G(-i\beta m/\pi, B) \dfrac{m^{2n-1}}{e^{2\beta m}-1}$

$$= \frac{1}{4n}\left\{ \alpha^n B_0 B_{2n}(A) - (-\beta)^n B_0(A) B_{2n} - \delta(n) a_0 \right\} ,$$

where $\delta(1) = 1$ and $\delta(n) = 0$ for $n > 1$.

We remark that the only reason that we have inserted $B_0 = 1$ in (11.1) is to exhibit a slight amount of more symmetry in the formula.

$\quad$ Proof.$\quad$ In the periodic Poisson formula (3.2), let $f(x) = x^{2n-1}/(e^{2\alpha x}-1)$, where n is a positive integer. We let $c = 0$ and $d = \infty$. Throughout the entire section, the justification for letting d tend to ∞ is easily accomplished by two integrations by parts, and so we shall not further comment on this. For $n \geq 1$, we have the well-known evaluation [39, p. 1076],

(11.2) $\quad \displaystyle\int_0^{\infty} \dfrac{x^{2n-1}}{e^{2\alpha x}-1}\, dx = (\dfrac{\pi}{\alpha})^{2n} \dfrac{(-1)^{n-1}}{4n} B_{2n}$.

Also, for $n \geq 1$ and $b > 0$, we have [39, p. 494]

(11.3) $\quad \displaystyle\int_0^{\infty} \dfrac{x^{2n-1}\cos(bx)}{e^x - 1}\, dx = (-1)^{n-1} \dfrac{\partial^{2n-1}}{\partial b^{2n-1}} \left\{ \dfrac{\pi}{2}\coth(b\pi) - \dfrac{1}{2b} \right\}$

$$= (-1)^{n-1}\frac{\pi}{2}\frac{\partial^{2n-1}}{\partial b^{2n-1}}\left\{ 1 + \frac{2}{e^{2b\pi}-1} \right\} + \frac{(-1)^{n-1}(2n-1)!}{2b^{2n}}$$

$$= (-1)^n \pi(2\pi)^{2n-1} \sum_{m=1}^{\infty} m^{2n-1} e^{-2bm\pi} + \frac{(-1)^{n-1}(2n-1)!}{2b^{2n}} .$$

Hence, from (11.3), for $j > 0$,

(11. 4) $I_j(n, \alpha) \equiv \displaystyle\int_0^\infty \frac{x^{2n-1}}{e^{2\alpha x}-1} \cos(2\pi jx/k)dx$

$$= (2\alpha)^{-2n} \int_0^\infty \frac{y^{2n-1}}{e^y - 1} \cos(\pi jy/\alpha k)dy$$

$$= \frac{1}{2}(-\beta/\alpha)^n \sum_{m=1}^\infty m^{2n-1} e^{-2\beta mj/k} + \frac{1}{2}(-1)^{n-1}(2n-1)! \left(\frac{k}{2\pi j}\right)^{2n} ,$$

where we have used the fact that $\alpha\beta = \pi^2$. Hence,

(11. 5) $\qquad\qquad 2 \displaystyle\sum_{j=1}^\infty b_j I_j(n, \alpha)$

$$= (-\beta/\alpha)^n \sum_{m=1}^\infty m^{2n-1} \sum_{j=1}^\infty b_j e^{-2\beta mj/k} + (-1)^{n-1}(2n-1)! \left(\frac{k}{2\pi}\right)^{2n} \zeta(2n;B) .$$

In (11. 5), put $j = \mu k - \nu$, $1 \le \mu < \infty$, $0 \le \nu \le k - 1$, use (2. 12), and recall that B is even. We then get

(11. 6) $\qquad\qquad 2 \displaystyle\sum_{j=1}^\infty b_j I_j(n, \alpha)$

$$= (-\beta/\alpha)^n \sum_{m=1}^\infty m^{2n-1} G(-i\beta m/\pi, B)/(e^{2\beta m} - 1) + B_{2n}(A)/(4n) .$$

Now if $n = 1$, $f(0) = 1/(2\alpha)$; if $n > 1$, $f(0) = 0$. Using these values,
(11. 2), and (11. 6) in (3. 2), we obtain (11. 1) after a small amount of manipulation.

If $A = I$ and $\alpha\beta = \pi^2$, (11. 1) yields

(11. 7) $\qquad \alpha \displaystyle\sum_{m=1}^\infty \frac{m}{e^{2\alpha m}-1} + \beta \sum_{m=1}^\infty \frac{m}{e^{2\beta m}-1} = \frac{\alpha+\beta}{24} - \frac{1}{4}$

and

$$(11.8) \quad \alpha^n \sum_{m=1}^{\infty} \frac{m^{2n-1}}{e^{2\alpha m}-1} - (-\beta)^n \sum_{m=1}^{\infty} \frac{m^{2n-1}}{e^{2\beta m}-1} = \frac{B_{2n}}{4n}\left\{\alpha^n - (-\beta)^n\right\},$$

where $n > 1$. If $\alpha = \beta = \pi$, (11.7) and (11.8) reduce, respectively, to

$$(11.9) \qquad \sum_{m=1}^{\infty} \frac{m}{e^{2\pi m}-1} = \frac{1}{24} - \frac{1}{8\pi}$$

and

$$(11.10) \qquad \sum_{m=1}^{\infty} \frac{m^{2n-1}}{e^{2\pi m}-1} = \frac{B_{2n}}{4n} \qquad (n > 1),$$

where in (11.10) n is odd.

We shall give a short history of the previous four formulas. A more complete history may be found in [17]. Formulas (11.7) and (11.9) are found in Ramanujan's Notebooks [61, vol. I, p. 257, no. 9; vol. II, p. 170, Cor. 1]. Furthermore, Ramanujan once stated (11.9) as a problem [60]. For these reasons, (11.7) and (11.9) are normally attributed to Ramanujan. However, proofs were given considerably earlier by 0. Schlömilch [63], [64, p. 157]. Formulas (11.8) and (11.10) are also found in Ramanujan's Notebooks [61, vol. I, p. 259, no. 14] and [61, vol. II, p. 171, Cor. iv], respectively. Hardy [43], [44, p. 537] attributes both of these results to Ramanujan. However, Glaisher [38, p. 81] seems to have been the first to prove (11.10). The first published proof of (11.8) appears to be by S. L. Malurkar [55]. There are many additional proofs of (11.7)-(11.10) in the literature, and the reader should consult [17] for an account of these.

Let $A = \{(-1)^j\}$. Since $b_0 = 0$ and $b_1 = 1$, $G(-i\beta m/\pi, B) = e^{\beta m}$. Since also $B_{2n}(A) = (2^{2n}-1)B_{2n}$, $n \geq 0$, (11.1) yields, for $n \geq 1$,

(11.11) $\alpha^n \sum_{m=1}^{\infty} (-1)^m \dfrac{m^{2n-1}}{e^{2\alpha m}-1} - \dfrac{1}{2} (-\beta)^n \sum_{m=1}^{\infty} \dfrac{m^{2n-1}}{\sinh(\beta m)}$

$$= \dfrac{\alpha^n}{4n} (2^{2n} - 1) B_{2n} - \dfrac{1}{4} \delta(n) ,$$

where $\alpha\beta = \pi^2$. The case $\alpha = \beta = \pi$ in (11.11) yields the following interesting formulas

(11.12) $\displaystyle\sum_{m=1}^{\infty} \dfrac{m^{2n-1}}{e^{\pi m} - (-1)^m} = \dfrac{1}{4n} (2^{2n} - 1) B_{2n} - \dfrac{1}{4\pi} \delta(n) ,$

where $n \geq 1$. For $n > 1$, formula (11.12) was first established by Glaisher [38, p. 82].

Next, let $A = \chi$, where $\chi(n)$ is even, primitive, and of modulus k . From (2.9), $G(-i\beta m/\pi, B) = G(\chi) G(-i\beta m/\pi, \overline{\chi})/k$. Thus, from (11.1) and (8.15), we get, for $n \geq 1$,

(11.13) $\alpha^n G(\overline{\chi}) \displaystyle\sum_{m=1}^{\infty} \chi(m) \dfrac{m^{2n-1}}{e^{2\alpha m}-1} - (-\beta)^n \sum_{m=1}^{\infty} G(-i\beta m/\pi, \overline{\chi}) \dfrac{m^{2n-1}}{e^{2\beta m}-1}$

$$= \dfrac{\alpha^n}{4n} G(\overline{\chi}) B_{2n}(\chi) .$$

If $\alpha = \beta = \pi$, (11.13) is given in [17, Theorem 4.22].

Theorem 11.2. Let A be odd, and let α and β be positive numbers such that $\alpha\beta = \pi^2$. Then if $n \geq 1$ is an integer,

(11.14) $\alpha^{n+\frac{1}{2}} \displaystyle\sum_{m=1}^{\infty} a_m \dfrac{m^{2n}}{e^{2\alpha m}-1} + i(-1)^n \beta^{n+\frac{1}{2}} \sum_{m=1}^{\infty} G(-i\beta m/\pi, B) \dfrac{m^{2n}}{e^{2\beta m} - 1}$

$$= \dfrac{\alpha^{n+\frac{1}{2}}}{2(2n+1)} B_{2n+1}(A) .$$

Proof. We employ (3.3) with $f(x) = x^{2n}/(e^{2\alpha x} - 1)$, $c = 0$, and $d = \infty$. From [39, p. 494] and a calculation akin to that in (11.3), we have for $b > 0$,

$$(11.15) \qquad \int_0^\infty \frac{x^{2n} \sin(bx)}{e^x - 1} dx = (-1)^n \frac{\partial^{2n}}{\partial b^{2n}} \left\{ \frac{\pi}{2} \coth(b\pi) - \frac{1}{2b} \right\}$$

$$= (-1)^n \pi(2\pi)^{2n} \sum_{m=1}^\infty m^{2n} e^{-2bm\pi} + \frac{(-1)^{n+1}(2n)!}{2b^{2n+1}},$$

where the first equality holds for $n \geq 0$, while the second holds for $n > 0$. Hence, for $n > 0$, from (11.15),

$$(11.16) \qquad \int_0^\infty \frac{x^{2n} \sin(2\pi jx/k)}{e^{2\alpha x} - 1} dx$$

$$= \frac{1}{2}(-1)^n (\beta/\alpha)^{n+\frac{1}{2}} \sum_{m=1}^\infty m^{2n} e^{-2\beta mj/k} + \frac{1}{2}(-1)^{n+1}(2n)! \left(\frac{k}{2\pi j}\right)^{2n+1},$$

since $\alpha\beta = \pi^2$. The remainder of the proof is similar to the previous proof. For $n \geq 1$, we substitute (11.16) into (3.3), use (2.12), change summation indices as before, and arrive at (11.14) with no difficulty.

If $A = \chi$, where $\chi(n)$ is an odd, primitive character, and $\alpha = \beta = \pi$, then (11.14) yields a result found in the author's paper [17, Theorem 4.22]. In particular, if $\chi = \chi_4$, we obtain the following representation for Euler numbers due to Chowla [24]:

$$E_{2n} = (-1)^n \sum_{m=1}^\infty \frac{m^{2n}}{\cosh(\pi m/2)} - 4 \sum_{m=0}^\infty \frac{(-1)^m (2m+1)^{2n}}{e^{2\pi(2m+1)} - 1} \qquad (n \geq 1).$$

Theorem 11.3. Let A be even, let α and β be positive numbers such that $\alpha\beta = \pi^2$, and let $0 < r < 1$. Then for $n \geq 1$,

(11.17) $\alpha^n \left\{ \sum_{m=1}^{\infty} a_m \dfrac{m^{2n-1} \cos(2\pi mr/k)}{e^{2\alpha m} - 1} + \dfrac{1}{4\alpha} \delta(n) + \dfrac{1}{2} \varphi(r, 0, 1-2n; A) \right\}$

$= \dfrac{1}{2} (-\beta)^n \sum_{m=1}^{\infty} m^{2n-1} \left\{ b_0 e^{-2\beta mr/k} + 2G(-i\beta m/\pi, B) \dfrac{\cosh(2\beta mr/k)}{e^{2\beta m} - 1} \right\}$,

where $\varphi(x, a, s; A)$ is given in Definition 2.4.

Proof. The proof is similar to that of Theorem 11.1. In (3.2), put $c = 0, d = \infty$, and $f(x) = x^{2n-1} \cos(2\pi xr/k)/(e^{2\alpha x} - 1)$, where $n \geq 1$, $\alpha > 0$, and $0 < r < 1$. Note that we shall need to use (11.4) with j replaced by r there. Using the identity $\cos(2\pi xr/k) \cos(2\pi xj/k) = \dfrac{1}{2} \{\cos(2\pi x(j+r)/k) + \cos(2\pi x(j-r)/k)\}$, we see that we shall need to employ (11.4) twice again with j replaced by $j + r$ and $j - r$, respectively, in the two applications. Hence, we find that

(11.18) $R(\alpha, r) \equiv b_0 \displaystyle\int_0^{\infty} f(x)dx + 2 \sum_{j=1}^{\infty} b_j \int_0^{\infty} f(x) \cos(2\pi xj/k)dx$

$= \dfrac{1}{2} (-\beta/\alpha)^n \sum_{m=1}^{\infty} m^{2n-1} \left\{ b_0 e^{-2\beta mr/k} + \sum_{j=1}^{\infty} b_j \left\{ e^{-2\beta m(j+r)/k} + e^{-2\beta m(j-r)/k} \right\} \right\}$

$+ \dfrac{1}{2} (-1)^{n-1} (2n-1)! \left(\dfrac{k}{2\pi}\right)^{2n} \left\{ \varphi(0, r, 2n; B) + \varphi(0, 1-r, 2n; B_1^*) \right\}$.

Now put $j = \mu k - \nu$, $1 \leq \mu < \infty$, $0 \leq \nu \leq k - 1$, sum the resulting geometric series, and apply the functional equation (7.1). Upon doing all of this, we find that (11.18) becomes

$R(\alpha, r) = \dfrac{1}{2} (-\beta/\alpha)^n \sum_{m=1}^{\infty} m^{2n-1} \left\{ b_0 e^{-2\beta mr/k} + 2G(-i\beta m/\pi, B) \dfrac{\cosh(2\beta mr/k)}{e^{2\beta m} - 1} \right\}$

$- \dfrac{1}{2} \varphi(r, 0, 1-2n; A)$.

179

With the use of the above in (3.2), equation (11.17) readily follows.

Theorem 11.3 in the case $A = I$ was first established by the author [17, Proposition 3.7]. In particular, if we let $r = 1/2$, the identity further reduces to a result of Glaisher [38] for $n > 1$ and a result of Schlömilch [63], [64, p. 157] for $n = 1$. More complete details are found in [17].

The Eulerian numbers $H_n[\lambda]$, $n \geq 0$, $\lambda \neq 1$, are defined by

$$\frac{1-\lambda}{e^x - \lambda} = \sum_{n=0}^{\infty} H_n[\lambda] \frac{x^n}{n!} \ ,$$

where $|x|$ is sufficiently small. We indicated in [17] that for $n \geq 1$,

$$(11.19) \qquad \varphi(r, 0, -n; I) = \frac{(-1)^n H_n[e^{2\pi i r}]}{e^{-2\pi i r} - 1} \ .$$

Thus, $\varphi(r, 0, 1-2n; A)$ in (11.17) may be calculated in terms of Eulerian numbers.

Let us examine the special case $A = \{(-1)^j\}$. It appears best to use (11.18) in our calculation. Since $b_0 = 0$ and $b_1 = 1$, we find that for $n \geq 1$,

$$(-1)^{n-1}(2n-1)! \ \pi^{-2n} \left\{ \varphi(0, r, 2n; B) + \varphi(0, 1-r, 2n; B_1^*) \right\}$$

$$= (-1)^{n-1}(2n-1)! \ (2\pi)^{-2n} \left\{ \varphi(0, (r+1)/2, 2n; I) + \varphi(0, 1-(r+1)/2, 2n; I) \right\}$$

$$= -\varphi((r+1)/2, 0, 1-2n; I) = -\frac{H_{2n-1}[-e^{\pi i r}]}{e^{-\pi i r} + 1} \ ,$$

where we have employed (7.1) and (11.19). Thus, for $\alpha\beta = \pi^2$ and $n > 1$, Theorem 11.3 gives

$$\alpha^n \left\{ \sum_{m=1}^{\infty} \frac{(-1)^m m^{2n-1} \cos(\pi mr)}{e^{2\alpha m} - 1} + \frac{H_{2n-1}[-e^{\pi ir}]}{2(e^{-\pi ir} + 1)} \right\}$$

$$= \frac{1}{2} (-\beta)^n \sum_{m=1}^{\infty} \frac{m^{2n-1} \cosh(\beta mr)}{\sinh(\beta m)} .$$

Since $H_1[\lambda] = -1/(1-\lambda)$, a brief calculation shows that, with the use of (11.19),

$$\varphi((r+1)/2, 0, -1; I) = -\frac{1}{4} \csc^2 \{\pi(r+1)/2\} = -\frac{1}{4} \sec^2(\pi r/2) .$$

Thus, in the case $n = 1$, Theorem 11.3 yields

$$2\alpha \sum_{m=1}^{\infty} \frac{(-1)^m m \cos(\pi mr)}{e^{2\alpha m} - 1} + \beta \sum_{m=1}^{\infty} \frac{m \cosh(\beta mr)}{\sinh(\beta m)}$$

$$= \frac{\alpha}{4} \sec^2(\pi r/2) - \frac{1}{2} .$$

Theorem 11.4. Let A be even, let α and β be positive numbers such that $\alpha\beta = \pi^2$, and let $0 < r < 1$. Then

(11.20) $$\alpha^{\frac{1}{2}} \left\{ \sum_{m=1}^{\infty} a_m \frac{\sin(2\pi mr/k)}{e^{2\alpha m} - 1} + \frac{kb_0}{4\pi r} - \frac{kr}{2\pi} \sum_{m=1}^{\infty} \frac{b_m}{m^2 - r^2} \right\}$$

$$= \beta^{\frac{1}{2}} \left\{ -\sum_{m=1}^{\infty} G(-i\beta m/\pi, B) \frac{\sinh(2\beta mr/k)}{e^{2\beta m} - 1} + \frac{1}{4} b_0 \coth(\beta r/k) - \frac{ra_0}{2k} \right\} ,$$

and for $n \geq 1$,

$$\alpha^{n+\frac{1}{2}} \left\{ \sum_{m=1}^{\infty} a_m \frac{m^{2n} \sin(2\pi mr/k)}{e^{2\alpha m} - 1} - \frac{i}{2} \varphi(r, 0, -2n; A) \right\}$$

$$= \frac{1}{2} (-1)^n \beta^{n+\frac{1}{2}} \sum_{m=1}^{\infty} m^{2n} \left\{ b_0 e^{-2\beta mr/k} - 2G(-i\beta m/\pi, B) \frac{\sinh(2\beta mr/k)}{e^{2\beta m} - 1} \right\} .$$

Proof. The proof is similar to the other proofs in this section. We use (3.2) with $c = 0$, $d = \infty$, and $f(x) = x^{2n} \sin(2\pi xr/k)/(e^{2\alpha x} - 1)$, where $n \geq 0$. In the calculations, we shall need to employ the first equation of (11.15) three times if $n = 0$, and (11.16) three times if $n > 0$.

If $A = I$, (11.20) reduces to the curious identity

(11.21) $$\alpha^{\frac{1}{2}} \sum_{m=1}^{\infty} \frac{\sin(2\pi mr)}{e^{2\alpha m} - 1} + \beta^{\frac{1}{2}} \sum_{m=1}^{\infty} \frac{\sinh(2\beta mr)}{e^{2\beta m} - 1}$$

$$= \frac{1}{4} \left(\beta^{\frac{1}{2}} \coth(\beta r) - \alpha^{\frac{1}{2}} \cot(\pi r) \right) - \frac{1}{2} r\beta^{\frac{1}{2}} ,$$

where $\alpha\beta = \pi^2$ and $0 < r < 1$. Formula (11.21) was first established by Schlömilch [63], [64, p. 156] and is also found in Ramanujan's Notebook [61, vol. I, pp. 257, 259, no. 13; vol. II, p. 169, no. 8 (i)]. See also [17, Proposition 3.12].

Let $A = \{(-1)^j\}$. Then (11.20) yields the relation

(11.22) $$\alpha^{\frac{1}{2}} \sum_{m=1}^{\infty} \frac{(-1)^m \sin(\pi mr)}{e^{2\alpha m} - 1} + \frac{1}{2} \beta^{\frac{1}{2}} \sum_{m=1}^{\infty} \frac{\sinh(\beta mr)}{\sinh(\beta m)}$$

$$= \frac{\alpha^{\frac{1}{2}}}{4} \tan(\pi r/2) - \frac{\beta^{\frac{1}{2}} r}{4} ,$$

where $\alpha\beta = \pi^2$ and $0 < r < 1$. Formula (11.22) might be useful in the numerical computation of $\tan(\pi r/2)$.

The proof of the next theorem is along the same lines as the foregoing proofs.

182

Theorem 11.5. Let A be odd, let α and β be positive numbers with $\alpha\beta = \pi^2$, and let $0 < r < 1$. Then for $n \geq 1$,

(11.23)
$$\alpha^n \left\{ \sum_{m=1}^{\infty} a_m \frac{m^{2n-1} \sin(2\pi mr/k)}{e^{2\alpha m} - 1} - \frac{i}{2} \varphi(r, 0, 1-2n; A) \right\}$$

$$= -i(-\beta)^n \sum_{m=1}^{\infty} G(-i\beta m/\pi, B) \frac{m^{2n-1} \sinh(2\beta mr/k)}{e^{2\beta m} - 1}$$

and

$$\alpha^{n+\frac{1}{2}} \left\{ \sum_{m=1}^{\infty} a_m \frac{m^{2n} \cos(2\pi mr/k)}{e^{2\alpha m} - 1} + \frac{1}{2} \varphi(r, 0, -2n; A) \right\}$$

$$= i(-1)^{n+1} \beta^{n+\frac{1}{2}} \sum_{m=1}^{\infty} G(-i\beta m/\pi, B) \frac{m^{2n} \cosh(2\beta mr/k)}{e^{2\beta m} - 1} .$$

Let us let $A = \chi_4$ and $n = 1$ in (11.23). After some calculation, we find that

$$\alpha \sum_{m=1}^{\infty} \frac{(-1)^m (2m+1) \sin((2m+1)\pi r/2)}{e^{2\alpha(2m+1)} - 1} + \frac{\beta}{4} \sum_{m=1}^{\infty} \frac{m \sinh(\beta mr/2)}{\cosh(\beta m/2)}$$

$$= \frac{\alpha}{4} \tan(\pi r/2) \sec(\pi r/2) .$$

REFERENCES

1. Milton Abramowitz and Irene A. Stegun, editors, Handbook of mathematical functions, Dover, New York, 1965.

2. N. C. Ankeny, E. Artin, and S. Chowla, The class-number of real quadratic fields, Ann. of Math. (2) 56 (1952), 479-493.

3. Tom M. Apostol, Dirichlet L-functions and character power sums, J. Number Theory 2 (1970), 223-234.

4. Raymond Ayoub, An introduction to the analytic theory of numbers, American Mathematical Society, Providence, 1963.

5. R. Ayoub, S. Chowla and H. Walum, On sums involving quadratic characters, J. London Math. Soc. 42 (1967), 152-154.

6. Bruce C. Berndt, Identities involving the coefficients of a class of Dirichlet series. I, Trans. Amer. Math. Soc. 137 (1969), 345-359.

7. _____, Identities involving the coefficients of a class of Dirichlet series. III, Trans. Amer. Math. Soc. 146 (1969), 323-348.

8. _____, Identities involving the coefficients of a class of Dirichlet series. V, Trans. Amer. Math. Soc. 160 (1971), 139-156.

9. _____, Identities involving the coefficients of a class of Dirichlet series. VII, Trans. Amer. Math. Soc. 201 (1975), 247-261.

10. _____, The Voronoï summation formula, Proceedings of a conference on the theory of arithmetic functions, Kalamazoo, April, 1971, Springer-Verlag, Berlin, 1972, 21-36.

11. _____, On the Hurwitz zeta-function, Rocky Mountain J. Math. 2 (1972), 151-157.

12. _____, Two new proofs of Lerch's functional equation, Proc. Amer. Math. Soc. 32 (1972), 403-408.

13. _____, An elementary proof of some character sum identities of Apostol, Glasgow Math. J. 14 (1973), 50-53.

14. _____, On Gaussian sums and other exponential sums with periodic coefficients, Duke Math. J. 40 (1973), 145-156.

15. _____, The evaluation of infinite series by contour integration, Publ. Elektro. Fak. Univ. u Beogradu, No. 435 (1973), 119-122.

16. _____, Character analogues of the Poisson and Euler-Maclaurin summation formulas with applications, J. Number Theory (to appear).

17. Bruce C. Berndt, Modular transformations and generalizations of several formulae of Ramanujan, (submitted for publication).

18. _____, Classical theorems on quadratic residues, (in preparation).

19. Bruce C. Berndt and Lowell Schoenfeld, Periodic analogues of the Euler-Maclaurin and Poisson summation formulas with applica-tions to number theory, Acta Arith. (to appear).

20. George Boole, A treatise on the calculus of finite differences, third ed., G. E. Stechert & Co., New York, 1946.

21. L. Carlitz, Some sums connected with quadratic residues, Proc. Amer. Math. Soc. 4 (1953), 12-15.

22. K. Chandrasekharan and Raghavan Narasimhan, Hecke's function-al equation and arithmetical identities, Ann. of Math. 74 (1961), 1-23.

23. _____ and _____, Functional equations with multiple gamma factors and the average order of arithmetical func-tions, Ann. of Math. 76 (1962), 93-136.

24. S. D. Chowla, Some infinite series, definite integrals and asymptotic expansions, J. Indian Math. Soc. 17 (1927/28), 261-288.

25. _____, Some applications of the Riemann zeta, and allied functions, Tohoku Math. J. 30 (1928), 202-225.

26. _____, On a problem of analytic number theory, Proc. Nat. Inst. Sci. India 13 (1947), 231-232.

27. _____, On the signs of certain generalized Bernoulli numbers, K. Norske Vid. Selsk. Forh. 34 (1961), 102-104.

28. _____, On some formulae resembling the Euler-Maclaurin sum formula, K. Norske Vid. Selsk. Forh. 34 (1961), 107-109.

29. Kai-Lai Chung, Note on a theorem on quadratic residues, Bull. Amer. Math. Soc. 47 (1941), 514-516.

30. Harold Davenport, Multiplicative number theory, Markham, Chicago, 1967.

31. D. Davies and C. B. Haselgrove, The evaluation of Dirichlet L-functions, Proc. Royal Soc. London Ser. A 264 (1961), 122-132.

32. L. E. Dickson, History of the theory of numbers, vol. III, G. E. Stechert and Co., New York, 1934.

33. G. Lejeune Dirichlet, Sur l'usage des intégrales définies dans la sommation des séries finies ou infinies, J. Reine Angew. Math. 17 (1837), 57-67.

34. _____, Recherches sur diverses applications de l'analyse infinitésimale à la théorie des nombres, J. Reine Angew. Math. 19 (1839), 324-369.

35. Gustav Doetsch, Summatorische Eigenschaften der Besselschen Funktionen und andere Funktionalrelationen, die mit der linearen Transformationsformel der Thetafunktion äquivalent sind, Comp. Math. 1 (1934), 1-13.

36. A. Erdélyi, Gewisse Reihentransformationen, die mit der linearen Transformationsformel der Theta-funktion zusammenhängen, Comp. Math. 4 (1937), 406-423.

37. N. J. Fine, On a question of Ayoub, Chowla and Walum concerning character sums, Illinois J. Math. 14 (1970), 88-90.

38. J. W. L. Glaisher, On the series which represent the twelve elliptic and the four zeta functions, Mess. Math. 18 (1889), 1-84.

39. I. S. Gradshteyn and I. M. Ryzhik, Table of integrals, series and products, fourth ed., Academic Press, New York, 1965.

40. A. P. Guinand, On Poisson's summation formula, Ann. of Math. 42 (1941), 591-603.

41. _____, Gauss sums and primitive characters, Quart J. Math. Oxford Ser. 16 (1945), 59-63.

42. Hans Hamburger, Über einige Beziehungen, die mit der Funktionalgleichung der Riemannschen ζ-Funktion äquivalent sind, Math. Ann. 85 (1922), 129-140.

43. G. H. Hardy, A formula of Ramanujan, J. London Math. Soc.
3 (1928), 238-240.

44. _____, Collected papers, vol. IV, Clarendon Press,
Oxford, 1969.

45. G. H. Hardy and J. E. Littlewood, Notes on the theory of series
(XX): On Lambert series, Proc. London Math. Soc. 41 (1936),
257-270.

46. H. Hasse, Sulla generalizzazione di Leopoldt dei numeri di
Bernoulli e sua applicazione alla divisibilità del numero delle
classi nei corpi numerici abeliani, Rend. Mat. e Appl. (5) 21 (1962),
9-27.

47. Takeshi Kano, On the Bessel-series expression for $\Sigma \frac{1}{n} \sin \frac{x}{n}$,
Math. J. Okayama Univ. 16 (1974), 129-136.

48. Konrad Knopp, Theory and application of infinite series, Blackie
and Sons Ltd., London, 1951.

49. H. Kober, Transformationsformeln gewisser Besselscher Reihen.
Beziehungen zu Zeta-funktionen, Math. Z. 39 (1934), 609-624.

50. Edmund Landau, Vorlesungen über Zahlentheorie, Zweiter Band,
Chelsea, New York, 1947.

51. Heinrich-Wolfgang Leopoldt, Eine Verallgemeinerung der
Bernoullischen Zahlen, Abh. Math. Sem. Univ. Hamburg 22 (1958),
131-140.

52. M. Lerch, Essais sur le calcul du nombre des classes de formes
quadratiques binaires aux coefficients entiers, Acta Math. 29
(1905), 333-424.

53. E. Lindelöf, Calcul des résidus, Chelsea, New York, 1947.

54. G. G. Macfarlane, The application of Mellin transforms to the
summation of slowly convergent series, Philos. Mag. ser. 7
40 (1949), 188-197.

55. S. L. Malurkar, On the application of Herr Mellin's integrals to
some series, J. Indian Math. Soc. 16 (1925/26), 130-138.

56. L. J. Mordell, Some applications of Fourier series in the analytic theory of numbers, Proc. Cambridge Philos. Soc. 24 (1928), 585-596.

57. _____, Poisson's summation formula and the Riemann zeta-function, J. London Math. Soc. 4 (1929), 285-291.

58. Leo Moser, A theorem on quadratic residues, Proc. Amer. Math. Soc. 2 (1951), 503-504.

59. Niels Nielsen, Handbuch der Theorie der Cylinderfunktionen, B. G. Teubner, Leipzig, 1904.

60. S. Ramanujan, Question 387, J. Indian Math. Soc. 4 (1912), 120.

61. _____, Notebooks of Srinivasa Ramanujan (2 volumes), Tata Institute of Fundamental Research, Bombay, 1957.

62. J. Barkley Rosser and Lowell Schoenfeld, Approximation of the Riemann zeta-function (to appear).

63. O. Schlömilch, Ueber einige unendliche Reihen, Berichte Verh. K. Sächsischen Gesell. Wiss. Leipzig 29 (1877), 101-105.

64. _____, Compendium der höheren Analysis, Zweiter Band, 4th ed., Friedrich Vieweg und Sohn, Braunschweig, 1895.

65. Hermann Schmidt, Über die Werte der Riemannschen ζ-Funktion und verwandter Funktionen für ganzzahlige Argumente, S. -B. Bayer. Akad. Wiss. Math. - Natur. Kl. (1972), 87-99.

66. S. L. Segal, On Σ $1/n$ $\sin(x/n)$, J. London Math. Soc. (2) 4 (1972), 385-393.

67. I. N. Sneddon, The use of integral transforms, McGraw-Hill, New York, 1972.

68. J. Steinig, The changes of sign of certain arithmetical error-terms, Comm. Math. Helv. 44 (1969), 385-400.

69. E. C. Titchmarsh, Introduction to the theory of Fourier integrals, second ed., Clarendon Press, Oxford, 1948.

70. _____, The theory of the Riemann zeta-function, Clarendon Press, Oxford, 1951.

71. G. N. Watson, Some self-reciprocal functions, Quart. J. Math. Oxford ser. 2 (1931), 298-309.

72. _____, Theory of Bessel functions, second ed., University Press, Cambridge, 1966.

73. A. L. Whiteman, Theorems on quadratic residues, Math. Mag. 23 (1949/50), 71-74.

74. E. T. Whittaker and G. N. Watson, A course of modern analysis, 4th ed., University Press, Cambridge, 1962.

75. Y. Yamamoto, unpublished manuscript.

Problems and Prospects for Basic Hypergeometric Functions
George E. Andrews

1. Introduction.

The ordinary basic hypergeometric function is

$$
{}_m\phi_n\left[\begin{array}{c} a_1, \ldots, a_m ; q, z \\ b_1, \ldots, b_n \end{array}\right] = \sum_{j \geq 0} \frac{(a_1)_j (a_2)_j \cdots (a_m)_j z^j}{(b_1)_j (b_2)_j \cdots (b_n)_j (q)_j} ,
$$

where $(a)_n = (a;q)_n = (1-a)(1-aq)\ldots(1-aq^{n-1})$, $|z| < 1$, $|q| < 1$. In a recent survey paper, "Applications of basic hypergeometric functions" (Andrews (1974a), henceforward referred to ABHF) the applications of basic hypergeometric functions to partitions, number theory, finite vector spaces, combinatorial identities, and physics were discussed. Since that paper was written, a number of advances have been made (especially in combinatorial identities and the theory of partitions), which suggest that the theory of basic hypergeometric functions can be greatly extended and that the resulting theory will have considerable impact on the subjects that originally suggested the extensions. We might have called this paper "Multiple Basic Hypergeometric Series" since most of the extensions we shall consider relate to multiple series; however, such a title would not really convey the intended spirit of this paper.

In Section 2, we shall describe the recent work in the theory of partitions which suggests the need for an extended theory of basic hypergeometric functions. In Section 3 we shall present an extension of Watson's q-analog of Whipple's theorem (Watson (1929)) and we shall show how this result implies many new analytic identities of the Rogers-

191

Ramanujan type including some recently discovered ones (Andrews (1974b)); we also describe the recent work of Holman, Biedenharn, and Louck (1975) which suggests alternative analogs of Whipple's theorem (Whipple (1926)). In Section 4, we shall study basic Appell and Lauricella functions in order to provide generalizations of the q-analogs of the Chu-Vandermonde summation and Saalschütz's theorem. In Section 5, we shall study the Dyson conjecture as a higher dimensional analog of Dixon's summation, and we shall inquire about possible q-analogs; also a possible q-analog of MacMahon's Master Theorem (MacMahon (1894), (1915)) will be considered. In Section 6, we examine a partition theorem of Göllnitz (1967) which leads to the consideration of nonterminating Saalschützian series; also q-inversion theorems are briefly discussed in this section.

In each section of the paper we shall begin with the statement of one or more problems. The main body of each section will then be devoted to the "prospects" for solution (s) and sometimes a related problem will be stated. In some instances such as Section 4, we shall be able to describe a number of known results.

2. Partitions identities.

PROBLEM 1. What finite linear homogeneous ordinary q-difference equations with coefficients that are polynomials in x and q have multiple basic hypergeometric series as solutions?

Let us begin by showing why this problem is important. For more details we refer the reader to Andrews (1972a), (1974c). Let S denote the set of all $\pi = (1^{f_1} 2^{f_2} 3^{f_3} \ldots)$, the partitions of non-negative integers. The set S forms a distributive lattice under the partial ordering $(1^{f_1} 2^{f_2} 3^{f_3} \ldots) \leq (1^{g_1} 2^{g_2} 3^{g_3} \ldots)$, whenever $f_i \leq g_i$ for all i; the "meet" and "join" operations are defined by

$$(1^{f_1} 2^{f_2} \ldots) \cap (1^{g_1} 2^{g_2} \ldots) = (1^{\min(f_1, f_1)} 2^{\min(f_2, g_2)} \ldots)$$

$$(1^{f_1} 2^{f_2} \ldots) \cup (1^{g_1} 2^{g_2} \ldots) = (1^{\max(f_1, g_1)} 2^{\max(f_2, g_2)} \ldots) .$$

A subset C of S which has the property that whenever $\pi \in C$ and $\pi' \leq \pi$ then $\pi' \in C$ is called a partition ideal.

Let C be a partition ideal, and denote by $C^{(m)}$ that subset of C that consists of those $\pi = (1^{f_1} 2^{f_2} 3^{f_3} \ldots)$ with $f_1 = f_2 = \ldots = f_m = 0$, i.e. $C^{(m)}$ consists of those partitions in C that have all their parts $> m$. Define $\varphi: S \to S^{(1)}$ by $\varphi((1^{f_1} 2^{f_2} 3^{f_3} \ldots)) = (1^0 2^{f_1} 3^{f_2} \ldots)$; intuitively φ adds 1 to each part of each partition. We shall say that C has modulus $m > 0$ if m is the least positive integer such that $\varphi^m C = C^{(m)}$; it is not difficult to show that if C has a modulus m then $\varphi^M C = C^{(M)}$ if and only if $m \mid M$.

We shall say that C satisfies the finiteness property if for each $j \geq 1$ there exists M_j such that if $\pi = (1^{f_1} 2^{f_2} 3^{f_3} \ldots) \in C$ then $f_j \leq M_j$ for each j.

We shall add partitions in the obvious manner:

$$(1^{f_1} 2^{f_2} 3^{f_3} \ldots) \oplus (1^{f'_1} 2^{f'_2} 3^{f'_3} \ldots)$$

$$= (1^{f_1 + f'_1} 2^{f_2 + f'_2} 3^{f_3 + f'_3} \ldots) .$$

Now if C has a modulus m, then it is easy to see that for each $\pi \in C$, there exists $\pi_0, \pi_1, \pi_2, \ldots \in L_C$ (not necessarily distinct) such that

$$(2.1) \qquad \pi = \pi_0 \oplus \varphi^m \pi_1 \oplus \varphi^{2m} \pi_2 \oplus \varphi^{3m} \pi_3 \oplus \ldots ,$$

where $L_C = \left\{ \left(1^{f_1} 2^{f_2} 3^{f_3} \ldots \right) \in C \mid f_i = 0 \text{ for } i > m \right\}$.

We shall say that C is a $\underline{\text{linked partition ideal}}$ if (1) C has a modulus, (2) C satisfies the finiteness property, (3) for each $\pi' \in L_C$ there exists a positive integer $\ell(\pi')$ (called the $\underline{\text{span}}$ of π') and a subset $L_C(\pi')$ of L_C (called the $\underline{\text{linking set}}$ of π') such that for each $\pi \in C$ the representation (2.1) fulfills the requirement that if $\pi_j = \pi'$, then $\pi_{j+1} = \pi_{j+2} = \ldots = \pi_{j+\ell(\pi')-1} = (1^0 2^0 3^0 \ldots)$ and $\pi_{j+\ell(\pi')} \in L_C(\pi')$.

We define

$$f_C(x;q) = \sum_{\pi \in C} x^{f_1+f_2+f_3+\ldots} q^{f_1 \cdot 1+f_2 \cdot 2+f_3 \cdot 3+\ldots} .$$

Theorem 1. If C is a linked partition ideal, then $f_C(x;q)$ satisfies a finite homogeneous linear q-difference equation whose coefficients are polynomials in x and q.

We refer the reader to Andrews (1974c) for a proof of this theorem. We shall content ourselves with an example of Theorem 1 for which Problem 1 can also be answered affirmatively. As a result we shall be abel to derive a well-known partition identity of I. J. Schur (1926). Let

$$C = \{\pi = (1^{f_1} 2^{f_2} 3^{f_3} \ldots) \,|\, f_i + f_{i+1} + f_{i+2} \leq 1, \ f_{3i} + f_{3i+1} + f_{3i+2} + f_{3i+3} \leq 1 \} ;$$

then

$$L_C = \{\pi_0 = (1^0 2^0 3^0 4^0 5^0 \ldots), \ \pi_1 = (1^1 2^0 3^0 4^0 5^0 \ldots) ,$$

$$\pi_2 = (1^0 2^1 3^0 4^0 5^0 \ldots), \ \pi_3 = (1^0 2^0 3^1 4^0 5^0 \ldots) \} ;$$

$$\ell(\pi_j) = 1 \text{ for } j = 0,1,2,3, \text{ and}$$

$$L_C(\pi_0) = \{\pi_0, \pi_1, \pi_2, \pi_3\} , \quad L_C(\pi_1) = \{\pi_0, \pi_1, \pi_2, \pi_3\}$$

$$L_C(\pi_2) = \{\pi_0, \pi_2, \pi_3\} , \quad L_C(\pi_3) = \{\pi_0\} .$$

The fact that C is linked with modulus 3 makes it a simple matter to deduce several relations among generating functions related to C. Namely if $\pi_i \in L_C$, let

(2.2) $$H_i(x) = \Sigma' \, x^{f_1+f_2+\ldots} q^{f_1 \cdot 1+f_2 \cdot 2+\ldots}$$

194

where Σ' is over those $\{f_j\} \in C$ such that $\{f_1, f_2, \ldots, f_m, 0, 0, 0, \ldots\}$ $= \pi_i$, then

(2. 3) $H_0(x) = H_0(xq^3) + H_1(xq^3) + H_2(xq^3) + H_3(xq^3)$

(2. 4) $H_1(x) = xq(H_0(xq^3) + H_1(xq^3) + H_2(xq^3) + H_3(xq^3))$

(2. 5) $H_2(x) = xq^2(H_0(xq^3) + H_2(xq^3) + H_3(xq^3))$

(2. 6) $H_3(x) = xq^3 H_0(xq^3)$.

Each of these equations is a special case of the following general identity for linked partition ideals (proved in Andrews (1974c))

(2. 7) $H_i(x) = x^{\#(\pi_i)} q^{\sigma(\pi_i)} \sum_{\pi_\ell \in L_C(\pi_i)} H_\ell\left(xq^{\ell(\pi_i)}\right)$;

here $\#(\pi_i) = \sum f_j$, $\sigma(\pi_i) = \sum f_j \cdot j$, where $\pi_i = \{f_j\}$.

We may eliminate H_1, H_2 and H_3 from (2. 3)-(2. 6) and in this way we deduce that

$$H_0(x) = (1 + xq^4 + xq^5) H_0(xq^3) + xq^6(1 - xq^6) H_0(xq^6) .$$

Since clearly

$$H_0(x) = f_C(xq^3;q) ,$$

we see that

(2. 8) $f_C(x;q) = (1 + xq + xq^2) f_C(xq^3;q) + xq^3(1 - xq^3) f_C(xq^6;q)$.

It is not difficult to show (see Andrews (1968a) for details) that given the boundary condition $f_C(0;q) = 1$, the unique solution of (2. 8) is

195

$$(2.9) \qquad f_C(x;q) = \sum_{m \geq 0} \sum_{n \geq 0} \frac{(-q;q^3)_n(-q^2;q^3)_n(-1)^m x^{n+m} q^{3/2\, m(m-1)}}{(q^3;q^3)_m (q^3;q^3)_n} .$$

Thus we have a solution of (2.8) of the type envisaged in Problem 1.

Applying a simple identity of Euler to the outer sum in (2.9), we see that

$$f_C(x;q) = (x;q^3)_\infty \sum_{n \geq 0} \frac{(-q;q^3)_n(-q^2;q^3)_n x^n}{(q^3;q^3)_n} .$$

Therefore if B(n) denotes the number of partitions of n in C (i.e. the number of partitions of n with difference between all parts at least 3 and at least 6 between multiples of 3), then

$$(2.10) \qquad \sum_{n=0}^{\infty} B(n)q^n = f_C(1;q)$$

$$= \lim_{x \to 1^-} (x;q^3)_\infty \sum_{n \geq 0} \frac{(-q;q^3)_n(-q^2;q^3)_n x^n}{(q^3;q^3)_n}$$

$$= (q^3;q^3)_\infty \frac{(-q;q^3)_\infty(-q^2;q^3)_\infty}{(q^3;q^3)_\infty}$$

(by Abel's theorem, Andrews (1971b))

$$= \prod_{\substack{n=1 \\ 3 \nmid n}}^{\infty} (1+q^n) = \prod_{\substack{n=1 \\ 3 \nmid n}}^{\infty} \frac{(1-q^{2n})}{(1-q^n)} = \prod_{\substack{n=1 \\ 3 \nmid n, \\ \text{and } 2 \nmid n}}^{\infty} \frac{1}{1-q^n}$$

$$= \prod_{n=0}^{\infty} (1-q^{6n+1})^{-1}(1-q^{6n+5})^{-1} = \sum_{n=0}^{\infty} A(n)q^n ,$$

where $A(n)$ is the number of partitions of n into parts $\equiv 1$ or 5 (modulo 6). Comparing coefficients of q^n in the extremes of (2.10), we deduce

Theorem 2 (Schur (1926)). The partitions of n into parts $\equiv 1$ or 5 (mod 6) are equinumerous with the partitions of n in which all parts differ by at least 3 and all parts that are multiples of 3 differ by at least 6.

The above proof of Schur's theorem illustrates how useful a general solution to Problem 1 would be. We conclude this section with one more example of a known case of Problem 1. A. Selberg (1936) has shown that there exist polynomials $a_j(x, q)$ in x and q such that if

$$(2.11) \qquad f(x) = \sum_{n \geq 0} \frac{(-1)^n x^{kn} q^{\frac{1}{2}((2k+1)n^2 + n)} (1 - x^k q^{(2n+1)k})}{(q)_n (xq^{n+1})_\infty},$$

then

$$(2.12) \qquad \sum_{j=0}^{k} a_j(x, q) f(xq^j) = 0.$$

Recently (Andrews (1974b)) a multiple series representation of $f(x)$ has been found:

$$(2.13) \qquad f(x) = \sum_{n_1, \ldots, n_{k-1} \geq 0} \frac{x^{N_1 + \ldots + N_{k-1}} q^{N_1^2 + N_2^2 + \ldots + N_{k-1}^2}}{(q)_{n_1} (q)_{n_2} \cdots (q)_{n_{k-1}}},$$

where $N_j = n_j + n_{j+1} + \ldots + n_{k-1}$.

We may directly deduce from (2.11) and (2.13) that

$$(2.14) \qquad \sum_{n_1, \ldots, n_{k-1} \geq 0} \frac{q^{N_1^2 + \ldots + N_{k-1}^2}}{(q)_{n_1} (q)_{n_2} \cdots (q)_{n_{k-1}}} = \prod_{\substack{n=1 \\ n \not\equiv 0, \pm k (\mathrm{mod}\ 2k+1)}} (1 - q^n)^{-1}$$

simply applying Jacobi's triple product identity (Andrews (1971b)) to (2.11) when $x = 1$.

The analytic identity (2.14) (and $k-1$ related identities) are closely related to the following partition theorem of B. Gordon (1961) (see Andrews (1966), (1974b) for a more complete discussion):

Theorem 3. Let $A_{k,a}(n)$ denote the number of partitions of n into parts $\not\equiv 0$, $\pm a$ (mod $2k+1$). Let $B_{k,a}(n)$ denote the number of partitions of n of the form $n = b_1 + b_2 + \ldots + b_s$, $b_j \geq b_{j+1}$, $b_j - b_{j+k-1} \geq 2$, $b_{s-a+1} > 1$. Then $A_{k,a}(n) = B_{k,a}(n)$ for all n provided $1 \leq a \leq k$.

The cases $k = 2$, $a = 1$, 2 are the celebrated Rogers-Ramanujan identities (Andrews 1971b)).

2. Identities for Multiple Hypergeometric Series.

PROBLEM 2. What are possible multiple series generalizations of the q-analog of Whipple's theorem?

In a way Problem 2 seems quite specialized in comparison with Problem 1; indeed, it is. However we remark that Watson's (1929) q-analog of Whipple's theorem:

(3.1)
$$
{}_8\phi_7 \left[\begin{array}{l} a,\ q\sqrt{a},\ -q\sqrt{a},\ b,\ c,\ d,\ e,\ q^{-N}; q,\ \dfrac{a^2 q^{N+2}}{bcde} \\[2mm] \sqrt{a},\ -\sqrt{a},\ aq/b,\ aq/c,\ aq/d,\ aq/e,\ aq^{N+1} \end{array} \right]
$$

$$
= \dfrac{(a)_N (aq/ef)_N}{(aq/e)_N (aq/f)_N}\ {}_4\phi_3 \left[\begin{array}{l} aq/cd,\ e,\ f,\ q^{-N}; q,\ q \\[2mm] efq^{-N}/a,\ aq/c,\ aq/d \end{array} \right]
$$

has been utilized to prove numerous partition identities (see Andrews (1971a)) including the Rogers-Ramanujan identities (Watson (1929)).

198

As an example of a generalization of (3.1) we prove the following identity. We shall show its relationship to (2.14), and we shall derive several new analytic identities of the Rogers-Ramanujan type.

Theorem 4. For $k \geq 1$, N a nonnegative integer,

$$
{}_{2k+4}\phi_{2k+3}\left[\begin{array}{c} a, q\sqrt{a}, -q\sqrt{a}, b_1, c_1, b_2, c_2, \ldots, b_k, c_k, q^{-N}; q, \dfrac{a^k q^{k+N}}{b_1 \cdots b_k c_1 \cdots c_k} \\[2mm] \sqrt{a}, -\sqrt{a}, aq/b_1, aq/c_1, aq/b_2, aq/c_2, \ldots, aq/b_k, aq/c_k, aq^{N+1} \end{array}\right]
$$

$$
= \frac{(aq)_N (aq/b_k c_k)_N}{(aq/b_k)_N (aq/c_k)_N} \sum_{m_1, \ldots, m_{k-1} \geq 0} \frac{(aq/b_1 c_1)_{m_1} (aq/b_2 c_2)_{m_2} \cdots (aq/b_{k-1} c_{k-1})_{m_{k-1}}}{(q)_{m_1} (q)_{m_2} \cdots (q)_{m_{k-1}}}
$$

$$
\cdot \frac{(b_2)_{m_1} (c_2)_{m_1} (b_3)_{m_1+m_2} (c_3)_{m_1+m_2} \cdots (b_k)_{m_1+\ldots+m_{k-1}}}{(aq/b_1)_{m_1} (aq/c_1)_{m_1} (aq/b_2)_{m_1+m_2} (aq/c_2)_{m_1+m_2} \cdots (aq/b_{k-1})_{m_1+\ldots m_{k-1}}}
$$

$$
\cdot \frac{(c_k)_{m_1+\ldots+m_{k-1}}}{(aq/c_{k-1})_{m_1+\ldots+m_{k-1}}} \cdot \frac{(q^{-N})_{m_1+m_2+\ldots+m_{k-1}}}{(b_k c_k q^{-N}/a)_{m_1+m_2+\ldots+m_{k-1}}}
$$

$$
\cdot \frac{(aq)^{m_{k-2}+2m_{k-3}+\ldots+(k-2)m_1} \; q^{m_1+m_2+\ldots+m_{k-1}}}{(b_2 c_2)^{m_1} (b_3 c_3)^{m_1+m_2} \cdots (b_{k-1} c_{k-1})^{m_1+m_2+\ldots+m_{k-2}}} .
$$

Remark. Since this identity is between two terminating series, we can replace each b_i by q^{b_i}, each c_i by q^{c_i}, a by q^a and then let $q \to 1$; this procedure will provide a multiple series identity that contains Whipple's (1926) theorem in the case $k = 2$. Also we should point out that this result is closely related to some work of V. N. Singh (1957); for a complete discussion of this relationship see Andrews (1974b).

Proof. We proceed by induction on k. When $k = 1$, the theorem reduces to the terminating, limiting case of Jackson's theorem (Slater (1966), p. 247. eq. (IV. 7), $d = q^{-N}$). When $k = 2$, the result is Watson's q-analog of Whipple's theorem (Slater (1966), p. 100, eq. (3.4. 1.5)). Let us therefore assume that the theorem is true for each integer smaller than a given $k \geq 3$.

Now by the q-analog of Saalschutz's theorem (Slater (1966), p. 247, eq. (IV. 4))

$$(3.2) \quad \frac{a^n q^n (b_1)_n (c_1)_n}{b_1^n c_1^n (aq/b_1)_n (aq/c_1)_n} = \sum_{m_1=0}^{n} \frac{(aq/b_1 c_1)_{m_1} (aq^n)_{m_1} (q^{-n})_{m_1} q^{m_1}}{(q)_{m_1} (aq/b_1)_{m_1} (aq/c_1)_{m_1}}.$$

Hence

$$_{2k+4}\phi_{2k+3}\left[\begin{array}{c} a, q\sqrt{a}, -q\sqrt{a}, b_1, c_1, \ldots, b_k, c_k, q^{-N}; q, \dfrac{a^k q^{k+N}}{b_1 \cdots b_k c_1 \cdots c_k} \\ \sqrt{a}, -\sqrt{a}, aq/b_1, aq/c_1, \ldots, aq/b_k, aq/c_k, aq^{1+N} \end{array}\right]$$

$$= \sum_{n \geq 0} \frac{(a)_n (1-aq^{2n}) (b_2)_n (c_2)_n \cdots (b_k)_n (c_k)_n}{(q)_n (1-a)(aq/b_2)_n (aq/c_2)_n \cdots (aq/b_k)_n (aq/c_k)_n} \frac{(q^{-N})_n}{(aq^{N+1})_n}$$

$$\left(\frac{a^{k-1} q^{k-1+N}}{b_2 \cdots b_k c_2 \cdots c_k}\right)^n \sum_{m_1 \geq 0} \frac{(aq/b_1 c_1)_{m_1} (aq^n)_{m_1} (q^{-n})_{m_1} q^{m_1}}{(q)_{m_1} (aq/b_1)_{m_1} (aq/c_1)_{m_1}} \qquad \text{(by (3.2))}$$

$$= \sum_{m_1 \geq 0} \frac{(aq/b_1 c_1)_{m_1} q^{m_1 + \binom{m_1}{2}} (-1)^{m_1}}{(q)_{m_1} (aq/b_1)_{m_1} (aq/c_1)_{m_1}} \cdot \sum_{n \geq m_1} \frac{(a)_{n+m_1} (1-aq^{2n})}{(q)_{n-m_1} (1-a)}$$

$$\frac{(b_2)_n (c_2)_n \cdots (b_k)_n (c_k)_n (q^{-N})_n}{(aq/b_2)_n (aq/c_2)_n \cdots (aq/b_k)_n (aq/c_k)_n (aq^{N+1})_n} \left(\frac{a^{k-1} q^{k-1+N-m_1}}{b_2 \cdots b_k c_2 \cdots c_k}\right)^n$$

$$= \sum_{m_1 \geq 0} \frac{(aq/b_1 c_1)_{m_1} q^{\binom{m_1+1}{2}} (-1)^{m_1} (b_2)_{m_1} (c_2)_{m_1} \cdots (b_k)_{m_1} (c_k)_{m_1} (q^{-N})_{m_1}}{(q)_{m_1} (aq/b_1)_{m_1} (aq/c_1)_{m_1} (aq/b_2)_{m_1} (aq/c_2)_{m_1} \cdots (aq/b_k)_{m_1} (aq/c_k)_{m_1}}$$

$$\left(\frac{a^{k-1} q^{k-1+N-m_1}}{b_2 \cdots b_k c_2 \cdots c_k}\right)^{m_1} \frac{(aq)_{2m_1}}{(aq^{1+N})_{m_1}} \sum_{n \geq 0} \frac{(aq^{2m_1})_n (1-aq^{2m_1} q^{2n})(b_2 q^{2m_1})_n}{(q)_n (1-aq^{2m_1})(aq^{m_1+1}/c_1)_n}$$

$$\frac{(c_2 q^{m_1})_n \cdots (b_k q^{m_1})_n (c_k q^{m_1})_n (q^{-N+m_1})_n}{(aq^{m_1+1}/b_2)_n \cdots (aq^{m_1+1}/b_k)_n (aq^{m_1+1}/c_k)_n (aq^{N+m_1+1})_n} \left(\frac{a^{k-1} q^{k-1+N-m_1}}{b_2 \cdots b_k c_2 \cdots c_k}\right)^n$$

$$= (aq)_N \sum_{m_1 \geq 0} \frac{(aq/b_1 c_1)_{m_1} q^{\binom{m_1+1}{2}} (-1)^{m_1} (b_2)_{m_1} (c_2)_{m_1} \cdots (b_k)_{m_1} (c_k)_{m_1} (q^{-N})_{m_1}}{(q)_{m_1} (aq/b_1)_{m_1} (aq/c_1)_{m_1} \cdots (aq/b_k)_{m_1} (aq/c_k)_{m_1}}$$

$$\left(\frac{a^{k-1} q^{k-1+N-m_1}}{b_2 \cdots b_k c_2 \cdots c_k}\right)^{m_1} \frac{1}{(aq^{2m_1+1})_{N-m_1}} \; {}_{2k+2}\phi_{2k+1}\left[\begin{array}{c} aq^{2m_1}, q\sqrt{aq^{2m_1}}, -q\sqrt{aq^{2m_1}}, \\[4pt] \sqrt{aq^{2m_1}}, -\sqrt{aq^{2m_1}}, \end{array}\right.$$

$$\left.\begin{array}{c} b_2 q^{m_1}, c_2 q^{m_1}, \ldots, b_k q^{m_1}, c_k q^{m_1}, q^{-N+m_1} \\[4pt] \dfrac{aq^{m_1+1}}{b_2}, \dfrac{aq^{m_1+1}}{c_2}, \ldots, \dfrac{aq^{m_1+1}}{b_k}, \dfrac{aq^{m_1+1}}{c_k}, aq^{N+m_1+1} \end{array}\; ;q, \; \frac{a^{k-1} q^{k-1+N-m_1}}{b_2 \cdots b_k c_2 \cdots c_k}\right]$$

$$= (aq)_N \sum_{m_1 \geq 0} \frac{(aq/b_1 c_1)_{m_1} q^{\binom{m_1+1}{2}} (-1)^{m_1} (b_2)_{m_1} (c_2)_{m_1} \cdots (b_k)_{m_1} (c_k)_{m_1} (q^{-N})_{m_1}}{(q)_{m_1} (aq/b_1)_{m_1} (aq/c_1)_{m} \cdots (aq/b_k)_{m_1} (aq/c_k)_{m_1}}$$

$$\left(\frac{a^{k-1}q^{k-1+N-m_1}}{b_2 \cdots b_k c_2 \cdots c_k}\right)^{m_1} \frac{(aq/b_k c_k)_{N-m_1}}{(\frac{aq}{b_k})_{N-m} (\frac{aq}{c_k})_{N-m_1}}$$

$$\sum_{m_2, \ldots, m_{k-1} \geq 0} \frac{(aq/b_2 c_2)_{m_2} \cdots (aq/b_{k-1} c_{k-1})_{m_{k-1}}}{(q)_{m_2} \cdots (q)_{m_{k-1}}}$$

$$\frac{(b_3 q^{m_1})_{m_2} (c_3 q^{m_1})_{m_2} \cdots (b_k q^{m_1})_{m_2+\ldots+m_{k-1}} (c_k q^{m_1})_{m_2+\ldots+m_{k-1}}}{(\frac{aq}{b_2})_{m_2}^{m_1+1} (\frac{aq}{c_2})_{m_2}^{m_1+1} \cdots (\frac{aq}{b_{k-1}})_{m_2+\ldots+m_{k-1}}^{m_1+1} (\frac{aq}{c_{k-1}})_{m_2+\ldots+m_{k-1}}^{m_1+1}}$$

$$\frac{(q^{-N+m_1})_{m_2+\ldots+m_{k-1}} (aq)^{m_{k-2}+2m_{k-3}+\ldots+(k-3)m_2} q^{m_2+\ldots+m_{k-1}}}{(b_k c_k q^{-N+m_1}/a)_{m_2+\ldots+m_{k-1}} (b_3 c_3)^{m_2} (b_4 c_4)^{m_2+m_3} \cdots (b_{k-1} c_{k-1})^{m_2+\ldots+m}}$$

$$= \frac{(aq)_N (aq/b_k c_k)_N}{(\frac{aq}{b_k})_N (\frac{aq}{c_k})_N} \sum_{m_1, m_2, \ldots, m_{k-1} \geq 0} \frac{(aq/b_1 c_1)_{m_1} (aq/b_2 c_2)_{m_2} \cdots (aq/b_{k-1} c_{k-1})_{m_k}}{(q)_{m_1} (q)_{m_2} \cdots (q)_{m_{k-1}}}$$

$$\frac{(b_2)_{m_1} (c_2)_{m_1} (b_3)_{m_1+m_2} (c_3)_{m_1+m_2} \cdots (b_k)_{m_1+\ldots+m_{k-1}}}{(aq/b_1)_{m_1} (aq/c_1)_{m_1} (aq/b_2)_{m_1+m_2} (aq/c_2)_{m_1+m_2} \cdots (aq/b_k)_{m_1+\ldots+m_{k-1}}}$$

$$\frac{(c_k)_{m_1+\ldots+m_{k-1}}}{(aq/c_k)_{m_1+\ldots+m_{k-1}}} \cdot \frac{(q^{-N})_{m_1+\ldots+m_{k-1}} \left(\frac{a^{k-1}q^{k-1+N-m_1}}{b_2 \cdots b_k c_2 \cdots c_k}\right)^{m_1} q^{\binom{m_1}{2}+1} (-1}{(1-\frac{aq}{b_k c_k})^N \cdots (1-\frac{aq}{b_k c_k})^{N-m_1+1})(\frac{b_k c_k q^{-N+m_1}}{d})_{m_2+\ldots+m}}$$

$$\cdot \; \frac{(aq)^{m_{k-2}+2m_{k-3}+\ldots+(k-3)m_2}\; q^{m_2+\ldots+m_{k-1}}}{(b_3 c_3)^{m_2}(b_4 c_4)^{m_2+m_3}\ldots(b_{k-1}c_{k-1})^{m_2+\ldots+m_{k-2}}}$$

$$= \frac{(aq)_N (aq/b_k c_k)_N}{(aq/b_k)_N (aq/c_k)_N}\sum_{m_1,\ldots,m_{k-1}\geq 0}\frac{(aq/b_1 c_1)_{m_1}\ldots(aq/b_{k-1}c_{k-1})_{m_{k-1}}}{(q)_{m_1}(q)_{m_2}\ldots(q)_{m_{k-1}}}$$

$$\frac{(b_2)_{m_1}(c_2)_{m_1}(b_3)_{m_1+m_2}(c_3)_{m_1+m_2}\ldots(b_k)_{m_1+\ldots+m_{k-1}}}{(aq/b_1)_{m_1}(aq/c_1)_{m_1}(aq/b_2)_{m_1+m_2}(aq/c_2)_{m_1+m_2}\ldots(aq/b_k)_{m_1+\ldots+m_{k-1}}}$$

$$\frac{(c_k)_{m_1+\ldots+m_{k-1}}}{(aq/c_k)_{m_1+\ldots+m_{k-1}}} \quad \frac{(q^{-N})_{m_1+\ldots+m_{k-1}}}{(b_k c_k q^{-N}/a)_{m_1+\ldots+m_{k-1}}}$$

$$\frac{(aq)^{m_{k-2}+2m_{k-3}+\ldots+(k-2)m_1}\; q^{m_1+\ldots+m_{k-1}}}{(b_2 c_2)^{m_1}(b_3 c_3)^{m_1+m_2}\ldots(b_{k-1}c_{k-1})^{m_1+\ldots+m_{k-1}}}$$

which is the desired formula. Hence Theorem 4 is established. □

Corollary 4.1.

$$\sum_{n\geq 0}\frac{(a)_n(1-aq^{2n})(b)_n(c)_n(-1)^n(aq^k/bc)^n q^{(2k-1)n(n-1)/2}}{(q)_n(1-a)(aq/b)_n(aq/c)_n}$$

$$= \frac{(aq)_\infty(aq/ac)_\infty}{(aq/b)_\infty(aq/c)_\infty}\sum_{m_1,\ldots,m_{k-1}\geq 0}\frac{(b)_{m_1+\ldots+m_{k-1}}(c)_{m_1+\ldots+m_{k-1}}}{(q)_{m_1}(q)_{m_2}\ldots(q)_{m_{k-1}}}$$

$$(q/bc)^{m_1+m_2+\ldots+m_{k-1}} \cdot a^{(k-1)m_1+(k-2)m_2+\ldots+2m_{k-2}+m_{k-1}}$$

$$q^{m_1^2+(m_1+m_2)^2+\ldots+(m_1+\ldots+m_{k-2})^2} \ .$$

Proof. Set $b_k = b$, $c_k = c$ in Theorem 4; then let $b_1, c_1, \ldots,$ b_{k-1}, c_{k-1} and N all tend to infinity. An appeal to a multiple series analog of Tannery's theorem (see Slater (1966), p. 104 for the case $k = 2$) justifies taking these limits. □

Corollary 4.2. Equation (2.14) is valid.

Proof. Let b and c tend to ∞ in Corollary 4.1; then set $a = 1$, and utilize the fact that

$$1 + \sum_{n=1}^{\infty} (-1)^n q^{\frac{1}{2}((2k+1)n^2 - n)} (1+q^n) = \prod_{\substack{n=1 \\ n \equiv 0,\ \pm k\,(\mathrm{mod}\,2k+1)}}^{\infty} (1-q^n)$$

a specialization of Jacobi's triple product identity. □

As a final corollary we generalize an identity of L. J. Slater (Slater (1952), eq. (36)) thus obtaining the analytic counterpart of the generalization of the Göllnitz-Gordon identities (see Andrews (1967b) in the case $k = a$).

Corollary 4.3.

$$\sum_{m_1,\ldots,m_{k-1}\geq 0} \frac{(-q;q^2)_{M_1}\ q^{M_1^2+M_2^2+\ldots+M_{k-1}^2}}{(q^2;q^2)_{m_1}(q^2;q^2)_{m_2}\ldots(q^2;q^2)_{m_{k-1}}}$$

$$= \prod_{\substack{n=1 \\ n\not\equiv 2\,(\mathrm{mod}\,4) \\ n\not\equiv 0,\ \pm(2k-1)\,(\mathrm{mod}\,4k)}}^{\infty} (1-q^n)^{-1}\ ,$$

where $M_j = m_j + m_{j+1} + \ldots + m_{k-1}$.

Proof. In Corollary 4.1 replace q by q^2, then set $b = q$, $a = 1$ and let $c \to \infty$. In the resulting identity, we observe that

$$1 + \sum_{n=1}^{\infty} (-1)^n q^{2kn^2-n}(1+q^{2n}) = \prod_{\substack{n=1 \\ n \equiv 0, \pm(2k-1)(\bmod 4k)}}^{\infty} (1-q^n) \; ,$$

another specialization of Jacobi's triple product identity.   □

Numerous other q-series identities can be derived from Theorem 4. We have chosen a small sampling to illustrate the possibilities.

In closing this section we wish to stress that Theorem 4 may well not be the only useful multiple series generalization of Watson's q-analog of Whipple's theorem.

In particular Holman, Biedenharn and Louck (1975) have proved a generalization of Whipple's theorem that is definitely not in the same direction as Theorem 4 above. Their result arises from work in physics and relies for its proof on various combinatorial and recurrent series arguments. Significant work must be undertaken before we can determine whether q-analogs exist for their results; past experience suggests that such q-analogs may well exist. If they do exist, their importance in the theory of partitions may well depend on whether they provide a solution to the following question.

PROBLEM 3. Are there multiple series q-analogs of well-poised hypergeometric series that specialize to either cases of the quituple product identity, or Winquist's identity or some other multiple theta series that sums to an infinite product?

The quintuple product identity is proved neatly by Carlitz and Subbarao (1972b):

$$\prod_{n=1}^{\infty} (1-q^n)(1-zq^n)(1-z^{-1}q^{n-1})(1-z^2q^{2n-1})(1-z^{-2}q^{2n-1})$$

$$= \sum_{n=-\infty}^{\infty} q^{(3n^2+n)/2}(z^{3n}-z^{-3n-1}) \quad .$$

205

Winquist's (1969) identity has also been nicely proved and gener-alized by Carlitz and Subbarao (1972a),

$$\prod_{n=1}^{\infty} (1-q^n)^2 (1-yq^{n-1})(1-zq^{n-1})(1-yz^{-1}q^{n-1})(1-yzq^{n-1})$$

$$\cdot (1-y^{-1}q^n)(1-z^{-1}q^n)(1-y^{-1}zq^n)(1-y^{-1}z^{-1}q^n)$$

$$= \sum_{i=0}^{\infty} \sum_{j=-\infty}^{\infty} (-1)^{i+j} \left\{ (y^{-3i}-y^{3i+3})(z^{-3j}-z^{3j+1}) \right.$$

$$\left. + (z^{-3j+1}-z^{3j+2})(z^{3i+2}-z^{-3i-1}) \right\} q^{\frac{3}{2}i(i+1)+\frac{1}{2}j(3j+1)} .$$

As we have seen in special cases of Theorem 4, the Jacobi triple product identity is the one invoked when the specializations of the well-poised $_{2k+4}\phi_{2k+3}$ are made in Corollaries 4.2 and 4.3.

As we have shown in ABHF, p. 466, the quintuple product identity may be deduced by specializing parameters in a bilateral well-poised basic hypergeometric series, and this observation suggests that a more extensive study of the $K_{\lambda,k,i}(a_0,a_1,\ldots,a_\lambda;a;q)$ (see ABHF, p. 453) is in order.

In closing this section, we remark that the work of Andrews (1968b) may be used to show that the $_{2k+4}\phi_{2k+3}$ appearing in Theorem 4 is the unique solution of a certain k-th order q-difference equation on the variable a of the type described in Problem 1.

4. Basic Appell and Lauricella Series.

In view of the directions indicated by Sections 2 and 3 it is quite natural to ask if there are theories of multiple basic hypergeometric series identities comparable to results known for the ordinary multiple hyper-geometric series, namely the Appell and more generally Lauricella func-tions. To illustrate the possibilities we pose the following problem to which we shall give some reasonably satisfactory answers.

PROBLEM 4. Are there multiple q-series analogs of the Chu-Vandermonde summation and Saalschütz's theorem?

We begin by defining the basic Lauricella function which is the q-analog of the fourth ordinary Lauricella function (see Slater (1966), p. 288, eq. (8.6.4))

$$\Phi_D[a;b_1,b_2,\ldots,b_n;c;x_1,\ldots,x_n]$$

$$= \sum_{m_1,\ldots,m_n \geq 0} \frac{(a)_{m_1+\ldots+m_n}(b_1)_{m_1}\cdots(b_n)_{m_n} x_1^{m_1}\ldots x_n^{m_n}}{(q)_{m_1}(q)_{m_2}\cdots(q)_{m_n}(c)_{m_1+m_2+\ldots+m_n}}.$$

Theorem 5 (Andrews (1972), p. 621).

$$\Phi_D[a;b_1,b_2,\ldots,b_n;c;x_1,\ldots,x_n]$$

$$= \frac{(a)_\infty(b_1x_1)_\infty(b_2x_2)_\infty\cdots(b_nx_n)_\infty}{(c)_\infty(x_1)_\infty(x_2)_\infty\cdots(x_n)_\infty} {}_{n+1}\phi_n\left[\begin{array}{c} c/a, x_1, x_2,\ldots,x_n;q,a \\ b_1x_1, b_2 x_2,\ldots,b_nx_n \end{array}\right]$$

Proof.

$$\Phi_D[a;b_1,b_2,\ldots,b_n;c;x_1,x_2,\ldots,x_n]$$

$$= \frac{(a)_\infty}{(c)_\infty} \sum_{m_1,\ldots,m_n \geq 0} \frac{(b_1)_{m_1}\cdots(b_n)_{m_n}}{(q)_{m_1}\cdots(q)_{m_n}} x_1^{m_1}\ldots x_n^{m_n} \frac{(cq^{m_1+m_2+\ldots+m_n})_\infty}{(aq^{m_1+m_2+\ldots+m_n})_\infty}$$

$$= \frac{(a)_\infty}{(c)_\infty} \sum_{m_1,\ldots,m_n \geq 0} \frac{(b_1)_{m_1}\cdots(b_n)_{m_n}}{(q)_{m_1}\cdots(q)_{m_n}} x_1^{m_1}\ldots x_n^{m_n} \sum_{r \geq 0} \frac{(c/a)_r}{(q)_r} a^r q^{(m_1+\ldots+m_n)r}$$

(by Slater (1966), p. 92, eq. (3.2.2.11))

207

$$= \frac{(a)_\infty (b_1 x_1)_\infty \cdots (b_n x_n)_\infty}{(c)_\infty (x_1)_\infty \cdots (x_n)_\infty} \sum_{r \geq 0} \frac{(c/a)_r (x_1)_r \cdots (x_n)_r a^r}{(q)_r (b_1 x_1)_r \cdots (b_n x_n)_r}$$

(by Slater (1966), p. 92, eq. (3.2.2.11))

$$= \frac{(a)_\infty (b_1 x_1)_\infty \cdots (b_n x_n)_\infty}{(c)_\infty (x_1)_\infty \cdots (x_n)_\infty} \; _{n+1}\phi_n \left[\begin{array}{c} c/a, x_1, \ldots, x_n ; q, a \\ b_1 x_1, \ldots, b_n x_n \end{array} \right]. \qquad \square$$

Corollary 5.1.

$$\Phi_D [a; b_1, x_1/x_2, x_2/x_3, \ldots, x_{n-1}/x_n ; c; x_1, \ldots, x_n]$$

$$= \frac{(a)_\infty}{(c)_\infty (x_n)_\infty} \sum_{r \geq 0} \frac{(c/a)_r (x_n)_r (b_1 x_1 q^r)_\infty a^r}{(q)_r}.$$

Proof. Set $b_2 = x_1/x_2$, $b_3 = x_2/x_3, \ldots, b_n = x_{n-1}/x_n$ in Theorem 5. $\qquad \square$

We can now easily deduce an analog of the Chu-Vandermonde summation

Corollary 5.2. If N is a nonnegative integer

$$\Phi_D [a; q^{-N+1}/x_1, x_1/x_2, x_2/x_3, \ldots, x_{n-2}/x_{n-1}, x_{n-1}/q; c; x_1, x_2, \ldots, x_{n-1}, q]$$

$$= \frac{a^N (c/a)_N}{(c)_N}.$$

Remark. When $n = 1$, Corollary 5.2 asserts that

$$_2\phi_1 \left[\begin{array}{c} a, q^{-N} ; q, q \\ c \end{array} \right] = \frac{a^N (c/a)_N}{(c)_N},$$

which is one of the forms of the q-analog of the Chu-Vandermonde summation [Slater (1966), p. 97, eq. (3.3.2.7)).

Proof. By Corollary 5.1,

$$\Phi_D[a;q^{-N+1}/x_1, x_1/x_2, x_2/x_3, \ldots, x_{n-2}/x_{n-1}, x_{n-1}/q;c;x_1, x_2, \ldots, x_{n-1}, q]$$

$$= \frac{(a)_\infty}{(c)_\infty (q)_\infty} \sum_{r \geq 0} (c/a)_r (q^{-N+r+1})_\infty a^r$$

$$= \frac{(a)_\infty}{(c)_\infty (q)_\infty} \sum_{r \geq 0} (c/a)_{r+N} (q^{r+1})_\infty a^{r+N}$$

$$= \frac{a^N (a)_\infty (c/a)_N}{(c)_\infty} \sum_{r \geq 0} \frac{(cq^N/a)_r a^r}{(q)_r}$$

$$= \frac{a^N (a)_\infty (c/a)_N (cq^N)_\infty}{(c)_\infty (a)_\infty} \qquad \text{(by Slater (1966), p. 92, eq. (3.2.2.11))}$$

$$= \frac{a^N (c/a)_N}{(c)_N} \quad .$$

Corollary 5.3 (Al-Salam (1965), eq. (9)).

$$\Phi_D(\beta\beta'/\alpha;\beta, \beta';\beta\beta';x, y)$$

$$= \frac{(x\beta'/\alpha)_\infty (y\beta/\alpha)_\infty}{(x)_\infty (y)_\infty} \Phi_D[\alpha;\beta, \beta';\beta\beta';\beta'x/\alpha, \beta y/\alpha] \quad .$$

Proof. Let us recall an identity of Newman Hall for the $_3\phi_2$-function (Hall (1936)):

$$(4.1) \qquad _3\phi_2 \begin{bmatrix} a, b, c;q, ef/abc \\ e, f \end{bmatrix} = \frac{(a)_\infty (ef/ab)_\infty (ef/ac)_\infty}{(e)_\infty (f)_\infty (ef/abc)_\infty}$$

$$_3\phi_2 \begin{bmatrix} e/a, f/a, ef/abc;q, a \\ ef/ab, ef/ac \end{bmatrix} \quad .$$

In (4.1), we set $a = \alpha$, $b = x$, $c = y$, $e = \beta x$, $f = \beta' y$, and then we multiply the resulting identity by

$$\frac{(\beta\beta'/\alpha)_\infty (\beta x)_\infty (\beta'y)_\infty}{(\beta\beta')_\infty (x)_\infty (y)_\infty} \; .$$

Hence

(4.2)

$$\frac{(\beta\beta'/\alpha)_\infty (\beta x)_\infty (\beta'y)_\infty}{(\beta\beta')_\infty (x)_\infty (y)_\infty} \; {}_3\phi_2 \left[\begin{array}{c} \alpha, x, y; q, \gamma/\alpha \\ \\ \beta x, \beta'y \end{array} \right]$$

$$= \frac{(\alpha)_\infty (y\beta\beta'/\alpha)_\infty (x\beta\beta'/\alpha)_\infty}{(\beta\beta')_\infty (x)_\infty (y)_\infty} \; {}_3\phi_2 \left[\begin{array}{c} \beta x/\alpha, \beta'y/\alpha, \beta\beta'/\alpha; q, \alpha \\ \\ y\beta\beta'/\alpha, x\beta\beta'/\alpha \end{array} \right]$$

$$= \frac{(x\beta'/\alpha)_\infty (y\beta/\alpha)_\infty}{(x)_\infty (y)_\infty}$$

$$\cdot \frac{(\alpha)_\infty (x\beta\beta'/\alpha)_\infty (y\beta\beta'/\alpha)_\infty}{(\beta\beta')_\infty (x\beta'/\alpha)_\infty (y\beta/\alpha)_\infty} \; {}_3\phi_2 \left[\begin{array}{c} \beta\beta'/\alpha, \beta x/\alpha, \beta'y/\alpha; q, \alpha \\ \\ x\beta\beta'/\alpha, y\beta\beta'/\alpha \end{array} \right] \; .$$

Hence by Theorem 5 applied to (4.2), we see that

$$\Phi_D[\beta\beta'/\alpha; \beta, \beta'; \beta\beta'; x, y] = \frac{(x\beta'/\alpha)_\infty (y\beta/\alpha)_\infty}{(x)_\infty (y)_\infty}$$

$$\cdot \Phi_D[\alpha; \beta, \beta'; \beta\beta'; \beta'x/\alpha, \beta y/\alpha] \; ,$$

which is the desired result. □

Next we derive Al-Salam's Saalschützian theorem for double q-series.

Corollary 5.4 (Al-Salam (1965), eq. (7)).

$$\sum_{r, s \geq 0} \frac{(q^{-n})_r (q^{-m})_s (\alpha)_{r+s} (\beta)_r (\beta')_s}{(q)_r (q)_s (\gamma)_{r+s} (\delta)_r (\delta')_s} \; q^{r+s} = \frac{(\beta\beta'/\alpha)_{n+m} (\beta)_m (\beta')_n}{(\beta\beta')_{n+m} (\beta/\alpha)_m (\beta'/\alpha)_n}$$

provided $\delta = q^{1-n}\alpha/\beta'$, $\delta' = q^{1-m}\alpha/\beta$, $\gamma = \beta\beta'$.

Proof. We note the simple identity

$$\sum_{n \geq 0} \frac{(A)_n t^n}{(q)_n} = \frac{(At)_\infty}{(t)_\infty} \qquad \text{(Slater (1966), p. 92, eq. (3.2.2.11))} .$$

Hence we may compare coefficients of $x^m y^n$ on both sides of the identity in Corollary 5.3:

$$\frac{(\beta\beta'/\alpha)_{m+n}(\beta)_m(\beta')_n}{(\beta\beta')_{m+n}(q)_n(q)_n} = \sum_{r,\,s \geq 0} \frac{(\alpha)_{r+s}(\beta)_r(\beta')_s(\beta'/\alpha)_{m-r}(\beta/\alpha)_{n-s}}{(q)_r(q)_s(\beta\beta')_{r+s}(q)_{m-r}(q)_{n-s}} (\frac{\beta'}{\alpha})^r (\frac{\beta}{\alpha})^s .$$

Therefore

$$\frac{(\beta\beta'/\alpha)_{m+n}(\beta)_m(\beta')_n}{(\beta\beta')_{m+n}(\beta/\alpha)_m(\beta'/\alpha)_n} = \sum_{r,\,s \geq 0} \frac{(\alpha)_{r+s}(\beta)_r(\beta')_s(q^{-m})_r(q^{-n})_s}{(q)_r(q)_s(\beta\beta')_{r+s}(q^{1-n}\alpha/\beta')_r(q^{1-m}\alpha/\beta)_s} q^{r+s} ,$$

which is the desired result. $\qquad\square$

Corollary 5.5 (Carlitz (1963), eq. (9)).

$$\sum_{r,\,s \geq 0} \frac{[-m]_r[-n]_s[\alpha]_{r+s}[\beta]_r[\beta']_s}{r!\,s!\,[\gamma]_{r+s}[\delta]_r[\delta']_s} = \frac{[\beta+\beta'-\alpha]_{m+n}[\beta']_m[\beta]_n}{[\beta+\beta']_{m+n}[\beta'-\alpha]_m[\beta-\alpha]_n} ,$$

where $[A]_r = A(A+1)\ldots(A+r-1)$, $\gamma = \beta+\beta'$, $\delta = \alpha-\beta'-m+1$, $\delta' = \alpha-\beta-n+1$.

Proof. Replace α by q^α, β by q^β and γ by q^γ in Corollary 5.4; then let $q \to 1^-$. The limiting process is clearly legitimate since all series terminate. $\qquad\square$

Corollary 5.6. If $x_1 = c/ab$, then

$$\Phi_D[a;b, x_1/x_2, x_2/x_3, \ldots, x_{n-1}/x_n; c; x_1, x_2, \ldots, x_n] = \frac{(c/a)_\infty(ax_n)_\infty}{(c)_\infty(x_n)_\infty} .$$

Proof. By Corollary 5.1

$$\Phi_D[a;b, x_1/x_2, x_2/x_3, \ldots, x_{n-1}/x_n ;c;x_1, x_2, \ldots, x_n]$$

$$= \frac{(a)_\infty (bx_1)_\infty}{(c)_\infty (x_n)_\infty} \sum_{r=0} \frac{(c/a)_r (x_n)_r a^r}{(q)_r (bx_1)_r}$$

$$= \frac{(a)_\infty (bx_1)_\infty}{(c)_\infty (x_n)_\infty} \sum_{r=0}' \frac{(x_n)_r a^r}{(q)_r}$$

$$= \frac{(a)_\infty (bx_1)_\infty (x_n a)_\infty}{(c)_\infty (x_n)_\infty (a)_\infty} = \frac{(bx_1)_\infty (x_n a)_\infty}{(c)_\infty (x_n)_\infty} . \qquad \square$$

The above result is an extension of the q-analog of Gauss's theorem to basic Lauricella functions. We may directly deduce an analog of Gauss's summation:

Corollary 5.7. If $\operatorname{Re} c \geq \operatorname{Re} a$, $\operatorname{Re}(a-c+nb) < 0$, $\operatorname{Re}((1-n)(a-c)-nb -nx_2) < 0$ and for $2 \leq i \leq n-1$ $\operatorname{Re}(a-c+nx_i -nx_{i+1}) < 0$, then

$$F_D[a;b, c-a-b+x_2, x_2-x_3, \ldots, x_{n-1}-x_n ;c;1, 1, 1, \ldots, 1]$$

$$= \frac{\Gamma(c)\Gamma(x_n)}{\Gamma(c-a)\Gamma(a+x_n)} .$$

Remark. When $n = 1$ this is just Gauss's summation. The case $n = 2$ has been treated by Appell and Kampe de Feriet (1926), and in fact, their method can also be easily generalized to prove the above result. B. C. Carlson (1974) gives the terminating form of this result as his Theorem 3.1.

Proof. Replace a by q^a, b by q^b, c by q^c, and x_i by q^{x_i} for $1 \leq i \leq n$ in Corollary 5.6 and let $q \to 1^-$. The inequalities prescribed for the parameter guarantee convergence of the Lauricella function since the general term in absolute value is

$$\left| \frac{\Gamma(c)}{\Gamma(a)\Gamma(b)\Gamma(c-a-b+x_2)\Gamma(x_2-x_3)\cdots\Gamma(x_{n-1}-x_n)} \right|$$

$$\cdot \left| \frac{\Gamma(a+m_1+\cdots+m_n)\Gamma(b+m_1)\Gamma(c-a-b+x_2+m_2)\Gamma(x_2-x_3+m_3)\cdots\Gamma(x_{n-1}-x_n+m_n)}{\Gamma(c+m_1+\cdots+m_n)\Gamma(1+m_1)\Gamma(1+m_2)\cdots\Gamma(1+m_n)} \right|$$

$$\sim K(a,b,c,x_1,\ldots,x_n)(m_1+\cdots+m_n)^{Re(a-c)} m_1^{Reb-1}$$

$$m_2^{Re(c-a-b+x_2)-1} \cdots m_n^{Re(x_{n-1}-x_n)-1}$$

$$\leq n^{Re(a-c)} K(a,b,c,x_1,\ldots,x_n) m_1^{\frac{1}{n}Re(a-c)+Reb-1} m_2^{Re((1-\frac{1}{n})(c-a)-b+x_2)-1}$$

$$m_3^{Re(\frac{1}{n}(a-c)+x_2-x_3)-1} \cdots m_n^{Re(\frac{1}{n}(a-c)+x_{n-1}-x_n)-1},$$

where the last inequality is just the arithmetic-geometric mean inequality and can be used here since $Re(a-c) \leq 0$. The remaining inequalities stated in the hypothesis now imply convergence by the comparison test. □

As we have remarked many of the results of this section were originally derived in different ways. It is hoped that the unification provided by Theorem 5 may be one of a class of general multiple q-series identities useful in dealing with some of our earlier problems.

5. MacMahon's Master Theorem and the Dyson Conjecture.

PROBLEM 5. Are there q-analogs of MacMahon's Master Theorem and the Dyson Conjecture?

First let us recall:

MacMahon's Master Theorem (MacMahon (1894), (1915)). The coefficient of $X_1^{P_1} X_2^{P_2} \cdots X_n^{P_n}$ in

$$\prod_{i=1}^{n} (A_{i1}X_1 + A_{i2}X_2 + \ldots + A_{in}X_n)^{P_i}$$

is the same as the corresponding coefficient in

$$(\det(\delta_{ij} - A_{ij}X_j))^{-1}$$

where $\delta_{ij} = 1$ if $i = j$ and $\delta_{ij} = 0$ if $i \neq j$.

This theorem has remarkable consequences in the theory of permutations and has been used to provide solutions to generalized Rencontre problems (MacMahon (1915)), the Menage problem (Percus (1971)), and numerous other results. Section 3 of MacMahon's book (MacMahon (1915)) as well as pages 18-31 of J. K. Percus (1971) provide ample evidence of this assertion.

I. J. Good (1962) utilized the Master Theorem to prove the following terminating form of Dixon's theorem (Dixon (1891), (1903)):

$$(5.1) \qquad \sum_{k=-\infty}^{\infty} (-1)^k \binom{b+c}{b+k}\binom{c+a}{c+k}\binom{a+b}{a+k} = \frac{(a+b+c)!}{a!b!c!} .$$

Good's observation is that the coefficient of $x^{b+c}y^{a+c}z^{a+b}$ in $(y-z)^{b+c}(z-x)^{c+a}(x-y)^{a+b}$ is just the left and side of (5.1); the Master Theorem may then be applied to produce the right hand side (5.1). D. Foata mentioned that the coefficient of $x^{b+c}y^{a+c}z^{a+b}$ in $(y-z)^{b+c}(z-x)^{c+a}(x-y)^{a+b}$ is precisely the same as the constant term in

$$(1 - \frac{x}{z})^a (1 - \frac{x}{y})^a (1 - \frac{y}{x})^b (1 - \frac{y}{z})^b (1 - \frac{z}{x})^c (1 - \frac{z}{y})^c ;$$

thus Foata observes that the above special case of the Master Theorem coincides with the three variable case of the following conjecture of F. J. Dyson (1962):

Dyson's Conjecture (Proved by Gunson (1962), Wilson (1962), and Good (1970)). The constant term in the expansion of

$$\prod_{1 \le i \ne j \le n} \left(1 - \frac{x_i}{x_j} \right)^{a_i}$$

is $(a_1 + a_2 + \ldots + a_n)! / (a_1! a_2! \ldots a_n!)$.

Proof. Good (1970) presents the most elegant proof of Dyson's conjecture; so we include it. Define $F_n(a_1, \ldots, a_n) = \prod_{1 \le i \ne j \le n} (1 - \frac{x_i}{x_j})^{a_i}$

and let $c_n(a_1, \ldots, a_n)$ be the constant term in $F_n(a_1, \ldots, a_n)$. Clearly $c_n(0, 0, \ldots, 0) = 1$. Furthermore if $a_h = 0$, then $c_n(a_1, \ldots, a_n) = c_{n-1}(a_1, \ldots, a_{h-1}, a_{h+1}, \ldots, a_n)$ since $a_h = 0$ implies that only nonpositive powers of x_h appear in $F_n(a_1, \ldots, a_n)$ and consequently the constant term in this case coincides with that in $F_{n-1}(a_1, \ldots, a_{h-1}, a_{h+1}, \ldots, a_n)$.

Finally we note that

(5.2)
$$1 = \sum_{j=1}^{n} \prod_{\substack{i=1 \\ i \ne j}}^{n} \frac{(x - x_j)}{(x_i - x_j)} \ ,$$

since the expression on the right is a polynomial in x of degree at most $n - 1$ that assumes the value 1 for each of $x = x_1, x_2, \ldots, x_n$ which is possible only if it is identically 1 . Setting $x = 0$ in (5.2), we see that

(5.3)
$$1 = \sum_{j=1}^{n} \prod_{\substack{i=1 \\ i \ne j}}^{n} \frac{1}{1 - \frac{x_i}{x_j}} \ .$$

Multiplying both sides of (5.3) by $F(a_1, \ldots, a_n)$, we find that

(5.4)
$$F_n(a_1, \ldots, a_n) = \sum_{j=1}^{n} F_n(a_1, \ldots, a_j - 1, \ldots, a_n) \ ,$$

and therefore

215

$$(5.5) \qquad c_n(a_1, \ldots, a_n) = \sum_{j=1}^{n} c_n (a_1, \ldots, a_j-1, \ldots, a_n) \ .$$

The two initial conditions on the $c_n(a_1, \ldots, a_n)$ and (5.5) uniquely define these numbers; since the multinomial coefficients $(a_1+\ldots+a_n)!/$
$(a_1! \ldots a_n!)$ also satisfy the same initial conditions and recurrence we
see that $c_n(a_1, \ldots, a_n) = (a_1+\ldots+a_n)!/(a_1! \ldots a_n!)$ as asserted in
Dyson's conjecture. $\qquad\qquad\qquad\qquad\qquad\qquad\qquad\qquad \Box$

On the question of q-analogs for the Master Theorem and the Dyson
conjecture, we make the following:

<u>Conjecture.</u> The constant term in the expansion of

$$\prod_{1 \le i \ne j \le n} (x_i \, \varepsilon_{ij}/x_j)_{a_i}$$

is $(q)_{a_1+a_2+\ldots+a_n}/((q)_{a_1}(q)_{a_2}\cdots(q)_{a_n})$, where $\varepsilon_{ij} = 1$ if $i < j$ and
$\varepsilon_{ij} = q$ if $i > j$.

This conjecture is trivial if $n = 2$ and when $n = 3$ it reduces to

$$(5.6) \qquad \sum_{k=-\infty}^{\infty} (-1)^k q^{\frac{1}{2}(3k+1)} \begin{bmatrix} b+c \\ c+k \end{bmatrix}\begin{bmatrix} c+a \\ a+k \end{bmatrix}\begin{bmatrix} a+b \\ b+k \end{bmatrix} = \frac{(q)_{a+b+c}}{(q)_a(q)_b(q)_c}$$

where $\begin{bmatrix} A \\ B \end{bmatrix} = (q)_A/(q)_B(q)_{A-B}$ for $0 \le B \le A$ and $\begin{bmatrix} A \\ B \end{bmatrix} = 0$ otherwise.
Identity (5.6) is the terminating form of the q-analog of Dixon's summa-
tion, a result due in part to F. H. Jackson (1941) and completed by L.
Carlitz (1969).

For completeness we present a short proof of (5.6) that relies only
on the q-analog of Saalschütz's theorem (Bailey (1935), p. 68); which we
state in the following form:

$$(5.7) \qquad q^{k^2}\begin{bmatrix} a+c \\ a+k \end{bmatrix}\begin{bmatrix} a+b \\ b+k \end{bmatrix} = \sum_{r=0}^{b} q^{(b-r)^2}\begin{bmatrix} b-k \\ r \end{bmatrix}\begin{bmatrix} c+k \\ c-b+r \end{bmatrix}\begin{bmatrix} a+c+r \\ c+b \end{bmatrix} \ .$$

216

Hence

$$\sum_{k=-\infty}^{\infty} (-1)^k q^{\frac{1}{2}k(3k+1)} \begin{bmatrix} b+c \\ c+k \end{bmatrix} \begin{bmatrix} c+a \\ a+k \end{bmatrix} \begin{bmatrix} a+b \\ b+k \end{bmatrix}$$

$$= \sum_{k=-\infty}^{\infty} (-1)^k q^{\frac{1}{2}k(k+1)} \sum_{r=0}^{b} \frac{q^{(b-r)^2} (q)_{a+c+r}}{(q)_r (q)_{b-k+r} (q)_{b+k-r} (q)_{a-b+r}}$$

$$= \sum_{r=0}^{b} q^{(b-r)^2} \begin{bmatrix} a+c+r \\ b+c \end{bmatrix} \frac{(q)_{b+c}}{(q)_r (q)_{2b-2r} (q)_{c-b+r}} \sum_{k=-\infty}^{\infty} (-1)^k q^{\frac{1}{2}k(k+1)} \begin{bmatrix} 2b-2r \\ b+k-r \end{bmatrix}$$

$$= \sum_{r=0}^{b} (-1)^{b-r} q^{(b-r)^2 + \frac{1}{2}(r+1-b)(r-b)} \begin{bmatrix} a+c+r \\ b+c \end{bmatrix} \frac{(q)_{b+c} (q^{r+1-b})_{2b-2r}}{(q)_r (q)_{2b-2r} (q)_{c-b+r}} \quad .$$

Now for $0 \leqq r < b$, $(q^{r+1-b})_{2b-2r}$ contains $1 - q^0 = 0$ as a factor and so vanishes. Hence the only nonvanishing term in the above sum occurs at $r = b$. Therefore

$$\sum_{k=-\infty}^{\infty} (-1)^k q^{\frac{1}{2}k(3k+1)} \begin{bmatrix} b+c \\ c+k \end{bmatrix} \begin{bmatrix} c+a \\ a+k \end{bmatrix} \begin{bmatrix} a+b \\ b+k \end{bmatrix}$$

$$= \begin{bmatrix} a+c+b \\ b+c \end{bmatrix} \frac{(q)_{b+c}}{(q)_b (q)_c} = \frac{(q)_{a+b+c}}{(q)_a (q)_b (q)_c} \quad ,$$

which proves (5.6).

If our conjecture is true in general it would suggest strongly the importance of this type of theorem in the general theory of basic hypergeometric functions.

217

6. Saalschützian Series and Inversion Theorems.

We turn now to a topic that is related to our work in Section 2.

PROBLEM 6. What q-inversion theorems imply or are closely related to partition identities?

Göllnitz (1967) proved the following partition theorem:

Theorem 6. Let $A(n)$ denote the number of partitions of n into parts $\equiv 2, 5$, or $11 \pmod{12}$. Let $B(n)$ denote the number of partitions of n into parts $\equiv 2, 4, 5 \pmod 6$. Let $C(n)$ denote the number of partitions of n of the form $n = b_1 + b_2 + \ldots + b_s$, where $b_i - b_{i+1} \geq 6$ with strict inequality if $b_i \equiv 0, 1, 3 \pmod 6$ and $b_s \neq 1, 3$. Then $A(n) = B(n) = C(n)$.

Göllnitz proved this theorem using a very intricate arithmetic argument patterned after Gleissberg's (1928) proof of Schur's (1926) theorem (Theorem 2 above).

Subsequently (Andrews (1969)) the following analytic theorem was proved which implies Theorem 6.

Theorem 7. If Λ_n is defined by

$$\prod_{n=0}^{\infty} (1+xq^{6n+2})(1+xq^{6n+4})(1+xq^{6n+5})(1-xq^{6n})^{-1} = \sum_{n=0}^{\infty} \frac{\Lambda_n x^n}{\prod_{j=1}^{n} (1-q^{6j})} ,$$

and if

$$d_m = 1 + \sum_{n=1}^{\infty} C_m(n) q^n$$

where $C_m(n)$ is the number of those partitions enumerated by $C(n)$ with the added restriction that each part is $\leq m$, then

$$(6.1) \qquad d_{6n-1} = \sum_{0 \leq 2j \leq n} q^{6nj - 3j^2 + 2j} {\begin{bmatrix} n-j \\ j \end{bmatrix}}_6 \Lambda_{n-2j} ,$$

where

$$[{}_n^m]_r = \begin{cases} \dfrac{(1-q^{rm})(1-q^{r(m-1)})\ldots(1-q^{r(m-n+1)})}{(1-q^{rn})(1-q^{r(n-1)})\ldots(1-q^r)}, & 0 \leq n \leq m, \\[20pt] 0 & \text{otherwise.} \end{cases}$$

H. Gould (1962), Gould and Hsu (1973) and L. Carlitz (1963b), (1973) have extensively studied inversion theorems clearly related to (6.1). In particular the methods of Carlitz (1963b), (1973) may be used to prove.

Theorem 8. For $\lambda \geq 2$ each of the following identities implies the other

$$(6.2) \quad v_n = \sum_{0 \leq \lambda s \leq n} \frac{(-1)^s q^{(\lambda-1)s(s-1)/2+(3-\lambda)s}(q)_{n-\lambda s+s} u_{n-\lambda s}}{(q^{\lambda-1};q^{\lambda-1})_s (q)_{n-\lambda s}} \quad \text{for all } n \geq 0 \text{ ;}$$

$$(6.3) \quad u_n = \sum_{0 \leq \lambda s \leq n} \frac{q^{(3-\lambda)s}(q)_n (1-q^{n-\lambda s+1}) v_{n-\lambda s}}{(q^{\lambda-1};q^{\lambda-1})_s (q)_{n-s+1}} \quad \text{for all } n \geq 0 \text{ .}$$

When $\lambda = 2$, it is easy to show (Andrews (1969)) that (6.1) is a special case of (6.2). A natural follow up on Problem 6 is: Are there other cases of Theorem 8 related to partition theorems besides the case $\lambda = 2$? We should remark that one of Schur's partition theorems (Theorem 2 above, Schur (1926)) may be proved (Andrews (1967a), Theorem 5) using an identity that is precisely q-binomial inversion (see Rota and Goldman (1970), eq. (5), p. 244).

Concerning Theorem 7, it has recently been shown that

(6.4) $$\sum_{n \geq 0} d_{6n-1} x^n = \frac{(q^6;q^6)_\infty (-xq^2;q^6)_\infty (-xq^4;q^6)_\infty (-xq^5;q^6)_\infty}{(x;q^6)_\infty}$$

$$_5\phi_4 \left[\begin{array}{c} x, ixq^{-1/2}, -ixq^{-1/2}, ixq^{5/2}, -ixq^{5/2};q^6, q^6 \\ -x^2 q^{-1}, -xq^2, -xq^4, -xq^5 \end{array} \right]$$

The series here is a nonterminating Saalschützian series since the product of the denominator parameters divided by the product of the numerator parameters is q^6, the base. From the above identity Göllnitz's theorem follows immediately since

$$\sum_{n \geq 0} C(n)q^n = \lim_{n \to \infty} d_{6n-1} = \lim_{x \to 1} (1-x) \sum d_{6n-1} x^n$$

$$= (-q^2;q^6)_\infty (-q^4;q^6)_\infty (-q^5;q^6)_\infty \qquad \text{(by (6.4))}$$

$$= \sum_{n \geq 0} B(n)q^n \ .$$

The fact that $A(n) = B(n)$ is a consequence of a simple manipulation of the above infinite product.

PROBLEM 7. Are there other partition identities related to certain nonterminating Saalschützian series like that in (6.4)?

7. Conclusion. I hope this paper has presented topics in basic hypergeometric functions of some interest and has raised some problems worthy of further research. It has been my desire to show that there is a fruitful interaction between ordinary and basic hypergeometric functions while at the same time demonstrating that basic series possess a special character and have particular applications which clearly make them not

just a subtopic or curious generalization in the study of hypergeometric functions.

In ABHF applications have been considered at length. Also a short conference "Eulerian Series and Applications" (Andrews et al. (1974)) further explored basic hypergeometric functions and considered a number of problems that are not covered in this paper. These works should be consulted for further information and references on basic hypergeometric functions.

REFERENCES

W. A. Al-Salam (1965) Saalschützian theorems for basic double series, J. London Math. Soc., 40, 455-458.

G. E. Andrews (1966) An analytic proof of the Rogers-Ramanujan-Gordon identities, Amer. J. Math., 88, 844-846.

G. E. Andrews (1967a) On Schur's second partition theorem, Glasgow Math. J., 8, 127-132.

G. E. Andrews (1967b) A generalization of the Göllnitz-Gordon partition theorems, Proc. Amer. Math. Soc., 18, 945-952.

G. E. Andrews (1968a) On partition functions related to Schur's second partition theorem, Proc. Amer. Math. Soc., 19, 441-444.

G. E. Andrews (1968b) On q-difference equations for certain well-poised basic hypergeometric series, Quart. J. Math., 19, 433-447.

G. E. Andrews (1969) On a partition theorem of Göllnitz and related formulae, J. für die reine und angew. Math., 236, 37-42.

G. E. Andrews (1971a) A new property of partitions with applications to the Rogers-Ramanujan identities, J. Comb. Th., 10, 266-270.

G. E. Andrews (1971b) Number Theory, W. B. Saunders, Philadelphia.

G. E. Andrews (1972a) Partition identities, Advances in Math., 9, 10-51.

G. E. Andrews (1972b) Summations and transformations for basic Appell series, J. London Math. Soc. (2), 4, 618-622.

G. E. Andrews (1974a) (ABHF) Applications of basic hypergeometric functions, S. I. A. M. Review, 16, 441-484.

G. E. Andrews (1974b) An analytic generalization of the Rogers-Ramanujan identities for odd moduli, Proc. Nat. Acad. Sci., $\underline{71}$, 4082-4085.

G. E. Andrews (1974c) A general theory of identities of the Rogers-Ramanujan type, Bull. Amer. Math. Soc., $\underline{80}$, 1033-1952.

G. E. Andrews, R. Askey, L. Carlitz, N. J. Fine, H. W. Gould, K. W. Kadell, D. P. Roselle, and A. Verma (1974) Eulerian Series and Applications, Abstracts from the Conference at the Pennsylvania State University, May, 1974.

P. Appell and J. Kampé de Fériet (1926) Fonctions hypergeometriques et hypersphériques, Gauthier Villars, Paris.

W. N. Bailey (1935) Generalized Hypergeometric Series, Cambridge University Press (Reprinted: Hafner, New York, 1964).

L. Carlitz (1963a) A Saalschützian theorem for double series, J. London Math. Soc., $\underline{38}$, 415-418.

L. Carlitz (1963b) Some inversion formulas, Rend. Circ. Mat. Palermo (2), $\underline{12}$, 1-17.

L. Carlitz (1967a) Summation of a double hypergeometric series, Le Matematiche, $\underline{22}$, 138-142.

L. Carlitz (1967b) A summation theorem for double hypergeometric series, Rend. Sem. Mat. Padova, $\underline{37}$, 230-233.

L. Carlitz (1969) Some formulas of F. H. Hackson, Monat. für Math., $\underline{73}$, 193-198.

L. Carlitz (1970) Some applications of Saalschutz's theorem, Rend. Sem. Mat. Padova, $\underline{44}$, 91-95.

L. Carlitz (1973) Some inverse relations, Duke Math. J., $\underline{40}$, 893-901.

L. Carlitz and M. V. Subbarao (1972a) On a combinatorial identity of Winquist and its generalization, Duke Math. J., $\underline{39}$, 165-172.

L. Carlitz and M. V. Subbarao (1972b) On the quituple product identity, Proc. Amer. Math. Soc., $\underline{32}$, 42-44.

B. C. Carlson (1974) Inequalities for Jacobi polynomials and Dirichlet averages, S. I. A. M. J. Math. Anal., $\underline{5}$, 586-596.

A. C. Dixon (1891) On the sum of the cubes of the coefficients in a certain expansion by the binomial theorem, Messenger of Math., 20, 79-80.

A. C. Dixon (1903) Summation of a certain series, Proc. London Math. Soc. (1), 35, 285-289.

F. J. Dyson (1962) Statistical theory of the energy levels of complex systems I, J. Math. Phys., 3, 140-156.

W. Gleissberg (1928) Über einen Satz von Herrn I. Schur, Math. Zeit., 28, 372-382.

H. Göllnitz (1967) Partitionen mit Differenzenbedingungen, J. für die reine und angew. Math., 225, 154-190.

I. J. Good (1962) Proofs of some 'binomial' identities by means of MacMahon's master theorem, Proc. Camb. Phil. Soc., 58, 161-162.

I. J. Good (1970) Short proof of a conjecture of Dyson, J. Math. Phys., 11, 1884.

B. Gordon (1961) A combinatorial generalization of the Rogers-Ramanujan identities, Amer. J. Math., 83, 393-399.

H. W. Gould (1962) A new convolution formula and some new orthogonal relations for inversion of series, Duke Math. J., 29, 393-404.

H. W. Gould and L. C. Hsu (1973) Some new inverse series relations, Duke Math. J., 40, 885-891.

J. Gunson (1962) Proof of a conjecture by Dyson in the statistical theory of energy levels, J. Math. Phys., 3, 752-753.

N. Hall (1936) An algebraic identity, J. London Math. Soc., 11, 276.

W. J. Holman, L. C. Biedenharn and J. D. Loucke (1975) On hypergeometric series well-poised in SU(n), SIAM J. Math. Anal. (to appear).

F. H. Jackson (1941) Certain q-identities, Quart. J. Math., 12, 167-172.

P. A. MacMahon (1894) A certain class of generating functions in the theory of numbers, Philos. Trans., 185, 111-160.

P. A. MacMahon (1902), The sums of powers of the binomial coefficients, Quart. J. Math., 33, 274-288.

P. A. MacMahon (1915) Combinatory Analysis, Vol. 1, Cambridge University Press, Cambridge (Reprinted: Chelsea, New York, 1960).

J. K. Percus (1971) Combinatorial Methods, Springer, New York.

G. C. Rota and J. Goldman (1970) On the foundations of combinatorial theory IV: Finite vector spaces and Eulerian generating functions, Studies in Appl. Math., 49, 239-258.

I. J. Schur (1926) Zur additiven Zahlentheorie, Sitzungsber. Akad. Wissensch. Berlin, Phys.-Math. Klasse, 488-495.

A. Selberg (1936) Über einige arithmetische Identitäten, Avh. Norsk. Vidensk. Akad. Oslo Mat. Naturvidensk. Kl. 8.

V. N. Singh (1957) Certain generalized hypergeometric identities of the Rogers-Ramanujan type, Pacific J. Math., 7, 1011-1014.

L. J. Slater (1952) Further identities of the Rogers-Ramanujan type, Proc. London Math. Soc. (2), 54, 147-167.

L. J. Slater (1966) Generalized Hypergeometric Functions, Cambridge University Press, Cambridge.

G. N. Watson (1929) A new proof of the Rogers-Ramanujan identities, J. London Math. Soc., 4, 4-9.

F. J. W. Whipple (1926) On well-poised series, generalized hypergeometric series having parameters in pairs, each pair with the same sum, Proc. London Math. Soc. (2), 24, 247-263.

K. Wilson (1962) Proof of a conjecture by Dyson, J. Math. Phys., 3, 1040-1043.

L. Winquist (1969) An elementary proof of $p(11m+6) \equiv 0 \pmod{11}$, J. Comb. Th., 6, 56-59.

Partially supported by National Science Foundation Grant MSP 74-97282.

An Introduction to Association Schemes and
Coding Theory
N. J. A. Sloane

ABSTRACT

Association schemes originated in statistics, but have recently
been used in coding theory and combinatorics by Delsarte, McEliece and
others to obtain strong upper bounds on the size of codes and other combinatorial objects, and to characterize those objects (such as perfect
codes) which meet these bounds. A central role is played by the eigenvalues of the association scheme, which in many cases come from a
family of orthogonal polynomials. In the most important case these are
the Krawtchouk polynomials. This paper gives an introduction to association schemes and the way they are used in coding theory and combinatorics.

§1 INTRODUCTION

Association schemes were first introduced by statisticians in connection with the design of experiements [6], [7], [39], [61], and have
since proved very useful in the study of permutation groups [9], [33]-[38]
and graphs [4], [10]. Recently, starting with the work of Delsarte [11]-
[15], association schemes have been applied with considerable success
in coding theory and in other combinatorial problems [16]-[21], [29].

One of the interesting features of this work is that to most of these
association schemes there corresponds one or more families of orthogonal
polynomials. For the Hamming association scheme, the most important

225

scheme for coding theory, these are the Krawtchouk polynomials $K_k(x;n)$, $k = 0, \ldots, n$, which are orthogonal on the set $\{0, 1, \ldots, n\}$ with respect to the weighting function $w(i) = \binom{n}{i}$.

There are two kinds of problems which are considered. The first is to find upper bounds on the size of a subset of an association scheme having certain desirable properties (a code, a t-design, an orthogonal array, etc). The second is to characterize those subsets which meet the bounds. The second problem includes the famous problem of finding all perfect codes - see §§2, 8 below.

The key result is a theorem published by Delsarte in 1972 [11] which states that certain linear combinations of the parameters of the subset must be nonnegative (Th 8 below). This theorem was independently discovered by McEliece, Rodemich, Rumsey and Welch [55] in the special case of error-correcting codes. Because of this result, the first problem can be stated as a linear programming problem. By converting to the dual problem, and using properties of Krawtchouk polynomials, Delsarte [11], [14] and McEliece et al. [55], [56] have obtained very good upper bounds on the size of codes – see §§2, 7 below. The same technique has been applied to other combinatorial problems including:

(i)   Codes of constant weight [14], [15]

(ii) t-designs and orthogonal arrays [14], [15]

(iii) Families of lines with a prescribed number of angles between them [21]

(iv) Bilinear and alternating bilinear forms over $GF(q)$ [19], [20]. This gives a natural setting for subcodes of the second-order Reed Muller codes, including Kerdock codes. (See also [27].)

The linear programming approach is also useful in the second problem. For example, in this way Delsarte [11], [15] proved Lloyd's theorem (Cor. 15 below) which states that if a perfect error-correcting code exists then a certain Krawtchouk polynomial must have distinct integer zeros in a certain interval.

This paper assumes no previous knowledge of coding theory or association schemes. Most of the results can be found in Delsarte [11]-[15]. Indeed an appropriate subtitle would be "an introduction to Delsarte's work".

§2 Error-Correcting Codes

Error-correcting codes were invented to transmit data more accurately. Suppose there is a telegraph wire from Madison to New York down which 0's and 1's can be sent. Usually when a 0 is sent it is received as a 0, but occasionally a 0 will be received as a 1, or a 1 as a 0 .

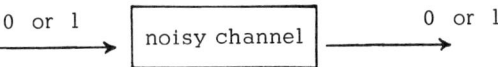

0 or 1 ⟶ noisy channel ⟶ 0 or 1

There are a lot of important messages to be sent down this wire, and they must be sent as quickly and reliably as possible. The messages are already written as strings of 0's and 1's — perhaps they are being produced by a computer.

The method we shall use is to encode the messages into codewords. Only codewords will be sent down the channel.

An example will make this clear. Suppose only two messages are to be sent, e.g. "Yes" or "No" . "Yes" will be encoded as 00000 and "No" as 11111 . This is a code with two codewords.

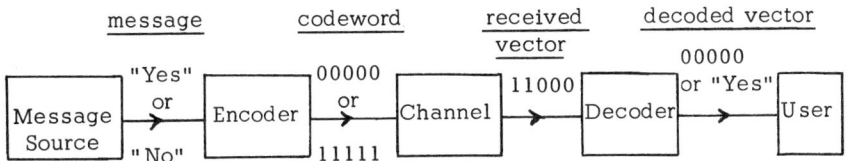

message — codeword — received vector — decoded vector

Message Source → "Yes" or "No" → Encoder → 00000 or 11111 → Channel → 11000 → Decoder → 00000 or "Yes" → User

Suppose 11000 is received at the far end. The decoder argues that 00000 was more likely to have been sent than 11111, because 11000 is somehow closer to 00000 than to 11111 .

To make this precise we define the Hamming distance dist(u, v) between two vectors $u = u_1 \ldots u_n$ and $v = v_1 \ldots v_n$ to be the number of places where they differ. Thus dist(00000, 11000) = 2,   dist(11111, 11000) = 3 .

It is easy to check that this is a metric.

Then the decoder's strategy is to decode the received vector as the closest codeword (in Hamming distance). This is because a digit is more likely to be correct than in error.

Notice that in this example the decoder was able to correct two errors. This is because the Hamming distance between the codewords 00000 and 11111 is 5 . The error correction procedure is perhaps best explained by a picture. Suppose we have a code in which any two distinct codewords differ in at least 3 places.

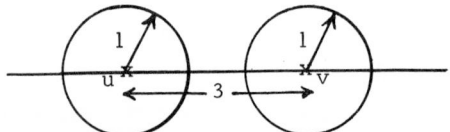

x = codewords

Then this code can correct one error. If the codeword u is transmitted and at most one error occurs, the received vector will still be within the "sphere" of radius 1 about u, and (by the triangle inequality) is closer to u than to any other codeword v . The "spheres" of radius 1 about each codeword are disjoint.

The same argument shows that a code in which any two distinct codewords have Hamming distance at least d apart can correct $\left\lfloor \frac{d-1}{2} \right\rfloor$ errors. The "spheres" of Hamming radius $\left\lfloor \frac{d-1}{2} \right\rfloor$ around the codewords are disjoint.

This motivates our definition. An (n, M, d) error-correcting code is a set of M vectors $u = u_1 \ldots u_n$ of 0's and 1's of length n (called codewords) such that any two distinct codewords differ in at least d places. n is called the length of the code, M is the size and d is the minimum distance. This is a $[\frac{1}{2}(d-1)]$-error-correcting code.

228

Examples of Codes

    1. The preceding example $\{00000, 11111\}$ is a $(5, 2, 5)$ code.

    2. A $(3, 4, 2)$ code: $\{000, 011, 101, 110\}$. More generally, the $(n, 2^{n-1}, 2)$ <u>even weight</u> code consists of all vectors of even Hamming weight, where the <u>Hamming weight</u> $\text{wt}(u)$ of a vector $u = u_1 \cdots u_n$ is the number of nonzero u_i. Clearly

$$\text{dist}(u, v) = \text{wt}(u - v) \ .$$

    3. The Hamming $(7, 16, 3)$ code:

0000000	1111111
1101000	0010111
0110100	1001011
0011010	1100101
0001101	1110010
1000110	0111001
0100011	1011100
1010001	0101110

The reader will recognize codewords 2 through 8 as forming the incidence matrix of

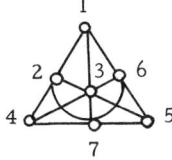

the projective plane of order 2.

    These examples are all <u>linear</u> codes. That is, the sum (taken componentwise, modulo 2) of two codewords is again a codeword. They are also <u>cyclic</u>: if $u_1 \cdots u_n$ is a codeword so is $u_2 \cdots u_n u_1$.

    The $(7, 16, 3)$ code has another nice property: it is perfect. An e-error-correcting code is called <u>perfect</u> if every binary vector of length n is within Hamming distance e of some codeword.

    To state this another way, let $Q_n = \{u_1 \cdots u_n : u_i = 0 \text{ or } 1\}$ be the set of all binary vectors of length n, i.e. the vertices of the unit n-cube.

An (n, M, d) code C is a subset of Q_n of size M, with the property that the spheres

$$S_e(v) = \{u \epsilon Q_n : \text{dist}(u, v) \leq e\}, \quad v \epsilon C \ ,$$

of radius $e = [\frac{1}{2}(d-1)]$ about the codewords are disjoint. In a perfect code these spheres include all the points of Q_n: they are both a packing and a covering of Q_n . We shall return to perfect codes in §8.

The Coding Theory Problem

In a good code n is small (for fast transmission), M is large (for efficiency), and d is large (to correct many errors).

The first thing one wants to know is how large M can be, for given values of n and d . Upper and lower bounds on the largest M (when n is large) due to Elias, Gilbert and Varshamov have been known for some time (see [1, Ch 13], [59, Ch 4]). Recently Levenshtein [47] and Sidel'nikov [64] have given small improvements on the Elias upper bound. One of the goals of this paper is to describe a new technique for obtaining upper bounds, discovered independently by Delsarte [11] and McEliece, Rodemich, Rumsey and Welch [55], and known as the linear programming method (see §7). Welch, McEliece and Rumsey [69] have recently used this method to improve the Elias bound when d/n is close to $\frac{1}{2}$. Finally, while this paper was being written, McEliece, Rodemich, Rumsey and Welch [56] have announced the following bound: if C is any (n, M, d) code, then

$$\frac{\log_2 M}{n} \lesssim H_2 \left(\frac{1}{2} - \sqrt{\frac{d}{n}\left(1 - \frac{d}{n}\right)} \right), \quad \text{as } n \to \infty \ ,$$

where $H_2(x) = -x \log_2 x - (1-x)\log_2(1-x)$. This is a considerable improvement on the Elias bound over most of the range of d/n, and was obtained by using the linear programming method together with asymptotic properties of Krawtchouk polynomials. One expects that further upper bounds will be obtained using this method. However, the true upper bound probably coincides with the Gilbert-Varshamov lower bound, and a proof of this seems a long way off.

For small values of n the story is similar. Upper and lower bounds are known [32], [41], [52], [67], and again the linear programming method has recently been used [55] to improve the upper bound in many cases.

The second main problem in coding theory is of course to find codes which come close to these bounds. This problem is still essentially unsolved, although a lot of constructions are known [1], [43], [48], [52], [59].

It is interesting to list a few of the best known codes and the mathematical techniques used to construct them:

Reed-Muller codes (Boolean functions, 1954),
Bose-Chaudhuri-Hocquenghem codes (Galois fields, 1959),
Quadratic residue codes (number theory, around 1960),
Geometry codes (finite geometries, 1967),
Goppa codes (alternating determinants, 1970 [30], [2], [31], [52]).

The latest technique to be introduced is that of association schemes (in 1973, by Delsarte [12]). This has led to the powerful linear programming bounds just mentioned, and to the other applications mentioned in the introduction.

§3 Association Schemes

This section gives some of the basic theory, mostly following Delsarte [14], and then the next three §'s contain examples.

Definition (Bose and Shimamoto [7]) An association scheme with n classes consists of a finite set X of v points, together with n + 1 symmetric relations $R_0, R_1, \ldots, R_n$ defined on X which satisfy

(i) For every x, y $\in$ X, (x, y) $\in$ R_i for exactly one i .

(ii) R_0 = {(x, x): x $\in$ X} is the identity relation.

(iii) If (x, y) $\in$ R_k, the number of z $\in$ X such that (x, z) $\in$ R_i and (y, z) $\in$ R_j is a constant $p_{i,j,k}$ depending on i, j, k but not on the particular choice of x and y .

231

Two points x and y are called <u>ith associates</u> if $(x, y) \in R_i$. In words, the definition states that if x and y are ith associates so are y and x; every pair of points are ith associates for exactly one i ; each point is its own zeroth associate while distinct points are never zeroth associates; and finally if x and y are kth associates then the number of points z which are both ith associates of x and jth associates of y is a constant p_{ijk} .

An association scheme can be described by a complete graph having v nodes (corresponding to the points of X), in which the edge joining nodes x and y is labeled by i if x and y are ith associates. The numbers p_{ijk} are called the <u>intersection numbers</u> of the scheme.

p_{ijk} is the number of triangles

in the graph which have a fixed base

The number of ith associates of any point x is

$$p_{ii0} = v_i \quad (\text{say}) ,$$

the <u>valency</u> of the ith relation. The p_{ijk} must satisfy the following identities:

$$p_{ijk} = p_{jik} , \tag{1}$$

$$p_{0jk} = \delta_{jk}, \quad v_k \, p_{ijk} = v_i \, p_{kji} ,$$

$$\sum_{j=0}^{n} p_{ijk} = v_i ,$$

$$\sum_{m=0}^{n} p_{ijm} \, p_{mk\ell} = \sum_{h=0}^{n} p_{ih\ell} \, p_{jkh} . \tag{2}$$

The last identity follows from counting the quadrilaterals

232

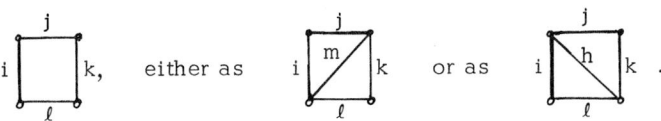

Let D_i be the <u>adjacency matrix</u> of R_i $(i = 0, \ldots, n)$. That is, D_i is the $v \times v$ matrix with rows and columns labeled by the points of X, defined by

$$(D_i)_{x,y} = \begin{cases} 1 & \text{if } (x,y) \in R_i \text{ ,} \\ 0 & \text{otherwise.} \end{cases}$$

The definition of an association scheme is equivalent to saying that the D_i are symmetric $(0,1)$-matrices which satisfy

(i) $\sum_{i=0}^{n} D_i = J$ (the all-ones matrix),   (3)

(ii) $D_0 = I$,

(iii) $D_i D_j = \sum_{k=0}^{n} p_{ijk} D_k$, $i, j = 0, \ldots, n$.   (4)

Indeed the $(x,y)^{\text{th}}$ entry of the left side of Eq. (4) is the number of paths $x \overset{i}{\circ}\!\!-\!\!\overset{}{\circ}\!\!-\!\!\overset{j}{\circ} y$ in the graph. Also

$$D_i J = J D_i = v_i J \text{ .}$$ (5)

The Bose-Mesner Algebra

Let $\mathcal{A}$ be the vector space consisting of all matrices of the form

$$\sum_{i=0}^{n} c_i D_i, \quad \text{where the } c_i \text{ are real.}$$

233

All the matrices in $\mathcal{A}$ are symmetric. Equation (3) implies that $D_0, \ldots,$ D_n are linearly independent, and the dimension of $\mathcal{A}$ is $n+1$. Furthermore Eq. (4) implies that $\mathcal{A}$ is closed under multiplication, so $\mathcal{A}$ is an algebra. Multiplication is commutative, from Eq. (1), and associative, since $\mathcal{A}$ is an algebra of matrices and matrix multiplication is associative. (Alternatively, associativity follows from Eq. (2).) We call this associative, commutative algebra $\mathcal{A}$ the <u>Bose-Mesner algebra</u> of the association scheme, after Ref. [6].

Since the matrices in $\mathcal{A}$ are symmetric and commute with each other, they can be simultaneously diagonalized ([54, p. 77]). I. e. there is a matrix S such that to each $A \in \mathcal{A}$ there is a diagonal matrix Λ_A with

$$S^{-1} AS = \Lambda_A . \tag{5'}$$

Therefore $\mathcal{A}$ is semisimple and has a unique basis of primitive idempotents $J_0, \ldots, J_n$. These are real $n \times n$ matrices satisfying (see [58], [8], [68a])

$$J_i^2 = J_i, \qquad i = 0, \ldots, n , \tag{6a}$$

$$J_i J_k = 0, \qquad i \neq k , \tag{6b}$$

$$\sum_{i=0}^{n} J_i = I . \tag{6c}$$

From Eqs. (3), (5), $\frac{1}{v} J$ is a primitive idempotent, so we shall choose

$$J_0 = \frac{1}{v} J .$$

Let D_k be expressed in terms of the basis $J_0, \ldots, J_n$ by

$$D_k = \sum_{i=0}^{n} P_k(i) J_i, \qquad k = 0, \ldots, n , \tag{7}$$

for some uniquely determined real numbers $P_k(i)$. Equations (6), (7) imply

$$D_k J_i = P_k(i) J_i \ .\tag{8}$$

Therefore the $P_k(i)$, $i = 0, \ldots, n$, are the eigenvalues of D_k . Also the columns of the J_i span the common eigenspaces of all the matrices in $\mathcal{Q}$. Let $\mu_i = \text{rank } J_i$ be the <u>multiplicity</u> of the ith eigenspace.

Conversely, to express the J_k in terms of the D_i, let P be the real $(n+1) \times (n+1)$ matrix

$$P = \begin{bmatrix} P_0(0) & P_1(0) & \cdots & P_n(0) \\ P_0(1) & P_1(1) & \cdots & P_n(1) \\ & \cdots & & \cdots \\ P_0(n) & P_1(n) & \cdots & P_n(n) \end{bmatrix}\tag{9}$$

and let

$$Q = v\, P^{-1} = \begin{bmatrix} Q_0(0) & Q_1(0) & \cdots & Q_n(0) \\ Q_0(1) & Q_1(1) & \cdots & Q_n(1) \\ & \cdots & & \cdots \\ Q_0(n) & Q_1(n) & \cdots & Q_n(n) \end{bmatrix} \quad (\text{say}).\tag{10}$$

We call P and Q the <u>eigenmatrices</u> of the association scheme. Then

$$J_k = \frac{1}{v} \sum_{i=0}^{n} Q_k(i)\, D_i, \qquad k = 0, \ldots, n \ .\tag{11}$$

235

Lemma 1

$$P_0(i) = Q_0(i) = 1, \quad P_k(0) = v_k, \quad Q_k(0) = \mu_k .$$

Proof   Only the last equation is not immediate.   Since J_k is an idempotent the diagonal entries of $S^{-1} J_k S$ (see Eq. (5')) are 0 and 1 . Therefore

$$\text{trace } S^{-1} J_k S = \text{trace } J_k = \text{rank } J_k = \mu_k .$$

Since trace $D_i = v \, \delta_{0i}$, (11) implies $\mu_k = Q_k(0)$.   Q. E. D.

Theorem 2

The eigenvalues $P_k(i)$ and $Q_k(i)$ satisfy the following orthogonality conditions.

$$\sum_{i=0}^{n} \mu_i \, P_k(i) \, P_\ell(i) = v \, v_k \, \delta_{k\ell} , \tag{13}$$

$$\sum_{i=0}^{n} v_i \, Q_k(i) \, Q_\ell(i) = v \, \mu_k \, \delta_{k\ell} . \tag{14}$$

Also

$$\mu_j \, P_i(j) = v_i Q_j(i), \qquad i, j = 0, \ldots, n . \tag{15}$$

Proof   To prove (14), expand $J_k J_\ell = J_k \delta_{k\ell}$ in the basis $D_0, \ldots, D_n$, and equate the coefficients of D_0 . To prove (13), write (14) as $Q^T A Q = v B$, where $A = \text{diag}(v_0, \ldots, v_n)$, $B = \text{diag}(\mu_0, \ldots, \mu_n)$, and T denotes transpose. Then $v A = P^T B P$, which is (13). Finally (from (10)) $A Q = P^T B$, which proves (15).   Q. E. D.

An Isomorphic Algebra of (n+1) × (n+1) Matrices

We briefly mention that there is an algebra of (n+1) × (n+1) matrices which is isomorphic to $\mathcal{A}$, and is often easier to work with. Let

$$
L_i = \begin{bmatrix}
p_{i00} & p_{i10} & \cdots & p_{in0} \\
p_{i01} & p_{i11} & \cdots & p_{in1} \\
\cdots & & \cdots & \\
p_{i0n} & p_{i1n} & \cdots & p_{inn}
\end{bmatrix}, \quad i = 0, \ldots, n .
$$

Then Eq. (2) implies

$$
L_i L_j = \sum_{k=0}^{n} p_{ijk} L_k .
$$

Thus the L_i multiply in the same manner as the D_i. Since $p_{ik0} = \delta_{ik}$, it follows that $L_0, \ldots, L_n$ are linearly independent. Therefore the algebra $\mathcal{B}$ consisting of all matrices $\sum_{i=0}^{n} c_i L_i$ (c_i real) is an associative commutative algebra, which is isomorphic to $\mathcal{A}$ under the mapping $D_i \to L_i$.

Equating eigenvalues on both sides of Eq. (4) gives

$$
P_i(\ell) P_j(\ell) = \sum_{k=0}^{n} p_{ijk} P_k(\ell), \quad \ell = 0, \ldots, n .
$$

This implies that

$$
P L_k P^{-1} = diag(P_k(0), \ldots, P_k(n)) .
$$

Thus the $P_k(i)$, $i = 0, \ldots, n$, are also the eigenvalues of L_i. (Alternatively, since $\mathcal{A}$ and $\mathcal{B}$ are isomorphic, D_i and L_i have the same minimal polynomials and therefore the same eigenvalues.)

Further general properties of association schemes are given by Delsarte [14], [15]. But now it is time for some examples.

237

§4 The Hamming Association Scheme

The <u>Hamming</u> or <u>hypercubic</u> association scheme is the most important example for coding theory (see Delsarte [14], [15]). In this scheme
$X = Q_n$, the set of binary vectors of length n, and two vectors u,
$v \in Q_n$ are ith associates if they are Hamming distance i apart.
Plainly conditions (i), (ii) of the definition of an association scheme are
satisfied. To verify (iii), let dist(u, v) = k . Without loss of generality
we may take u = 00...0, v = 11...100...0 . We show that the number
of $w \in Q_n$ such that dist(u, w) = i,   dist(v, w) = j is a constant p_{ijk}
independent of the choice of u, v, w . Consider the figure

$$\longleftarrow k \longrightarrow$$

$$u = 000 \quad 000 \quad 000 \quad 000$$

$$v = 111 \quad 111 \quad 000 \quad 000$$

$$w = \underset{a}{111} \quad \underset{b}{000} \quad \underset{c}{111} \quad \underset{d}{000}$$

Then a + c = i,   b + c = j,   a + b = k . Hence $a = \frac{1}{2}(i - j + k)$, $c = \frac{1}{2}(i + j - k)$
and so

$$
p_{ijk} = \begin{cases}
\dbinom{k}{\dfrac{i-j+k}{2}} \dbinom{n-k}{\dfrac{i+j-k}{2}} & \text{if} \quad i+j-k \text{ is even,} \\
\\
0 & \text{if} \quad i+j-k \text{ is odd.}
\end{cases}
$$

Also $v_i = \binom{n}{i}$. The matrices in the Bose-Mesner algebra $\mathcal{A}$ are $2^n \times 2^n$
matrices, with rows and columns labeled by vectors $u \in Q_n$. In partic-
ular the $(u, v)^{th}$ entry of D_k is 1 if and only if dist(u, v) = k .
For example, if n = 2, $Q_2 = \{00, 01, 10, 11\}$,   and we label the rows

and columns by $00, 01, 10, 11$. The graph

shows that $v_0 = p_{000} = 1$, $v_1 = p_{110} = 2$, $v_2 = p_{220} = 1$, $p_{111} = 2$, $p_{112} = 1$, etc. The adjacency matrices are

$$
D_0 = \begin{bmatrix} 1 & 0 & 0 & 0 \\ 0 & 1 & 0 & 0 \\ 0 & 0 & 1 & 0 \\ 0 & 0 & 0 & 1 \end{bmatrix}, \quad
D_1 = \begin{bmatrix} 0 & 1 & 1 & 0 \\ 1 & 0 & 0 & 1 \\ 1 & 0 & 0 & 1 \\ 0 & 1 & 1 & 0 \end{bmatrix}, \quad
D_2 = \begin{bmatrix} 0 & 0 & 0 & 1 \\ 0 & 0 & 1 & 0 \\ 0 & 1 & 0 & 0 \\ 1 & 0 & 0 & 0 \end{bmatrix}.
$$

We shall need the following simple result.

Lemma 3

$$
\sum_{u \in Q_n} (-1)^{u \cdot v} = 2^n \delta_{0, v} \; ,
$$

where $u \cdot v$ denotes the real scalar product $\sum_{i=1}^{n} u_i v_i$.

Lemma 4 The primitive idempotent J_k is the matrix which has $(u, v)^{th}$ entry equal to

$$
\frac{1}{2^n} \sum_{wt(w)=k} (-1)^{(u+v) \cdot w}, \qquad k = 0, \ldots, n \; . \tag{16}
$$

Proof Let A_k be the matrix (16). We show that the $n+1$ matrices A_k satisfy (6a), (6b) and (6c), and therefore are the primitive idempotents. The $(u, w)^{th}$ entry of $A_k A_\ell$ is

$$
\frac{1}{2^{2n}} \sum_{v \in Q_n} \sum_{wt(x)=k} (-1)^{(u+v) \cdot x} \sum_{wt(y)=\ell} (-1)^{(v+w) \cdot y}
$$

$$
= \frac{1}{2^{2n}} \sum_{wt(x)=k} \sum_{wt(y)=\ell} (-1)^{u \cdot x + v \cdot y} \sum_{v \in Q_n} (-1)^{v \cdot (x+y)}
$$

239

Have also to check that these belong to B-M algebra!
See Delsarte's thesis pp. 39-40.

$$= \frac{1}{2^n} \, \delta_{k,\ell} \sum_{wt(x)=k} (-1)^{(u+v) \cdot x}, \quad \text{by Lemma 3 },$$

which is the $(u, w)^{th}$ entry of $A_k \delta_{k,\ell}$ and proves (6a) and (6a). Equation (6c) follows from Lemma 3. Q. E. D.

For example, when $n = 2$,

$$J_0 = \frac{1}{4} \begin{bmatrix} 1 & 1 & 1 & 1 \\ 1 & 1 & 1 & 1 \\ 1 & 1 & 1 & 1 \\ 1 & 1 & 1 & 1 \end{bmatrix}, \quad J_1 = \frac{1}{2} \begin{bmatrix} 1 & 0 & 0 & - \\ 0 & 1 & - & 0 \\ 0 & - & 1 & 0 \\ - & 0 & 0 & 1 \end{bmatrix}, \quad J_2 = \frac{1}{4} \begin{bmatrix} 1 & - & - & 1 \\ - & 1 & 1 & - \\ - & 1 & 1 & - \\ 1 & - & - & 1 \end{bmatrix},$$

where - stands for -1. The ranks are $\mu_0 = 1$, $\mu_1 = 2$, $\mu_2 = 1$.

The eigenvalues $P_k(i)$ will turn out to be the values of Krawtchouk polynomials.

Definition For any positive integer n, the kth Krawtchouk polynomial ([44], [68]) is defined by

$$K_k(x;n) = \sum_{j=0}^{k} (-1)^j \binom{x}{j} \binom{n-x}{k-j}, \qquad k = 0, 1, 2, \ldots, \tag{17}$$

where x is an indeterminate, and

$$\binom{x}{m} = \begin{cases} \dfrac{x(x-1)\ldots(x-m+1)}{m!} & \text{if } m \text{ is a positive integer,} \\ 1 & \text{if } m = 0, \\ 0 & \text{otherwise.} \end{cases}$$

Thus $K_k(x;n)$ is a polynomial in x of degree k. The first few Krawtchouk polynomials are

$$K_0(x;n) = 1,$$
$$K_1(x;n) = n - 2x,$$
$$K_2(x;n) = \binom{n}{2} - 2nx + 2x^2.$$

Theorem 5

If u is any vector of weight i,

$$\sum_{wt(v)=k} (-1)^{u \cdot v} = K_k(i;n) \tag{18}$$

Proof   The figure

$$\longleftarrow i \longrightarrow$$

$$
\begin{array}{llll}
u = 111 & 111 & 000 & 000 \\
v = 111 & 000 & 111 & 000 \\
\quad\ j & & k\text{-}j &
\end{array}
$$

shows that the left hand side is

$$\sum_{j=0}^{i} (-1)^j \binom{i}{j}\binom{n-i}{k-j} = K_k(i;n) \ .$$

Q. E. D.

Theorem 6   The eigenvalues of the Hamming association scheme are
$P_k(i) = Q_k(i) = K_k(i;n)$,   for $i, k = 0, \ldots, n$.

Proof   Since the expansion (7) is unique, $P_k(i) = K_k(i;n)$ will follow if we show that

$$D_k = \sum_{i=0}^{n} K_k(i;n) J_i, \qquad k = 0, \ldots, n \ .$$

This is now a routine calculation using Lemmas 3, 4 and Th. 5.   Similarly for $Q_k(i)$.

Q. E. D.

E. G.,   when $n = 2$

$$P = Q = \begin{pmatrix} 1 & 2 & 1 \\ 1 & 0 & -1 \\ 1 & -2 & 1 \end{pmatrix} .$$

241

From Th. 2 we obtain

$$\sum_{i=0}^{n} \binom{n}{i} K_k(i;n) K_\ell(i;n) = 2^n \binom{n}{k} \delta_{k,\ell} . \tag{19}$$

Thus the Krawtchouk polynomials $K_k(x;n)$, $k = 0, \ldots, n$, are an orthogonal family on the set $\{0, 1, \ldots, n\}$, with respect to the weighting function $w(i) = \binom{n}{i}$.

A generating function for these polynomials is

$$(1+z)^{n-x}(1-z)^x = \sum_{k=0}^{\infty} K_k(x;n)z^k . \tag{20}$$

If i is an integer in the range $0 \le i \le n$ this becomes

$$(1+z)^{n-i}(1-z)^i = \sum_{k=0}^{n} K_k(i;n)z^k . \tag{21}$$

From (20) it is not difficult to obtain the three-term recurrence

$$(k+1)K_{k+1}(x;n) = (n-2x) K_k(x;n) - (n-k+1) K_{k-1}(x;n) \tag{22}$$

for $k = 1, 2, \ldots$, and the alternative expressions

$$K_k(x;n) = \sum_{j=0}^{k} (-2)^j \binom{x}{j}\binom{n-j}{k-j} ,$$

$$= \sum_{j=0}^{k} (-1)^j 2^{k-j} \binom{n-k+j}{j}\binom{n-x}{k-j} \tag{23}$$

Thus $K_k(x;n)$ is a polynomial in x of degree k, with leading coefficient $(-2)^k/k!$ and constant term $\binom{n}{k}$. Some other useful formulae are

$$\binom{n}{i} K_k(i;n) = \binom{n}{k} K_i(k;n) , \tag{24}$$

$$\sum_{i=0}^{n} K_k(i;n) K_i(\ell;n) = 2^n \delta_{k,\ell} \ , \tag{25}$$

$$\sum_{i=0}^{k} K_i(x;n) = K_k(x-1;n-1), \qquad k = 0, 1, \ldots \tag{26}$$

$$\sum_{i=0}^{n} \binom{n-i}{n-j} K_i(x;n) = 2^j \binom{n-x}{j}, \qquad j = 0, \ldots, n \ . \tag{27}$$

The usual classical definition of these polynomials is a bit more general than this (see [44], [45], [68], [24], [25]). Other generalizations have been used in coding theory (see [14]-[17], [53]).

§5 The Johnson Association Scheme

Upper bounds on the size of codes with a prescribed minimum distance and in which all codewords have the same weight have been given by Levenshtein [46] and Johnson [40]-[42]. Not only are these bounds important in themselves, but they often lead to improved bounds on unrestricted codes, see [40]-[42], [14, Th 3.7]. The appropriate association scheme is what we shall call the Johnson scheme, following Delsarte. (In the statistical literature this is called a triangular association scheme [57], [71]). As will be seen in §7, this scheme also has applications to t-designs.

Let V and n be fixed integers with $0 \le n \le \frac{1}{2} V$. In the Johnson scheme, X consists of all $\binom{V}{n}$ binary vectors of length V and weight n. Two vectors u, v are ith associates if $\text{dist}(u, v) = 2i$, for $i = 0, 1, \ldots, n$.

The figure

shows that this is an association scheme, with intersection numbers

$$p_{ijk} = \sum_{s=0}^{n-k} \binom{k}{n-i-s}\binom{k}{n-j-s}\binom{n-k}{s}\binom{V-n-k}{i+j+s-n} \ ,$$

and valencies

$$v_i = \binom{n}{i}\binom{V-n}{i} \ .$$

It can be shown that the eigenvalues are given by $P_k(i) = E_i(i;V,n)$, where $E_k(x;V,n)$ is the Eberlein polynomial ([26], [14], [15], [18]) defined by

$$E_k(x;V,n) = \sum_{j=0}^{k} (-1)^{k-j}\binom{n-j}{k-j}\binom{n-x}{j}\binom{V-n+j+x}{j} \ . \tag{28}$$

In fact $E_k(x;V,n)$ is a polynomial $\Phi_k(z;V,n)$, say, of degree k in the indeterminate $z = x(V+1-x)$.

Theorem 2 implies that the Eberlein polynomials $\Phi_k(z;V,n)$ are an orthogonal family on the set $\{z_i = i(V+1-i): i = 0, \ldots, n\}$ with respect to the weighting function $w(z_i) = \binom{V}{i} - \binom{V}{i-1}$.

The eigenvalues $Q_k(i) = Q_k(i;V,n)$ are also obtained from orthogonal polynomials $Q_k(x;V,n)$, where (see [14])

$$Q_0(x;V,n) = 1 \ ,$$

$$Q_1(x;V,n) = \frac{V-1}{n(V-n)} \{n(V-n)-Vx\} \ , \tag{29}$$

$Q_k(x;V,n)$ is a polynomial in x of degree k .

§6 Association Schemes Obtained from Graphs and Other Sources

Let Γ be a connected graph with v nodes, containing no loops or multiple edges. Let X be the set of nodes of Γ . The distance

244

$\rho(x, y)$ between nodes x and y is defined to be the number of edges on the shortest path joining them. The maximum distance n (say) between any two nodes is called the diameter of Γ .

Definition The graph Γ is called metrically regular [23] (or perfectly regular [36], or distance-regular [4], [10]) if the following condition is satisfied. For any pair of nodes x, y with $\rho(x, y) = k$, the number of nodes z such that $\rho(x, z) = i$ and $\rho(y, z) = j$ is a constant p_{ijk} depending on i, j, k but not on the particular choice of x and y .

Clearly we may obtain an association scheme with n classes from the nodes X of a metrically regular graph by calling x and y ith associates if $\rho(x, y) = i$, $i = 0, \ldots, n$. This example explains why the p_{ikj} are called intersection numbers: let $\Gamma_i(x) = \{y \in X : \rho(x, y) = i\}$ be the nodes at distance i from x . Then if $\rho(x, y) = k$,

$$p_{ijk} = \left| \Gamma_i(x) \cap \Gamma_j(y) \right| .$$

Association schemes that can be obtained from graphs in this way are called metric schemes. To construct the graph from the scheme, one defines x and y to be adjacent if and only if $(x, y) \in R_1$.

Examples of Metrically Regular Graphs

1. The Hamming and Johnson Schemes are metric schemes.

2. Metrically regular graphs of diameter 2 are known as strongly regular graphs, and have been extensively studied ([5], [28], [62], [63]). Any association scheme with two classes is metric, and corresponds to a strongly regular graph.

But not all association schemes are metric. As pointed out by Delsarte [14], an association scheme is metric if and only if (i) $k = i+j$ $\Longrightarrow p_{ijk} \neq 0$, and (ii) $p_{ijk} \neq 0 \Longrightarrow |i-j| \leq k \leq i + j$. Delsarte has also given an interesting characterization of metric schemes in terms of the eigenvalues $P_k(i)$, as follows.

Let $z_0, \ldots, z_n$ be distinct nonnegative real numbers, with $z_0 = 0$.

245

Suppose the entries of the eigenmatrix P can be written as

$$P_k(i) = \Phi_k(z_i), \quad i, k = 0, \ldots, n \ ,$$

where $\Phi_k(z)$ is a polynomial of degree k . Then the association scheme is called a <u>P-polynomial scheme</u> with respect to the z_i . A <u>Q-polynomial scheme</u> is defined similarly.

Theorem 2 implies that in a P-polynomial scheme, $\Phi_0(z), \ldots, \Phi_n(z)$ are an orthogonal family on the set $\{z_0, \ldots, z_n\}$ with respect to the weighting function $w(z_i) = \mu_i$. It can also be shown that the sum polynomials

$$\Psi_k(z) = \Phi_0(z) + \ldots + \Phi_k(z), \quad k = 0, \ldots, n\text{-}1 \tag{30}$$

form an orthogonal family on the set $\{z_1, \ldots, z_n\}$ with respect to the weighting function $w(z_i) = \mu_i z_i$. The Hamming and Johnson schemes are both P- and Q-polynomials schemes.

Theorem 7 (Delsarte [14])

An association scheme is metric if and only if it is a P-polynomial scheme.

(No such characterization is known for Q-polynomial schemes.)

In a metric scheme, D_1 is the ordinary node-node adjacency matrix of the graph. Also it is easy to see that D_i is a polynomial in D_1 of degree i . Thus the Bose-Mesner algebra of a metric association scheme is the algebra of polynomials in D_1 . This algebra and its eigenvalues have been studied by several authors (see for example Biggs [3], [4], [4a]).

An interesting subclass of metrically regular graphs are <u>distance-transitive</u> graphs, defined as follows [4], [67a]. A graph is distance-transitive if, for any nodes u, v, x, y with $\rho(u, v) = \rho(x, y)$, there is a permutation of the nodes which preserves adjacency and which takes u to x and v to y . It is easy to see that a distance-transitive graph is metrically regular. The graphs corresponding to the Hamming and

Johnson schemes are distance-transitive.

Association Schemes from Permutation Groups

Metrically regular and distance-transitive graphs are important in studying permutation groups. In fact, many of the sporadic simple groups can be obtained as automorphism groups of such graphs. But we have already strayed too far from coding theory, and so just refer the reader to the extremely interesting references Higman [33]-[38], Cameron [9] and Biggs [3]. Wielandt [70] also contains much relevant material, although without using the terminology of association schemes.

Association Schemes in Statistics

Here again, space does not permit us to describe the role of association schemes in statistics. See Bose and Shimamoto [7], James [39], Ogasawara [57], Ogawa [58], Raghavarao [61], and Yamamoto et al. [71].

§7 The Linear Programming Bound

The definition of an error-correcting code given in §2 amounts to saying that a code is a subset of $X = Q_n$ in the Hamming association scheme. More generally, let us define a code Y in any association scheme to be a nonempty subset of the points X . Elements of Y are called codewords. We shall use the term error-correcting code to refer to a code as defined in §2.

The distance distribution of the code Y is defined to be the (n+1)-tuple of rational numbers $(B_0, \ldots, B_n)$, where

$$B_i = \frac{1}{|Y|} |R_i \cap Y^2|$$

is the average number of codewords which are ith associates of a given codeword. The distance distribution of an error-correcting code in the Hamming scheme gives the minimum distance d of the code and other useful information (see [52]).

Note that $B_0 = 1$, $\displaystyle\sum_{i=0}^{n} B_i = |Y|$, and of course $B_i \geq 0$. A much stronger result is

Theorem 8 (Delsarte [11], [15])

The distance distribution of any code $Y \subset X$ satisfies

$$B_k' \triangleq \frac{1}{|Y|} \sum_{i=0}^{n} B_i Q_k(i) \geq 0 \ ,$$

for $k = 0, \ldots, n$. Note that $B_0' = 1$, $\sum_{k=0}^{n} B_k' = |X|/|Y|$.

Proof Let $u = (u_x)_{x \in X}$ be the characteristic vector of Y, defined by $u_x = 1$ if $x \in Y$, $= 0$ if $x \notin Y$. Then $B_i = \frac{1}{|Y|} u D_i u^T$. Also

$$B_k' = \frac{1}{|Y|^2} u \left(\sum_{i=0}^{n} Q_k(i) D_i \right) u^T = \frac{|X|}{|Y|^2} u J_k u^T, \quad \text{from (11)} \ .$$

Now J_k has eigenvalues 0 and 1 (see the proof of Lemma 1), so is positive semi-definite. Therefore $B_k' \geq 0$. Q. E. D.

Remark For a linear error-correcting code in the Hamming scheme it turns out that B_i is the number of codewords of Hamming weight i, and (by the MacWilliams theorem [51], [52, Ch 5]) B_i' is the number of codewords of Hamming weight i in the dual code. So in this case Th. 8 is trivial. But the importance of Th. 8 comes from the fact that it applies to nonlinear error-correcting codes and more generally to codes in arbitrary association schemes.

The simultaneous linear inequalities in Th. 8 suggest the use of linear programming. Let us say that a code Y in an association scheme has minimum distance d if no codeword is an ith associate of any other codeword, for $0 < i < d$. The distance distribution of Y must satisfy

$$B_1 = B_2 = \ldots = B_{d-1} = 0 \ .$$

This includes error-correcting codes with minimum distance d in the Hamming scheme as a special case.

The problem of finding the largest code of minimum distance d is related to:

Linear Programming Problem (I)   Choose the real variables B_d, B_{d+1}, ..., B_n so as to

$$\text{maximize } g = \sum_{i=d}^{n} B_i$$

subject to the inequalities

$$B_i \geq 0, \quad i = d, \ldots, n , \tag{31}$$

$$Q_k(0) + \sum_{i=d}^{n} B_i Q_k(i) \geq 0, \quad k = 1, \ldots, n . \tag{32}$$

An $(n+1)$-tuple $B = (B_0, \ldots, B_n)$ with $B_0 = 1$, $B_1 = \ldots = B_{d-1} = 0$ is called a feasible solution to Problem (I) if it satisfies (31) and (32). A feasible solution is optimal if g is maximized. Let g_{max} be the maximal value of g .

If a code Y with minimum distance d exists in this association scheme, its distance distribution $(B_0, B_1, \ldots, B_n)$ is (from Th. 8) a feasible solution to Problem (I). Therefore

$$|Y| \leq g_{max} + 1$$

is an upper bound on the size of the code. This is the first version of the linear programming bound for codes.

Associated with this linear programming problem is the:

Dual Problem (II)   Choose the real variables $\beta_1, \ldots, \beta_n$ so as to

$$\text{minimize } \gamma = \sum_{k=1}^{n} \gamma_k Q_k(0)$$

subject to the inequalities

249

$$\beta_k \geq 0, \qquad k = 1, \ldots, n \ , \tag{33}$$

$$1 + \sum_{k=1}^{n} \beta_k \, Q_k(i) \leq 0, \qquad i = d, \ldots, n \ . \tag{34}$$

An $(n+1)$-tuple $\beta = (\beta_0, \beta_1, \ldots, \beta_n)$ with $\beta_0 = 1$ is called a _feasible solution_ to Problem (II) if it satisfies (33) and (34), and an _optimal_ solution if γ is minimized.

We now invoke the following theorem from linear programming theory:

Theorem 9 ([65])

(a) If B is a feasible solution to Problem (I) and β is a feasible solution to Problem (II) then $g \leq \gamma$.

(b) Optimal solutions exist to both Problems, and the optimal values of g and γ are equal (to g_{max}) .

(c) If B is an optimal solution to (I) and β is an optimal solution to (II), then

$$\beta_k \left(Q_k(0) + \sum_{i=d}^{n} B_i Q_k(i) \right) = 0, \qquad k = 1, \ldots, n \tag{35}$$

and

$$B_i \left(\sum_{k=0}^{n} \beta_k Q_k(i) \right) = 0, \qquad i = d, \ldots, n \ . \tag{36}$$

(d) Conversely, if a pair of feasible solutions B, β satisfy (35) and (36), then they are optimal solutions.

Suppose the association scheme is such that $Q_k(x)$ is a polynomial in x of degree k . (This includes the Hamming and Johnson schemes, as we have seen.) The $Q_k(x)$ are a family of orthogonal polynomials. Th. 9 can now be used to obtain another upper bound on $|Y|$.

Theorem 10 <u>The second version of the linear programming bound for</u> <u>codes</u> (Delsarte [11]).

Suppose a polynomial $\beta(x)$ of degree at most n can be found with the following properties. Let us write

$$\beta(x) = \sum_{k=0}^{n} \beta_k \, Q_k(x) \ . \tag{37}$$

Then $\beta(x)$ should satisfy

$$\beta_0 = 1 \ , \tag{38a}$$

$$\beta_k \geq 0 \quad \text{for} \quad k = 1, \ldots, n \ , \tag{38b}$$

$$\beta(i) \leq 0 \quad \text{for} \quad i = d, \ldots, n \ .$$

Then if Y is any code with minimum distance d ,

$$|Y| \leq \beta(0) \ .$$

<u>Proof</u> If such a $\beta(x)$ can be found, then $(\beta_0, \ldots, \beta_n)$ is a feasible solution to Problem (II), since (33) and (34) hold, and $\gamma = \beta(0) - 1$. The theorem follows from Th. 9(a).   Q.E.D.

Application to Error-Correcting Codes

Next we describe how these bounds have been applied to error-correcting codes in the Hamming and Johnson schemes. In the Hamming scheme, McEliece et al. [55] have obtained excellent numerical results by using the simplex algorithm and a computer. Delsarte et al. [22] have obtained preliminary numerical results for the Johnson scheme.

For theoretical purposes the second version of the linear programming bound is easier to use. This turns out to be a very powerful technique. For example Delsarte [11], [14] has derived the Plotkin, Singleton, sphere-packing and Grey bounds in this way. We illustrate the technique with three examples.

251

Theorem 11   The Plotkin bound [60]

If C is an (n, M, d) error-correcting code with $n < 2d$,   then

$$M \le \frac{2d}{2d - n}$$

Proof   We use Th. 9a,   taking $\beta(x)$ to be a polynomial of degree 1 .   The polynomials $Q_k(x)$ in Eq. (37) are now Krawtchouk polynomials.   Thus $\beta(x)$ has the form

$$\beta(x) = 1 + \beta_1 K_1(x;n) = 1 + \beta_1(n-2x) \ .$$

The best choice for β_1 to satisfy (38) is to make $\beta(d) = 0$,   i. e., $\beta_1 = 1/(2d - n)$.   Then $\beta(x) = (2d - 2x)/(2d - n)$,   and $\beta(0) = 2d/(2d - n)$.   The theorem now follows from Th. 9a.   Q. E. D.

Theorem 12   (Johnson [40, Th. 3]; Delsarte [14])

If C is a $(V, M, 2\delta)$ error-correcting code in which every code-word has weight n,   then

$$M \le \frac{\delta V}{\delta V - n(V-n)} \ ,$$

provided $n(V-n) < \delta V$.

Proof C is a code in the Johnson scheme, with minimum distance δ .
The theorem is now proved in the same way as Th. 11,   taking $\beta(x)$ to have degree 1 and using Eq. (29).   Q. E. D.

Theorem 13   (The Singleton bound [66], [14])

If C is an (n, M, d) error-correcting code then

$$M \le 2^{n - d + 1} \ .$$

Proof   Let us use

$$\beta(x) = 2^{n-d+1} \prod_{i=d}^{n} \left(1 - \frac{x}{i}\right)$$

in Th. 10. Certainly (38c) is satisfied. Since the Krawtchouk polynomials are orthogonal, Eq. (25), the coefficients of the expansion

$$\beta(x) = \sum_{k=0}^{n} \beta_k K_k(x;n)$$

are given by

$$\beta_k = \frac{1}{2^n} \sum_{i=0}^{n} \beta(i) K_i(k;n)$$

$$= \frac{1}{2^{d-1}} \sum_{i=0}^{n} \binom{n-i}{n-d+1} K_i(k;n) \Big/ \binom{n}{d-1}$$

$$= \binom{n-k}{d-1} \Big/ \binom{n}{d-1}, \quad \text{from Eq. (27)} \ ,$$

for $k = 0, \ldots, n-d+1$, and $\beta_k = 0$ for $k = n-d+2, \ldots, n$. Thus (38a), (38b) are satisfied. The result follows from Th. 10, since $\beta(0) = 2^{n-d+1}$.

Q. E. D.

Applications to Designs

A code Y in an association scheme, with distance distribution $(B_0, \ldots, B_n)$, will be called a t-design if

$$B_1' = \ldots = B_t' = 0 \ .$$

Then t is called the strength of Y . The connection with the usual designs in statistics is given by:

Theorem 14 (Delsarte [14], [15])

A t-design Y in the Johnson association scheme is equivalent to

253

a classical t-(V, k, λ) design. (That is, a family of k-subsets, called blocks, of a V-set, such that any t-subset of the V-set is contained in exactly λ blocks.) Also a t-design in the Hamming association scheme is equivalent to an orthogonal array of strength t .

Again the linear programming approach has led to a number of bounds on t-designs (see [14], [15], [22]).

§8 Properties of Perfect Codes

Let Y be a code of minimum distance d in a metric association scheme (see §6). Then Y is called an e-perfect code if the spheres

$$S_e(y) = \{x \in X: \rho(x, y) \le e\}, \quad y \in Y ,$$

of radius $e = [\frac{1}{2}(d-1)]$ about the codewords include all points of X . (These spheres are disjoint by the definition of d .) This definition includes perfect error-correcting codes in the Hamming scheme as a special case. The distance distribution of an e-perfect code is an extremal solution to the linear programming Problem (I), and hence Th. 9c can be applied to give:

Theorem 15 (Delsarte [14])

If an e-perfect code exists in a metric association scheme, then the sum polynomial $\Psi_e(z)$ of Eq. (30) has e distinct zeros in the set $\{z_1, \ldots, z_n\}$.

Corollary 16 (Lloyd [50])

If a perfect (n, M, d=2e+1) error-correcting code exists, then the Krawtchouk polynomial $K_e(x-1;n-1)$ has e distinct zeros in the range [1, n] .

This theorem was the basis for the recent proof by Tietäväinen and Van Lint that we know all the perfect error-correcting codes, see [49] . Other applications of Th. 15 can be found in [14], [29], and [49]. Perfect codes in graphs have been studied by Biggs [3a], [4a].

Acknowledgments

I should like to thank Philippe Delsarte, Jean-Marie Goethals, Jessie MacWilliams and Colin Mallows for many helpful discussions.

REFERENCES

1. E. R. Berlekamp, Algebraic Coding Theory, McGraw-Hill, N. Y. 1968.

2. E. R. Berlekamp, Goppa codes, IEEE Trans. Info. Theory, 19 (1973), 590-592.

3. N. L. Biggs, Finite Groups of Automorphisms, London Math. Soc. Lecture Notes Series, No. 6, Cambridge Univ. Press, Cambridge, 1971.

3a. N. L. Biggs, Perfect codes in graphs, J. Combinatorial Theory, 15B (1973), 289-296.

4. N. L. Biggs, Algebraic Graph Theory, Cambridge Univ. Press, London, 1974.

4a. N. L. Biggs, Perfect codes and distance-transitive graphs, in Combinatorics, edited by T. P. McDonough and V. C. Mavron, London Math. Soc., Lecture Notes No. 13, Cambridge Univ. Press, Cambridge, 1974, pp. 1-8.

5. R. C. Bose, Strongly regular graphs, partial geometries and partially balanced designs, Pacif. J. Math., 13 (1963), 389-419.

6. R. C. Bose and D. M. Mesner, On linear associative algebras corresponding to association schemes of partially balanced designs, Ann. Math. Stat., 30 (1959), 21-38.

7. R. C. Bose and T. Shimamoto, Classification and analysis of partially balanced incomplete block designs with two associate classes, J. Amer. Stat. Assoc., 47 (1952), 151-184.

8. M. Burrow, Representation Theory of Finite Groups, Academic Press, N.Y., 1965.

9. P. J. Cameron, Suborbits in transitive permutation groups, in Combinatorics, M. Hall, Jr., and J. H. Van Lint, editors, Math. Centre Tracts 57 (1974), 98-129 (Math. Centre, Amsterdam).

10. R. M. Damerell, On Moore graphs, Proc. Comb. Phil. Soc., 74 (1973), 227-236.

11. P. Delsarte, Bounds for unrestricted codes, by linear programming, Philips Research Reports, 27 (1972), 272-289.

12. P. Delsarte, Four fundamental parameters of a code and their combinatorial significance, Information and Control 23 (1973), 407-438.

14. P. Delsarte, An algebraic approach to the association schemes of coding theory, Philips Research Reports Supplements, No. 10, 1973.

15. P. Delsarte, The association schemes of coding theory, in Combinatorics, edited by M. Hall, Jr., and J. H. Van Lint, Math. Centre Tracts, 55 (1974), 139-157 (Math. Centre, Amsterdam).

16. P. Delsarte, Association schemes in certain lattices, to appear.

17. P. Delsarte, Association schemes and t-designs in regular semi-lattices, to appear.

18. P. Delsarte, Properties and applications of the recurrence $F(i+1, k+1, n+1) = q^{k+1} F(i, k+1, n) - q^k F(i, k, n)$, SIAM J. Appl. Math. to appear.

19. P. Delsarte, The association scheme of bilinear forms over GF(q), to appear.

20. P. Delsarte and J. M. Goethals, Alternating bilinear forms over GF(q), J. Combinatorial Theory, to appear.

21. P. Delsarte, J. M. Goethals, and J. J. Seidel, Bounds for systems of lines, and Jacobi polynomials, to appear.

22. P. Delsarte, W. Haemers and C. Weug, Unpublished.

23. M. Doob, On graph products and association schemes, Utilitas Mathematica, 1 (1972), 291-302.

24. C. F. Dunkl, A Krawtchouk polynomial addition theorem and wreath products of symmetric groups, to appear.

25. C. F. Dunkl and D. E. Ramirez, Krawtchouk polynomials and the symmetrization of hypergroups, SIAM J. Math. Anal., 5 (1974), 351-366.

26. P. J. Eberlein, A two parameter test matrix, Math. Comp. 18 (1964), 296-298.

27. J. M. Goethals, Two dual families of nonlinear binary codes, Electronics Letters, 10 (1974), 471-472.

28. J. M. Goethals and J. J. Seidel, Strongly regular graphs derived from combinatorial designs, Canad. J. Math., 22 (1970), 597-614.

29. J. M. Goethals and H. C. A. van Tilborg, Uniformly packed codes, to appear.

30. V. D. Goppa, A new class of linear correcting codes, Problems of Information Transmission 6 (1970), 207-212.

31. H. J. Helgert, Alternant codes, Information and Control 26 (1974), 369-380.

32. H. J. Helgert and R. D. Stinaff, Minimum-distance bounds for binary linear codes, IEEE Trans. Info. Theory, 19 (1973), 344-356.

33. D. G. Higman, Finite permutation groups of rank 3, Math. Zeit., 86 (1964), 145-156.

34. D. G. Higman, Intersection matrices for finite permutation groups, J. Algebra 4 (1967), 22-42.

35. D. G. Higman, Characterization of families of rank 3 permutation groups by the subdegrees I and II, Archiv. Math. 21 (1970), 151-156 and 353-361.

36. D. G. Higman, Combinatorial Considerations about Permutation Groups, Lecture notes, Math. Inst., Oxford, 1972.

37. D. G. Higman, Invariant relations, coherent configurations, and generalized polygons, in Combinatorics, edited by M. Hall, Jr., and J. H. van Lint, Math. Centre Tracts, 57 (1974), 27-43 (Math. Centre, Amsterdam).

38. D. G. Higman, Coherent configurations, part I: ordinary representation theory, Geometriae Dedicata, to appear.

39. A. T. James, The relationship algebra of an experimental design, Ann. Math. Stat., 28 (1957), 993-1002.

40. S. M. Johnson, A new upper bound for error-correcting codes, IEEE Trans. Info. Theory, 8 (1962), 203-207.

41. S. M. Johnson, On upper bounds for unrestricted binary error-correcting codes, IEEE Trans. Info. Theory, 17 (1971), 466-478.

42. S. M. Johnson, Upper bounds for constant weight error correcting codes, Discrete Math., 3 (1972), 109-124.

43. J. Justesen, A class of constructive asymptotically good algebraic codes, IEEE Trans. Info. Theory, 18 (1972), 652-656.

44. M. Kratchouk, Sur une généralisation des polynômes d'Hermite, Comptes Rendus, 189 (1929), 620-622.

45. M. Krawtchouk, Sur la distribution des racines des polynomes orthogonaux, Comptes Rendus, 196 (1933), 739-741.

46. V. I. Levenshtein, Upper bounds for fixed-weight codes, Problems of Information Transmission, 7 (1971), 281-287.

47. V. I. Levenshtein, On minimal redundancy of binary error-correcting codes (In Russian), Problemy Peredachi Informatsii, 10 (No. 2, 1974), 26-42.

48. J. H. van Lint, Coding Theory, Lecture Notes in Math. 201, Springer-Verlag, New York, 1971.

49. J. H. van Lint, Recent results on perfect codes and related topics, in Combinatorics, edited by M. Hall, Jr., and J. H. van Lint, Math. Centre Tracts, 55 (1974), 158-178 (Math. Centre, Amsterdam).

50. S. P. Lloyd, Binary block coding, Bell Syst. Tech. J., 36 (1957), 517-535.

51. F. J. MacWilliams, A theorem on the distribution of weights in a systematic code, Bell Syst. Tech. J., 42 (1963), 79-94.

52. F. J. MacWilliams and N. J. A. Sloane, Combinatorial Coding Theory, North-Holland, Amsterdam (in preparation).

53. F. J. MacWilliams, N. J. A. Sloane, and J. M. Goethals, The MacWilliams identities for nonlinear codes, Bell Syst. Tech. J., 51 (1972), 803-819.

258

ASSOCIATION SCHEMES AND CODING THEORY

54. M. Marcus and H. Minc, A Survey of Matrix Theory and Matrix Inequalities, Allyn and Bacon, Boston, 1964.

55. R. J. McEliece, E. R. Rodemich, H. C. Rumsey, Jr., and L. R. Welch, unpublished.

56. R. J. McEliece, E. R. Rodemich, H. C. Rumsey, Jr., and L. R. Welch, private communication, March 14, 1975.

57. M. Ogasawara, A necessary condition for the existence of regular and symmetrical PBIB designs of T_m type, Inst. of Stat. Mimeo Series No. 418, Univ. of N. Carolina, Chapel Hill, N.C., Feb. 1965.

58. J. Ogawa, The theory of the association algebra and the relationship algebra of a partially balanced incomplete block design, Institute of Statistics, Mimeograph Series No. 224, Univ. of North Carolina, Chapel Hill, N.C., April 1959.

59. W. W. Peterson and E. J. Weldon, Jr., Error-Correcting Codes, 2nd Ed., MIT Press, Cambridge, Mass., 1972.

60. M. Plotkin, Binary codes with specified minimum distances, IEEE Trans. Info. Theory, 6 (1960), 445-450.

61. D. Raghavarao, Constructions and Combinatorial Problems in Design of Experiments, Wiley, N.Y., 1971.

62. J. J. Seidel, Strongly regular graphs, pp. 185-198 of Recent Progress in Combinatorics, W. T. Tutte, ed., Academic Press, N.Y., 1969.

63. J. J. Seidel, Graphs and two-graphs, Proc. 5th Southeastern Conference on Combinatorics, Graph Theory, Computing, Boca Raton, Florida, 1974.

64. V. M. Sidel'nikov, Upper bounds for number of words in a binary code with given minimal distance (In Russian), Problemy Peredachi Informatsii, 10 (No. 2, 1974), 43-51.

65. M. Simonnard, Linear Programming, Prentice-Hall, N.J., 1966 (Chapter 5).

259

66. R. C. Singleton, Maximum distance q-nary codes, IEEE Trans. Info. Theory, 10 (1964), 116-118.

67. N. J. A. Sloane, A survey of constructive coding theory, and a table of binary codes of highest known rate, Discrete Math., 3 (1972), 265-294.

67a. D. H. Smith, Distance-transitive graphs, in Combinatorics, edited by T. P. McDonough and V. C. Mavron, London Math. Soc., Lecture Notes No. 13, Cambridge Univ. Press, Cambridge, 1974, pp. 145-153.

68. G. Szegö, Orthogonal Polynomials, Vol. 23, Amer. Math. Soc., Providence, R. I., 1967.

68a. J. H. M. Wedderburn, Lectures on Matrices, Dover, N. Y., 1964.

69. L. R. Welch, R. J. McEliece, and H. Rumsey, Jr., A low-rate improvement on the Elias bound, IEEE Trans. Info. Theory, 20 (1974), 676-678.

70. H. Wielandt, Finite Permutation Groups, Academic Press, N. Y., 1964.

71. S. Yamamoto, Y. Fujii and N. Hamada, Composition of some series of association algebras, J. Sci. Hiroshima Univ., Ser. A-I, 29 (1965), 181-215.

Bell Laboratories
Murray Hill, New Jersey 07974

Linear Growth Models with Many Types and Multidimensional Hahn Polynomials
S. Karlin and J. McGregor

In this paper we formulate some genetics models in which the state space is multi-dimensional and discrete, and for which the basic system of differential equations satisfied by the transition probabilities can be solved by separation of the variables. The models can be viewed as discrete analogues of multi-dimensional diffusion models with the variables separable in spherical polar coordinates. The spherical harmonics are here replaced by functions of discrete variables which can be expressed explicitly in terms of Hahn polynomials.

1. <u>Multi-allele Moran mutation models</u>. We consider a population consisting of a fixed number N of haploid individuals. Each individual may be of any one of r possible types (type k, $k = 1, 2, \ldots, r$), and the state of the system is specified by giving the number of individuals of each type. If there are m_k of type k, $k = 1, \ldots, r$ we say the system is in state $\overline{m} = (m_1, \ldots, m_r)$. The set of all states, denoted by Δ_r, is thus the set of all r-tuples $\overline{m} = (m_1, \ldots, m_r)$ of non-negative integers with

$$m_1 + \ldots + m_r = N .$$

Events occur at random times. The time intervals $T_1, T_2, \ldots$ between successive events are independent random variables all with the same exponential distribution

(1.1) $$P\{T_i > t\} = \exp(-N\mu t), \quad 0 \leq t < \infty$$

where μ is a positive constant.

When an event occurs two members, possibly the same, are independently chosen at random from the population. The first chosen individual dies and is replaced by a new individual, whose type is determined according to certain probabilistic laws from the type of the second chosen individual. Hence, the state of the system either remains unchanged, or else the number of individuals of one type decreases by one and the number of individuals of some other type increases by one.

Suppose the system is in state $\overline{m} = (m_1, m_2, \ldots, m_r)$ and an event occurs. The first chosen individual, who dies, is of type i with probability $\dfrac{m_i}{N}$. Let $p_j(\overline{m})$ be the probability that the new individual is of type j. Then the probability that the event results in a loss of one individual of type i and a gain of one individual of type j is

$$(1.2) \qquad p_{i,j}(\overline{m}) = \frac{m_i}{N} p_j(\overline{m}) \ .$$

Hence the probability that the event does not result in a change of state is

$$q(\overline{m}) = \sum_{i=1}^{r} \frac{m_i}{N} p_i(\overline{m}) = \sum_{i=1}^{r} p_{i,i}(\overline{m}) \ .$$

The functions $p_j(\overline{m})$ can be defined in various ways to account for different phenomena (random sampling, mutation, migration, etc.).

Mutation models. Assume the new individual is created by the following mechanism. The second chosen individual duplicates, producing a progeny of his own type, and the progeny is then subject to mutation. The probability that a progeny of type k mutates and yields a new individual of type j is $\beta_{k,j}$ where $\beta_{k,j} \geq 0$, $\sum_j \beta_{k,j} = 1$. Since the probability that the second chosen individual, and hence the progeny, is of type k is m_k/N, we have

$$(1.3) \qquad p_j(\overline{m}) = \sum_{k=1}^{r} \frac{m_k}{N} \beta_{k,j} \ .$$

An important analytical simplification occurs if the mutation probabilities $\beta_{k,j}$ do not depend on k, more precisely if

(1.4)
$$\begin{cases} \beta_{k,j} = \beta_j & \text{when } j \neq k \ , \\ \beta_{k,k} = 1 - \sum_{j \neq k} \beta_j \ . \end{cases}$$

For this case (1.3) can be written in the form

(1.5)
$$p_j(\overline{m}) = [1-(\beta_1 + \beta_2 + \cdots + \beta_r)] \frac{m_j}{N} + \beta_j \ .$$

Migration models. Alternatively we may assume the type of the new individual does not depend on the second chosen individual. We imagine the new individual to be an immigrant from an infinite external population in which the relative frequencies $\gamma_1, \gamma_2, \ldots, \gamma_r$ of the r types are fixed $(\gamma_i \geq 0, \sum_i \gamma_i = 1)$. Then the new individual is of type j with probability γ_j, that is

(1.6)
$$p_j(\overline{m}) = \gamma_j \ .$$

Here $p_j(\overline{m})$ does not depend on $\overline{m}$, a substantial analytic simplification which appears in (1.5) only in the special case $\beta_1 + \cdots + \beta_r = 1$.

Mutation-migration models. More generally the new individual may arise by mutation with probability θ and by immigration with probability $1 - \theta$. In this case

(1.7)
$$p_j(\overline{m}) = \theta \sum_{k=1}^{r} \frac{m_k}{N} \beta_{k,j} + (1-\theta)\gamma_j \ ,$$

or, with the assumptions (1.4).

(1.8)
$$p_j(\overline{m}) = \theta[1 - (\beta_1 + \cdots + \beta_r)] \frac{m_j}{N} + \theta\beta_j + (1-\theta)\gamma_j \ .$$

We note that (1.5) and (1.6) are special cases of (1.8).

Let $X_i(t)$, $i = 1, 2, \ldots, r$ be the number of individuals of type i in the population at time t and let $\overline{X}(t) = (X_1(t), X_2(t), \ldots, X_r(t))$. Then $\overline{X}(t)$ is a temporally homogeneous Markoff process with state space Δ_r. The transition probabilities are defined by

$$P_{\overline{m}, \overline{n}}(t) = P\{\overline{X}(t) = \overline{n} \,|\, \overline{X}(0) = \overline{m}\}, \quad t \geq 0, \overline{m}, \overline{n} \in \Delta_r .$$

For $\overline{m} = (m_1, m_2, \ldots, m_r) \in \Delta_r$ and $1 \leq i, j \leq r, i \neq j$ we define $(\overline{m})_{ij} = (m_1', m_2', \ldots, m_r')$ as follows,

$$m_i' = m_i - 1 ,$$

$$m_j' = m_j + 1 ,$$

$$m_k' = m_k \quad \text{if} \quad k \neq i, j .$$

Then the transition probabilities satisfy the differential equations

$$(1.9) \qquad \frac{1}{N\mu} \frac{dP_{\overline{m}, \overline{n}}}{dt} = \sum_{\substack{i \leq i, j \leq r \\ i \neq j}} p_{ij}(\overline{m})[P_{(\overline{m})_{i, j}, \overline{n}} - P_{\overline{m}, \overline{n}}] ,$$

and the initial conditions

$$(1.10) \qquad P_{\overline{m}, \overline{n}}(0) = \begin{cases} 1, & \overline{m} = \overline{n} , \\ 0, & \overline{m} \neq \overline{n} . \end{cases}$$

The transition probabilities form a matrix $P = (P_{\overline{m}, \overline{n}})$, indexed by the set Δ_r, and (1.9) can be written in matrix form

$$(1.11) \qquad \frac{1}{N\mu} \frac{dP}{dt} = AP$$

where $A = (a_{\overline{m}, \overline{\ell}})$ is a matrix defined by

$$
a_{\overline{m},\overline{\ell}} = \begin{cases} p_{i,j}(\overline{m}) & \text{if } \overline{\ell} = (\overline{m})_{i,j} \text{ for some } i,j \text{ with } i \neq j \\[2mm] - \displaystyle\sum_{\substack{1 \leq i,\,j \leq r \\ i \neq j}} p_{i,j}(\overline{m}) & \text{if } \overline{\ell} = \overline{m}\, , \\[6mm] 0 & \text{otherwise} \end{cases}
$$

The unique solution of (1.11) satisfying (1.10) is $P(t) = \exp(N\mu At)$ but this symbolic formula is not sufficiently explicit to be useful. Our aim will be to find a more explicit and informative representation of the solution. It is convenient to assume $\theta > 0$ and distinguish three cases:

$$
\tag{1.12} \beta_1 + \ldots + \beta_r < 1 \, ,
$$

$$
\tag{1.13} \beta_1 + \ldots + \beta_r = 1 \, ,
$$

$$
\tag{1.14} \beta_1 + \ldots + \beta_r > 1 \, .
$$

The case (1.13) is essentially simpler than the other two. This simpler case is excluded from the present discussion.

2. Representation of $P(t)$. In this section we will assume $\theta > 0$,

$$
\theta\beta_j + (1-\theta)\gamma_j > 0, \qquad j = 1, 2, \ldots, r
$$

and we will consider only the case (1.12). We can then write

$$
\tag{2.1} p_{ij}(\overline{m}) = \frac{\theta[1 - (\beta_1 + \ldots + \beta_r)]}{N^2} \, m_i(m_j + \alpha_j + 1)
$$

where

$$
\tag{2.2} \alpha_j = \frac{N}{\theta[1 - (\beta_1 + \ldots + \beta_r)]} [\theta\beta_j + (1-\theta)\gamma_j] - 1 \, ,
$$

so that $\alpha_j > -1$.

The matrix A in (1.11) is symmetrizable, in the sense that there is a system of positive weights $\pi(\overline{m})$, $\overline{m} \in \Delta_r$, such that

$$(2.3) \qquad a_{\overline{m},\overline{\ell}} \, \pi(\overline{m}) = a_{\overline{\ell},\overline{m}} \, \pi(\overline{\ell}), \quad \overline{\ell}, \overline{m} \in \Delta_r \ .$$

The weights are determined uniquely by (2.3) except for a multiplicative constant. We choose this constant so that $\sum_{\overline{m}} \pi(\overline{m}) = 1$, and then the weights are given by

$$(2.4) \qquad \pi(\overline{m}) = \binom{m_1 + \alpha_1}{m_1}\binom{m_2 + \alpha_2}{m_2} \cdots \binom{m_r + \alpha_r}{m_r} \bigg/ \binom{N + \alpha_1 + \ldots + \alpha_r + r - 1}{N} \ .$$

In the vector space of all complex valued functions $f = f(\overline{m})$ defined for $\overline{m} \in \Delta_r$ we introduce the inner product

$$(f, g) = \sum_{\overline{m} \in \Delta_r} f(\overline{m}) \, \overline{g(\overline{m})} \, \pi(\overline{m}) \ .$$

The matrix A determines a linear transformation of the vector space $f \to Af$ where

$$(Af)(\overline{m}) = \sum_{\overline{n} \in \Delta_r} a_{\overline{m},\overline{n}} \, f(\overline{n}) \ .$$

Because of (2.3) the linear transformation is Hermitian, that is $(Af, g) = (f, Ag)$ for all f and g. There is therefore a complete orthogonal set of eigenvectors $\phi_1, \phi_2, \ldots \phi_d$ with eigenvalues $\lambda_1, \lambda_2, \ldots, \lambda_d$,

$$(2.5) \qquad A\phi_\nu = -\lambda_\nu \phi_\nu \qquad \nu = 1, 2, \ldots, d$$

where d is the dimension of the vector space, that is

266

$$d = \text{number of points in } \Delta_r$$

$$= \begin{pmatrix} N + r - 1 \\ r - 1 \end{pmatrix} .$$

A minus sign has been incorporated in (2.5) to make the eigenvalues λ_ν all non-negative.

When the eigenvalues λ_ν and eigenvectors ϕ_ν are known the formula

$$(2.6) \qquad P_{\overline{m}, \overline{n}}(t) = \pi(\overline{n}) \sum_{\nu=0}^{d} e^{-N\mu\lambda_\nu t} \phi_\nu(\overline{m}) \phi_\nu(\overline{n}) \rho_\nu ,$$

where

$$\frac{1}{\rho_\nu} = \sum_{m \in \Delta_r} [\phi_\nu(\overline{m})]^2 \pi(\overline{m}) ,$$

is an explicit formula for $P_{\overline{m}, \overline{n}}(t)$.

The constant $\theta\{1 - (\beta_1 + \ldots + \beta_r)\}/N^2$ in (2.1) can be conveniently absorbed in the eigenvalue. If we let

$$(2.7) \qquad z_\nu = \lambda_\nu N^2/\theta\{1 - (\beta_1 + \ldots + \beta_r)\}$$

then the equation to be satisfied by the eigenfunctions assumes the canonical form

$$(2.8) \qquad -z_\nu \phi_\nu(\overline{m}) = \sum_{\substack{i, j \\ i \neq j}} m_i(m_j + \alpha_j + 1)[\phi_\nu((\overline{m})_{i, j}) - \phi_\nu(\overline{m})] .$$

3. <u>Relation with multi-dimensional linear growth</u>. A one-dimensional linear growth process is a birth and death process whose birth rates and death rates are linear functions of the state variable. We will

267

consider the following special case. A population of variable size n is subject to changes of size by way of events which occur at random times. If the population is of size n then the probability that in the next h units of time a death occurs, resulting in a decrease to population size n - 1, is $\mu n h + o(h)$, the probability that a birth occurs, resulting in an increase to size n + 1, is $\lambda n h + o(h)$, the probability that a new individual arrives by immigration is $\gamma h + o(h)$, and the probability that more than one of these events occurs is $o(h)$. The transition probabilities $P_{n,m}(t)$ for the system are found to satisfy

$$(3.1) \quad \frac{dP_{n,m}(t)}{dt} = n\mu P_{n-1,m}(t) - [n(\lambda+\mu)+\gamma]P_{n,m}(t) + (n\lambda+\gamma)P_{n+1,m}$$

for $n, m = 0, 1, \ldots$ and $P_{n,m}(0) = \delta_{n,m}$. The constants λ, μ are called the birth rate and death rate per individual, and γ is the immigration rate.

The system (3.1) can be solved in the sense that explicit formulas for $P_{n,m}(t)$ can be given [1]. Here we discuss only the special case $\mu = \lambda$.

The explicit representation for the solution (3.1), given in [1], depends on the polynomials $Q_n(x)$ which satisfy the recurrence relation

$$(3.2) \quad Q_0(x) = 1, \quad Q_{-1}(x) = 0,$$

$$-xQ_n(x) = n\frac{\lambda}{\mu}Q_{n-1} - (n\frac{\lambda+\mu}{\mu}+\frac{\gamma}{\mu})Q_n + (n\frac{\lambda}{\mu}+\frac{\gamma}{\mu})Q_{n+1}.$$

We introduce the renormalized Laguerre polynomials

$$(3.3) \quad \ell_n^{(\alpha)}(x) = L_n^{(\alpha)}(x)/L_n^{(\alpha)}(0)$$

which satisfy the recurrence relation

$$-x\ell_n^{(\alpha)}(x) = n\ell_{n-1}^{(\alpha)} - (2n+\alpha+1)\ell_n^{(\alpha)} + (n+\alpha+1)\ell_{n+1}^{(\alpha)}.$$

If $\lambda = \mu$ the solution of (3.2) is $Q_n(x) = \ell_n^{(\alpha)}(x)$ where $\alpha = \frac{\gamma}{\mu} - 1$, and the representation for $P_{n,m}(t)$ is

(3.4) $\qquad P_{n,m}^{(\alpha)}(t) = \binom{m+\alpha}{m} \int_0^\infty e^{-\mu x t} Q_n(x) Q_m(x) \frac{x^\alpha e^{-x}}{\Gamma(\alpha+1)} dx$.

If $\lambda \neq \mu$ there is a representation with Meixner polynomials [1].

Now consider a direct product of several, say r linear growth processes, all with the same parameter $\lambda = \mu$, but with possibly different immigration rates $\gamma_1, \gamma_2, \ldots, \gamma_r$. Let $\alpha_r = (\gamma_r/\mu) - 1$. A state of the direct product process is an r-tuple of non-negative integers $\overline{n} = (n_1, n_2, \ldots, n_r)$ and its transition probability function is a product of r factors like (3.4),

(3.5) $\qquad P_{\overline{n}, \overline{m}}(t) = P_{n_1, m_1}^{(\alpha_1)}(t) \, P_{n_2, m_2}^{(\alpha_2)}(t) \ldots P_{n_r, m_r}^{(\alpha_r)}(t)$.

We adopt the following notation. If $\overline{n} = (n_1, \ldots, n_r)$ is an r-tuple of integers and $1 \leq i \leq r$ then

$$(\overline{n})_{i,} = (n_1', n_2', \ldots, n_r')$$

where

$$n_i' = n_i - 1, \quad n_\nu' = n_\nu \quad \text{if } \nu \neq i \ ,$$

and

$$(\overline{n})_{,i} = (n_1'', n_2'', \ldots, n_r'')$$

where

$$n_i'' = n_i + 1, \quad n_\nu'' = n_\nu \quad \text{if } \nu \neq i \ .$$

Then the differential equation satisfied by (3.5) is

269

(3.6)

$$\frac{1}{\mu}\frac{d}{dt}P_{\overline{n},\overline{m}} = \sum_{i=1}^{r} n_i[P_{(\overline{n})_{i,}}, \overline{m} - P_{\overline{n},\overline{m}}]$$

$$+ \sum_{i=1}^{r} (n_i+\alpha_i+1)[P_{(\overline{n})_{,i}}, \overline{m} - P_{\overline{n},\overline{m}}] .$$

To start from (3.6) and find a representation for the function $P_{\overline{n},\overline{m}}(t)$, by separating the variables, leads to consideration of the recurrence relation

(3.7)

$$-y\,\Psi(\overline{n}) = \sum_{i} n_i[\Psi((\overline{n})_{i,}) - \Psi(\overline{n})]$$

$$+ \sum_{i} (n_i+\alpha_i+1)[\Psi((\overline{n})_{,i}) - \Psi(\overline{n})]$$

and this has solutions which are products of the r-variables $n_1, \ldots, n_r$, namely, in the notation (3.3),

(3.8)

$$\Psi(\overline{n}) = \ell_{n_1}^{(\alpha_1)}(x_1)\ell_{n_2}^{(\alpha_2)}(x_2)\ldots\ell_{n_r}^{(\alpha_r)}(x_r)$$

with $x_1+x_2+\ldots+x_r = y$. This solution of (3.7) can be used to construct the representation formula (3.4), (3.5).

In (3.6) the coefficients are linear functions of the n_i as opposed to the quadratic functions in (2.1). Moreover in this process, when an event occurs either one of the n_i decreases by one unit or one of them increases by one unit or all the n_i remain unchanged. Approximately, the probabilities of a decrease or an increase in a small time dt are respecively

$$\sum_i n_i\mu dt = n\mu dt$$

and

$$\sum_i (n_i+\alpha_i+1)\mu dt = (n+\alpha+r)\mu dt$$

270

where $n = \sum_i n_i$, $\alpha = \sum_i \alpha_i$. Since these quantities depend only on n, not on the individual n_i, we see that the total population size n is a one-dimensional linear growth process. It is therefore strongly suggested that $P_{\overline{n}, \overline{m}}(t)$ may have another natural representation distinct from (3.4), (3.5), a representation in which the variable $n = \sum_i n_i$ is more prominent.

A solution of (3.7) which depends on $n = \sum_i n_i$ but not on the separate n_i exists, namely

(3.9)
$$\Psi(\overline{n}) = \ell_{n_1 + \ldots + n_r}^{(\alpha_1 + \ldots + \alpha_r + r - 1)}(y)$$

where again the notation (3.3) is used. We wish to find additional solutions of (3.7) which together with (3.9) can be used to generate a representation for the direct product process. It is reasonable to look for solutions of the form

(3.10)
$$\Psi(\overline{n}) = R_{n_1 + \ldots + n_r}(y) \, \phi(\overline{n}) = R_n(y) \phi(\overline{n}) \; .$$

analogous to solutions of the Helmholtz equation of the form $R(r) \, H(\theta, \phi)$ in spherical polar coordinates. In our case the role of the radial coordinate r is played by the variable $n = n_1 + \ldots + n_r$ and the function $H(\theta, \phi)$ of the angular variables is replaced by the factor $\phi(\overline{n})$. At any rate, let us substitute (3.10) into (3.7), to obtain

$$-y R_n(y) \phi(\overline{n}) = R_{n-1}(y) \sum_i n_i \phi((\overline{n})_{i,})$$

$$+ R_{n+1}(y) \sum_i (n_i + \alpha_i + 1) \phi((\overline{n})_{,i})$$

$$- R_n(y) \sum_i (2n_i + \alpha_i + 1) \phi(\overline{n}) \; .$$

Consequently (3.10) will be a solution of (3.7) provided auxilliary functions $a(n)$ and $b(n)$ can be found such that $R_n(y)$ and $\phi(\overline{n})$ satisfy

(3.11a)
$$\sum_i n_i \phi((\overline{n})_{i,}) = a(n) \, \phi(\overline{n}), \quad n \geq 1 \; ,$$

(3.11b) $$\sum_i (n_i + \alpha_i + 1)\, \phi(\bar{n})_{,i} = b(n)\,\phi(\bar{n}), \qquad n \geq 0 \ ,$$

(3.11c) $$-y\,R_n(y) = a(n)\,R_{n-1} + b(n)\,R_{n+1} - (2n+\alpha+r)\,R_n, \qquad n \geq 0 \ .$$

Equations (3.11a) and (3.11b) lead to a consistency condition as follows. Multiply (3.11b) by $a(n+1)$ and transform the resulting left member by using (3.11a). This gives an identity (A). Again, multiply (3.11a) by $b(n-1)$ and transform the resulting left member by using (3.11b). This gives an identity (B). Now if (B) is subtracted from (A), most of the terms cancel, and we are left with

$$[a(n+1)\,b(n) - a(n)\,b(n-1)]\,\phi(\bar{n}) = (2n+\alpha+r)\,\phi(\bar{n}), \quad n \geq 1 \ ,$$

and for non-trivial $\phi(\bar{n})$ this requires that

$$a(n+1)\,b(n) - a(n)\,b(n-1) = 2n + \alpha + r \ ,$$

and hence

(3.12) $$a(n+1)\,b(n) = (n+1)\,(n+\alpha+r) - x$$

where x is a constant. Now using (3.11a, b) and (3.12) we find $\phi(\bar{n})$ must satisfy

(3.13) $$-x\phi(\bar{n}) = \sum_{\substack{i,j \\ i \neq j}} n_i(n_j+\alpha_j+1)\,[\phi((\bar{n})_{i,j}) - \phi(\bar{n})]$$

which is the same as (2.8). For the present purposes we need solutions of (2.8) for different n which are connected by first order relations of the form (3.11a) and (3.11b).

4. **The case** $r = 2$ **and the Hahn polynomials.** The Moran model for the case of two alleles was discussed in detail by the authors in [2], [3]. The functions $\phi(\bar{m}) = \phi(m_1, m_2)$ can be expressed most conveniently in terms of the Hahn polynomials [4], the relevant properties of which are outlined below.

272

The Hahn polynomials are terminating ${}_3F_2$'s . Recall that

$$
{}_3F_2(a_1, a_2, a_3; b_1, b_2; z) = \sum_{k=0}^{\infty} \frac{(a_1)_k (a_2)_k (a_3)_k}{(b_1)_k (b_2)_k} \cdot \frac{z^k}{k!}
$$

where $(a)_0 = 1$, $(a)_k = a(a+1)\ldots(a+k-1)$ for $k \geq 1$. The Hahn polynomial $Q_i(\xi; \alpha, \beta, N)$ is a polynomial in the variable ξ, depending on parameters α, β, N, and defined by

(4.1) $$ Q_i(\xi; \alpha, \beta, N) = {}_3F_2(-i, -\xi, i+\alpha+\beta+1; \alpha+1, -N+1; 1) . $$

Always N is a positive integer, and the above formula for the ith polynomial $i = 0, 1, \ldots, N-1$ then makes sense provided $(\alpha+1)_i \neq 0$. This provision will obtain if α is not a negative integer, or if $\alpha < -N + 1$.

The Hahn polynomials satisfy the differen e equation

$$
-i(i+\alpha+\beta+1)Q_i(\xi) = \xi(N+\beta-\xi) Q_i(\xi-1) + (N-1-\xi)(\alpha+1+\xi) Q_i(\xi+1)
$$

(4.2)
$$
- [\xi(N+\beta-\xi) + (N-1-\xi)(\alpha+1+\xi)]Q_i(\xi)
$$

for $i = 0, 1, \ldots, N-1$ and all complex ξ. Hence if m_1, m_2 are non-negative integers and we define

(4.3) $$ \phi_i(\overline{m}) = \phi_i((m_1, m_2)) = Q_i(m_1; \alpha_1, \alpha_2, m_1+m_2+1) $$

then this function satisfies

$$
-z_i \phi_i(\overline{m}) = m_1(m_2+\alpha_2+1) \phi_i((m_1-1, m_2+1)) + m_2(m_1+\alpha_1+1) \phi_i((m_1+1, m_2-1))
$$

(4.4)
$$
- [m_1(m_2+\alpha_2+1) + m_2(m_1+\alpha_1+1)]\phi_i(\overline{m})
$$

where

(4.5) $$ z_i = i(i+\alpha_1+\alpha_2+1) . $$

273

Since (4.4) and (2.8) are the same when $r = 2$ it follows that (4.3) defines $N + 1$ solutions $(N = m_1 + m_2)$ of (2.8) for $r = 2$.

From the orthogonality relation of the Hahn polynomials [4], we obtain for the functions (4.3) the relation

$$(4.6) \qquad \sum_{\overline{m} \in \Delta_2(N)} \phi_i(\overline{m}) \phi_j(\overline{m}) \; \frac{\binom{m_1 + \alpha_1}{m_1} \binom{m_2 + \alpha_2}{m_2}}{\binom{m_1 + m_2 + \alpha_1 + \alpha_2 + 1}{m_1 + m_2}} = \frac{\delta_{ij}}{\rho_i}$$

where

$$\rho_0 = 1, \quad \rho_i =$$

$$(4.7) \qquad \frac{\binom{N}{i}}{\binom{N + \alpha_1 + \alpha_2 + i + 1}{i}} \cdot \frac{\Gamma(\alpha_2 + 1)}{\Gamma(\alpha_1 + 1)\Gamma(\alpha_1 + \alpha_2 + 1)} \cdot \frac{\Gamma(\alpha_1 + i + 1)\Gamma(\alpha_1 + \alpha_2 + i + 1)}{\Gamma(\alpha_2 + i + 1)\Gamma(i + 1)} \cdot \frac{2i + \alpha_1 + \alpha_2 + 1}{\alpha_1 + \alpha_2 + 1}$$

and $N = m_1 + m_2$ is the size parameter of Δ_r . (Note that the Hahn polynomials which figure in (4.3) and (4.6) belong to the parameter value $N + 1$ according to (4.1)).

At this stage the representation (2.6), for the case $r = 2$, is completely determined. The cardinality of $\Delta_2(N)$ is $d = N + 1$ and the orthogonal system of $N + 1$ solutions $\phi_i(\overline{m})$, $i = 0, 1, \ldots$ of (2.5) are given by (4.3). The eigenvalues λ_i are determined by (4.5) and (2.7).

A generating function for the Hahn polynomials is

$$(4.8) \qquad \sum_{m=0}^{N-1} \binom{N-1}{m} Q_i(m; \alpha, \beta, N) w^m = (1+w)^{N-1} \; \frac{P_i^{\alpha, \beta}\left(\frac{1-w}{1+w}\right)}{P_i^{\alpha, \beta}(1)} \; ,$$

where $P_i^{\alpha, \beta}$ is the usual Jacobi polynomial. In this generating function only the factor $(1+w)^{N-1}$ depends on N . Hence for $i \leq N - 2$

274

$$\sum_{m=0}^{N-1} \binom{N-1}{m} Q_i(m;\alpha, \beta, N)w^m = (1+w) \sum_{m=0}^{N-1} \binom{N-2}{m} Q_i(m;\alpha, \beta, N-1)w^m \;.$$

By equating coefficients of w^m we obtain a first order recurrence relation

(4.9) $(N-1)Q_i(m;\alpha, \beta, N) = mQ_i(m-1;\alpha, \beta, N-1) + (N-1-m) Q_i(m;\alpha, \beta, N-1)$

which, for the functions (4.3) is equivalent to

(4.10) $(m_1+m_2)\phi_i((m_1, m_2)) = m_1\phi_i((m_1-1, m_2) + m_2\phi_i((m_1, m_2-1))$

valid for $i \le m_1 + m_2 - 1$. This is of the form (3.11a) with $a(m) = m$.

Another first order relation can be obtained by combining (4.9) and (4.2) or by combining (4.10) and (4.3). The two forms which result are

(4.11) $\dfrac{(N-i)(N+i+\alpha+\beta+1)}{N} \; Q_i(m;\alpha, \beta, N)$

$= (m+\alpha+1) Q_i(m+1;\alpha, \beta, N+1) + (N-m+\beta) Q_i(m;\alpha, \beta, N+1)$,

(4.12) $\dfrac{(m_1+m_2+1-i)(m_1+m_2+1+i+\alpha_1+\alpha_2+1)}{m_1+m_2+1} \; \phi_i((m_1, m_2))$

$= (m_1+\alpha_1+1) \phi_i((m_1+1, m_2)) + (m_2+\alpha_2+1) \phi_i((m_1, m_2+1))$.

The last equation is the same as (3.11b) with

$$b(m) = \frac{(m+1-i)(m+i+\alpha_1+\alpha_2+2)}{m+1} \;.$$

We can now determine, for the case $r = 2$, the polynomials $R_{n_1+n_2}$ in (3.10). When the above values of $a(n)$ and $b(n)$ are substituted in (3.11c) that equation becomes $(\alpha = \alpha_1+\alpha_2+1)$

$$-yR_n(y) = nR_{n-1}(y) + \frac{(n+1-i)(n+i+\alpha+1)}{n+1} R_{n+1}(y) - (2n+\alpha+1)R_n(y) \;.$$

A solution of this can be expressed in terms of Laguerre polynomials, by comparing with (3.2),

$$(4.13) \qquad R_n(y) = \begin{cases} 0 & \text{if } n < i \ , \\[2ex] \binom{n}{n-i} \ell_{n-i}^{(\alpha+2i+1)}(y) & \text{if } n \geq i \ . \end{cases}$$

5. <u>Moran model with r types</u>. Consider the r type Moran model with $p_j(\overline{m})$ given by (1.8). We continue to assume $\beta_1 + \ldots + \beta_r < 1$. The probabilities for an increase by one or a decrease by one, of type r , in a small time dt are respectively given by (to first approximation)

$$N\mu dt \sum_{i=1}^{r-1} \frac{m_i}{N} \, p_r(\overline{m}) = \mu dt (m_1 + \ldots + m_{r-1}) p_r(\overline{m}) \ ,$$

and

$$N\mu dt \, \frac{m_r}{N} \sum_{i=1}^{r-1} p_i(\overline{m}) \ .$$

These quantities depend on $m_1, \ldots, m_{r-1}$ only in the combination $m_1 + \ldots + m_{r-1}$. Consequently if we observe only the number m_r of type r and the number $m_1 + \ldots m_{r-1}$ not of type r, then we have a Moran model of two types. This suggests that the eigenfunctions $\phi(\overline{m})$ in (2.6) might be obtained in a factored form

$$\phi(\overline{m}) = S(m_1 + \ldots + m_{r-1}, \, m_r) H(m_1, \ldots, m_{r-1}) \ .$$

The functions $\phi_\nu(\overline{m})$ which satisfy (2.8) are functions of the variables $\overline{m} = (m_1, \ldots, m_r)$, depend on the mutation constants $\overline{\alpha} = (\alpha_1, \ldots, \alpha_r)$ and are parametrized by an index ν which must range over some set with the cardinality of Δ_r . It will be convenient to use as parameter space the set of all $(r-1)$-tuples $\overline{\nu} = (\nu_1, \ldots, \nu_{r-1})$ of non-negative integers ν_k with $\nu_1 + \ldots + \nu_{r-1} \leq N$. We will define a family of functions

$$(5.1) \qquad \phi\left(\begin{array}{c}\overline{m} \\ \overline{\alpha}\end{array}\middle|\ \overline{\nu}\right) = \phi\left(\begin{array}{c}m_1, \ldots, m_r \\ \alpha_1, \ldots, \alpha_r\end{array}\middle|\ \nu_1, \ldots, \nu_{r-1}\right)$$

which satisfy (2.8) with the eigenvalues

$$(5.2) \qquad z(\overline{\nu}) = (\nu_1 + \ldots + \nu_{r-1})(\nu_1 + \ldots + \nu_{r-1} + \alpha_1 + \ldots + \alpha_r + r - 1) \ .$$

For $r = 2$ the function is a Hahn polynomial,

$$(5.3) \qquad \phi\left(\begin{array}{c}\overline{m} \\ \overline{\alpha}\end{array}\middle|\ \overline{\nu}\right) = Q_{\nu_1}(m_1; \alpha_1, \alpha_2, m_1 + m_2 + 1)$$

in agreement with (4.3). For $r > 2$ the definition will be inductive with the induction on r. When $\overline{m} = (m_1, \ldots, m_r)$, $\overline{\alpha} = (\alpha_1, \ldots, \alpha_r)$, $\overline{\nu} = (\nu_1, \ldots, \nu_{r-1})$ we let

$$(5.4) \qquad m = m_1 + \ldots + m_r, \quad \alpha = \alpha_1 + \ldots + \alpha_r, \quad \nu = \nu_1 + \ldots + \nu_{r-1} \ ,$$

$$(5.5) \qquad \left\{\begin{array}{l} \overline{m}' = (m_1, \ldots, m_{r-1}), \quad \overline{\alpha}' = (\alpha_1, \ldots, \alpha_{r-1}), \quad \overline{\nu}' = (\nu_1, \ldots, \nu_{r-2}) \ , \\[2mm] m' = m_1 + \ldots + m_{r-1}, \quad \alpha' = \alpha_1 + \ldots + \alpha_{r-1}, \quad \nu' = \nu_1 + \ldots + \nu_{r-2} \ . \end{array}\right.$$

The inductive definition of (5.1) for $r > 2$ is

$$(5.6) \qquad \phi\left(\begin{array}{c}\overline{m} \\ \overline{\alpha}\end{array}\middle|\ \overline{\nu}\right) = \frac{\binom{m'}{\nu'}}{\binom{m}{\nu}} Q_{\nu_{r-1}}(m' - \nu'; \alpha' + 2\nu' + r - 2, \alpha_r, m - \nu' + 1)\ \phi\left(\begin{array}{c}\overline{m}' \\ \overline{\alpha}'\end{array}\middle|\ \overline{\nu}'\right) .$$

The polynomial $Q_{\nu_{r-1}}$ is a Hahn polynomial. In order to show that these functions satisfy (2.8) it will be necessary to show that they also satisfy first order relations like (3.11a, b). We begin by constructing a generating function.

Let $\bar{w} = (w_1, w_2, \ldots, w_r)$, $\bar{w}' = (w_1, \ldots, w_{r-1})$, $w = w_1 + \ldots + w_r$, $w' = w_1 + \ldots + w_{r-1}$. It will be shown that the generating function

$$(5.7) \quad G_{r,N}\left(\begin{array}{c}\bar{w}\\\bar{\alpha}\end{array}\bigg|\ \bar{\nu}\right) = \sum_{m\epsilon\,\bar{\Delta}_r(N)} \phi\left(\begin{array}{c}\bar{m}\\\bar{\alpha}\end{array}\bigg|\ \bar{\nu}\right) \frac{N!}{m_1!\ldots m_r!}\, w_1^{m_1} w_2^{m_2} \ldots w_r^{m_r}$$

is of the form

$$(5.8) \quad G_{r,N}\left(\begin{array}{c}\bar{w}\\\bar{\alpha}\end{array}\bigg|\ \bar{\nu}\right) = w^N\, H_r\left(\begin{array}{c}\bar{w}\\\bar{\alpha}\end{array}\bigg|\ \bar{\nu}\right)$$

where H_r does not depend on N. From (4.8) we have

$$G_{2,N}\left(\begin{array}{c}w_1, w_2\\\alpha_1, \alpha_2\end{array}\bigg|\ \nu_1\right) = w^N\, \frac{P_{\nu_1}^{\alpha_1, \alpha_2}\left(\dfrac{w_2-w_1}{w_1+w_2}\right)}{P_{\nu_1}^{\alpha_1, \alpha_2}(1)}$$

so the result holds for $r = 2$. The procedure is by induction on r, using (5.6). We have, for $r > 2$,

$$G_{r,N}\left(\begin{array}{c}\bar{w}\\\bar{\alpha}\end{array}\bigg|\ \bar{\nu}\right)$$

$$= \sum_{(m', m_r)\epsilon\,\Delta_2(N)} \frac{\binom{m'}{\nu'}}{\binom{m}{\nu'}}\, Q_{\nu_{r-1}}(m'-\nu'; \alpha'+2\nu'+r-2, \alpha_r, m-\nu'+1) \binom{N}{m'} w_r^{m_r}\, \sigma$$

where

$$\sigma = \sum_{\bar{m}'\,\epsilon\,\bar{\Delta}_r(m')} \phi\left(\begin{array}{c}\bar{m}'\\\bar{\alpha}'\end{array}\bigg|\ \nu'\right) \frac{m'!}{m_1!\ldots m_{r-1}!}\, w_1^{m_1} \ldots w_{r-1}^{m_{r-1}}$$

$$= w'^{m'}\, H_{r-1}\left(\begin{array}{c}\bar{w}'\\\bar{\alpha}'\end{array}\bigg|\ \bar{\nu}'\right) \quad \text{by assumption}$$

278

and hence (note that $N = m$)

$$G_{r, N}\left(\left.\begin{array}{c}\overline{w}\\ \overline{\alpha}\end{array}\right|\ \overline{\nu}\right)$$

$$= w^{N-\nu'}\ w'^{\nu'}\ H_{r-1}\left(\left.\begin{array}{c}\overline{w'}\\ \overline{\alpha'}\end{array}\right|\ \overline{\nu'}\right)\ \frac{P^{\alpha'+2\nu'+r-2,\ \alpha_r}_{\nu_{r-1}}\left(\dfrac{w_r-w'}{w_r+w'}\right)}{P^{\alpha'+2\nu'+r-2,\ \alpha_r}_{\nu_{r-1}}}\quad (1)$$

Since this is of the form (5.8) the result is proved.

Just as in section 4, from (5.8) follows the first order relation

$$(5.9) \qquad m\ \phi\left(\left.\begin{array}{c}\overline{m}\\ \overline{\alpha}\end{array}\right|\ \overline{\nu}\right) = \sum_i m_i\ \phi\left(\left.\begin{array}{c}(\overline{m})_{i,}\\ \overline{\alpha}\end{array}\right|\ \overline{\nu}\right)$$

valid for $\nu \le m - 1$.

We will next show that (5.6) satisfies the difference equation (2.8). This is already known for $r = 2$ and we proceed by induction on r. Substitution of (5.6) into the right hand member of (2.8), and use of (5.9) and the induction assumption, yields

$$\sum_{\substack{(i,j)\\ i\ne j}} m_i(m_j+\alpha_j+1)\left[\phi\left(\left.\begin{array}{c}(\overline{m})_{ij}\\ \overline{\alpha}\end{array}\right|\ \overline{\nu}\right) - \phi\left(\left.\begin{array}{c}\overline{m}\\ \overline{\alpha}\end{array}\right|\ \overline{\nu}\right)\right]$$

$$= -\nu'(\nu'+\alpha'+r-2)\ \phi\left(\left.\begin{array}{c}\overline{m}\\ \overline{\alpha}\end{array}\right|\ \overline{\nu}\right)$$

$$+ (m_r+\alpha_r+1)\ \frac{\binom{m'-1}{\nu'}}{\binom{m}{\nu'}}\ Q_{\nu_{r-1}}(m'-\nu'-1;\alpha'+2\nu'+r-2,\ \alpha_r,\ m-\nu'+1)\cdot m'\ \phi\left(\left.\begin{array}{c}\overline{m'}\\ \overline{\alpha'}\end{array}\right|\ \overline{\nu'}\right)$$

$$+ m_r\ \frac{\binom{m'+1}{\nu'}}{\binom{m}{\nu'}}\ Q_{\nu_{r-1}}(m'-\nu'+1;\alpha'+2\nu'+r-2,\ \alpha_r,\ m-\nu'+1)\cdot \frac{(m'+1-\nu')(m'+1+\alpha'+\nu'+r-2)}{m'+1}$$

279

$$\otimes \quad \phi\left(\begin{array}{c|c} \overline{m} \\ \overline{\alpha'} \end{array} \middle| \begin{array}{c} \overline{\nu'} \end{array}\right)$$

$$- [m_r(m'+\alpha'+1) + (m_r+\alpha_r+1)m'] \; \phi\left(\begin{array}{c|c} \overline{m} \\ \overline{\alpha} \end{array} \middle| \begin{array}{c} \overline{\nu} \end{array}\right)$$

$$= \frac{\begin{pmatrix} m' \\ \nu' \end{pmatrix}}{\begin{pmatrix} m \\ \nu' \end{pmatrix}} \; \phi\left(\begin{array}{c|c} \overline{m'} \\ \overline{\alpha'} \end{array} \middle| \begin{array}{c} \overline{\nu'} \end{array}\right) \cdot T \; ,$$

where

$$T = -[\nu'(\nu'+\alpha'+r-2) + m_r(m'+\alpha'+r-1)$$

$$+ m'(m_r+\alpha_r+1)] Q_{\nu_{r-1}} (m'-\nu';\alpha'+2\nu'+r-2, \alpha_r, m-\nu'+1)$$

$$+ (m'-\nu')(m_r+\alpha_r+1) Q_{\nu_{r-1}} (m'-\nu'-1;\alpha'+2\nu'+r-2, \alpha_r, m-\nu'+1)$$

$$+ m_r(m'+1+\alpha'+\nu'+r-2) Q_{\nu_{r-1}} (m'-\nu'+1;\alpha'+2\nu'+r-2, \alpha_r, m-\nu'+1)$$

$$= -\nu(\nu+\alpha+r-1) Q_{\nu_{r-1}} (m'-\nu';\alpha'+2\nu'+r-2, \alpha_r, m-\nu'+1)$$

by (4.2). Thus (5.6) satisfies (2.8) with $z = \nu(\nu+\alpha+r-1)$.

From (5.9) and (2.8) it now follows that (5.6) satisfies another first order relation, namely

(5.10) $\quad \dfrac{(m+1-\nu)(m+\alpha+\nu+r)}{m+1} \; \phi\left(\begin{array}{c|c} \overline{m} \\ \overline{\alpha} \end{array} \middle| \begin{array}{c} \overline{\nu} \end{array}\right) = \displaystyle\sum_{j=1}^{r} (m_j+\alpha_j+1) \; \phi\left(\begin{array}{c|c} (\overline{m}), j \\ \overline{\alpha} \end{array} \middle| \begin{array}{c} \overline{\nu} \end{array}\right) \;$.

We denote by $\sum_r(N)$ the set of all non-negative integral parameter vectors $\overline{\nu} = (\nu_1, \dots, \nu_{r-1})$ with

$$\nu_1 + \dots + \nu_{r-1} \leq N \;\; .$$

This set has the same cardinality as $\Delta_r(N)$, and for each $\bar{\nu} \in \Delta_r(N)$ we have a solution (5.6) of the difference equation (2.8). We show that these solutions form a complete orthogonal system of functions on $\Delta_r(N)$ relative to the weights (2.4).

Formulas (4.6), (4.7) can be written

$$(5.11) \quad \sum_{(m_1, m_2) \in \Delta_2(N)} \phi\begin{pmatrix} m_1, m_2 \\ \alpha_1, \alpha_2 \end{pmatrix}\begin{vmatrix} \\ \nu_1 \end{vmatrix}\phi\begin{pmatrix} m_1, m_2 \\ \alpha_1, \alpha_2 \end{vmatrix}\lambda_1 \end{pmatrix} \frac{\begin{pmatrix} m_1+\alpha_1 \\ m_1 \end{pmatrix}\begin{pmatrix} m_2+\alpha_2 \\ m_2 \end{pmatrix}}{\begin{pmatrix} N+\alpha_1+\alpha_2+1 \\ N \end{pmatrix}}$$

$$= \left[\frac{\begin{pmatrix} N \\ \nu_1 \end{pmatrix}}{\begin{pmatrix} N+\alpha_1+\alpha_2+\nu_1+1 \\ \nu_1 \end{pmatrix}} \theta_2(\alpha_1, \alpha_2; \nu_1) \right]^{-1} \delta_{\nu_1, \lambda_1}$$

where θ_2 does not depend on N, and is

$$(5.12) \quad \theta_2 = \begin{pmatrix} \alpha_1+\alpha_2+\nu_1 \\ \nu_1 \end{pmatrix} \frac{\begin{pmatrix} \alpha_1+\nu_1 \\ \nu_1 \end{pmatrix}}{\begin{pmatrix} \alpha_2+\nu_1 \\ \nu_1 \end{pmatrix}} \cdot \frac{\alpha_1+\alpha_2+2\nu_1+1}{\alpha_1+\alpha_2+1}.$$

We will show that for $r \geq 2$ and $\bar{\nu}, \bar{\lambda} \in \sum_r(N)$

$$(5.13) \quad \sum_{\bar{m} \in \Delta_r(N)} \phi\begin{pmatrix} \bar{m} \\ \bar{\alpha} \end{pmatrix}\begin{vmatrix} \\ \bar{\nu} \end{vmatrix}\phi\begin{pmatrix} \bar{m} \\ \bar{\alpha} \end{vmatrix}\bar{\lambda} \end{pmatrix} \frac{\begin{pmatrix} m_1+\alpha_1 \\ m_1 \end{pmatrix} \cdots \begin{pmatrix} m_r+\alpha_r \\ m_r \end{pmatrix}}{\begin{pmatrix} N+\alpha+r-1 \\ N \end{pmatrix}} = \frac{\delta_{\bar{\nu}, \bar{\lambda}}}{\rho(\bar{\nu})}$$

where

$$(5.14) \quad \rho(\bar{\nu}) = \frac{\begin{pmatrix} N \\ \nu \end{pmatrix}}{\begin{pmatrix} N+\alpha+\nu+r-1 \\ \nu \end{pmatrix}} \theta_r(\bar{\alpha}; \bar{\nu})$$

and $\theta_r(\overline{\alpha}; \overline{\nu})$ is positive and does not depend on N. The proof will lead to a recurrence relation from which θ_r is easily computed.

For $r = 2$ the result is contained in (5.11) and (5.12). We proceed by induction on r. The left hand member of (5.13) can be written, using (5.6), as

$$(5.15) \quad \sum_{(m', \, m_r) \in \Delta_2(N)}$$

$$Q_{\nu_{r-1}}(m'-\nu';\alpha+2\nu'+r-2, \alpha_r, m-\nu'+1) \, Q_{\lambda_{r-1}}(m'-\lambda';\alpha'+2\lambda'+r-2, \alpha_r, m-\lambda'+1) \cdot S$$

where

$$(5.16) \quad S = \frac{\dbinom{m'}{\nu'}\dbinom{m'}{\lambda'}}{\dbinom{N}{\nu'}\dbinom{N}{\lambda'}} \cdot \frac{\dbinom{m_r+\alpha_r}{m_r}\dbinom{m'+\alpha'+r-2}{m'}}{\dbinom{N+\alpha+r-1}{N}}$$

$$\otimes \sum_{\overline{m}' \in \Delta_{r-1}(\overline{m}')}' \phi\left(\frac{\overline{m}}{\overline{\alpha}} \,\middle|\, \overline{\nu}'\right) \phi\left(\frac{\overline{m}}{\overline{\alpha}} \,\middle|\, \overline{\lambda}'\right) \frac{\dbinom{m_1+\alpha_1}{m_1} \cdots \dbinom{m_{r-1}+\alpha_{r-1}}{m_{r-1}}}{\dbinom{m'+\alpha'+r-2}{m'}}$$

$$= \delta_{\overline{\nu}', \overline{\lambda}'} \frac{\dbinom{m'}{\nu'}^2}{\dbinom{N}{\nu'}^2} \cdot \frac{\dbinom{m_r+\alpha_r}{m_r}\dbinom{m'+\alpha'+r-2}{m'}}{\dbinom{N+\alpha+r-1}{N}} \cdot \frac{\dbinom{m'+\alpha'+\nu'+r-2}{\nu'}}{\dbinom{m'}{\nu'}} \theta_{r-1}(\overline{\alpha}'; \overline{\nu}')$$

by the induction assumption. This is substituted into (5.15). Since $\overline{\lambda}' = \overline{\nu}'$ implies $\lambda' = \nu'$ the left member of (5.13) can be written as (note that $\dbinom{m'}{\nu'}$ is zero if $m' - \nu' < 0$),

(5.17) $\delta_{\overline{\nu'},\overline{\lambda'}}$ $\sum Q_{\nu_{r-1}}(m'-\nu';\alpha'+2\nu'+r-2,\alpha_r,m-\nu'+1)Q_{\lambda_{r-1}}(m'-\nu';\dots)$

$(m'-\nu',m_r)\epsilon\,\Delta_2(N-\nu')$

$\otimes\,\dfrac{\dbinom{m'+\alpha'+\nu'+r-2}{m'-\nu'}\dbinom{m_r+\alpha_r}{m_r}}{\dbinom{N+\alpha+\nu'+r-1}{N-\nu'}}\cdot K$

where

$$K=\frac{\dbinom{\alpha'+2\nu'+r-2}{2\nu'}\dbinom{2\nu'}{\nu'}\dbinom{N+\alpha+\nu'+r-1}{N-\nu'}}{\dbinom{N}{\nu'}^{2}\dbinom{N+\alpha+r-1}{N}\theta_{r-1}(\overline{\alpha'};\overline{\nu'})}$$

Since K does not depend on m' or m_r we can use (5.11) to evaluate (5.17). As a result the left member of (5.13) is found to be

$\delta_{\overline{\nu'},\overline{\lambda'}}\,\delta_{\nu_{r-1},\lambda_{r-1}}\cdot K\cdot\dfrac{\dbinom{N+\alpha+\nu+r-1}{\nu_{r-1}}}{\dbinom{N-\nu'}{\nu_{r-1}}\theta_2(\alpha'+2\nu'+r-2,\alpha_r;\nu_{r-1})}$

$=\delta_{\overline{\nu},\overline{\lambda}}\,\dfrac{\dbinom{N+\alpha+\nu+r-1}{\nu}\dbinom{\alpha'+2\nu'+r-2}{2\nu}}{\dbinom{N}{\nu}\dbinom{\alpha+2\nu'+r-1}{2\nu'}\theta_2(\alpha'+2\nu'+r-2,\alpha_r;\nu_{r-1})\theta_{r-1}(\overline{\alpha'};\overline{\nu'})}\,.$

This is of the required form and thus (5.13) is proved. The positive function θ_r is defined for $r=2$ by (5.12) and for $r>2$ defined recursively by

(5.18) $\theta_r(\overline{\alpha};\overline{\nu})=\dfrac{\dbinom{\alpha+2\nu'+r-1}{2\nu'}}{\dbinom{\alpha'+2\nu'+r-2}{2\nu'}}\,\theta_2(\alpha'+2\nu'+r-2,\alpha_r;\nu_{r-1})\theta_{r-1}(\overline{\alpha'};\overline{\nu'})\,.$

For each $\bar{v} \in \sum_r(N)$, formula (5.6) defines a non-zero function ϕ on $\Delta_r(N)$, and this collection of functions has been shown to be an orthogonal system relative to the positive weights $\pi(\bar{m})$. Since $\sum_r(N)$ and $\Delta_r(N)$ are of the same cardinality, it follows that the orthogonal system is complete.

The representation formula for the transition probabilities of the r-type Moran model can now be written down. Let

$$(5.19) \qquad c = \mu \, \theta[1 - (\beta_1 + \ldots + \beta_r)]/N \ .$$

Then the transition probabilities are

$$P_{\bar{m},\bar{n}}(t) = \pi(\bar{n}) \sum_{\bar{v} \in \sum_r(N)} \exp(-cz_v t) \, \phi\left(\begin{array}{c}\bar{m} \\ \bar{\alpha}\end{array}\middle|\ \bar{v}\right) \phi\left(\begin{array}{c}\bar{n} \\ \bar{\alpha}\end{array}\middle|\ \bar{v}\right) \frac{\binom{N}{v}\theta_r(\bar{\alpha};\bar{v})}{\binom{N+\alpha+v+r-1}{v}}$$

$$(5.20)$$

$$= \pi(\bar{n}) \sum_{v=0}^{N} \exp(-cz_v t) \, \frac{\binom{N}{v}}{\binom{N+\alpha+v+r-1}{v}} \sum_{\bar{v} \in \Delta_{r-1}(v)} \phi\left(\begin{array}{c}\bar{m} \\ \bar{\alpha}\end{array}\middle|\ \bar{v}\right) \phi\left(\begin{array}{c}\bar{n} \\ \bar{\alpha}\end{array}\middle|\ \bar{v}\right) \theta_r(\bar{\alpha};\bar{v}$$

where in the latter form terms belonging to the same eigenvalue z_v are grouped together.

6. <u>Linear growth model with r types.</u> The first order relations (5.9) and (5.10) are of the form (3.11a) and (3.11b) with

$$a(n) = n, \quad b(n) = (n+1-v)(n+\alpha+v+r)/(n+1) \ .$$

These values may now be substituted in (3.11c). The solution of the resulting recurrence relation, unique for $n \geq v$ except for a constant factor, is

$$(6.1) \qquad R_n(y) = \binom{n}{v} \ell_{n-v}^{(\alpha+2v+r-1)}(y) \ .$$

where $\ell_k^{(\beta)}$ is defined in (3.3).

We interpret the right member of (6.1) as zero when $n < \nu$. With this
interpretation formula (6.1) defines a function which satisfies the recurrence relation (3.11c) for all $n \geq 0$, since $b(\nu-1) = 0$.

Combining (6.1) and (5.1) as prescribed in (3.10) we obtain a collection of functions

$$(6.2) \qquad \Psi(\overline{m}) = \Psi(\overline{m};\overline{\nu}, y) = \binom{m}{\nu} \ell_{m-\nu}^{(\alpha+2\nu+r-1)}(y) \, \phi\left(\begin{matrix} \overline{m} \\ \overline{\alpha} \end{matrix} \middle| \overline{\nu} \right)$$

defined for $\overline{m} \in \Delta_r = \bigcup_{0 \leq N < \infty} \Delta_r(N)$ and depending on parameters $\overline{\nu}$, y ,

$\overline{\nu} \in \Sigma_r$, $y \in [0, \infty)$.

The function $\Psi(\overline{m};\overline{\nu}, y)$ is zero if $\nu < m$. From the orthogonality relations (5.13) and

$$(6.3) \qquad \int_0^\infty \ell_m^{(\gamma)}(y) \, \ell_n^{(\gamma)}(y) \, \frac{y^\gamma e^{-y}}{\Gamma(\gamma+1)} \, dy = \frac{\delta_{m,n}}{\binom{m+\gamma}{m}}$$

we find by a routine calculation that the functions (6.2) satisfy an orthogonality relation as follows.

Let

$$(6.4) \qquad w_\nu(y) = e^{-y} y^{\alpha+2\nu+r-1} / \Gamma(\alpha+2\nu+r) \ .$$

Then

$$\int_0^\infty \Psi(\overline{m};\overline{\nu}, y) \, \Psi(\overline{n};\overline{\nu}, y) w_\nu(y) dy$$

$$(6.5)$$

$$= \frac{\delta_{m,n}}{\binom{m+\alpha+r-1}{m}} \cdot \frac{\binom{m}{\nu}}{\binom{m+\alpha+\nu+r-1}{\nu}} \binom{\alpha+2\nu+r-1}{2\nu}\binom{2\nu}{\nu} \phi\left(\begin{matrix} \overline{m} \\ \overline{\alpha} \end{matrix} \middle| \overline{\nu}\right) \phi\left(\begin{matrix} \overline{n} \\ \overline{\alpha} \end{matrix} \middle| \overline{\nu}\right) \ .$$

From (6.5) and (5.13) we find

$$-\sum_{\nu \in \Sigma_r} \int_0^\infty \Psi(\overline{m};\overline{\nu}, y)\, \Psi(\overline{n};\overline{\nu}, y) w_\nu(y) dy \; \frac{\theta_r(\overline{\alpha};\overline{\nu})}{\binom{\alpha+2\nu+r-1}{2\nu}\binom{2\nu}{\nu}}$$

(6.6)

$$= \frac{\delta_{\overline{m},\,\overline{n}}}{K(\overline{m})}$$

where

(6.7) $$K(\overline{m}) = \binom{m_1+\alpha_1}{m_1}\binom{m_2+\alpha_2}{m_2} \cdots \binom{m_r+\alpha_r}{m_r} \; .$$

The transition probability $P_{\overline{m},\,\overline{n}}(t)$ of the r-dimensional linear growth process, which is given explicitly by (3.4) and (3.5) and which satisfies (3.6), is also given by the alternative explicit formula

(6.8) $$P_{\overline{m},\,\overline{n}}(t) = K(\overline{n}) \sum_{\nu \in \Sigma_r} \int_0^\infty e^{-yt}\, \Psi(\overline{m};\overline{\nu}, y)\Psi(\overline{n};\overline{\nu}, y) \; \frac{w_\nu(y) dy\, \theta_r(\overline{\alpha};\overline{\nu})}{\binom{\alpha+2\nu+r-1}{2\nu}\binom{2\nu}{\nu}} \; .$$

Formulas (3.5) and (3.6) express $P_{\overline{m},\,\overline{n}}(t)$ as a product of Laplace transforms, and this can be written as the transform of a convolution. Interesting identities can be found by comparing with (6.8).

7. <u>The eigenfunctions when</u> $\beta_1 + \ldots + \beta_r > 1$. The previous analysis must be modified to deal with the case (1.14) when $\beta_1 + \ldots + \beta_r > 1$. Note that from (1.4) we have for any k

$$\sum_{j \neq k} \beta_j \leq 1 \; .$$

We let $\sigma = \beta_1 + \ldots + \beta_r > 1$ and $\sigma_j = \sigma - \beta_j$. The probability $p_j(\overline{m})$ in (1.8) now assumes the form

286

$$p_j(\overline{m}) = \left[\sum_{k \neq j} \frac{m_k}{N} \beta_j + \frac{m_j}{N} (1-\sigma_j) \right] + (1-\theta)\gamma_j$$

$$= \frac{\theta(\sigma-1)}{N} \left[\frac{N\beta_j}{\sigma-1} - m_j \right] + (1-\theta)\gamma_j$$

and hence

(7.1) $$p_{i,j}(\overline{m}) = \frac{\theta(\sigma-1)}{N^2} m_i (N-m_j+a_j+1)$$

where

(7.2) $$a_j + 1 = \frac{N(1-\sigma_j)}{\sigma-1} + \frac{1-\theta}{\theta} \frac{N}{\sigma-1} \gamma_j .$$

We assume $\theta > 0$.

The recurrence relation satisfied by the eigenfunctions is therefore not (2.8) but rather

(7.3) $$-z'_\nu \phi_\nu(\overline{m}) = \sum_{\substack{i,j \\ i \neq j}} m_i (N-m_j+a_j+1) [\phi_\nu(\overline{m})_{i,j} - \phi_\nu(\overline{m})] .$$

The Hahn polynomials are defined by formula (4.1) as analytic functions of the complex variables α, β which are regular except if α has one of the values $-1, -2, \ldots, -N$. Various functional relations satisfied by the functions $\phi\left(\frac{\overline{m}}{\alpha} \middle| \overline{\nu}\right)$ when the $\alpha_i + 1$ are all positive can therefor be extended to other values of the α_i by analytic continuation.

We let

(7.4) $$\alpha_j = -(N - a_j + 2), \qquad j = 1, 2, \ldots, r$$

and determine the resulting form of the recurrence relation (2.8) satisfied by the functions ϕ . It is exactly (7.3) with

(7.5) $$z'_\nu = \nu(rN+a+r+1-\nu)$$

287

where $a = a_1 + \ldots + a_r$. Hence the functions $\phi\left(\dfrac{\overline{m}}{\alpha} \middle| \overline{\nu}\right)$ with α_j determined by (6.4) are eigenfunctions for the problem at hand.

REFERENCES

[1]   Karlin, S. and McGregor, J. L., "Linear Growth, Birth and Death Processes", J. Math. and Mech. 7 (1958), 643-662.

[2]   Karlin, S. and McGregor, J. L., "On a Genetics Model of Moran", Proc. Camb. Phil. Soc. 58 (1962), 299-311.

[3]   Karlin, S. and McGregor, J. L., 'On Some Stochastic Models in Genetics', Stochastic Models in Medicine and Biology Univ. of Wisconsin Press (1964), 245-279.

[4]   Karlin, S. and McGregor, J. L., "The Hahn Polynomials, Formulas and an Application", Scripta Math., XXVI (1961), 33-46.

Orthogonal Polynomials Revisited

K. M. Case

I. Introduction

Here we wish to summarize some recent work on the theory of orthogonal polynomials. The essential point is to study the spectral properties of the matrices associated with the recursion relations which the polynomials satisfy. We were led to this approach somewhat indirectly. In a primarily pedagogical article Kac and I[1] studied a discrete version of the inverse scattering problem. There it turned out that the properties of orthogonal polynomials throw considerable light on the methods used to solve that problem. It seemed natural to turn the question around and ask what insight is gained from looking at orthogonal polynomials from the viewpoint of scattering theory.

We find a certain unification of various well known results about orthogonal polynomials. It turns out that almost all properties are contained in the limiting value of a particular solution of the recursion relations. (In analogy with the corresponding function in scattering theory we call this the Jost function.) An explicit construction of this function in terms of the weight function is readily given. While this function is mentioned in the literature its simple origin and significance seems not to have been emphasized. In particular from the interpretation we are led to an infinite set of sum rules of which only the first one or two were noticed before.

In Section II the problem of polynomials defined on a finite portion of the real axis is discussed. Following this we give some applications. Namely we show how the asymptotic properties of the polynomials are simply obtained and how the new sum rules are produced. Section III is devoted to a sketch of the results for polynomials defined on the unit circle. While rather similar there are a number of amusing differences.

As indicated this is a review. Very little in the way of proofs will be given. They can be found in the references.[2]

II. Polynomials on the Real Axis

Let us collect some well known properties of orthogonal polynomials.[3]

Suppose we are given some non-decreasing function $\rho(\lambda)$ defined on the real axis. We are to find polynomials $\psi(\lambda, n)$ such that

(i) $\psi(\lambda, n)$ is a polynomial of exact degree n and its leading coefficient is positive

(ii) The orthonormality relations hold:

$$\int_{-\infty}^{\infty} \psi(\lambda, n)\, \psi(\lambda, m)\, d\rho(\lambda) = \delta(m, n) \quad . \tag{II.1}$$

The construction is a straightforward application of the Hilbert-Schmidt procedure. Indeed, if we denote the moments of ρ by s_m, i.e.,

$$s_m = \int_{-\infty}^{\infty} \lambda^m d\rho(\lambda) \quad , \tag{II.2}$$

the result is

$$\psi(\lambda, n) = \frac{1}{\sqrt{D_{n-1} D_n}}
\begin{vmatrix}
s_0 & s_1 & \cdots & s_n \\
s_1 & s_2 & \cdots & s_{n+1} \\
\cdots & \cdots & \cdots & \cdots \\
\cdots & \cdots & \cdots & \cdots \\
s_{n-1} & s_n & \cdots & s_{2n-1} \\
1 & \lambda & \cdots & \lambda^n
\end{vmatrix}
\qquad n = 1, 2, 3 \tag{II.3}$$

where

$$D_n = \begin{vmatrix} s_0 & s_1 & \cdots & s_n \\ s_1 & s_2 & \cdots & s_{n+1} \\ \cdots\cdots\cdots\cdots\cdots\cdots\cdots \\ s_n & s_{n+1} & \cdots & s_{2n} \end{vmatrix} . \qquad (II.4)$$

(The Eq. (II. 3) also holds for $n \geq 0$, if we <u>define</u> $D_{-1} = s_0^2$).

While we thus have an explicit formal solution, it is not very useful. However, from it one readily shows that the $\psi(\lambda, n)$ satisfy the 3-term recursion relation:

$$a(n+1)\psi(\lambda, n+1) + b(n)\psi(\lambda, n) + a(n)\psi(\lambda, n-1) = \lambda\psi(\lambda, n), \quad n = 1, 2, 3 \ldots (II.5)$$

where

$$b(n) = \int_{-\infty}^{\infty} \lambda\psi^2(\lambda, n)d\rho(\lambda) \quad ,$$

and

$$a(n+1) = \int_{-\infty}^{\infty} \lambda\psi(\lambda, n)\psi(\lambda, n+1)d\rho(\lambda) = \sqrt{\frac{D_{n+1}D_{n-1}}{D_n^2}} \quad . (II.6)$$

(Note: This also holds for $n = 0$ provided we define $a(0)\psi(\lambda, -1) = 0$.)

Now we will take Eq. (II. 5) plus the initial conditions

$$a(0)\psi(\lambda, -1) = 0, \quad \psi(\lambda, 0) = C = 1/\sqrt{s_0} \quad , \qquad (II.7)$$

as fundamental for our discussion of orthogonal polynomials.

Here we will restrict attention to the case when $a(\infty)$, $b(\infty)$ exist and the limits are approached at least as fast as $1/n^2$. We will see that in this case the support of $d\rho(\lambda)$ is compact.

We define as "regular" those solutions of Eq. (II. 5) with the given initial conditions which are bounded as $n \to \infty$. Under our conditions such exist for all λ such that

291

$$b(\infty) - 2a(\infty) \le \lambda \le b(\infty) + 2a(\infty) \ . \tag{II.8}$$

These are conveniently described by z such that

$$\lambda = b(\infty) + a(\infty)(z + z^{-1}) \ . \tag{II.9}$$

The statement then is that the Jacobi matrix formed from the $a(n)$, $b(n)$ has a continuous spectrum for λ in the interval described by Eq. (II.8) or, alternatively for z lying on the unit circle $(z = e^{i\theta})$. In addition there may be some discrete eigenvalues λ_i corresponding to square summable solutions. It is readily shown that these are real, simple, finite in number and outside of or at the edge of the continuum. In z they are real, within the unit circle or at $z = \pm 1$. (We remark that these results will be seen to imply that the $\rho(\lambda)$ used to form our orthogonal polynomials has only a finite number of jumps outside the interval plus a continuous part within the interval.)

As in scattering theory it is convenient to introduce two other solutions of Eq. (II.5), $\psi_{\pm}(z, n)$ by the conditions

$$\lim_{n \to \infty} |\psi_{\pm}(z, n) - z^{\pm n}| = 0, \quad \begin{matrix} |z| \le 1 \\ |z| \ge 1 \end{matrix} \ . \tag{II.10}$$

Then we will define $f_{\pm}(z)$ by:

$$f_{\pm}(z) = a(0) \, \psi_{\pm}(z, -1) \ ,$$

i.e. we use Eq. (II.5) at $n = 0$ to define the $f_{\pm}$. We call $f_{+}(z)$ the "Jost function" since it will be seen to play the same fundamental role for orthogonal polynomials as $f_{+}(k)$ does in scattering theory. Thus:

1) f_{+} determines the asymptotic behavior of the continuum functions. Proof: Using the analog of the Wronskian theorem for solutions of Eq. (II.5), i.e.

$$W\left[\psi^{(1)}, \psi^{(2)}\right] = a(n+1)\left\{\psi^{(1)}(\lambda, n+1)\psi^{(2)}(\lambda, n) - \psi^{(2)}(\lambda, n+1)\psi^{(1)}(\lambda, n)\right\}$$
$$= \text{constant} \tag{II.11}$$

we can, using the boundary conditions, express the continuum $\psi(\lambda, n)$ in terms of the $\psi_{\pm}$ as:

$$\psi(\lambda, n) = \frac{C}{a(\infty)2i \sin \theta} \{f_-(z)\psi_+(z, n) - f_+(z)\psi_-(z, n)\} , \quad z = e^{i\theta} . \quad \text{(II.12)}$$

Then the asymptotic behavior is:

$$\psi(\lambda, n) \xrightarrow[n \to \infty]{} \frac{C|f_+(z)|\sin(n\theta+\delta)}{a(\infty) \sin \theta} , \quad \text{(II.13)}$$

with $\delta = -\arg f_+$.

2) The zeros of $f_+(z)$ within the unit circle determine the discrete eigenvalues, i. e. obviously

$$f_+(z_i) = 0 \to \psi(\lambda_i, n) = \frac{C\psi_+(z_i, n)}{\psi_+(z_i, 0)} . \quad \text{(II.14)}$$

3) The "orthogonality" of solutions corresponding to different λ's in Eq. (II.15) can be shown as usual by a Green's type identity and use of the boundary condition. The result is:

$$\sum_{n=0}^{\infty} \psi(\lambda, n) \psi(\lambda', n) = \frac{\delta(\lambda-\lambda')}{\rho'(\lambda)} , \quad \lambda, \lambda' \varepsilon \text{ continuum},$$

$$\rho'(\lambda) = \frac{a(\infty)\sin \theta}{\pi C^2 |f_+|^2} , \quad \text{(II.15)}$$

$$\sum_{n=0}^{\infty} \psi(\lambda_i, n) \psi(\lambda_j, n) = \frac{\delta(\lambda_i, \lambda_j)}{\rho_i} , \quad \lambda_i, \lambda_j \text{ discrete},$$

$$\rho_i = \psi_+(z_i, 0)/C^2 \left(\frac{df_+}{dz}\right)_{\lambda_i} ,$$

and

$$\sum_{n=0}^{\infty} \psi(\lambda_i, n) \psi(\lambda, m) = 0, \quad \lambda\varepsilon \text{ continuum} .$$

(We see $df_+/d\lambda_i \neq 0$, i.e. the eigenvalues are simple). These "orthogonal relations" are really the completeness relations for our orthogonal polynomials. It is perhaps remarkable that such a simple proof of completeness follows from the present approach.

The true orthogonality relations are, of course, those of Eq. (II.1). We have apparently begged the question here. Our polynomials are indeed "complete" as derived here with ρ related to f_+ as indicated in Eq. (II.15). However, we really need to show that this is the $\rho(\lambda)$ with which we originally defined our orthogonal polynomials. Thus we need to show that if

$$d\rho(\lambda) = \rho'(\lambda)d\lambda, \quad b(\infty) - 2a(\infty) \leq \lambda \leq b(\infty) + 2a(\infty) \quad ,$$

$$= \sum_i \rho_i \, \delta(\lambda, \lambda_i)d\lambda, \quad \lambda \text{ not in the interval}, \tag{II.16}$$

where $\rho'(\lambda)$, ρ_i are as written in Eqs. (II.15), then Eq. (II.1) holds.

Let us sketch the proof. First note some analytic properties.

(i) $\psi(\lambda, n)$ is analytic within the unit circle, except for a pole at $z = 0$. (It is a polynomial in λ).

(ii) $\psi_+(z, n)$ is analytic within the unit circle. The proof of this is more complicated. We construct the discrete analog of an integral equation, solve this by iteration, and prove the appropriate convergence by using the assumed bounds on $a(n)$ and $b(n)$.

(iii) $f_+(z)$ is analytic within the unit circle except for a simple pole at $z = 0$. Indeed from the definition we have

$$f_+(z) = [\lambda - b(0)] \psi_+(z, 0) - a(1) \psi_+(z, 1) \quad , \tag{II.17}$$

and by direct calculation from Eq. (II.5)

$$\left. \text{Res } f_+(z) \right|_{z=0} = a(\infty) \prod_{i=1}^{\infty} \frac{a(\infty)}{a(i)} \neq 0 \quad . \tag{II.18}$$

Second, consider the integral

$$I = \frac{1}{2\pi i} \oint G(\lambda, n;m)d\lambda \, , \qquad\qquad (II.19)$$

where

$$G(\lambda, n;m) = \begin{cases} -\psi(\lambda, n) \, \psi_+(\lambda, m), & n \le m \\ -\psi_+(\lambda, n) \, \psi(\lambda, m), & n \ge m \, , \end{cases} \qquad (II.20)$$

and the path is such that z goes around the unit circle.

Now we evaluate I in two ways.

a) We transform it to an integral over the continuum range of λ .

b) We calculate it by residues.

Equating the two results yields Eqs. (II.1) with the $\rho(\lambda)$ related to f_+ as given by Eqs. (II.15).

Note that we have paid for the simplicity of the completeness proof with a more complicated orthogonality proof.

Our conclusion now is that f_+ contains most of the essential information about our polynomials. The central problem is then given $\rho(\lambda)$ to construct f_+ . This is simply done. We are to construct a function analytic within the unit circle given its absolute value on the circle and its zeros (z_i) within. A simple variant of the Poisson-Jensen formula yields:

$$f_+(z) = \frac{\underset{+}{\Pi} (z_i - z) \underset{-}{\Pi} (z - z_i)}{z \, \Pi_i (1 - z_i z)} \exp J \, , \qquad (II.21)$$

with

$$J = -\frac{(z - z^{-1})}{4\pi} \int\limits_{b(\infty) - 2a(\infty)}^{b(\infty) + 2a(\infty)} \frac{d\lambda'}{\sin \theta'(\lambda' - \lambda)} \, \ell n \left[\frac{\pi C^2 \rho'(\lambda')}{a(\infty) \sin \theta'} \right] \, .$$

(Here $\underset{+}{\Pi}(\underset{-}{\Pi})$ means product over positive (negative) z_i).

295

III.   Applications

A) Asymptotic formulae.

The problem we consider is the following: Given $d\rho(\lambda)$, what is the asymptotic form of our polynomials as $n \to \infty$. The solution is very straightforward. Given $d\rho(\lambda)$ we compute f_+ from Eqs. (II. 21). Then the asymptotic form is given by Eq. (II. 13).

Consider, as an example, the simple case of the Legendre polynomials. Then:

$$d\rho(\lambda) = d\lambda, \quad -1 \le \lambda \le 1$$
$$= 0, \quad |\lambda| > 1$$
$$\rho_i = 0 \quad \text{(no jumps)} \ . \tag{III. 1}$$

Thus

$$b(\infty) = 0, \quad a(\infty) = 1/2, \quad C = 1/\sqrt{\int d\rho} = 1/\sqrt{2} \ . \tag{III. 2}$$

Inserting in Eqs. (II. 21) yields

$$f_+(z) = \left(\frac{z^{-2}-1}{2\pi} \right)^{1/2} , \tag{III. 3}$$

and then Eq. (II. 13) gives

$$\psi(\lambda, n) \to \left(\frac{2}{\pi \sin \theta} \right)^{1/2} \cos\left[(n+1/2)\theta - \frac{\pi}{4} \right] . \tag{III. 4}$$

This is the conventional result since our polynomials are <u>normalized</u> such that

$$\psi(\lambda, n) = \sqrt{\frac{2n+1}{2}} \ P_n(\lambda) \ . \tag{III. 5}$$

B) Sum Rules

These relate sums of products of the $a(n)$, $b(n)$ to integrals of the logarithm of the weight function. Their existence becomes obvious when we remember that $f_+(z) = a(0) \, \psi_+(z, -1)$. Since f_+ has only a simple pole at $z = 0$ it can be expanded in a Laurent series there. The coefficients can be obtained as integrals involving $\ell n \rho'$ and the z_i by expanding Eqs. (II. 21). Alternatively we can find these coefficients by solving Eq. (II. 5) subject to the boundary condition of Eq. (II. 10) for small z . The coefficients will then be expressed in terms of sums of powers of the $a(n), b(n)$. Equating the two expressions yields the sum rules. Explained in this way, the calculations would seem to be very tedious and the relations messy. Remarkably enough they turn out to be very simple and have an elegant closed form.

First some notation. Let

$$\alpha(n) = a(n)/a(\infty), \quad \beta(n) = \frac{b(n) - b(\infty)}{a(\infty)}, \quad \mu = z + z^{-1} = \frac{\lambda - b(\infty)}{a(\infty)} \quad \text{(III. 6)}$$

Then Eq. (II. 5) becomes

$$(L \psi)(\mu, n) = \mu \psi(\mu, n)$$

with

$$(L \psi)(\mu, n) = \alpha(n+1) \, \psi(\mu, n+1) + \beta(n) \, \psi(\mu, n) + \alpha(n) \, \psi(\mu, n-1) . \quad \text{(III. 7)}$$

(Effectively the general problem has been reduced to the case $a(\infty) = 1$, $b(\infty) = 0$.)

Similarly we introduce a comparison operator L_0 defined by

$$(L_0 \psi)(\mu, n) = \psi(\mu, n+1) + \psi(\mu, n-1) \quad \text{(III. 8)}$$

The general result is:

$$\gamma_n = \sum_{m=1}^{n} \frac{\sigma_{mn}}{m} \, \text{Tr} \left[L_0^m - L^m \right], \quad n = 1, 2, \ldots \quad \text{(III. 9)}$$

Here:

$$\gamma_n = \frac{1}{\pi} \int_0^{\pi} \cos n\theta' \, \ell n \left| \frac{a(\infty)\sin\theta'}{\pi C^2_{\rho'}} \right| d\theta'$$

$$+ \sum_i \frac{\left[(z_i)^n - (z_i)^{-n} \right]}{n} \,. \qquad \text{(III.10)}$$

The σ_{mn} are universal and simple. Thus:

$$\sigma_{mn} = 0, \qquad m > n$$

$$\sigma_{mn} = 0, \quad \text{unless} \ n - m \ \text{is even and otherwise}$$

$$\sigma_{mn} = \begin{pmatrix} \frac{n+m}{2} \\ m-1 \end{pmatrix} (-1)^{\frac{n-m}{2}} \,. \qquad \text{(III.11)}$$

There is one more rule. This is obtained by equating the two expressions for the residue of $f_+(z)$ at 0. Thus:

$$a(\infty) \prod_{i=1}^{\infty} a(\infty)/a(i) = \prod_i |z_i| \exp\int_0^{\pi} \ell n \left[\frac{a(\infty)\sin\theta'}{C^2_{\rho'}} \right] d\theta' = R \quad \text{(III.12)}$$

Some remarks:

(i) On taking the logarithm of Eq. (III.12) we see that it is in some sense the extension of Eqs. (III.9, 10) to $n = 0$.

(ii) In a distantly related form Eq. (III.12) has been known before. Thus from this and the relation of the D_n to the $a(n)$ Eq. (II.6) we can derive the relation for the Hankel forms D_n that

$$\lim_{n\to\infty} (D_n)^{1/n} a^n(\infty) = a^3(\infty)/D_0 R^2 \,. \qquad \text{(III.13)}$$

The analogous result for Toeplitz forms has been given by Szegö.[4]

As an example let us again consider the case of the Legendre polynomials.

Then Eq. (III.12) becomes

$$\frac{1}{2} \prod_{i=1}^{\infty} 1/\alpha(i) = (2\pi)^{-\frac{1}{2}} \qquad \text{(III.14)}$$

and since

$$\gamma_n = \frac{1}{\pi} \int_0^{\pi} \cos n\theta' \, \ln \frac{\sin \theta'}{\pi} \, d\theta'$$

$$= \begin{cases} -\dfrac{1}{n} & n \text{ even} \\[2ex] 0 & n \text{ odd} \end{cases} \qquad \text{(III.15)}$$

we have for the first few non-trivial[5] sum rules

$$\frac{1}{2} = \sum_{n=1}^{\infty} [\alpha^2(n) - 1]$$

$$5 = \sum_{n=1}^{\infty} \left\{ 2[\alpha^4(n) - 1] + 4[\alpha^2(n)\alpha^2(n+1) - 1] \right\} ,$$

$$11 = \sum_{n=0}^{\infty} [\alpha^6(n+1) - 1] + 3 \sum_{n=0}^{\infty} \left\{ \alpha^2(n+1)[\alpha^2(n+1)\alpha^2(n+2) \right.$$

$$\left. + \alpha^4(n+2) + \alpha^2(n+2) \, \alpha^2(n+3)] - 3 \right\} . \qquad \text{(III.16)}$$

If we insert the known form for the coefficients

$$\alpha^2(n) = n^2/(n^2 - \tfrac{1}{4}) \qquad \text{(III.17)}$$

we can check the relations by direct, if extremely tedious, calculation.

IV. Polynomials on the Unit Circle[6]

Here we are given a positive function $\rho'(\theta)$ defined on $-\pi \le \theta \le \pi$. We are to find polynomials $\phi(z, n)$ $(z = e^{i\theta})$ such that

$$\frac{1}{2\pi} \int_{-\pi}^{\pi} \rho'(\theta) \, \phi(z, n) \, \overline{\phi(z, m)} \, d\theta = \delta(m, n), \quad n, m = 0, 1, 2, \ldots \quad \text{(IV.1)}$$

Again the standard orthogonalization techniques give an explicit (but not very useful) answer: Thus if

$$c_n = \frac{1}{2\pi} \int_{-\pi}^{\pi} \rho'(\theta) \, e^{-in\theta} \, d\theta, \quad n = 0, \pm 1, \pm 2, \ldots \quad \text{(IV.2)}$$

we have

$$\phi(z, n) = (D_{n-1} D_n)^{\frac{1}{2}} \begin{vmatrix} c_0 & c_{-1} & \cdots & c_{-n} \\ c_1 & c_0 & \cdots & c_{-n+1} \\ \cdots\cdots\cdots\cdots\cdots\cdots\cdots \\ c_{n-1} & c_{n-2} & \cdots & c_{-1} \\ 1 & z & \cdots & z^n \end{vmatrix}, \quad \text{(IV.3)}$$

where

$$D_n = \begin{vmatrix} c_0 & c_{-1} & c_{-2} & \cdots & c_{-n} \\ \cdots\cdots\cdots\cdots\cdots\cdots\cdots \\ c_n & \cdots\cdots\cdots\cdots & c_0 \end{vmatrix} \quad n = 1, 2, 3, \ldots \quad \text{(IV.4)}$$

(The Eq. (IV.3) also holds for $n = 0$ provided we <u>define</u> $D_{-1} = c_0^2$) .

Instead of working with these solutions we merely note that they imply the recursion relations

$$\phi(z, n+1) = \frac{\kappa(n+1)}{\kappa(n)} \, z\phi(z, n) + \frac{\alpha(n+1)}{\kappa(n)} \, z^n \, \overline{\phi}\left(\frac{1}{z}, n\right)$$

$$\overline{\phi}(1/z, n+1) = \frac{\kappa(n+1)}{\kappa(n)} \, z^{-1} \, \overline{\phi}(1/z, n) + \frac{\overline{\alpha}(n+1)}{\kappa(n)} z^{-n} \phi(z, n) , \quad \text{(IV.5)}$$

where

$$\kappa(n) = (D_{n-1} D_n^{-1})^{\frac{1}{2}} , \tag{IV. 6}$$

$\alpha(n) = \phi(0, n),$ and we have the relation

$$|\alpha(n+1)|^2 = \kappa^2(n+1) - \kappa^2(n) . \tag{IV. 7}$$

(Here the notation is used that $\overline{\phi(z, n)}$ is the complex conjugate, while $\overline{\phi}(z, n)$ is the polynomial obtained by using the complex conjugate coefficients).

It is convenient to group the $\phi(z, n)$ and $\overline{\phi}(1/z, n)$ into one two-component quantity $\Psi(z, n)$ such that

$$\Psi(z, n) = \begin{pmatrix} \phi(z, n) \\ \overline{\phi}(\frac{1}{z}, n) \end{pmatrix} . \tag{IV. 8}$$

Then our two recursion relations can be written

$$\Psi(z, n+1) = A(n)\Psi(z, n) . \tag{IV. 9}$$

with

$$A(n) = \begin{pmatrix} \dfrac{\kappa(n+1)}{\kappa(n)} z & \dfrac{\alpha(n+1)}{\kappa(n)} z^n \\ \dfrac{\alpha(n+1)}{\kappa(n)} z^{-n} & \dfrac{\kappa(n+1)}{\kappa(n)} z^{-1} \end{pmatrix} . \tag{IV. 10}$$

We take Eqs. (IV. 9, 10) with the initial condition $\phi(z, 0) = 1/\sqrt{c_0}$ to be basic.

Now let us assume that $\lim\limits_{n \to \infty} \kappa(n)$ exists. We notice that

$$\lim_{n \to \infty} A(n) = \begin{pmatrix} z & 0 \\ 0 & z^{-1} \end{pmatrix} . \tag{IV. 11}$$

301

Accordingly we define 2 solutions of Eqs. (IV. 9, 10)

$$\Psi^{\pm}(z, n) = \begin{pmatrix} \phi_{\pm}(z, n) \\ \phi'_{\pm}(z, n) \end{pmatrix}$$ (IV. 12)

by

$$\lim_{n \to \infty} \left| \Psi^{+}(z, n) - \begin{pmatrix} z^n \\ 0 \end{pmatrix} \right| = 0, \quad |z| \le 1 \quad ,$$

$$\lim_{n \to \infty} \left| \Psi^{-}(z, n) - \begin{pmatrix} 0 \\ z^{-n} \end{pmatrix} \right| = 0, \quad |z| \ge 1 \quad .$$ (IV. 13)

We again have the analog of a Wronskian theorem. Namely, if Ψ, Ψ' are two solutions of Eqs. (IV. 9, 10) we find that

$$W[\Psi, \Psi'] = \tilde{\Psi}(z, n) \begin{pmatrix} 0 & -1 \\ 1 & 0 \end{pmatrix} \Psi'(z, n)$$ (IV. 14)

is independent of n . In particular

$$W[\Psi^{-}, \Psi^{+}] = 1 \quad .$$ (IV. 15)

Hence Ψ^{+} and Ψ^{-} are linearly independent and we can write

$$\Psi(z, n) = f^{-}(z)\Psi^{+}(z, n) + f^{+}(z)\Psi^{-}(z, n), \quad |z| = 1 \quad ,$$ (IV. 16)

where

$$f^{\pm}(z) = \tilde{\Psi}(z, n) \begin{pmatrix} 0 & -1 \\ 1 & 0 \end{pmatrix} \Psi^{\pm}(z, n) \quad .$$ (IV. 17)

·Now since these functions are independent of n it is simplest to evaluate them in the limit $n \to \infty$, since there $\Psi^{\pm}$ are simple. In particular

302

$$f^+(z) = \lim_{n\to\infty} z^n \phi(1/z, n), \quad |z| \leq 1 \ . \tag{IV.18}$$

We will again call $f^+(z)$ the Jost function – since it has such similar properties to that encountered in Section II. Thus:

(i) $f^+(z)$ is analytic in the unit circle

(ii) f^+ is non-zero in the unit circle

(iii) The completeness relation for our polynomials can be shown to be

$$\frac{1}{2} \sum_{n=0}^{\infty} \tilde{\Psi}(z, n) \begin{pmatrix} 1+\bar{z}'z & 0 \\ 0 & 1+z'\bar{z} \end{pmatrix} \overline{\Psi(z', n)}$$

$$= \frac{1}{2} (1+\bar{z}'z) \sum_{n=0}^{\infty} \overline{\phi(z', n)}\phi(z, n)$$

$$+ (1+z'\bar{z}) \sum_{n=0}^{\infty} \phi(z', n) \overline{\phi}(z, n) = |f^+(z)|^2 \delta(\theta_1-\theta_2) \tag{IV.19}$$

while (iv) the orthogonality relations are

$$\frac{1}{2\pi} \int_{-\pi}^{\pi} \frac{\phi(z, n)\overline{\phi(z, m)}\,d\theta}{|f^+(z)|^2} = \delta(n, m) \tag{IV.20}$$

Thus we see

$$\rho'(\theta) = |f^+(z)|^{-2}, \quad (z = e^{i\theta}) \ . \tag{IV.21}$$

Since we have seen that the Jost function tells us most of the interesting properties we would like an explicit expression in terms of the weight function $\rho'(\theta)$. Again this is obtained from the Poisson-Jensen formula with the result:

$$f^+(z) = \exp -\frac{1}{4\pi} \int_{-\pi}^{\pi} [\ln \rho'(\theta)] \frac{[e^{i\theta}+z]}{[e^{-i\theta}-z]} \, d\theta, \quad |z| < 1 \ . \tag{IV.22}$$

As a simple application we note that we again have an infinite number of sum rules. Thus, let us expand $f^+(z)$ around $z = 0$. (It is now regular there). From Eqs. (IV. 9, 10) we can find the coefficients as sums of powers of the $\kappa(n)$ and $\alpha(n)$. On the other hand if we expand Eq. (IV. 22) we have expressions for the expansion coefficients as the Fourier coefficients of $\ell n \rho'$.

V. Conclusion

It is hoped that the advantages of the approach adopted here has been demonstrated. We consider as basic the description of orthogonal polynomials by their recursion relations. These are regarded as eigenvalue equations and the methods of scattering theory are applied. Previously known and previously unknown results are obtained in a unified fashion.

FOOTNOTES

1. K. M. Case and M. Kac, J. Math. Phys., 14, 594 (1973).
2. K. M. Case, J. Math. Phys., 15, 2166 (1974).
   K. M. Case, J. Math. Phys., (to be published).
   K. M. Case and J. Geronimo, J. Math. Phys., (In preparation).
3. See, for example, N. I. Akhiezer, "The Classical Moment Problem", (Oliver and Boyd, Edinburgh, 1965).
4. G. Szegö, Communications du Seminaire Mathematique de l'Université de Lund, tome supplémentaire dédié a M. Riesz, 228, (1952).
5. We say non-trivial since in the case $\beta(n) \equiv 0$, as it is in the Legendre case, the odd sum rules are merely the statement $0 = 0$.
6. This Section describes work done in collaboration with J. Geronimo, cf. Reference (2).

This work was supported in part by the Air Force Office of Scientific Research, Grant 722187.

Symmetry, Separation of Variables, and Special Functions
Willard Miller, Jr.

In this survey we describe a method whereby one can associate a Lie symmetry algebra $\mathcal{L}$ of Lie derivatives, hence a local Lie symmetry group G, with each of the principal linear homogeneous second order partial differential equations of mathematical physics, $T\Phi = 0^{(\dagger)}$. The elements of $\mathcal{L}$ are just those Lie derivatives which map all solutions of the differential equation into other solutions. Similarly we define the symmetry vector space $\mathcal{S}$ of all first and second order linear partial differential operators which map solutions of (†) into solutions. We then show that every coordinate system for which variables separate in (†), hence all separated solutions of (†), can be characterized by a set $\mathcal{L}_K = \{L_1, \ldots, L_K\}$ of commuting operators in $\mathcal{S}$. The separated solutions are exactly those solutions of (†) which are simultaneous eigenfunctions of the elements of $\mathcal{L}_K$. Here, the K eigenvalues are the separation constants.

For equations of especially high symmetry, e. g., the Laplace, wave, heat, and free-particle Schrödinger equations, the elements of $\mathcal{S}$ belong to the enveloping algebra of $\mathcal{L}$ and the problem of relating separated solutions of (†) becomes a special case of the representation theory of $\mathcal{L}$ and G. For equations of lower symmetry, e. g., the Schrödinger equation for the Kepler problem, $\mathcal{S}$ contains elements not in the enveloping algebra of $\mathcal{L}$. However, even in this case $\mathcal{L}$ acts on $\mathcal{S}$ via the adjoint representation and splits $\mathcal{S}$ into $\mathcal{L}$-orbits which can be associated with the separable coordinate systems. Once again the representation theory of $\mathcal{L}$ provides much information about the separable solutions of (†) and their relationships.

The methods discussed here were first explicitly advocated in a few special cases by Winternitz and collaborators in 1965. The general application of these methods to derive new results in special function theory is due to C. Boyer, E. G. Kalnins, J. Patera, P. Winternitz and the author.

The survey will be illustrated by many examples showing that group theory is a powerful tool permitting derivation of new results and rational classification of old results not only for functions of hypergeometric type, but for all functions obtained via separation of variables, e.g., the Mathieu, Lamé, Ince, and spheroidal functions.

Section 1. The Helmholtz equation.

Let $Q\psi = 0$ be a linear homogeneous second-order partial differential equation where $\psi = \psi(x_1, \ldots, x_n) = \psi(x)$ and

$$(1.1) \qquad Q = \sum_{i,j=1}^{n} A_{ij}(x) \partial_{x_i x_j} + \sum_{i=1}^{n} B_i(x) \partial_{x_i} + C(x) .$$

Here the A_{ij}, B_i, C and ψ are analytic functions of $x_1, \ldots, x_n$ in some common domain $\mathcal{D}$. The <u>symmetry algebra</u> $\mathcal{G}$ of this differential equation is the set of all linear differential operators

$$(1.2) \qquad L = \sum_{i=1}^{n} D_i(x) \partial_{x_i} + E(x)$$

with analytic coefficients D_i, E, such that $QL\psi = 0$ whenever $Q\psi = 0$, i.e., $\mathcal{G}$ is the set of all first-order operators L which map the null space $\mathfrak{I}$ of Q into itself. It follows from this definition that $L \in \mathcal{G}$ if and only if

$$(1.3) \qquad [L, Q] = R_L(x)Q$$

where $[L, Q] = LQ - QL$ is the commutator of L and Q, and R_L is an analytic function.

306

It is now easy to show that $\mathcal{L}$ is a Lie algebra, i. e., $a_1 L_1 + a_2 L_2 \in \mathcal{L}$ and $[L_1, L_2] \in \mathcal{L}$ for all $L_1, L_2 \in \mathcal{L}$ and all constants a_1, a_2. For $n \geq 3$ this Lie Algebra is always finite-dimensional whereas for $n = 2$, e. g., $\Delta_2 \psi = 0$ and $(\partial_{tt} - \partial_{xx})\psi = 0$, $\mathcal{L}$ may be infinite-dimensional. When $\mathcal{L}$ is finite-dimensional, it follows from standard results in Lie theory that we can construct a local Lie group G of operators $\exp aL$, $L \in \mathcal{L}$, which also map solutions into solutions. Here,

(1. 4) $[\exp(aL)f](x) = \nu(x, a) \, g(x(a))$

where f is any analytic function and $x(a)$, $\nu(x, a)$ are uniquely determined by the differential equations

$$\frac{dx_i}{da}(a) = D_i(x(a)), \quad i = 1, \ldots, n, \quad \frac{d}{da} \, \nu(x, a) = \nu(x, a) \, E(x(a)) \ ,$$

(1. 5)

$$x_i(0) = x_i, \quad \nu(x, 0) = 1 \ ,$$

see [1]. We call G the (local Lie) <u>symmetry group</u> of the equation $Q\psi = 0$.

We also associate with Q the vector space $\mathbf{g}'$ consisting of all first and second order operators S, of the general form (1. 1), which map solutions into solutions. It follows that $S \in \mathbf{g}'$ if and only if

(1. 6) $[S, Q] = U_S Q$

where U_S is a first order differential operator, i. e., an operator of the form (1. 2). It is obvious that $f(x)Q \in \mathbf{g}'$ for any analytic function f. Let $\mathcal{J}$ be the space of all such trivial symmetries fQ, and denote by $\mathbf{g}$ the factor space $\mathbf{g}'/\mathcal{J}$ of non-trivial symmetries. (Note that fQ acts like the zero operator on $\mathcal{J}$.) Here, $\mathbf{g}$ is in general not a Lie algebra since the commutator $[S_1, S_2]$ of two second-order symmetries may be third order. However, $[L, S] \in \mathbf{g}$ for all $L \in \mathcal{L}$, $S \in \mathbf{g}$ so $\mathcal{L}$, hence G, acts on $\mathbf{g}$ via the adjoint representation.

307

Note that the null space $\mathfrak{J}$ of Q is a representation space for $\mathscr{L}$. If $S \in \mathfrak{S}$ and S belongs to the enveloping algebra of $\mathscr{L}$, i. e., S can be expressed as a second-order polynomial in the elements of $\mathscr{L}$, then the problem of computing eigenfunctions of S in $\mathfrak{J}$ reduces to a problem in the representation theory of $\mathscr{L}$. We call an equation $Q\psi = 0$ in which every second-order symmetry S belongs to the enveloping algebra of $\mathscr{L}$, an equation of <u>class I</u>. If there is an $S \in \mathfrak{S}$ which doesn't belong to the enveloping algebra of $\mathscr{L}$ then the equation is of <u>class II</u>.

The most important equations of mathematical physics, e. g., the wave, Laplace, heat, and free-particle Schrödinger equations, are of class I. For these equations all question involving second-order symmetries reduce to problems concerning the enveloping algebra of $\mathscr{L}$. Furthermore, many class II equations can be shown to arise from class I equations by partial separation of variables. We will show that for all such equations, the problem of separation of variables is equivalent to a problem in the representation theory of $\mathscr{L}$.

The most important class II equation which doesn't arise directly from one of class I is the Schrödinger equation for the two-body Kepler problem (hydrogen atom). Here too, the representation theory of $\mathscr{L}$ is important for the separation of variable problem but one must also examine explicitly the elements of $\mathfrak{S}$ which don't lie in the enveloping algebra of $\mathscr{L}$, see [2].

To understand the relationship between elements of $\mathfrak{S}$ and separation of variables we consider a simple example, the Helmholtz equation

(1.7)
$$(\Delta_2 + k^2)\psi = 0 \ .$$

Here, $\Delta_2 = \partial_{xx} + \partial_{yy}$ and k is a non-zero real constant. Let $u(x, y)$, $v(x, y)$ be another real coordinate system defined, analytic, and with nonzero Jacobian in some open set $\mathfrak{O}$ of the real plane. We say that variables separate in (1.7) in terms of the u, v coordinates if the equation (1.7) for $\psi(u, v) = U(u)V(v)$ is equivalent to the two ordinary differential equations.

(1.8)

$$AU \equiv (A_1(u) \frac{d^2}{du^2} + A_2(u)\frac{d}{du}) U = -(k_1\Phi_{11}(u) + k_2\Phi_{12}(u))U$$

$$BV \equiv (B_1(v) \frac{d^2}{dv^2} + B_2(v)\frac{d}{dv}) V = -(k_1\Phi_{21}(v) + k_2\Phi_{22}(v))V \ .$$

Here $k_1 = k^2$ and k_2 is the (arbitrary) separation constant. We assume

$$\det \begin{pmatrix} \Phi_{11} & \Phi_{12} \\ \Phi_{21} & \Phi_{22} \end{pmatrix} = \Phi \neq 0$$

in the region $\emptyset$. The relationship between the separated equations (1.8) and (1.7) is well-known, [3]. Indeed,

(1.9) $$\Phi^{-1}(\Phi_{22}A - \Phi_{12}B) + k^2 = \Delta_2 + k^2 = Q \ .$$

The separable coordinate systems (u, v) are just those coordinate systems in which $\Delta_2 + k^2$ can be expressed in the form of the left-hand side of (1.9). Setting $\psi = UV$ we see from (1.9) that

(1.10) $$U^{-1}\Phi_{12}^{-1}\{AU + k^2\Phi_{11}U\} = V^{-1}\Phi_{22}^{-1}\{BV + k^2\Phi_{21}V\} = -k_2$$

which gives the explicit separation of the variables. Note also that the operator

(1.11) $$S = \Phi^{-1}(\Phi_{21}A - \Phi_{11}B)$$

has the property

(1.12) $$S(UV) = k_2(UV), \quad Q(UV) = 0 \ ,$$

i.e., the separated solutions UV are eigenfunctions of S with eigenvalue k_2, the separation constant corresponding to U and V .

309

(1.17)
$$(P_1^2 + P_2^2 + k^2)\psi = 0 \ .$$

A basis for $\mathfrak{s}$ is provided by the second-order operators

(1.18)
$$P_2^2, \ P_1 P_2, \ M^2, \ \{M, P_1\}, \ \{M, P_2\}$$

in addition to the three operators (1.15). (Here, $\{A, B\} = AB + BA$) . It follows that (1.7) is a class I equation.

Let $\mathfrak{O}$ be the five-dimensional subspace of $\mathfrak{s}$ spanned by the second-order operators (1.18). It is easy to show that $\mathcal{E}(2)$, hence the Euclidean group $E(2)$ acts on $\mathfrak{O}$ via the adjoint representation. Thus if $g \in E(2)$ and $S \in \mathfrak{O}$, then $S' = gSg^{-1} \in \mathfrak{O}$. If S and cS', c a nonzero real constant, are related to separable coordinate systems $\{u, v\}$ and $\{u', v'\}$ via (1.13) it follows that one system can be obtained from the other by the Euclidean transformation g . We regard two such coordinate systems as equivalent. Thus distinct separable coordinate systems are associated with orbits in $\mathfrak{O}$ under the adjoint action of $E(2)$.

In reference [4] it is shown that $\mathfrak{O}$ consists of four orbit types with representatives

(1.19)
$$P_2^2, \ M^2, \ \{M, P_2\}, \ M^2 - a^2 P_2^2, \ a \neq 0 \ .$$

Each $S \in \mathfrak{O}$ lies on the same $E(2)$-orbit as a scalar multiple of exactly one of these four operators. Furthermore, the orbits correspond exactly to the four coordinate systems in which variables separate in (1.7). The correspondence was first noted in reference [4] and is given explicitly in Table 1.

Table 1

Operator S	Coordinates	Separated solutions
1. $\quad P_2^2$	Cartesian x, y	product of exponential functions
2. $\quad M^2$	polar $x=r\cos\theta,\ y=r\sin\theta$	Bessel times exponential function
3. $\quad \{M, P_2\}$	parabolic cylinder $x=\frac{1}{2}(\xi^2-\eta^2),\ y=\xi\eta$	product of parabolic cylinder functions
4. $\quad M^2-a^2P_2^2$	elliptic $x=a\cosh\alpha\cos\beta,$ $y=a\sinh\alpha\sin\beta$	product of Mathieu functions

(Although the subspace $\mathfrak{Q}$ of purely second-order operators in $\mathfrak{g}$ suffices to explain the separable coordinate systems for the Helmholtz equation, it may be necessary to use orbits in the full space $\mathfrak{g}$ to describe the separable coordinates for other equations.)

Now we sketch one method for exploiting the above correspondence to derive properties of the separable solutions. This method involves use of the Fourier transform to introduce a Hilbert space structure on the solution space of the Helmholtz equation. By proceeding formally it is easy to show that ψ is a solution of $(\Delta_2 + k^2)\psi = 0$ provided

$$(1.20) \qquad \psi(x, y) = \int_0^{2\pi} e^{ik(x\cos\varphi + y\sin\varphi)} h(\varphi)d\varphi = I(h) \ .$$

We parametrize $E(2)$ in the form

$$(1.21) \qquad g(\theta, a, b) = \begin{pmatrix} \cos\theta & -\sin\theta & a \\ \sin\theta & \cos\theta & b \\ 0 & 0 & 1 \end{pmatrix} \in E(2), \quad \theta, a, b \in R$$

and define the action of $E(2)$ on the functions ψ by the operators

$$(1.22) \qquad \underline{T}(a, b, \theta) = \exp(aP_1 + bP_2)\exp(\theta M)$$

where P_1, P_2 and M are given by (1.15). It follows from (1.20) that these operators induce operators $\underline{T}(g)$ acting on the functions $h(\varphi)$ and defined by

(1.23) $\qquad \underline{T}(a, b, \theta) \, h(\varphi) = \exp[ik(a \cos \varphi + b \sin \varphi)] h(\varphi - \theta)$.

If we restrict the functions h to the Hilbert space $L_2[-\pi, \pi]$ of Lebesgue square integrable functions on the interval $[-\pi, \pi]$ with inner product

(1.24) $\qquad \langle h_1, h_2 \rangle = \int_{-\pi}^{\pi} h_1(\varphi) \overline{h_2(\varphi)} d\varphi$,

the operators (1.23) define an irreducible unitary representation of $E(2)$ on $L_2[-\pi, \pi]$. (We assume h is defined on the real line by the periodicity condition $h(\varphi) = h(\varphi + 2\pi)$.) The induced Lie algebra representation is defined by operators

(1.25) $\qquad P_1 = ik \cos \varphi, \quad P_2 = ik \sin \varphi, \quad M = -\partial_\varphi$.

We now restrict ourselves to solutions ψ of (1.7) with the form $\psi = I(h)$, $h \in L_2[-\pi, \pi]$. The space $\mathscr{Y}$ of such solutions is a Hilbert space with inner product.

(1.26) $\qquad (\psi_1, \psi_2) \equiv \langle h_1, h_2 \rangle, \quad \psi_j = I(h_j), \quad j = 1, 2$.

Note that we can also consider each $\psi \in \mathscr{Y}$ as an inner product

$$\psi(x, y) = I(h) = \langle h, H(x, y, \cdot) \rangle$$
(1.27)
$$H(x, y, \varphi) = \exp[-ik(x \cos \varphi + y \sin \varphi)] \in L_2[-\pi, \pi]$$.

Now we see that the Lie algebra elements (1.25) can be extended to skew-symmetric operators on $L_2[-\pi, \pi]$ and the quadratic elements (1.19) to essentially self-adjoint operators. Let S be the closure of one of these operators (1.19). It follows from the spectral theorem [5] that

313

there exists an orthonormal basis for $L_2[-\pi, \pi]$ consisting of (generalized) eigenvectors $\{f_\lambda\}$ of S :

(1.28) $$Sf_\lambda = \lambda f_\lambda, \quad \langle f_\lambda, f_\mu \rangle = \delta(\lambda-\mu) \quad .$$

Furthermore, it follows that the functions $\psi_\lambda = I(f_\lambda)$ form an orthonormal basis for $\mathscr{K}$ consisting of (generalized) eigenvectors of S :

(1.29) $$S\psi_\lambda = \lambda\psi_\lambda, \quad (\psi_\lambda, \psi_\mu) = \delta(\lambda-\mu) \quad .$$

Here, S is the operator on $\mathscr{K}$ which corresponds to the operator (1.28) on $L_2[-\pi, \pi]$. Since S on $\mathscr{K}$ corresponds to a separable coordinate system $\{u, v\}$ it follows that ψ_λ is a separable solution of the Helmholtz equation in the coordinates u, v and that λ is the separation constant.

Let S' be another one of the operators (1.19) on $L_2[-\pi, \pi]$ and let $\{f'_\mu\}$ be the corresponding basis of eigenfunctions. For $g \in E(2)$ we can expand the function $\underline{T}(g)f'_\mu$ in term of the basis $\{f_\lambda\}$:

(1.30) $$\underline{T}(g)f'_\mu = \int_{-\infty}^{\infty} \langle \underline{T}(g)f'_\mu, f_\lambda \rangle f_\lambda \, d\lambda \quad .$$

(Note that $\underline{T}(g)f'_\mu$ is an eigenfunction of the operator $\underline{T}(g)S'\underline{T}(g^{-1})$ with eigenvalue μ . This operator lies on the same orbit as S' . Note also that for $g = g(0, 0, 0)$ eqn. (1.30) is an expansion of the $\{f'_\mu\}$ basis in terms of the $\{f_\lambda\}$ basis.

Now it follows immediately from (1.26) or (1.27) that

(1.31) $$\underline{T}(g)\,\psi'_\mu(x, y) = \int_{-\infty}^{\infty} \langle \underline{T}(g)f'_\mu, f_\lambda \rangle \, \psi_\lambda(x, y)d\lambda$$

where convergence is pointwise and $\psi_\mu = I(f'_\mu)$. Here (1.31) is an expansion of one separable solution of (1.7) in terms of a basis of other separable solutions. This formula is especially significant because our $L_2[-\pi, \pi]$ model is so convenient for computational purposes. We will see that it is relatively easy to compute the functions $\{f_\lambda\}$, $\{f'_\mu\}$ and

314

the so-called "mixed basis matrix elements" $\langle T(g)f'_\mu, f_\lambda \rangle$ explicitly. Furthermore the integrals $\psi_\lambda(x, y) = I(f_\lambda)$, $\psi'_\mu(x, y) = I(f'_\mu)$ are easy to compute because we know that they are separable solutions of the Helmholtz equation in the appropriate variables. Thus the general integrals are determined to within four multiplicative constants which can be computed by evaluating the integrals for special values of x and y.

We give some examples involving the first three operators in Table 1. (Details concerning the fourth operator can be found in [6]).

1. The operator P_2^2.

On $L_2[-\pi, \pi]$, $P_2 = ik \cos \varphi$. Thus a basis of generalized eigenfunctions for P_2 is

(1.32)
$$f_\lambda^{(1)}(\varphi) = \delta(\varphi - \lambda), \quad P_2 f_\lambda^{(1)} = ik \cos \lambda \, f_\lambda^{(1)}, \quad -\pi \leq \lambda < \pi$$

$$\langle f_\lambda^{(1)}, f_{\lambda'}^{(1)} \rangle = \delta(\lambda - \lambda') .$$

Clearly the spectrum of P_2 is continuous and covers the interval $[-ik, ik]$ on the imaginary axis with multiplicity two. Furthermore

(1.33)
$$\psi_\lambda^{(1)} = I(f_\lambda^{(1)}) = \exp[ik(x \cos \lambda + y \sin \lambda)]$$

which is separable in Cartesian coordinates.

2. The operator M^2.

Here $M = -\partial_\varphi$ on $L_2[-\pi, \pi]$ and an orthonormal basis of eigenfunctions is

$$f_n^{(2)}(\varphi) = \frac{e^{-in\varphi}}{\sqrt{2\pi}}, \quad Mf_n^{(2)} = inf_n^{(2)}, \quad n = 0, \pm1, \pm2, \dots$$

$$\langle f_n^{(2)}, f_{n'}^{(2)} \rangle = \delta_{nn'} .$$

In this case the spectrum is discrete and

(1.34)
$$\psi_n^{(2)} = I(f_n^{(2)}) = \sqrt{2\pi} \, e^{in(\pi/2 - \theta)} J_n(kr)$$

315

where r, θ are polar coordinates and J_n is a Bessel function [7].

3. The operator $\{M, P_2\}$.

In this case $-\{M, P_2\} = 2i \sin \varphi \partial_\varphi + i \cos \varphi$ and a basis for $L_2[-\pi, \pi]$ of generalized eigenfunctions is

(1.35)
$$f_{\lambda+}^{(3)}(\varphi) = \begin{cases} \dfrac{1}{\sqrt{2\pi}}(1+\cos\varphi)^{+i\lambda/2-1/4}(1-\cos\varphi)^{-i\lambda/2-1/4}, & 0 < \varphi \le \pi \\ \\ 0 & -\pi \le \varphi < 0 \end{cases}$$

$$f_{\lambda-}^{(3)}(\varphi) = f_{\lambda+}^{(3)}(-\varphi), \quad -\{M, P_2\}f_{\lambda\pm}^{(3)} = 2\lambda f_{\lambda\pm}^{(3)}, \quad -\infty < \lambda < \infty$$

$$\langle f_{\lambda\pm}^{(3)}, f_{\lambda'\pm}^{(3)}\rangle = \delta(\lambda-\lambda'), \quad \langle f_{\lambda\pm}^{(3)}, f_{\lambda'\mp}^{(3)}\rangle = 0.$$

Here, the spectrum of $\{M, P_2\}$ is continuous and covers the real axis with multiplicity two. Also

$$\psi_\lambda^{(3)}(x, y) = \frac{1}{\sqrt{2}\cos(i\lambda\pi)}[D_{i\lambda-\frac{1}{2}}(\sigma\xi)D_{-i\lambda-\frac{1}{2}}(\sigma\eta) + D_{i\lambda-\frac{1}{2}}(-\sigma\xi)D_{-i\lambda-\frac{1}{2}}(-\sigma\eta)]$$

(1.36)

$$\psi_{\lambda-}^{(3)}(x, y) = \psi_{\lambda+}^{(3)}(x, -y), \quad \sigma = e^{i\pi/4}\sqrt{2k}, \quad x = \frac{\xi^2-\eta^2}{2}, \quad y = \xi\eta.$$

where $D_\nu(\xi)$ is a parabolic cylinder function [7].

Using these expressions we can compute some mixed basis matrix elements of interest. For example,

$$\langle f_n^{(2)}, f_\nu^{(1)}\rangle = f_n^{(2)}(\nu), \quad \langle f_{\lambda\pm}^{(3)}, f_\nu^{(1)}\rangle = f_{\lambda\pm}^{(3)}(\nu)$$

(1.37)

$$\langle f_n^{(2)}, f_{\lambda+}^{(3)}\rangle = \frac{e^{\frac{\pi}{2}(\frac{i}{2}+\lambda)}}{\pi\sqrt{2}}\Gamma(n+\tfrac{1}{2})\left[\frac{(-1)^n\Gamma(\frac{1}{2}-i\lambda)}{\Gamma(\frac{1}{2}-i\lambda+n)}\,{}_2F_1\left(\begin{matrix}\frac{1}{2}-i\lambda, \frac{1}{2}+n\\ 1+n-i\lambda\end{matrix}\,\bigg|\,-1\right)\right.$$

$$\left. -\frac{i\Gamma(\frac{1}{2}+i\lambda)}{\Gamma(1+i\lambda+n)}\,{}_2F_1\left(\begin{matrix}\frac{1}{2}+i\lambda, \frac{1}{2}+n\\ 1+n+i\lambda\end{matrix}\,\bigg|\,-1\right)\right]$$

$$\langle f_n^{(2)}, f_{\lambda-}^{(3)}\rangle = \langle f_{-n}^{(2)}, f_{\lambda+}^{(3)}\rangle.$$

Substituting these expressions into (1. 31), in the special case $g = g(0, 0, 0)$ we find

$$\exp[ik(x\cos\lambda + y\sin\lambda)] = \sum_{n=-\infty}^{\infty} e^{in[\lambda - \theta + \frac{\pi}{2}]} J_n(kr)$$

(1. 38)

$$\exp[ik(x\cos\varphi + y\sin\varphi] = \frac{(\sin\varphi)^{-\frac{1}{2}}}{\sqrt{2\pi}} \int_{-\infty}^{\infty} (\cot\frac{\varphi}{2})^{i\lambda} \psi_{\lambda+}^{(3)}(x, y)d\lambda, \quad 0 < \varphi < \pi,$$

the expansions of a plane wave in terms of cylindrical waves and parabolic cylindrical waves, and

$$\sqrt{2\pi} \; e^{in(\frac{\pi}{2} - \theta)} J_n(kr) = \int_{-\infty}^{\infty} \left[\langle f_n^{(2)}, f_{\lambda+}^{(3)} \rangle \psi_{\lambda+}^{(3)}(x, y) \right.$$

(1. 39)

$$\left. + \langle f_n^{(2)}, f_{\lambda-}^{(3)} \rangle \psi_{\lambda-}^{(3)}(x, y) \right] d\lambda \; ,$$

the expansion of a cylindrical wave in terms of parabolic cylindrical waves. Other more complicated formulas follow from (1. 31) for g other than the identity element.

Section 2. The Klein-Gordon equation.

The method of the preceding section yields dramatically different results when applied to the Klein-Gordon equation in two-dimensional space time,

(2.1)
$$(\partial_{tt} - \partial_{xx} + k^2) \Phi(x, t) = 0 \; .$$

A basis for the symmetry algebra $\mathcal{E}(1, 1)$ is

(2.2)
$$P_1 = \partial_t, \quad P_2 = \partial_x, \quad M = -t\partial_x - x\partial_t$$

with commutation relations

(2.3)
$$[P_1, P_2] = 0, \quad [M, P_1] = P_2, \quad [M, P_2] = P_1 \; .$$

317

In terms of these generators, eqn. (2.1) reads

(2.4) $$(P_1^2 - P_2^2 + k^2)\Phi = 0 \ .$$

Again, a basis for $\mathfrak{s}$ is provided by the five second-order operators (1.18) as well as the three first-order operators (2.2). It follows that (2.1) is a class I equation. The symmetry group of this equation is E(1,1), isomorphic to the group of matrices

(2.5) $$A(\theta, a, b) = \begin{pmatrix} \cosh \theta & \sinh \theta & a \\ \sinh \theta & \cosh \theta & b \\ 0 & 0 & 1 \end{pmatrix} , \qquad -\infty < \theta, a, b < \infty \ ,$$

which acts on the pseudo-Euclidean plane via the transformation $\underline{z} \to A\underline{z}$ where

$$\underline{z} = \begin{pmatrix} x \\ t \\ 1 \end{pmatrix} \ .$$

As in the preceding section we take $\mathfrak{D}$ to be the five-dimensional space of second-order operators in $\mathfrak{s}$. In [8] Kalnins showed that $\mathfrak{D}$ consists of twelve orbit types (under the action of E(1,1) extended by time inversion $t \to -t$ and space inversion $x \to -x$). Furthermore, he showed that (2.1) separates in eleven coordinate systems, related to eleven of the orbit types. The remaining orbit doesn't correspond to a separable system. A list of representatives of the orbit types and related coordinates is presented in Table 2.

In [9] Kalnins and the author introduce a Hilbert space structure on the solution space of (2.1) to derive properties of the separable solutions. Indeed, in analogy with (1.20) we consider solutions of the form

(2.6) $$\psi(x, t) = \int_{-\infty}^{\infty} \exp[ik(t \cosh y + x \sinh y)] h(y) dy = I(h)$$

Table 2

Operator S	Coordinates	Separated solutions
1. P_2^2	t, x (Cartesian)	product of exponentials
2. $P_1 P_2$	t, x	" " "
3. $(P_1 + P_2)^2$	t, x	" " "
4. M^2	$t = r \cosh \theta$, $x = r \sinh \theta$	exponential times Bessel function
5. $\{M, P_1\}$	$t = \xi\eta$, $x = (\eta^2 + \xi^2)/2$	product of parabolic cylinder functions
6. $M^2 - (P_1 + P_2)^2$	$t = \dfrac{u^2 + u^2 v^2 + v^2}{2uv}$, $x = \dfrac{u^2 - u^2 v^2 + v^2}{2uv}$	Bessel times Macdonald function
7. $M^2 + (P_1 + P_2)^2$	$t = \dfrac{u^2 - u^2 v^2 - v^2}{2uv}$, $x = \dfrac{u^2 + u^2 v^2 - v^2}{2uv}$	product of Macdonald functions
8. $\{M, P_1 - P_2\} + (P_1 + P_2)^2$	$t = -\frac{1}{4}(x_1 - x_2)^2 + \frac{1}{2}(x_1 + x_2)$, $x = -\frac{1}{4}(x_1 - x_2)^2 - \frac{1}{2}(x_1 + x_2)$	product of Airy functions
9. $M^2 - P_1 P_2$	$t + x = \cosh \frac{1}{2}(x_1 - x_2)$, $t - x = \sinh \frac{1}{2}(x_1 + x_2)$	product of Mathieu functions
10. $M^2 + P_2^2$	$t = \sinh x_1 \cosh x_2$, $x = \cosh x_1 \sinh x_2$	product of Mathieu functions
11. $M^2 - P_2^2$	$t = \cosh x_1 \cosh x_2$, $x = \sinh x_1 \sinh x_2$	product of Mathieu functions
12. $\{M, P_1 - P_2\}$		no separable solutions

(The mutually commuting operators of types 1-3 are associated with the same coordinate system)

319

where h belongs to the Hilbert space $L_2(R)$ with inner product

(2.7) $$\langle h_1, h_2 \rangle = \int_{-\infty}^{\infty} h_1(y)\overline{h}_2(y)dy \ .$$

The induced action of $E(1,1)$ on $L_2(R)$ is

(2.8) $\underline{T}(a, b, \theta)h(y) = \exp[ik(a \cosh y + b \sinh y)]h(y + \theta),$ $h \in L_2(R)$

and the operators $\underline{T}(a, b, \theta)$ define an irreducible unitary representation of $E(1,1)$. The corresponding Lie algebra action is defined by operators

(2.9) $$P_1 = ik \cosh y, \quad P_2 = ik \sinh y, \quad M = \partial_y \ .$$

Just as in the preceding section, it is possible to use the simpler $L_2(R)$ model to derive properties of the separated solutions listed in Table 2. The details are presented in references [9] and [10].

Section 3. The complex Helmholtz equation.

Next we consider the equation

(3.1) $$(\partial_{xx} + \partial_{yy} + k^2) \, \psi(x, y) = 0$$

where now x and y are complex variables and Φ is an analytic function in these variables. (Note that (1.7) and (2.1) can be considered as different real forms of the same complex equation (3.1).) The operators (1.15) form a basis for the complex symmetry algebra and the operators (1.15), (1.18) form a basis for the complex vector space $\mathfrak{g}^c$. The complex five-dimensional space $\mathfrak{2}^c$ spanned by the operators (1.18) can be decomposed into eight orbits under the action of the complex Euclidean group. (This is the group of all matrices (1.21) where a, b, θ take complex values.) These orbits correspond to the possible complex analytic coordinate systems which permit separation of variables in (3.1), as shown in Table 3.

Table 3

Operator S	Coordinates	Separated solutions
1. P_1^2	x, y	product of exponentials
2. $(P_1+iP_2)^2$	x, y	" " "
3. M^2	$x=r\cos\theta,\ y=r\sin\theta$	exponential times Bessel function
4. $M^2-a^2 P_2^2,\ a\neq 0$	$x=a\cosh\alpha\cos\beta,$ $y=a\sinh\alpha\sin\beta$	product of Mathieu functions
5. $\{M, P_2\}$	$x=\frac{1}{2}(\xi^2-\eta^2),\ y=\xi\eta$	product of parabolic cylinder functions
6. $M^2+(P_1+iP_2)^2$	$x=\dfrac{u^2-u^2v^2-v^2}{2uv},\ iy=\dfrac{u^2+u^2v^2-v^2}{2uv}$	product of Bessel functions
7. $\{M, P_1+iP_2\} + (P_1-iP_2)^2$	$x=\ -\frac{1}{4}(x_1-x_2)^2+\frac{1}{2}(x_1+x_2)$ $iy=-\frac{1}{4}(x_1-x_2)^2-\frac{1}{2}(x_1+x_2)$	product of Airy functions
8. $\{M, P_1+iP_2\}$		no separable solutions

Note that there are no coordinate systems for this equation which are not analytic continuations of coordinates listed in tables 1 and 2. However, systems which are real-inequivalent for the Helmholtz and Klein-Gordon equations become equivalent upon complexification of these equations.

Although the solution space of (3.1) doesn't appear to admit a Hilbert space structure, one can employ a method due to Louis Weisner to derive identities for the separated solutions. Weisner's method focuses on the solutions on orbit 3 . On introducing polar coordinates $x = r\cos\theta$, $y = r\sin\theta$ in (3.1), one obtains the equation

$$(3.2) \qquad (\partial_{rr}+r^{-1}\partial r+r^{-2}\partial_{\varphi\varphi}+k^2)\psi = 0 .$$

This equation admits separable solutions of the form

$$(3.3) \qquad J_{\pm m}(kr)s^m, \quad s = e^{i\varphi}, \quad m \in \mathscr{C}$$

where $J_\nu(kr)$ is a Bessel function. Furthermore a function $\psi = u(r)s^m$ is a solution of (3.2) if and only if u satisfies Bessel's equation

$$(3.4) \qquad (\partial_{rr} + r^{-1}\partial_r - r^{-2}m^2 + k^2)u = 0 .$$

Now suppose ψ is a solution of (3.2) such that $(rs)^{-m}\psi(r, s)$ is analytic in a neighborhood of $r = s = 0$ for some $m \in \not{\!C}$. It follows easily that

$$(3.5) \qquad \psi(r, s) = \sum_{n=0}^{\infty} c_n J_{m+n}(kr)s^{m+n}$$

where the power series converges in a neighborhood of $(r, s) = (0, 0)$ and the c_n are complex constants, i.e., ψ is a generating function for Bessel functions. If ψ is known, one can ordinarily compute the constants by evaluating the right-hand side of (3.5) for special choices of r or by using operator identities satisfied by ψ to obtain recurrence relations for the c_n . In order to make this procedure effective one needs an efficient method for obtaining explicit analytic solutions of (3.1). From our point of view an obvious answer to this problem is to take the separated solutions corresponding to each of the six non-trivial orbits in Table 3. Indeed, if Φ is a separated solution corresponding to the operator S in Table 3, $S\Phi = \lambda\Phi$, and g belongs to the complex Euclidean group, then the separated solution $\psi = T(g)\Phi$ corresponds to the operator $S' = T(g) ST(g^{-1})$ on the same orbit, $S'\psi = \lambda\psi$. Then ψ can be substituted into (3.5) to yield a generating function for the $J_\nu(r)$. Although Weisner nowhere explicitly states the relationship between symmetry operators and separation of variables, in his paper [11] he gives operator characterizations of separated solutions from all orbits except 4 to construct generating functions for Bessel functions.

Weisner's method applied to (3.1) provides generating functions only for Bessel functions. However, in [9] and [10] the author and E. G. Kalnins have also derived some generating functions for products of

322

parabolic cylinder functions and products of Bessel functions, corresponding to orbits 5 and 6 on Table 3.

Section 4. The Schrödinger equation.

Next we consider the free-particle Schrödinger equation

(4.1) $\qquad (\partial_{xx} + \partial_t) \, \psi(x, t) = 0, \qquad x, t \in R \;.$

This equations admits a real six-dimensional Lie algebra $\mathcal{Y}_1$ with basis

(4.2)
$$K_2 = -t^2 \partial_t - tx\partial_x - t/2 + ix^2/4, \quad K_1 = -t\partial_x + ix/2$$
$$K_0 = i, \quad K_{-1} = \partial_x, \quad K_{-2} = \partial_t, \quad K^0 = x\partial_x + 2t\partial_t + \frac{1}{2}$$

and commutation relations

(4.3)
$$[K^0, K_j] = jK_j, \quad j = 0, \pm 1, \pm 2, \quad [K_{-1}, K_1] = \frac{1}{2} K_0$$
$$[K_{-1}, K_2] = K_1, \quad [K_{-2}, K_1] = -K_{-1}, \quad [K_{-2}, K_2] = -K^0 \;.$$

Here, K_0 spans the center of $\mathcal{Y}_1$. (Note that in this case we cannot neglect multiplication by a constant since K_0 appears as a commutator.) The algebra $\mathcal{Y}_1$ is called the Schrödinger algebra and the corresponding local Lie group G_1 is called the Schrödinger group. For details on the structure of this group, see [12].

One can verify directly that (4.1) is a class I equation. Moreover, it can be shown that in this case the orbits in $\mathcal{Y}_1$ alone, under the adjoint action of G_1, suffice to describe the separable solutions of (4.1). There is no need to consider second-order operators in the enveloping algebra.

For this equation it becomes useful (and necessary) to consider R-separable solutions. To explain this concept we denote eqn. (4.1) as $X\psi = 0$. Suppose (u, v) is a real coordinate system and $R(u, v) = \exp iQ(u, v)$ is a function which cannot be factored in the form $R = a(u)b(v)$.

323

Furthermore, suppose the equation $X'\psi' = 0$ separates in the system (u, v) where $X' = R^{-1}XR$. Then the second equation admits separable solutions of the form $\psi'(u, v) = U(u) V(v)$ and the original equation admits R-separable solutions $\psi(u, v) = [\exp i Q(u, v)]\, U(u) V(v)$. Roughly speaking, R-separable solutions of $X\psi = 0$ correspond to separable solutions of equations equivalent to (4.1) under transformation by a function $R(u, v)$.

In [12], Kalnins and the author have shown that $\mathcal{J}_1$ modulo its center contains five orbit types: K_{-2}-K_2, K^0, $K_2 + K_{-1}$, K_{-2}, and K_{-1}. The association between orbits and separable coordinates is given in Table 4.

Table 4

Operator S	Coordinates	multiplier e^{iQ}	solutions
1. K_{-1}	$x=u$, $t=v$	$Q = 0$	product of exponentials
2. K_{-2}	$x=u$, $t=v$	$Q = 0$	product of exponentials
3a. K_{-2}-K_{-1}	$x=u+\frac{1}{2}v^2$, $t=v$	$Q = \frac{1}{2}uv$	Airy times exponential function
3b. $K_2 + K_{-1}$	$x=uv +\frac{1}{2v}$, $t=v$	$Q = \frac{1}{4}(u^2 v - \frac{u}{v})$	Airy times exponential function
4a. K^0	$x=u\sqrt{v}$, $t=v$	$Q = 0$	parabolic cylinder times exponential function
4b. $K_2 + K_{-2}$	$x=u\|1-v^2\|^{\frac{1}{2}}$, $t=v$	$Q=\pm\frac{u^2 v}{4}(+$if$\|v\|>1$, $-$if$\|v\|<1)$	parabolic cylinder times exponential function
5. K_2-K_{-2}	$x=u\sqrt{1+v^2}$, $t=v$	$Q = \frac{1}{4}u^2 v$	Hermite times exponenential function

Orbits 1 and 2 give equivalent results because K_{-1} and K_{-2} commute and can be simultaneously diagonalized. On orbits 3 and 4 we have chosen two different operators. Indeed, from the viewpoint of Galilean symmetry alone each of these orbits divides into two suborbits corresponding to distinct coordinate systems. However, from the viewpoint of the full Schrödinger symmetry group we see that the systems 3a, 3b and 4a, 4b are equivalent.

The significance of these systems is easy to understand. Indeed choosing coordinates 3a and choosing an appropriate multiplier $R(u, v)$ we can transform equation (4.1) to the form

(4.4)
$$(\partial_{uu} + \frac{u}{2} + i\partial_v)\Phi = 0 \ .$$

This is the Schrödinger equation with a linear potential where v now corresponds to the time coordinate and u to the space coordinate. Similarly, choosing coordinates 5 we can transform (4.1) to

(4.5)
$$(\partial_{uu} - \frac{u^2}{4} + i\partial_\tau)\Phi = 0, \quad v = \tan \tau$$

i. e., to the Schrödinger equation for the harmonic oscillator. By choosing coordinates 4b we can transform (4.1) to the form

(4.6)
$$(\partial_{uu} + \frac{u^2}{4} + i\partial_\mu)\Phi = 0, \quad \mu = \mu(v)$$

i. e., to the Schrödinger equation for the repulsive oscillator. It follows that these four Schrödinger equations are equivalent and correspond to the four non-trival orbits in the Schrödinger group. Thus, at one stroke we have solved the separation of variables problem for all these equations simultaneously.

The solutions of (4.1) admit a Hilbert space structure, as is well-known. Indeed, consider the Hilbert space $L_2(R)$ with inner product (2.7). If $h \in L_2(R)$ then

$$\psi(x, t) = \exp(t\mathcal{K}_{-2}) h(x) = l.\,i.\,m. \ \frac{1}{\sqrt{4\pi it}} \int_{-\infty}^{\infty}$$

(4.7)
$$\exp[-(x-y)^2/4it\,]h(y)dy$$

is a solution of (4.1) such that

(4.8) $$(\psi, \psi') \equiv \int_{-\infty}^{\infty} \psi(x, t)\overline{\psi'}(x, t)dx = \langle h, h' \rangle$$

independent of t, for solutions ψ, ψ' corresponding to $h, h' \in L_2(R)$.
As shown in [12] the symmetry operators (4.2) acting on the solution
space of (4.1) induce operators

(4.9)
$$\mathcal{K}_2 = ix^2/4, \quad \mathcal{K}_1 = ix/2, \quad \mathcal{K}_0 = i, \quad \mathcal{K}_{-1} = \partial_x$$

$$\mathcal{K}_{-2} = i\partial_{xx}, \quad \mathcal{K}^3 = x\partial_x + \frac{1}{2}$$

acting on $L_2(R)$. (Formally the script operators are obtained from (4.2)
by replacing ∂_t with $i\partial_{xx}$ and setting $t = 0$.) It is easy to check that
the operators (4.9) are essentially skew-adjoint and generate a unitary
irreducible representation of the Schrödinger group on $L_2(R)$. Furthermore the block and script operators are related by the unitary equivalence

(4.10) $$(\exp t\mathcal{K}_{-2}) \mathcal{K}_j(\exp -t\mathcal{K}_{-2}) = K_j \quad .$$

Note: This transformation is well-known to physicists as the transformation between the Heisenberg and Schrödinger pictures, [13].

Just as in the previous sections, one can use the Hilbert space
$L_2(R)$ and the operators (4.9) to compute the spectral resolutions of the
operators listed in Table 4 and then use (4.7) to map these results to the
Hilbert space of square integrable solutions of (4.1). The detailed results together with mixed-basis matrix elements are presented in [12].
The corresponding results relating the symmetry group of the heat equation

(4.11) $$(\partial_{xx} - \partial_t) \psi(x, t) = 0$$

to separation of variables are also given in [12]. The theory of the heat
equation is not as simple as that of the free-particle Schrödinger equation because there is no convenient Hilbert space structure for the solution space. Nevertheless, one can still discuss pointwise convergent

expansions of one basis in terms of another, e. g. , the expansions in heat polynomials as given in [14].

For the complex heat equation, i. e. , equation (4.11) with x and t complex, the symmetry algebra is easily seen to be the complexification $\mathcal{J}_1^C$ of the Lie algebra (4.3). $\mathcal{J}_1^C$ modulo its center contains four orbits which correspond to the first four orbits listed in Table 4. (Under $\mathcal{J}_1^C$ orbit 5 becomes equivalent to orbit 4 .) Furthermore it is not difficult to show that the complex heat equation admits R-separation in exactly three coordinate systems, orbits 1 and 2 corresponding to products of exponentials, orbit 3 to an Airy times an exponential function, and orbit 4 to a parabolic cylinder times an exponential function. Just as in Section 3, one can apply Weisner's method to these results to expand analytic solutions of (4.11) in series of parabolic cylinder functions. In fact, Weisner did just this in his paper [15] although he didn't mention that the equation he was studying was equivalent to the heat equation.

The separation of variables problem for the free-particle Schrödinger equation in three-dimensional space-time,

(4.12) $$(\partial_{xx} + \partial_{yy} + i\partial_t)\, \psi(x, y, t) = 0 \ ,$$

is considerably more complicated than that for equation (4.1). The real symmetry algebra $\mathcal{J}_2$ for this equation is nine-dimensional with basis

$$K_2 = -t^2 \partial_t - t(x\partial_x + y\partial_y) - t + \frac{i}{4}(x^2 + y^2), \quad K_{-2} = \partial_t$$

(4.13) $$P_1 = \partial_x, \quad P_2 = \partial_y, \quad B_1 = -t\partial_x + ix/2, \quad B_2 = -t\partial_y + iy/2$$

$$M = x\partial_y - y\partial_x, \quad E = i, \quad D = x\partial_x + y\partial_y + 2t\partial_t + 1$$

and commutation relations

$$[D, K_{\pm2}] = \pm 2K_{\pm2}, \quad [D, B_j] = B_j, \quad [D, P_j] = -P_j$$

$$[D, M] = 0, \quad [M, K_{\pm2}] = 0, \quad [P_j, M] = (-1)^{j+1} P_\ell \ ,$$

(4.14) $\qquad [B_j, M] = (-1)^{j+1} B_\ell, \quad [K_2, K_{-2}] = D, \quad [K_2, B_j] = 0$

$$[K_{-2}, B_j] = -P_j, \quad [K_{-2}, P_j] = 0, \quad [P_j, K_2] = B_j \ ,$$

$$[P_j, B_j] = \frac{1}{2}E, \quad [P_j, B_\ell] = 0, \quad j, \ell = 1, 2, \quad j \neq \ell \ ,$$

with E in the center of $\mathcal{L}_2$. Again one can show that (4.12) is a class I equation. The structure of the Schrödinger algebra $\mathcal{L}_2$ and its associated Lie symmetry group are studied in detail in [6]. Moreover, the possible R-separable coordinate systems for (4.12) and their relationship to $\mathcal{L}_2$ are listed for the first time in this reference.

There are now two constants of integration associated with each R-separable system and these constants can be interpreted as the eigenvalues of a pair of commuting operators K, S belonging to the enveloping algebra of $\mathcal{L}_2$. It turns out that one of these operators, K, always belongs to $\mathcal{L}_2$ while S is a second order homogeneous polynomial in the elements of $\mathcal{L}_2$. The twenty-six R-separable systems $\{u, v, w\}$ are listed in Table 5.

Table 5

Operators K, S	Coordinates	Multiplier e^{iQ}	Solutions
1a. K_2, B_1^2	x=uw, y=vw	$Q = \frac{1}{4}(u^2+v^2)w$	exponential exponential
1b. K_{-2}, P_1^2	x=u, y=v	0	exponential exponential
2a. K_2, M^2	x=uw cos v, y=uw sin v	$\frac{1}{4}u^2 w$	Bessel exponential
2b. K_{-2}, M^2	x=u cos v, y=u sin v	0	Bessel exponential
3a. $K_2, \{B_2, M\}$	x=$\frac{1}{2}$w(u^2-v^2), y=uvw	$\frac{1}{16}(u^2+v^2)^2 w$	parabolic cylinder parabolic cylinder

Table 5 Continued

Operators K, S	Coordinates	Multiplier e^{iQ}	Solutions
3b. $K_{-2}, \{P_2, M\}$	$x=\frac{1}{2}(u^2-v^2)$ $y=uv$	0	parabolic cylinder parabolic cylinder
4a. $K_2, M^2-B_2^2$	$x=w\cosh u \cos v$ $y=w\sinh u \sin v$	$\frac{w}{4}(\sinh^2 u + \cos^2 v)$	associated Mathieu Mathieu
4b. $K_{-2}, M^2-P_2^2$	$x=\cosh u \cos v$ $y=\sinh u \sin v$	0	associated Mathieu Mathieu
5a. $K_2-2aP_1-2bP_2$	$x=uw + a/w$	$\frac{w}{4}(u^2+v^2) - \frac{1}{2w}(au+bv)$	Airy
$B_2^2 + 2bEP_2$	$y=vw + b/w$		Airy
5b. $K_{-2}+2aB_1+2bB_2$	$x=u + aw^2$	$(au+bv)w$	Airy
$P_1^2-2aEB_1$	$y=v + bw^2$		Airy
6a. $K_2 - aP_1$	$x=\frac{w}{2}(u^2-v^2) + a/w$	$\frac{w}{16}(u^2+v^2)^2$	($*$)
$\{B_2, M\}-aP_2^2$	$y=uvw$	$-\frac{a}{4w}(u^2-v^2)$	($*$)
6b. $K_{-2} - 2aB_1$	$x=\frac{1}{2}(u^2-v^2)+aw^2$	$\frac{aw}{2}(u^2-v^2)$	($*$)
$\{P_2, M\}+2aB_2^2$	$y=uv$		($*$)
7. $K_{-2} - K_2$	$x=u\sqrt{1+w^2}$	$\frac{1}{4}(u^2 + v^2)w$	Hermite
$P_1^2 + B_1^2$	$y=v\sqrt{1+w^2}$		Hermite
8. $K_{-2} - K_2$	$x=u\sqrt{1+w^2}\cos v$	$\frac{1}{4}u^2 w$	Laguerre
M^2	$y=u\sqrt{1+w^2}\sin v$		exponential
9. $K_{-2} - K_2$	$x=\sqrt{1+w^2}\cosh u \cos v$	$\frac{1}{4}(\sinh^2 u+\cos^2 v)w$	Ince
$M^2 - P_2^2-B_2^2$	$y=\sqrt{1+w^2}\sinh u \sin v$		Ince

Table 5 Continued

Operators K, S	Coordinates	Multiplier e^{iQ}	Solutions
10a. $D, \{B_1, P_1\}$	$x = u\sqrt{w}$ $y = v\sqrt{w}$	0	parabolic cylinder parabolic cylinder
10b. $K_{-2} + K_2$ $P_1^2 - B_1^2$	$x = u\|w^2 - 1\|^{\frac{1}{2}}$ $y = v\|w^2 - 1\|^{\frac{1}{2}}$	$\frac{\varepsilon}{4}(u^2 + v^2)w$	" " " "
11a. D, M^2	$x = u\sqrt{w}\cos v$ $y = u\sqrt{w}\sin v$	0	Whittaker exponential
11b. $K_{-2} + K_2$ M^2	$x = \|w^2 - 1\|^{\frac{1}{2}}u\cos v$ $y = \|w^2 - 1\|^{\frac{1}{2}}u\sin v$	$\frac{\varepsilon}{4}u^2 w$	Whittaker exponential
12a. D $M^2 - \frac{1}{2}\{B_2, P_2\}$	$x = \sqrt{w}\cosh u\cos v$ $y = \sqrt{w}\sinh u\sin v$	0	finite Ince " "
12b. $K_{-2} + K_2$ $M^2 - P_2^2 + B_2^2$	$x = \|w^2 - 1\|^{\frac{1}{2}}\cosh u\cos v$ $y = \|w^2 - 1\|^{\frac{1}{2}}\sinh u\sin v$	$\frac{\varepsilon}{4}(\sinh^2 u + \cos^2 v)$	" " " "
13. P_1 $B_2^2 - 2bEP_2$	$x = u$ $y = vw + b/w$	$\frac{1}{4}wv^2 - bv/2w$	exponential Airy
14. P_1 $P_2^2 - 2aEB_2$	$x = u$ $y = v + aw^2$	avw	exponential Airy
15. $P_1, P_2^2 + B_2^2$	$x = u$ $y = v\sqrt{1 + w^2}$	$\frac{1}{4}wu^2$	exponential Hermite
16. $P_1, \{B_2, P_2\}$	$x = u$ $y = v\sqrt{w}$	0	exponential parabolic cylinder

Table 5 Continued

Operators K, S	Coordinates	Multiplier e^{iQ}	Solutions
17. P_1, $P_2^2 - B_2^2$	$x = u$	$\frac{\varepsilon}{4} v^2 w$	exponential
	$y = v\|w^2 - 1\|^{\frac{1}{2}}$		parabolic cylinder

In each case $w = t$ and the separated solution in the variable w is an exponential function. In the last column of Table 5 we list first the form of the separated solution in u followed by the separated solution in v. The symbol $\varepsilon = \pm 1$ denotes the sign of $1 - w^2$ and the functions (*) are solutions of a differential equation of the form

$$(4.15) \qquad f''(u) + (\lambda u^2 + \alpha u^4 - \beta) \, f(u) = 0, \qquad \alpha, \beta \in R \ .$$

From the viewpoint of the Galilean symmetry group alone there are twenty-six coordinate systems as listed in this table. However, from the viewpoint of the full Schrödinger symmetry group some of these systems become equivalent and we finally have seventeen orbits of commuting operators which correspond to the possible Schrödinger-inequivalent coordinate systems.

The significance of some of these systems is easy to understand. Indeed by appropriate choices of coordinates from our list and appropriate choices of multipliers $R(u, v, w)$ we can transform equation (4.12) to any one of the forms

$$(4.16) \qquad \begin{aligned} & (\partial_{uu} + \partial_{vv} + \frac{u}{2} + i\partial_w)\Phi = 0 \ , \\ & (\partial_{uu} + \partial_{vv} - \frac{u^2}{4} - \frac{v^2}{4} + i\partial_w)\Phi = 0 \ , \\ & (\partial_{uu} + \partial_{vv} + \frac{u^2}{4} + \frac{v^2}{4} + i\partial_w)\Phi = 0 \ , \end{aligned}$$

i.e., to the Schrödinger equations corresponding to linear, harmonic oscillator, or repulsive oscillator potentials. Ordinary separation of variables for any one of these equations (4.12), (4.16) leads to R-separation for the rest since the four equations are R-equivalent.

As is well-known from quantum theory, the solutions of (4.12) admit a Hilbert space structure. Indeed, let $L_2(R_2)$ be the space of Lebesgue square integrable functions in the plane, with inner product

(4.17) $\qquad \langle h, h' \rangle = \int\int_{-\infty}^{\infty} h(\underline{x}) \overline{h'}(\underline{x}) dx_1 dx_2, \quad h, h' \in L_2(R_2)$.

If $h \in L_2(R_2)$ then

(4.18) $\qquad \psi(\underline{x}, t) = \exp(t\mathcal{K}_{-2}) h(x) = \text{l. i. m.} \dfrac{1}{4\pi i t}$

$$\int\int_{-\infty}^{\infty} \exp\{[-(x_1 - s_1)^2 - (x_2 - s_2)^2]/4it\} h(s_1, s_2) ds_1 ds_2$$

is a solution of (4.12) such that

(4.19) $\qquad (\psi, \psi') \equiv \int\int_{-\infty}^{\infty} \psi(\underline{x}, t) \overline{\psi}'(\underline{x}, t) dx_1 dx_2 = \langle h, h' \rangle$,

independent of t . Furthermore, the symmetry operators (4.13) acting on the solution space of (4.12) induce operators

$$\mathcal{K}_2 = \frac{i}{4}(x^2 + y^2), \quad \mathcal{K}_{-2} = i(\partial_{xx} + \partial_{yy}), \quad \mathcal{P}_1 = \partial_x ,$$

(4.20) $\qquad \mathcal{P}_2 = \partial_y, \quad \mathcal{B}_1 = ix/2, \quad \mathcal{B}_2 = iy/2, \quad \mathcal{M} = x\partial_y - y\partial_x ,$

$$\mathcal{E} = i, \quad \mathcal{D} = x\partial_x + y\partial_y + 1 ,$$

acting on $L_2(R_2)$. These operators generate a unitary irreducible representation of the Schrödinger group on $L_2(R_2)$ and the corresponding block and script operators are related by the unitary equivalence

(4.21) $\qquad (\exp t\mathcal{K}_{-2}) \mathcal{K} (\exp{-t\mathcal{K}_{-2}}) = K$.

In [6] and [26], C. Boyer, E. G. Kalnins and the author have derived the above results and used the Hilbert space $L_2(R_2)$ to compute the spectral resolutions of all the operators listed in Table 5. We have also mapped the eigenfunctions in $L_2(R_2)$ to solutions of the Schrödinger equation via (4.18) and derived expansions relating various separated solutions.

For example, we treat the system 4a in a slightly altered form. In the $L_2(R_2)$ model the eigenfunction equations are

(4.22) $\qquad \mathcal{K}_2 h = \frac{i\gamma^2}{4} h, \quad (\mathcal{M}^2 + 4\beta_1^2 - 4\beta_2^2)h = -\lambda h$

where $\frac{i\gamma^2}{4}$, $-\lambda$ are eigenvalues of the defining operators. (Note that $\beta_1^2 + \beta_2^2 = i\mathcal{K}_2$.) Setting $x_1 = r \cos \theta$, $x_2 = r \sin \theta$ in R_2, we find the following basis of generalized eigenfunctions:

(4.23) $\qquad h_{\gamma, \lambda_n}(r, \theta) = \frac{\delta(r-\gamma)}{\sqrt{r}} \begin{cases} ce_{\lambda_n} (\theta, \gamma) \\ se_{\lambda_n} (\theta, \gamma) \end{cases}$

$$0 < \gamma, \ \lambda_0 < \lambda_1 < \cdots$$

where ce_λ and se_λ are the even and odd Mathieu functions, [16]. Furthermore,

(4.24) $\qquad \langle h_{\gamma, \lambda_n}, h_{\gamma', \lambda_m} \rangle = \delta(\gamma-\gamma')\delta_{\lambda_n, \lambda_m}$

which follows from the orthogonality relations for Mathieu functions.

Necessarily, the solutions $\psi_{\gamma, \lambda_n}(\underline{x}, t) = (\exp t\mathcal{K}_{-2})h_{\gamma, \lambda_n}(\underline{x})$ of (4.12) separate in the coordinates 4a. Indeed we find

333

$$\psi_{\gamma, \lambda_n}(\underline{x}, t) = k_{\gamma, n} \frac{\sqrt{\gamma}}{4\pi i w} \exp\left[i w(\cos^2\sigma + \sinh^2\rho + \gamma^2)\right]$$

(4.25)

$$\begin{cases} ce_{\lambda_n}(\sigma, \gamma) \ Ce_{\lambda_n}(\rho, \gamma) \\ se_{\lambda_n}(\sigma, \gamma) \ Se_{\lambda_n}(\rho, \gamma) \end{cases}$$

$$x_1 = -2 \, s \cosh \rho \cos \sigma, \quad x_2 = -2s \sinh \rho \sin \sigma, \quad t = w$$

where Ce_λ, Se_λ are associated Mathieu functions and $k_{\lambda, n}$ is a normalization constant. It follows immediately that for fixed t, $\{\psi_{\gamma, \lambda_n}(\underline{x}, t)\}$ is a basis for $L_2(R_2)$ and

(4.26)
$$(\psi_{\gamma, \lambda_n}, \psi_{\gamma', \lambda_m}) = \delta(\gamma - \gamma')\delta_{\lambda_n \lambda_m} \ .$$

As a second example, consider system 11a. Here, the eigenfunction equations in $L_2(R_2)$ are

(4.27)
$$\mathcal{B}h = \rho h, \quad \mathcal{M}^2 h = -m^2 h \ .$$

With $x_1 = r \cos \theta$, $x_2 = r \sin \theta$ we obtain the basis

$$h_{\rho m}(r, \theta) = \frac{1}{2\pi} r^{i\rho - 1} e^{im\theta}, \quad -\infty < \rho < \infty \ ,$$

(4.28)
$$m = 0, \pm 1, \pm 2, \ldots,$$

$$\langle h_{\rho m}, h_{\rho' m'} \rangle = \delta(\rho - \rho')\delta_{mm'} \ .$$

The corresponding separated solutions of (4.12) are

$$\psi_{\rho m}(\underline{x}, t) = (\exp t \mathcal{X}_{-2}) h_{\rho m}(\underline{x}) = \frac{2}{\pi i \sqrt{w}} (2 \sqrt{iw})^{1+i\rho}$$

$$\frac{\Gamma\left(\dfrac{m}{2} + \dfrac{1+i\rho}{2}\right)}{m!} u^{-1} e^{iu^2/8} e^{imv} M_{\frac{i\rho}{2}, \frac{m}{2}} (iu^2/4)$$

(4.29)

$$x_1 = \sqrt{w} \, u \cos v, \quad x_2 = \sqrt{w} \, u \sin v, \quad t = w$$

$$(\psi_{\rho, m}, \psi_{\rho', m'}) = \delta(\rho - \rho') \delta_{mm'}$$

where $M_{\mu, \lambda}(z)$ is a Whittaker function.

The overlap function relating these two bases is easily computed:

(4.30)
$$\langle h_{\gamma, \lambda_n}, h_{\rho m} \rangle = \gamma^{i\rho - 3/2} A_m^{(\lambda_n)} .$$

Here $A_m^{(\lambda_n)}$ is the coefficient of $e^{im\theta}$ in the Fourier series expansion

of $\left\{ \begin{array}{c} ce_{\lambda_n}(\theta, \gamma) \\ se_{\lambda_n}(\theta, \gamma) \end{array} \right\}$. Recall that the Mathieu functions are frequently de-

fined in terms of the coefficients $A_m^{(\lambda_n)}$. It follows immediately that

$$(\psi_{\gamma, \lambda_m}, \psi_{\rho m}) = \gamma^{i\rho - 3/2} A_m^{(\lambda_n)} ,$$

which allows us to expand either one of the sets of basis functions as sums and integrals over the other set. For details see [6].

Closely related to the above discussion is the separation of variables problem for the heat equation

(4.31)
$$\partial_t \Phi = (\partial_{xx} + \partial_{yy}) \Phi .$$

The separable coordinate systems and their relationship to the symmetry group of (4.31) are listed in [17].

The problem of analytic separation of variables for (4.31) where x, y, t are complex variables has not yet been studied in detail. Weisner's method applied to this equation leads to generating functions for Laguerre polynomials. Partial results are contained in [18], and [19], Chapters 4 and 5.

In [20], Boyer studied the Schrödinger equation

$$(4.32) \qquad (i\partial_t + \partial_{xx} + \partial_{yy} - \alpha/x^2 - \beta/y^2)\psi = 0$$

which is class II. He showed that (4.32) R-separates in 25 coordinate systems for $\alpha = 0$, $\beta \neq 0$ and in 15 coordinate systems for $\alpha \neq 0, \beta \neq 0$. Moreover, he found that each separable coordinate system corresponded to a pair of commuting second order symmetry operators of (4.32). Although this equation is class II it is still highly tractable since it can be obtained by a partial variable separation from the free-particle Schrödinger equation (class I) in a higher dimensional space.

Section 5. The wave equation.

We next study the wave equation

$$(5.1) \qquad (\partial_{00} - \partial_{11} - \partial_{22})\psi(x) = 0, \quad x = (x_0, x_1, x_2) \ .$$

The symmetry algebra of (5.1) is ten-dimensional (aside from the trivial symmetry $E = i$) and isomorphic to $so(3,2)$. A basis is provided by the momentum operators

$$(5.2) \qquad P_\alpha = \partial_\alpha, \quad \alpha = 0, 1, 2 \ ,$$

the generators of homogeneous Lorentz transformations

$$(5.3) \qquad M_{12} = x_1\partial_2 - x_2\partial_1, \quad M_{01} = x_0\partial_1 + x_1\partial_0, \quad M_{02} = x_0\partial_2 + x_2\partial_0 \ ,$$

the generator of dilations

336

(5.4)
$$D = -(\tfrac{1}{2} + x_0\partial_0 + x_1\partial_1 + x_2\partial_2) \ ,$$

and the generators of special conformal transformations

(5.5)
$$K_0 = -x_0 + (x\cdot x - 2x_0^2)\partial_0 - 2x_0x_1\partial_1 - 2x_0x_2\partial_2 \ ,$$
$$K_1 = x_1 + (x\cdot x + 2x_1^2)\partial_1 + 2x_1x_0\partial_0 + 2x_1x_2\partial_2 \ ,$$
$$K_2 = x_2 + (x\cdot x + 2x_2^2)\partial_2 + 2x_2x_0\partial_0 + 2x_2x_1\partial_1 \ ,$$

where

$$x\cdot y = x_0y_0 - x_1y_1 - x_2y_2 = x_0y_0 - \underline{x}\cdot\underline{y} \ .$$

One can verify directly that (5.1) is class I. Indeed, $\mathfrak{s}$ is a forty-five dimensional vector space with thirty-five independent second order elements in the enveloping algebra of $so(3,2)$.

As is well-known [21], we can introduce a Hilbert space structure on the solution space of (5.1) by considering only positive energy solutions

(5.6)
$$\psi(x) = \frac{1}{4\pi} \iint_{-\infty}^{\infty} e^{ik\cdot x} f(\underline{k})\, d\mu(\underline{k}) = I(f) \ ,$$

$$k = (k_0, k_1, k_2), \quad k_0 = \sqrt{k_1^2 + k_2^2}, \quad d\mu(k) = dk_1 dk_2/k_0 \ .$$

Here f belongs to the Hilbert space $\mathcal{H}$ of all Lebesgue square integrable functions in the plane $(\iint |f|^2 d\mu(\underline{k}) < \infty)$ with inner product

(5.7)
$$\langle f_1, f_2 \rangle = \iint_{-\infty}^{\infty} f_1(\underline{k})\, \overline{f_2}(\underline{k})\, d\mu(\underline{k}) \ .$$

The solutions ψ_1, ψ_2 of (5.1) related to $f_1, f_2 \in \mathcal{H}$ by (5.6) satisfy

(5.8) $(\psi_1, \psi_2) \equiv \langle f_1, f_2 \rangle = 4i \iint\limits_{x_0=t} \psi_1(x) \, \partial_0 \overline{\psi_2}(x) dx_1 dx_2$

$= -4i \iint\limits_{x_0=t} \overline{\psi_2}(x) \, \partial_0 \psi_1(x) dx_1 dx_2$

where the integral is independent of t. The operators (5.2)-(5.5) induce corresponding operators on $\mathcal{V}$.

(5.9) $P_0 = ik_0, \quad P_j = -ik_j, \quad j = 1, 2$

(5.10) $M_{12} = k_1 \partial_{k_2} - k_2 \partial_{k_1}, \quad M_{01} = k_0 \partial_{k_1}, \quad M_{02} = k_0 \partial_{k_2}$

(5.11) $D = \dfrac{1}{2} + k_1 \partial_{k_1} + k_2 \partial_{k_2}$

(5.12) $K_0 = ik_0(\partial_{k_1 k_1} + \partial_{k_2 k_2})$

$K_1 = i(k_1 \partial_{k_1 k_1} - k_1 \partial_{k_2 k_2} + 2k_2 \, \partial_{k_1 k_2} + \partial_{k_1})$

$K_2 = i(-k_2 \partial_{k_1 k_1} + k_2 \partial_{k_2 k_2} + 2k_1 \partial_{k_1 k_2} + \partial_{k_2})$.

It is straightforward to show that these operators generate a unitary irreducible representation of a covering group $\widetilde{SO}(3, 2)$ of $SO(3, 2)$ on $\mathcal{V}$. The action of $\widetilde{SO}(3, 2)$ on $\mathcal{V}$ is worked out in [22].

Equation (5.1) permits R-separation of variables in a great variety of coordinate systems, many of which are listed for the first time in references [22], [23] and [24] by E. G. Kalnins and the author. Every separable coordinate system we have found corresponds to a pair of commuting operators S, S' in $\mathcal{G}$. Each of the operators is purely first order or purely second order. As usual, we consider two coordinate systems as equivalent if they correspond to operators which lie on the same orbit under the adjoint action of $\widetilde{SO}(3, 2)$ on two-dimensional

338

commuting subspaces of $\mathfrak{S}$. However, it is sometimes useful to list more than one system on the same orbit if these systems lie on different orbits with respect to the Poincaré subgroup of $\widetilde{SO}(3,2)$.

Here we will discuss only those coordinate systems which are easiest to understand, the semi-subgroup coordinates, and refer the reader to the research literature for lists of the remaining systems known at this time. Semi-subgroup coordinates correspond to commuting operators S, S' such that we can choose either $S = A$ or $S = A^2$ where $A \epsilon$ so(3, 2) . If also $S' = B^2$, $B \epsilon$ so(3,2) then we have subgroup coordinates while if it is not possible to choose the operators S, S' such that either $S \epsilon$ so(3, 2) or S is a perfect square, the coordinates are non-subgroup. (As usual, the eigenvalues of S and S' are the separation constants for the associated coordinate system.)

In [22] we show that there are seven types of semi-subgroup systems corresponding to seven choices for S . The choices are:

1] $S = A^2$, $A = \frac{1}{2}(P_0 - K_0)$.

The eigenvalues of A are $i(\ell + \frac{1}{2})$, $\ell = 0,1,2,\ldots$. If ψ_ℓ is an eigenfunction of A, $A\psi_\ell = i(\ell + \frac{1}{2})\psi_\ell$, and a solution of (5.1), then (5.1) can be transformed to

(5.13) $\qquad (L_1^2 + L_2^2 + L_3^2)\psi_\ell = -\ell(\ell+1)\psi_\ell$

where

(5.14) $\qquad L_1 = \frac{1}{2}(P_1 + K_1)$, $L_2 = \frac{1}{2}(P_2 + K_2)$, $L_3 = M_{12}$

and $\{L_1, L_2, L_3\}$ form a basis for the Lie algebra so(3) . It can be shown that (5.13) is equivalent to the eigenvalue equation for the Laplace operator on the sphere. This last equation was studied by Patera and Winternitz [25], see also [26], [27]. There are exactly two separable coordinate systems which correspond to the two orbit types of second order symmetric operators in the enveloping algebra of so(3) under the adjoint action of the rotation group SO(3) :

339

1) $S' = L_3^2$ (spherical coordinates)

2) $S' = L_1^2 + a^2 L_2^2$, $a \neq 0$ (Lamé-type coordinates)

2] $S = A^2$, $A = P_0$

If ψ_λ is a solution of (5.1) with $P_0 \psi_\lambda = i\lambda\psi_\lambda$ we find $\psi_\lambda(x) = e^{i\lambda x_0}$ $\Phi(x_1, x_2)$ where

(5.15)
$$(\partial_{11} + \partial_{22} + \lambda^2)\Phi = 0 .$$

Thus the equation for the eigenfunctions reduces to the Helmholtz equation (1.7). The symmetry group of (5.15) is $E(2)$ and there are four separable coordinate systems corresponding to the following four choices for S' :

3) P_1^2 (Cartesian coordinates)

4) M_{12}^2 (polar coordinates)

5) $\{M_{12}, P_2\}$ (parabolic cylinder coordinates)

6) $M_{12}^2 + P_2^2$ (elliptic coordinates)

3] $S = A^2$, $A = P_2$

If ψ_γ is a solution of (5.1) with $P_2 \psi_\gamma = i\gamma\psi_\gamma$ we find $\psi_\gamma(x) =$ $e^{-i\gamma x_2} \Phi(x_0, x_1)$ where

(5.16)
$$(\partial_{00} - \partial_{11} + \gamma^2)\Phi = 0 .$$

Here, the eigenfunction equation has been reduced to the Klein-Gordon equation (2.1) with symmetry group $E(1, 1)$. There are nine distinct separable coordinate systems corresponding to the following choices for S' :

3') $P_0 P_1$

7) M_{01}^2

8) $\{M_{01}, P_0\}$

9) $M_{01}^2 - (P_0 + P_1)^2$

10) $M_{01}^2 + (P_0 + P_1)^2$

11) $\{M_{01}, P_0 - P_1\} - (P_0 + P_1)^2$

12) $M_{01}^2 - P_0 P_1$

13) $M_{01}^2 + P_1^2$

14) $M_{01}^2 - P_1^2$.

Case 3') yields the same coordinates as 3).

4] $S = A^2$, $A = D$

If ψ_ν is a solution of (5.1) with $D\psi_\nu = -i\nu\psi_\nu$ we have $\psi_\nu(x) = \rho^{i\nu - \frac{1}{2}} \Phi(s_0, s_1, s_2)$ where $x_\alpha = \rho s_\alpha$, $\rho \geq 0$, $s_0^2 - s_1^2 - s_2^2 = \varepsilon$ and $\varepsilon = +1, -1$ or 0 depending on whether $x \cdot x > 0$, < 0 or $= 0$. The eigenfunction equation for Φ then becomes

(5.17) $\qquad (M_{12}^2 - M_{01}^2 - M_{02}^2)\Phi(s) = (\nu^2 + \tfrac{1}{4})\Phi(s)$.

Here M_{12}, M_{01}, M_{02} form a basis for the Lie algebra $so(2,1)$, which is the symmetry algebra of (5.17). This equation separates in nine coordinate systems associated with nine symmetric second order operators in the enveloping algebra of $so(2,1)$. The details for $\varepsilon = -1$ are worked out in [27], [28], and [29], and the general case is treated in [23]. The choices for S' are

15) M_{12}^2 (spherical)

16) M_{01}^2 (equidistant)

7') $(M_{12} - M_{02})^2$ (horocyclic)

17) $M_{12}^2 + a^2 M_{01}^2$, $a \neq 0$ (elliptic)

341

18) $M_{01}^2 - a^2 M_{12}^2$, $0 < a < 1$ (hyperbolic)

19) $-\{M_{12}, M_{02}\} + a M_{01}^2$, $0 < a < \infty$ (semihyperbolic)

20) $a M_{01}^2 + M_{02}^2 + M_{12}^2 - \{M_{02}, M_{12}\}$, $0 < a$ (elliptic-parabolic)

21) $-a M_{01}^2 + M_{02}^2 + M_{12}^2 - \{M_{02}, M_{12}\}$, $0 < a$ (hyperbolic-paraboli

22) $\{M_{02}, M_{01}\} - \{M_{01}, M_{12}\}$, (semicircular-parabolic)

The coordinates 7') are equivalent to 7).

5] $S = P_0 + P_1$.

If ψ_β is a solution of (5.1) with $(P_0 + P_1)\psi_\beta = i\beta\psi_\beta$ we have $\psi_\beta(x) = e^{is\beta} \Phi(t, x_2)$ where $2s = x_0 + x_1$, $2t = x_1 - x_0$. Furthermore the equation for Φ reduces to the free-particle Schrödinger equation

(5.18)
$$(i\beta \partial_t + \partial_{22}) \Phi(t, x_2) = 0$$

which is essentially the same as (4.1). This equation admits as symmetries the operators

(5.19)
$$\mathcal{P} = P_2, \quad \mathcal{K}_{-2} = -P_0 + P_1, \quad \mathcal{E} = P_0 + P_1 ,$$
$$\mathcal{K}_1 = \frac{1}{2}(M_{02} - M_{12}), \quad \mathcal{K}_{-1} = -D - M_{01}, \quad \mathcal{K}_2 = -\frac{1}{4}(K_0 + K_1)$$

which satisfy the commutation relations (4.3) and form a basis for the Schrödinger algebra $\mathcal{J}_1$. As shown in [12] and Section 4, there are four separable coordinate systems corresponding to the following choices for S' :

3") P_2 (free particle)

23) $P_0 - P_1 - \frac{1}{2} K_0 - \frac{1}{4} K_1$ (oscillator)

24) $P_0 - P_1 + a M_{12} - a M_{02}$, $a \neq 0$ (linear potential)

25) $D + M_{01}$ (repulsive oscillator)

342

Coordinates 3") are the same as 3).

6] $S = A^2$, $A = M_{12}$.

If ψ_m is a solution of (5.1) and an eigenfunction of M_{12}, $M_{12}\psi_m$ = $im\psi_m$ then $\psi_m(x) = e^{im\varphi} \Phi(x_0, r)$ where $x_1 = r \cos \varphi$, $x_2 = r \sin \varphi$ and Φ satisfies the Euler-Poisson-Darboux (EPD) equation

(5.20)
$$(\partial_{00} - \partial_{rr} - \frac{1}{r}\partial r + \frac{m^2}{r^2})\Phi = 0$$

or

(5.21)
$$(L_3^2 - L_2^2 - L_1^2)\Phi = -(m + \tfrac{1}{2})(m - \tfrac{1}{2})\Phi$$

where

(5.22)
$$L_1 = D, \quad L_2 = \frac{1}{2}(P_0 + K_0), \quad L_3 = \frac{1}{2}(P_0 - K_0)$$

and these operators form a basis for the symmetry algebra $sl(2, R) \cong$ $so(2, 1)$ of the EPD equation. In analogy with case 4], the EPD equation separates in nine coordinate systems corresponding to nine orbits of symmetric second order operators in the enveloping algebra of $sl(2, R)$. The corresponding choices for S' are

1') L_3^2

4') $(L_3 + L_2)^2$

17') L_1^2

26) $2L_2^2 + \{L_2, L_3\}$

27) $2L_3^2 + \{L_3, L_2\}$

28) $L_2^2 + a\{L_3, L_1\}$, $a \neq 0$

29) $L_3^2 + aL_1^2$, $a \neq 0$

30) $L_1^2 + aL_2^2$, $a \neq 0$

31) $\{L_2 + L_3, L_1\}$.

Complete details are contained in [30].

7] $S = A^2$, $A = \frac{1}{2} M_{12} + \frac{1}{4} K_0 - \frac{1}{4} P_0$.

If ψ_κ is a solution of (5.1) and satisfies $A\psi_\kappa = i\kappa\psi_\kappa$ then $\psi_\kappa(x)$ = $\sqrt{\cos\sigma - \cos\varphi}\ e^{i\kappa\beta}\ \Phi(\sigma, \rho)$ where

(5.23) $\qquad x_0 = \frac{\sin\varphi}{\cos\sigma - \cos\varphi}$, $x_1 = \frac{\sin\sigma\cos\alpha}{\cos\sigma - \cos\varphi}$, $x_2 = \frac{\sin\sigma\sin\alpha}{\cos\sigma - \cos\varphi}$

$$\beta = \alpha + \varphi, \quad \rho = \alpha - \varphi$$

and the equation for Φ becomes

(5.24) $\qquad (H_3^2 - H_2^2 - H_1^2)\Phi = -(\kappa + \frac{1}{2})(\kappa - \frac{1}{2})\Phi$

with

(5.25) $\qquad H_1 = -\frac{1}{2} M_{02} + \frac{1}{4} P_1 - \frac{1}{4} K_1$, $H_2 = \frac{1}{2} M_{01} + \frac{1}{4} P_2 - \frac{1}{4} K_2$,

$$H_3 = \frac{1}{2} M_{12} + \frac{1}{4} P_0 - \frac{1}{4} K_0 .$$

The operators H_j satisfy the commutation relations for $sl(2, R)$ which is the symmetry algebra of (5.24). Separation is achieved in at most nine coordinate systems which correspond to the following choices for S':

1') H_3^2

32) H_2^2

33) $(H_2 + H_3)^2$

344

34) $2H_2^2 + \{H_2, H_3\}$

35) $2H_3^2 + \{H_3, H_2\}$

36) $H_2^2 + a\{H_3, H_1\}$, $a \neq 0$

37) $H_3^2 + aH_1^2$, $a \neq 0$

38) $H_1^2 + aH_2^2$, $a \neq 0$

39) $\{H_2 + H_3, H_1\}$

This completes our list of semi-subroup systems.   (It has not yet been settled which of the orbits 34)-39) correspond to separable systems.)

In analogy with our procedure in the earlier sections we can use the operators (5.9)-(5.12) acting on the Hilbert space $\mathcal{H}$ to derive expansion theorems for the separable solutions of the wave equation. Some results along this line can be found in [22] and [30].

In a classic work, [31], Bôcher has shown how to determine all (R-) separable orthogonal coordinate systems for the Laplace equations $\Delta_n \psi = 0$ whose coordinate curves are cyclides or their degenerate forms. Kalnins has adapted Bôcher's method to the wave equation (5.1) (where it becomes more complicated) and computed the orthogonal separable coordinate systems (Minkowski metric) for (5.1) whose coordinate curves are cyclides or their degenerate forms. The results, presented in our joint paper [23], show that there are 53 systems of this type, inequivalent with respect to the group $SO(3,2)$. As usual, each system corresponds to a pair of commuting operators in $\mathcal{G}$. The semi-subgroup systems of classes 1] - 4] and 6] are all orthogonal and of this type. (As a by-product of the analysis of [23] we find 53 separable orthogonal coordinate systems for the Klein-Gordon equation $(\partial_{tt} - \Delta_2)\psi = m^2\psi$.)

The semi-subgroup systems of classes 5] and 7] are non-orthogonal. General non-orthogonal separable systems for (5.1) are studied in [24] where we find a rather bewildering profusion of such systems. At this writing we cannot claim to have yet found all separable coordinate systems for the wave equation. (We can classify all systems whose variables separate in some well-defined fashion but we do not yet understand the most general way it is possible for variables to separate.) Furthermore, we have not yet classified all orbits of two-dimensional commuting subspaces in $\mathfrak{g}$ to see which orbits correspond to separable systems and which do not.

The symmetry algebra of the Laplace equation $\Delta_3 \psi = 0$ is isomorphic to $so(4,1)$. The known separable coordinate systems for this case are all orthogonal and can be obtained by the methods of Bôcher. However, at this writing the symmetry operators corresponding to each separable coordinate system have not been worked out. This equation is somewhat less interesting than the wave equation since the solution space appears to have no Hilbert space structure. The complex Laplace equation $\Delta_3 \psi = 0$ is related to Weisner's method for obtaining generating functions for the Gegenbauer polynomials. Some results are presented in [18], [30] and [32].

Work is just beginning on the wave, Laplace and complex Laplace equations in four-dimensional space. The complex Laplace equation is related to Weisner's method for obtaining generating functions for the hypergeometric functions $_2F_1$, [19], [33].

In references [18], [34], [35] and [36] the author has shown that the generalized hypergeometric functions $_pF_q$ and the Lauricella functions can be obtained by separation of variables in partial differential equations of higher order in the first case, and systems of partial differential equations in the second case. Weisner's method can then easily be applied to derive generating functions for the solutions by making use of the symmetry groups of these equations. A systematic analysis of separable coordinate systems for these equations has not been made.

We also mention the papers [37], [38] by Winternitz et al. in which the authors used methods similar to those described here to compute all time independent non-relativistic Schrödinger equations in two

346

and three dimensional space which can be solved by separation of variables. See also [39].

Finally we mention the only strictly class II equation which has been studied in detail: the time independent non-relativistic Schrödinger equation for the hydrogen atom. This equation has fascinated physicists for many years, partly because its first order symmetry algebra isn't large enough to explain the degeneracies of its eigenvalues. A detailed treatment from the point of view of this paper is contained in [2].

Section 6

The discerning reader will note that although the general features of our methods relating symmetry groups and separation of variables are now clear, many interesting problems remain. Among these problems are:

1) Can one give a clear, precise, definition of separation of variables which suffices for all examples?

2) We have shown how to find the set of commuting symmetry operators associated with a given separable system. Given a set of commuting symmetry operators in what sense can one characterize an associated separable system?

3) Why do some orbits of commuting symmetry operators correspond to separable coordinate systems while others do not?

REFERENCES

[1]   W. Miller, Jr., "Symmetry Groups and Their Applications",
       Academic Press, New York, 1972.

[2]   E. G. Kalnins, W. Miller, Jr., and P. Winternitz, The group
       0(4), separation of variables and the hydrogen atom, (to appear
       in SIAM Journal on Applied Mathematics).

[3]   P. Morse and H. Feshbach, "Methods of Theoretical Physics,
       Part I", McGraw-Hill, New York, 1953.

347

[4]   P. Winternitz and I. Fris, Invariant expansions of relativistic amplitudes and subgroups of the proper Lorentz group, Soviet Physics JNP, 1 (1965), pp. 636-643.

[5]   N. Dunford and J. Schwartz, "Linear Operators. Part II", Wiley-Interscience, New York, 1963.

[6]   C. Boyer, E. G. Kalnins, and W. Miller Jr. , Lie theory and separation of variables. 6. The equation $iU_t + \Delta_2 U = 0$, J. Math. Phys. $\underline{16}$, pp. 499-511, (1975).

[7]   A. Erdélyi et al. , "Higher Transcendental Functions. Vols. 1 and 2", McGraw-Hill, New York, 1953.

[8]   E. G. Kalnins, On the separation of variables for the Laplace equation $\Delta\psi + K^2\psi = 0$ in two and three dimensional Minkowski space, SIAM J. Math. Anal. $\underline{6}$, pp. 340-374, (1975).

[9]   E. G. Kalnins and W. Miller, Jr. , Lie theory and separation of variables. 3. The equation $f_{tt} - f_{ss} = \gamma^2 f$, J. Math. Phys. $\underline{15}$, pp. 1025-1032, (1974).

[10] W. Miller, Jr. , Lie theory and separation of variables. 1. Parabolic cylinder coordinates, SIAM J. Math. Anal. $\underline{5}$, pp. 626-643, (1974).

[11] L. Weisner, Generating functions for Bessel functions, Can. J. Math. $\underline{11}$, pp. 148-155, (1959).

[12] E. G. Kalnins and W. Miller, Jr. , Lie theory and separation of variables. 5. The equations $iU_t + U_{xx} = 0$ and $iU_t + U_{xx} - (c/x^2)U = 0$", J. Math. Phys. $\underline{15}$, pp. 1728-1737, (1974).

[13] A. Davydov, "Quantum Mechanics", Addison-Wesley, Reading, Massachusetts, 1965.

[14] P. C. Rosenbloom and D. V. Widder, Expansions in terms of heat polynomials and associated functions, Trans. Amer. Math. Soc. $\underline{92}$, pp. 220-266, (1959).

[15] L. Weisner, Generating functions for Hermite functions, Can. J. Math. $\underline{11}$, pp. 141-147, (1959).

[16] F. Arscott, "Periodic Differential Equations", Pergamon, Oxford, England, 1964.

[17] E. G. Kalnins and W. Miller, Jr., Symmetry and separation of variables for the heat equation, Proceedings of Conference on Symmetry, Similarity and Group-Theoretic Methods in Mechanics, Calgary, 1974.

[18] W. Miller, Jr., Lie algebras and generalizations of the hypergeometric function, Proceedings of the AMS Summer Institute (1972), "Harmonic Analysis on Homogeneous Spaces, Proceedings of Symposia in Pure Mathematics", Vol. XXVI, AMS., Providence, R.I., 1973.

[19] W. Miller, Jr., "Lie Theory and Special Functions", Academic Press, New York, 1968.

[20] C. Boyer, Lie theory and separation of variables for the equation $iU_t + \Delta_2 U - \left(\dfrac{\alpha}{x_1^2} + \dfrac{\beta}{x_2^2} \right) U = 0$ ", SIAM J. Math. Anal. (to appear).

[21] S. Schweber, "An Introduction to Relativistic Quantum Field Theory", Row, Peterson, Evanston, Illinois, 1961 (Chapter 3).

[22] E. G. Kalnins and W. Miller, Jr., Lie theory and separation of variables. 8. Semi-subgroup coordinates for $\psi_{tt} - \Delta_2 \psi = 0$, (submitted).

[23] E. G. Kalnins and W. Miller, Jr., Lie theory and separation of variables. 9. Orthogonal R-separable coordinate systems for the wave equation $\psi_{tt} - \Delta_2 \psi = 0$, J. Math. Phys. (to appear).

[24] E. G. Kalnins and W. Miller, Jr., Lie theory and separation of variables. 10. Nonorthogonal R-separable solutions of the wave equation $\partial_{tt} \psi = \Delta_2 \psi$, (submitted to the Journal of Mathematical Physics).

[25] J. Patera and P. Winternitz, A new basis for the representations of the rotation group. Lame and Heun polynomials. J. Math. Phys. $\underline{14}$, pp. 1130-1139, (1973).

[26] C. Boyer, E. G. Kalnins and W. Miller, Jr., Lie theory and separation of variables. 7. The harmonic oscillator in elliptic coordinates and Ince polynomials, J. Math. Phys. 16, pp. 512-517, (1975).

[27] E. G. Kalnins and W. Miller, Jr., Lie theory and separation of variables. 4. The groups S0(2,1) and S0(3), J. Math. Phys. 15, pp. 1263-1274, (1974).

[28] P. Winternitz, I. Lukac, and Y. Smorodinskii, Quantum numbers in the little groups of the Poincaré group, Soviet Physics JNP, 7, pp. 139-145, (1968).

[29] N. Macfadyen and P. Winternitz, Crossing symmetric expansions of physical scattering amplitudes; The 0(2,1) group and Lamé functions, J. Math. Phys. 12, pp. 281-293, (1971).

[30] E. G. Kalnins and W. Miller, Jr., Lie theory and separation of variables. 11. The Euler-Poisson-Darboux equation. (submitted).

[31] M. Bôcher, "Die Reihenentwickelungen der Potentialtheorie", Leipzig, 1894.

[32] B. Viswanathan, Generating functions for ultraspherical functions, Can. J. Math. 20, pp. 120-134, (1968).

[33] L. Weisner, Group-theoretic origin of certain generating functions, Pacific J. Math. 5, pp. 1033-1039, (1955).

[34] W. Miller, Jr., Lie theory and generalized hypergeometric functions, SIAM J. Math. Anal. 3, pp. 31-44, (1972).

[35] W. Miller, Jr., Lie theory and generalizations of the hypergeometric functions, SIAM J. Appl. Math. 25, pp. 226-235, (1973).

[36] W. Miller, Jr., Lie theory and the Lauricella functions F_D, J. Math. Phys. 13, pp. 1393-1399, (1972).

[37] P. Winternitz, Ya. A. Smorodinskii, M. Uhlir and I. Fris, Symmetry groups in classical and quantum mechanics, Soviet Physics JNP, 4, pp. 444-450, (1967).

[38] A. A. Makarov, J. A. Smorodinsky, K. Valiev and P. Winternitz, A systematic search for nonrelativistic systems with dynamical symmetries. Part I: The integrals of motion, Nuovo Cimento 52 A, pp. 1061-1984, (1967).

[39] C. Boyer and W. Miller, Jr., A classification of second order raising operators for Hamiltonians in two variables, J. Math. Phys. 15, pp. 1484-1489, (1974).

School of Mathematics

University of Minnesota

Nicholson–Type Integrals for Products of Gegenbauer
Functions and Related Topics
L. Durand

ABSTRACT

Nicholson's integral gives a generalization of the relation $\sin^2 x +$ $\cos^2 x = 1$ to the case of Bessel functions. This integral expresses the sum of the squares of the Bessel functions of the first and second kind, $[J_\nu^2(x) + Y_\nu^2(x)]$ as an integral over a hyperbolic Bessel function, with the integrand positive. We present an analogous result for general Gegenbauer and Legendre functions derived from a new representation for the product of two Gegenbauer functions of the second kind, $D_\lambda^{(\alpha)}(x)$. The result expresses the sum of the squares of the Gegenbauer functions on the cut, $[(C_\lambda^{(\alpha)}(x))^2 + (D_\lambda^{(\alpha)}(x))^2]$, as an integral over a Gegenbauer function of the second kind, with the integrand again positive. The result for the ordinary Legendre functions is quite simple,

$$\left[P_n(x) \right]^2 + \frac{4}{\pi^2}\left[Q_n(x) \right]^2 = \frac{4}{\pi^2} \int_1^\infty Q_n(x^2 + (1-x^2)z) \frac{dz}{\sqrt{z^2-1}}, \quad -1 < x < 1 \ .$$

We discuss various applications of the general result, including the derivation of an asymptotic expansion, bounds, and some monotonicity properties for the Gegenbauer functions, some new results on products of Bessel functions, and the derivation of an analogous integral for Hermite functions. The last can be used to establish higher monotonicity properties for the Hermite polynomials.

1. INTRODUCTION

A generalization of the familiar relation $\sin^2 x + \cos^2 x = 1$ appropriate for the case of Bessel functions was derived by J. W. Nicholson in 1910 [1]. The result, expressed by Nicholson's integral

$$J_\nu^2(x) + Y_\nu^2(x) = \frac{8}{\pi^2} \int_0^\infty K_0(2x \sinh t) \cosh 2\nu t \, dt \, , \tag{1}$$

has been of considerable importance in the theory of Bessel functions. It has been used by Watson [2] to derive bounds on the functions, obtain asymptotic expansions, and establish a number of interesting monotonicity properties. It follows, for example, from the analysis given by Watson [2, Sec. 13.74] that the function $x[J_\nu^2(x) + Y_\nu^2(x)]$ is completely monotonic for $\nu > \frac{1}{2}$, a result which has been used by Lorch and Szego [3] to prove a number of remarkable monotonicity properties of the nth differences of the zeros of Bessel functions, the areas under successive arches, etc.

One would expect expressions analogous to Nicholson's integral to exist for the classical orthogonal polynomials, as these all reduce to Bessel functions in appropriate confluent limits. However, despite the extensive literature on these polynomials [4, 5], no such results were known when I first learned about the problem from Professor Richard Askey in 1969. I was working at that time on scattering problems in physics which had necessitated a thorough study of Gegenbauer (or hyperspherical) functions of the second kind, and the derivation of some new addition and expansion formulas for those functions. With that background, I was able to write down the analog of the right hand side of Eq. (1) immediately, but was unclear as to how to define the function of the second kind which appears on the left, and let the matter drop. The problem was revived and solved for Gegenbauer functions in the fall of 1971, in conjunction with a series of seminars on special functions given by Professor Askey. The method of derivation used for the extended

"Nicholson formula" was based heavily on the use of the functions of the second kind, and is similar in that respect to Nicholson's work. I suspect, in fact, that Nicholson's results were derived in a similar context, as much of his work in physics at that time was also concerned with scattering problems. [1]

In the next section, I will give a sketch of the derivation of the basic result for Gegenbauer functions. Once its form is known, one can, of course, check its validity in different ways, for example, by using the third order differential equations satisfied by the product of two Gegenbauer functions [6]. However, the original method retains a certain group-theoretic flavor which may give some insight into generalizations for other cases. In the following sections, I will give applications of this result for Gegenbauer, Bessel, and Hermite functions. More details on these results will eventually appear in a series of papers in preparation.

2. DERIVATION OF A NICHOLSON-TYPE FORMULA FOR GEGEN-BAUER FUNCTIONS

a. Definitions

The Gegenbauer functions of the first and second kind, $C_\lambda^{(\alpha)}(z)$ and $D_\lambda^{(\alpha)}(z)$, are solutions of the differential equation

$$(1-z^2)\frac{d^2y}{dz^2} - (2\alpha+1)z\frac{dy}{dz} + \lambda(\lambda+2\alpha)y = 0 \ . \tag{2}$$

The definitions which we use for arbitrary λ, α are [6] [2]

$$C_\lambda^{(\alpha)}(z) = \frac{\Gamma(\lambda+2\alpha)}{\Gamma(2\alpha)\Gamma(\lambda+1)} \ _2F_1(-\lambda, \ \lambda+2\alpha;\alpha+\frac{1}{2}; \frac{1}{2}(1-z))$$

$$= \sqrt{\pi} \ 2^{-\alpha+\frac{1}{2}} \frac{\Gamma(\lambda+2\alpha)}{\Gamma(\alpha)\Gamma(\lambda+1)} (z^2-1)^{-\frac{1}{2}\alpha+\frac{1}{4}} P_{\lambda+\alpha-\frac{1}{2}}^{-\alpha+\frac{1}{2}}(z) \ , \tag{3}$$

$$D_\lambda^{(\alpha)}(z) = e^{i\pi\alpha} \frac{\Gamma(\lambda+2\alpha)}{\Gamma(\alpha)\Gamma(\lambda+\alpha+1)} (2z)^{-\lambda-2\alpha} {}_2F_1(\tfrac{1}{2}\lambda+\alpha, \tfrac{1}{2}\lambda+\alpha+\tfrac{1}{2}; \lambda+\alpha+1; z^{-2})$$

$$= \frac{1}{\sqrt{\pi}} e^{2\pi i(\alpha-\frac{1}{4})} 2^{-\alpha+\frac{1}{2}} \frac{\Gamma(\lambda+2\alpha)}{\Gamma(\alpha)\Gamma(\lambda+1)} (z^2-1)^{-\frac{1}{2}\alpha+\frac{1}{4}} Q_{\lambda+\alpha-\frac{1}{2}}^{-\alpha+\frac{1}{2}}(z) . \tag{4}$$

The functions $C_\lambda^{(\alpha)}(z)$ and $D_\lambda^{(\alpha)}(z)$ are analytic in the complex z plane cut respectively from -1 to $-\infty$, and from +1 to $-\infty$; $C_\lambda^{(\alpha)}(z)$ and $e^{-i\pi\alpha} D_\lambda^{(\alpha)}(z)$ are real and increase (decrease) monotonically with z for z real, $z > 1$, and λ and α real. $C_\lambda^{(\alpha)}(z)$ and $D_\lambda^{(\alpha)}(z)$ as defined above satisfy the same recurrence relations.

We will also need the functions "on the cut", $-1 < x < 1$. $C_\lambda^{(\alpha)}(z)$ is analytic in this region, so $C_\lambda^{(\alpha)}(x)$ will be defined simply as the restriction of $C_\lambda^{(\alpha)}(z)$ to this interval,

$$C_\lambda^{(\alpha)}(x) = \lim_{\varepsilon \to 0+} C_\lambda^{(\alpha)}(x\pm i\varepsilon)$$

$$= \sqrt{\pi}\, 2^{-\alpha+\frac{1}{2}} \frac{\Gamma(\lambda+2\alpha)}{\Gamma(\alpha)\Gamma(\lambda+1)} (1-x^2)^{-\frac{1}{2}\alpha+\frac{1}{4}} P_{\lambda+\alpha-\frac{1}{2}}^{-\alpha+\frac{1}{2}}(x), \quad -1 < x < 1 . \tag{5}$$

Here $P_\nu^\mu(x)$ is the usual Legendre function on the cut. If we note that

$$C_\lambda^{(\alpha)}(x) = \lim_{\varepsilon \to 0+} e^{-i\pi\alpha} [e^{i\pi\alpha} D_\lambda^{(\alpha)}(x+i\varepsilon) + e^{-i\pi\alpha} D_\lambda^{(\alpha)}(x-i\varepsilon)] , \tag{6}$$

a relation easily derived from the expressions in Eqs. (3) and (4), it is clear that we obtain a second independent function on the cut by defining $D_\lambda^{(\alpha)}(x)$ as

$$D_\lambda^{(\alpha)}(x) = \lim_{\varepsilon \to 0+} \frac{1}{i} e^{-i\pi\alpha} [e^{i\pi\alpha} D_\lambda^{(\alpha)}(x+i\varepsilon) - e^{-i\pi\alpha} D_\lambda^{(\alpha)}(x-i\varepsilon)]$$

$$= \frac{2}{\sqrt{\pi}} 2^{-\alpha+\frac{1}{2}} \frac{\Gamma(\lambda+2\alpha)}{\Gamma(\alpha)\Gamma(\lambda+1)} (1-x^2)^{-\frac{1}{2}\alpha+\frac{1}{4}} Q_{\lambda+\alpha-\frac{1}{2}}^{-\alpha+\frac{1}{2}}(x), \quad -1 < x < 1, \tag{7}$$

where Q_ν^μ is the usual Legendre function on the cut.

With the foregoing definitions,

$$\lim_{\varepsilon \to 0+} e^{-i\pi\alpha} D_\lambda^{(\alpha)}(x \pm i\varepsilon) = \frac{1}{2} e^{\mp i\pi\alpha}[C_\lambda^{(\alpha)}(x) \pm iD_\lambda^{(\alpha)}(x)] \ , \tag{8}$$

so, in particular,

$$[C_\lambda^{(\alpha)}(x)]^2 + [D_\lambda^{(\alpha)}(x)]^2 = \lim_{\varepsilon \to 0+} 4e^{-2\pi i\alpha} D_\lambda^{(\alpha)}(x+i\varepsilon) D_\lambda^{(\alpha)}(x-i\varepsilon) \ . \tag{9}$$

The expression on the left is clearly what we want in a generalization of the Nicholson-type formula to Gegenbauer functions. The problem is then to find an expression for the product of the two Gegenbauer functions of the second kind which appear on the right. [This situation is exactly parallel to that encountered in the classic derivation of Nicholson's result [2, Sec. 13.73] or in the relation $e^{ix} e^{-ix} = 1$.] A logical place to begin is apparently with the addition formulas for the Gegenbauer functions, which involve such products naturally.

b. Derivation

We begin with a non-standard addition theorem for the Gegenbauer functions of the second kind [7]

$$D_\lambda^{(\alpha)}(z_1 z_2 + \sqrt{z_1^2 - 1} \sqrt{z_2^2 - 1} \ x) = \frac{\Gamma(2\alpha-1)}{[\Gamma(\alpha)]^2} \sum_{n=0}^\infty \frac{4^n \Gamma(\lambda-n+1)[\Gamma(\alpha+n)]^2}{\Gamma(\lambda+2\alpha+n)}$$

$$\times (2n+2\alpha-1)(z_1^2-1)^{\frac{n}{2}}(z_2^2-1)^{\frac{n}{2}} D_{\lambda-n}^{(\alpha+n)}(z_1) C_{\lambda-n}^{(\alpha+n)}(z_2) C_n^{(\alpha-\frac{1}{2})}(x) \ . \tag{10}$$

This result, which converges in a region which includes the region z_1, z_2, x real with $z_1 > z_2 > 1$, $-1 \le x \le 1$, can be derived in a straightforward way from the extension of the standard group theoretical addition

357

theorem to general values of λ and α . Use of the orthogonality relations for the Gegenbauer polynomials [8, Sec. 3.15.1(16)] leads to the
expression

$$(z_1^2 -1)^{\frac{n}{2}} (z_2^2 -1)^{\frac{n}{2}} D_{\lambda-n}^{(\alpha+n)} (z_1) C_{\lambda-n}^{(\alpha+n)}(z_2)$$

$$= 2^{-2\alpha-2n+1} \frac{\Gamma(2\alpha-1)\Gamma(n+1)\Gamma(\lambda+2\alpha+n)}{[\Gamma(\alpha+n)]^2 \Gamma(n+2\alpha-1) \Gamma(\lambda-n+1)} \tag{11}$$

$$\times \int_{-1}^{1} D_{\lambda}^{(\alpha)} (z_1 z_2 + \sqrt{z_1^2 -1} \sqrt{z_2^2 -1} \; x) \; C_n^{(\alpha-\frac{1}{2})}(x) \; (1-x^2)^{\alpha-1} dx \; ,$$

$$z_1 > z_2 > 1, \quad -1 \le x \le 1 \; .$$

By comparing the asymptotic forms of the two sides of this equation for
$z_1 \to \infty$, z_2 fixed, we can establish a new integral representation for the
ordinary Gegenbauer functions,

$$(z^2 -1)^{\frac{n}{2}} C_{\lambda-n}^{(\alpha+n)}(z) = 2^{-n-2\alpha+1} \frac{\Gamma(2\alpha-1)\Gamma(\lambda+2\alpha)\Gamma(n+1)}{\Gamma(\alpha)\Gamma(n+2\alpha-1)\Gamma(\lambda-n+1)\Gamma(\alpha+n)}$$

$$\times (-1)^n \int_{-1}^{1} [z + x\sqrt{z^2 -1} \;]^{-\lambda-2\alpha} \; C_n^{(\alpha-\frac{1}{2})}(x) \, (1-x^2)^{\alpha-1} dx \; . \tag{12}$$

This reduces to a standard representation for $C_{\lambda}^{(\alpha)}(z)$ for $n = 0$ [8, Sec.
3.7(6)].

We can extract an expression for the product of two D's from Eq.
(11) as follows. We note that the contour of integration can be opened
up in the complex x (or z) plane as shown in Fig. 1. The integral over

Fig. 1

the right hand segment of the contour gives a function symmetric in z_1, z_2 which vanishes for $z_1, z_2 \to \infty$ as $(z_1 z_2)^{-\lambda - 2\alpha}$. This is just the behavior expected for the product of two D's. The function defined by the remaining segment of the contour is not of interest. After some calculation to determine the normalization constants, we obtain the following expression,

$$(z_1^2 - 1)^{\frac{n}{2}} (z_2^2 - 1)^{\frac{n}{2}} D_{\lambda-n}^{(\alpha+n)}(z_1) D_{\lambda-n}^{(\alpha+n)}(z_2)$$

$$= 2^{-2\alpha - 2n + 1} \frac{\Gamma(2\alpha-1)\Gamma(n+1)\Gamma(\lambda+2\alpha+n)}{[\Gamma(\alpha+n)]^2 \Gamma(n+2\alpha-1)\Gamma(\lambda-n+1)} \tag{13}$$

$$\times e^{i\pi\alpha} \int_1^{\infty} D_{\lambda}^{(\alpha)}(z_1 z_2 + z\sqrt{z_1^2 - 1}\sqrt{z_1^2 - 1}) C_n^{(\alpha - \frac{1}{2})}(z)(z^2 - 1)^{\alpha - 1} dz ,$$

$$\mathrm{Re}(\lambda - n + 1) > 0, \quad \mathrm{Re}\,\alpha > 0 .$$

A lengthy calculation shows that the right hand side of this expression divided by $(z_1^2 - 1)^{\frac{n}{2}} (z_2^2 - 1)^{\frac{n}{2}}$ gives solutions to the Gegenbauer equation of degree $\lambda - n$ and order $\alpha + n$, Eq. (2), in z_1 and z_2 separately. There is no restriction that n be an integer. The asymptotic behavior of the two sides of the equation for $z_1 \to \infty$ and $z_2 \to \infty$ then establishes the equality indicated. A careful analysis of the singularities of the integrand shows that Eq. (13) is valid in the form given for all z_1 and z_2 in the respective right half planes, with the usual cut of the D's extending in each case from $+1$ to $-\infty$, and is valid more generally for all z_1, z_2 such that $\arg(\sqrt{z_1^2 - 1}\sqrt{z_2^2 - 1}) < \pi$. The expression may be extended to other regions through the use of the reflection symmetry of the D's displayed by Eq. (4),

$$D_{\lambda}^{(\alpha)}(e^{\pm i\pi} z) = e^{\mp i(\lambda + 2\alpha)\pi} D_{\lambda}^{(\alpha)}(z) . \tag{14}$$

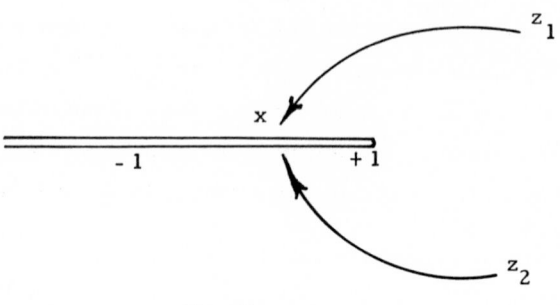

Fig. 2

Our analog of Nicholson's formula is now obtained by considering the limiting case in which the points z_1, z_2 approach a real point x, $-1 < x < 1$, as indicated in Eq. (9) and Fig. 2 . The integral in Eq. (13) is well-behaved in this limit, and we find that

$$(1-x^2)^n \left\{ [C_{\lambda-n}^{(\alpha+n)}(x)]^2 + [D_{\lambda-n}^{(\alpha+n)}(x)]^2 \right\}$$

$$= 2^{-2\alpha-2n+3} \, e^{-i\pi\alpha} \, \frac{\Gamma(2\alpha-1)\Gamma(n+1)\Gamma(\lambda+2\alpha+n)}{[\Gamma(\alpha+n)]^2 \Gamma(n+2\alpha-1)\Gamma(\lambda-n+1)}$$

$$\times \int_1^\infty D_\lambda^{(\alpha)}(x^2 + (1-x^2)z) \, C_n^{(\alpha-\frac{1}{2})}(z)(z^2-1)^{\alpha-1} dz \ ,$$

$$\text{Re}(\lambda-n+1) > 0, \quad \text{Re } \alpha > 0 \ . \tag{15}$$

This formula gives a generalization of the relation $\cos^2 x + \sin^2 x = 1$ to the case of general Gegenbauer functions. The generalizations of the expressions $\cos^2 x - \sin^2 x = \cos 2x$ and $2 \sin x \cos x = \sin 2x$ can be obtained by considering respectively the sums and differences of the expressions obtained for $z_1, z_2 \rightarrow x + i\varepsilon$, and $z_1, z_2 \rightarrow x - i\varepsilon$. There results are of less interest than that above, and will not be given here.

For completeness, we note also the analog for the D's of the integral representation for the C's given in Eq. (12),

$$(z^2-1)^{\frac{n}{2}} D_{\lambda-n}^{(\alpha+n)}(z) = e^{i\pi(\alpha+n)} 2^{-n-2\alpha+1} \frac{\Gamma(2\alpha-1)\Gamma(\lambda+2\alpha)\Gamma(n+1)}{\Gamma(\alpha)\Gamma(n+2\alpha-1)\Gamma(\lambda-n+1)\Gamma(\alpha+n)}$$

$$\times \int_1^\infty [z + x\sqrt{z^2-1}]^{-\lambda-2\alpha} C_n^{(\alpha-\frac{1}{2})}(x)(x^2-1)^{\alpha-1}dx \quad . \tag{16}$$

This result is obtained by considering the limit of Eq. (13) for $z_2 \to \infty$ with z_1 fixed. It reduces to a known result for $n = 0$ [8, Sec. 3.7(2)].

3. SOME APPLICATIONS FOR GEGENBAUER FUNCTIONS

a. <u>Bounds on</u> $C_\lambda^{(\alpha)}$ <u>and</u> $D_\lambda^{(\alpha)}$

We can obtain bounds on the functions $C_\lambda^{(\alpha)}(x)$ and $D_\lambda^{(\alpha)}(x)$ for real values of λ and α with $\alpha > 0$, $\lambda > -1$ by using Eq. (15) in the special case $n = 0$,

$$[C_\lambda^{(\alpha)}(x)]^2 + [D_\lambda^{(\alpha)}(x)]^2 = 2^{-2\alpha+3} e^{-i\pi\alpha} \frac{\Gamma(\lambda+2\alpha)}{[\Gamma(\alpha)]^2\Gamma(\lambda+1)}$$

$$\times \int_1^\infty D_\lambda^{(\alpha)}(x^2 + (1-x^2)z)(z^2-1)^{\alpha-1}dz \quad . \tag{17}$$

A change of variables from z to $u = x^2 + (1-x^2)z$ converts the integral to either of the equivalent forms

$$(1-x^2)^{-2\alpha+1} \int_1^\infty D_\lambda^{(\alpha)}(u)(u-1)^{\alpha-1}(u+1-2x^2)^{\alpha-1}du \tag{18a}$$

or

$$(1-x^2)^{-\alpha} \int_1^\infty D_\lambda^{(\alpha)}(u)(u-1)^{\alpha-1}(2 + \frac{u-1}{1-x^2})^{\alpha-1}du \quad . \tag{18b}$$

In the form (18a), the last factor in the integrand decreases mono-tonically as x increases from 0 to 1 if $\alpha - 1 > 0$, and increases monotonically for $0 < \alpha < 1$. It thus has its maximum (minimum) value at

361

any fixed value of u at $x^2 = 0$ if $\alpha > 1$ $(0 < \alpha < 1)$, and we obtain an upper (lower) bound on the integral by evaluating it at $x^2 = 0$. However, the value of the integral at $x^2 = 0$ multiplied by the numerical factor in Eq. (17) is just equal to $[C_\lambda^{(\alpha)}(0)]^2 + [D_\lambda^{(\alpha)}(0)]^2 \equiv L_0^2$, and we conclude that

$$[C_\lambda^{(\alpha)}(x)]^2 + [D_\lambda^{(\alpha)}(x)]^2 \le (1-x^2)^{-2\alpha+1} L_0^2, \quad \alpha \ge 1, \tag{19}$$

and

$$[C_\lambda^{(\alpha)}(x)]^2 + [D_\lambda^{(\alpha)}(x)]^2 \ge (1-x^2)^{-2\alpha+1} L_0^2, \quad 0 < \alpha \le 1. \tag{20}$$

The quantity L_0 is easily determined by evaluating $C_\lambda^{(\alpha)}(0)$ and $D_\lambda^{(\alpha)}(0)$ using representations for those functions in terms of hypergeometric functions of argument x^2,

$$L_0 = \frac{\Gamma(\frac{\lambda}{2} + \alpha)}{\Gamma(\alpha)\Gamma(\frac{\lambda}{2} + 1)}. \tag{21}$$

A second set of bounds can be obtained by using the form of the integral given in Eq. (18b). In this case, the last factor in the integrand has its minimum value at fixed u for $x^2 = 0$ if $\alpha > 1$, and its maximum value at $x^2 = 0$ if $0 < \alpha < 1$. We therefore obtain a lower (upper) bound on the integral for $\alpha > 1$ $(0 < \alpha < 1)$ by evaluating it at $x^2 = 0$, and find that

$$[C_\lambda^{(\alpha)}(x)]^2 + [D_\lambda^{(\alpha)}(x)]^2 \ge (1-x^2)^{-\alpha} L_0^2, \quad \alpha \ge 1, \tag{22}$$

and

$$[C_\lambda^{(\alpha)}(x)]^2 + [D_\lambda^{(\alpha)}(x)]^2 \le (1-x^2)^{-\alpha} L_0^2, \quad 0 < \alpha \le 1. \tag{23}$$

All of the bounds in Eqs. (19)-(23) hold for arbitrary real values of λ with $\lambda > -1$. The equalities hold only for $x^2 = 0$ except in the case

$\alpha = 1$, when they hold for all x $(C_\lambda^{(1)}(\cos\theta) = \sin(\lambda+1)\theta/\sin\theta$, $D_\lambda^{(1)}(\cos\theta)$ $= \cos(\lambda+1)\theta/\sin\theta)$. The result in Eq. (23) generalizes a standard inequality for $|C_\lambda^{(\alpha)}(x)|$, [4, Theorem 7.33.2]. The remaining inequalities apparently do not appear in standard references. Note that the separate inequalities for C and D implied by Eqs. (19) and (23) generalize to any real linear combination of C and D . Thus, from Eq. (23),

$$\left| a \, C_\lambda^{(\alpha)}(x) + b \, D_\lambda^{(\alpha)}(x) \right| \leq (1-x^2)^{-\alpha/2} \, L_0 \, \sqrt{a^2+b^2}, \quad 0 < \alpha \leq 1 \; . \quad (24)$$

One can also derive inequalities of another sort from Eq. (18a) by noting that the integral attains its minimum (maximum) value for $x^2 = 1$ if $\alpha > 1$ $(0 < \alpha < 1)$, and evaluating that bound by considering the limit for $x^2 \to 1$ of the left hand side of Eq. (17) multiplied by $(1-x^2)^{2\alpha-1}$,

$$[C_\lambda^{(\alpha)}(x)]^2 + [D_\lambda^{(\alpha)}(x)]^2 \geq \frac{1}{\pi} \left[\frac{\Gamma(\alpha-\frac{1}{2})}{\Gamma(\alpha)} \right]^2 (1-x^2)^{-2\alpha+1}, \; \alpha \geq 1 \; , \quad (25)$$

$$[C_\lambda^{(\alpha)}(x)]^2 + [D_\lambda^{(\alpha)}(x)]^2 \leq \frac{1}{\pi} \left[\frac{\Gamma(\alpha-\frac{1}{2})}{\Gamma(\alpha)} \right]^2 (1-x^2)^{-2\alpha+1}, \; \frac{1}{2} < \alpha \leq 1 \; . \quad (26)$$

The integral fails to converge and yields no useful bound for $\alpha \leq \frac{1}{2}$.

We conclude the discussion of bounds by noting that we can use the relation (which holds also for $D_\lambda^{(\alpha)}(x)$)

$$\left(\frac{d}{dx}\right)^n C_\lambda^{(\alpha)}(x) = 2^n \, \frac{\Gamma(\alpha+n)}{\Gamma(\alpha)} \, C_{\lambda-n}^{(\alpha+n)}(x) \; , \quad (27)$$

in conjunction with the general expression in Eq. (15) and manipulations similar to those above to obtain bounds on the nth derivatives of the C's and D's . Let

$$L_n = \frac{2^n \, \Gamma(\frac{\lambda}{2} + \alpha + \frac{n}{2})}{\Gamma(\alpha) \, \Gamma(\frac{\lambda}{2} - \frac{n}{2} + 1)} \; , \quad \lambda - n \geq -1 \; . \quad (28)$$

Then we find that

$$
[(\frac{d}{dx})^n C_\lambda^{(\alpha)}]^2 + [(\frac{d}{dx})^n D_\lambda^{(\alpha)}(x)]^2 \leq
\begin{cases}
L_n^2 (1-x^2)^{-\alpha-2n}, & 0 < \alpha \leq 1, \\
L_n^2 (1-x^2)^{-2\alpha-2n+1}, & \alpha \geq 1,
\end{cases}
$$

$$
[(\frac{d}{dx})^n C_\lambda^{(\alpha)}(x)]^2 + [(\frac{d}{dx})^n D_\lambda^{(\alpha)}(x)]^2 \geq
\begin{cases}
L_n^2 (1-x^2)^{-2\alpha-n+1}, & 0 < \alpha \leq 1, \\
L_n^2 (1-x^2)^{-\alpha-n}, & \alpha \geq 1. \quad (29)
\end{cases}
$$

These are perhaps most useful for $n = 1$, in which case they give bounds on the θ derivatives of $C_\lambda^{(\alpha)}(\cos\theta)$ and $D_\lambda^{(\alpha)}(\cos\theta)$.

b. Asymptotic Series

We can obtain an asymptotic expansion for the sum of the squares of $C_\lambda^{(\alpha)}$ and $D_\lambda^{(\alpha)}$ by expanding the last factor in Eq. (18b) and integrating term-by-term. For any finite interval $1 \leq u \leq u_{max}$,

$$
[1 + \frac{u-1}{2(1-x^2)}]^{\alpha-1} = \sum_{n=0}^{N-1} \frac{\Gamma(\alpha)}{\Gamma(n+1)\Gamma(\alpha-n)} 2^{-n}(1-x^2)^{-n}(u-1)^n
$$

$$(30)$$

$$
+ \frac{\Gamma(\alpha)}{\Gamma(N+1)\Gamma(\alpha-N)} 2^{-N}(1-x^2)^{-N}(u-1)^N [1 + \frac{\xi-1}{2(1-x^2)}]^{\alpha-N-1},
$$

$1 \leq \xi \leq u_{max}$. Let us suppose that λ is large. Then we can choose N so that $\lambda + \alpha + 1 > N + 1 > \alpha$. The last term in Eq. (30) is then a monotonically decreasing function of ξ which attains its maximum value at $\xi = 1$. Since setting $\xi = 1$ gives the next term in the Taylor expansion, the remainder in our asymptotic series will be bounded in magnitude and have the same sign as the first term neglected. Integration is straightforward [9, 7.133(2)], and we find that

$$[C_\lambda^{(\alpha)}(x)]^2 + [D_\lambda^{(\alpha)}(x)]^2 \sim \frac{1}{\sqrt{\pi}} 2^{-2\alpha+2}(1-x^2)^{-\alpha} \frac{\Gamma(\lambda+2\alpha)}{[\Gamma(\alpha)]^2\Gamma(\lambda+1)}$$

$$\times \sum_{n=0}^{N} (1-x^2)^{-n} \frac{\Gamma(\lambda+\alpha+n)\Gamma(n+\frac{1}{2})\Gamma(\alpha+n)}{\Gamma(\lambda+\alpha+n+1)\Gamma(n+1)\Gamma(\alpha-n)}, \quad \lambda+\alpha+1 > N+1 > \alpha .$$

(31)

c. Monotonicity Properties

It is relatively easy to prove the simplest monotonicity properties of the Gegenbauer functions. We will work with the functions considered as functions of θ with $x = \cos\theta$, and will make use of a representation for $C^2 + D^2$ obtained by letting $n = \nu$, $\alpha = \frac{1}{2}$ in Eq. (15). If we let $z = \cosh 2t$, and use the limit

$$\lim_{\alpha \to \frac{1}{2}} \frac{\Gamma(\nu+1)\Gamma(\alpha-\frac{1}{2})}{\Gamma(\nu+2\alpha-1)} C_\nu^{(\alpha-\frac{1}{2})}(z) = 2\cosh 2\nu t \tag{32}$$

and the relation

$$D_\lambda^{(\frac{1}{2})}(z) = \frac{i}{\pi} Q_\lambda(z) , \tag{33}$$

we find that

$$(\sin\theta)^{2\nu}\left\{[C_\lambda^{(\nu+\frac{1}{2})}(\cos\theta)]^2 + [D_\lambda^{(\nu+\frac{1}{2})}(\cos\theta)]^2\right\}$$

$$= 2^{-2\nu+3}\frac{1}{\pi} \frac{\Gamma(\lambda+2\nu+1)}{\Gamma(\lambda+1)[\Gamma(\nu+\frac{1}{2})]^2} \int_0^\infty Q_{\lambda+\nu}(\cos^2\theta+\sin^2\theta\cosh 2t)\cosh 2\nu t\, dt .$$

(34)

It is convenient to use instead of $C_\lambda^{(\alpha)}$ and $D_\lambda^{(\alpha)}$ functions $c_\lambda^{(\alpha)}$ and $d_\lambda^{(\alpha)}$ which are solutions of the Gegenbauer equation in the form

$$\left[\frac{d^2}{d\theta^2} + (\lambda+\alpha)^2 + \frac{\alpha(1-\alpha)}{\sin^2\theta}\right] y(\theta) = 0 \tag{35}$$

normalized to unit Wronskian,

365

$$c_\lambda^{(\alpha)}(\cos\theta) = 2^{\alpha-1} \, \Gamma(\alpha) \left[\frac{\Gamma(\lambda+1)}{\Gamma(\lambda+2\alpha)} \right]^{\frac{1}{2}} (\sin\theta)^\alpha \, C_\lambda^{(\alpha)}(\cos\theta) \ ,$$

(36)

$$d_\lambda^{(\alpha)}(\cos\theta) = 2^{\alpha-1} \, \Gamma(\alpha) \left[\frac{\Gamma(\lambda+1)}{\Gamma(\lambda+2\alpha)} \right]^{\frac{1}{2}} (\sin\theta)^\alpha \, D_\lambda^{(\alpha)}(\cos\theta) \ ,$$

$$W_\theta[d_\lambda^{(\alpha)}, \, c_\lambda^{(\alpha)}] = 1 \ .$$

Equation (34) then assumes the form

$$[c_\lambda^{(\nu+\frac{1}{2})}(\cos\theta)]^2 + [d_\lambda^{(\nu+\frac{1}{2})}(\cos\theta)]^2 =$$

(37)

$$= \frac{4}{\pi} \sin\theta \int_0^\infty Q_{\lambda+\nu}(\cos^2\theta + \sin^2\theta \cosh 2t) \cosh 2\nu t \, dt \ ,$$

where the left hand side is just the function $p(\theta)$ used by Lorch and Szego in their study of higher monotonicity properties of Sturm-Liouville functions [3].

The function $p(\theta)$ is positive, and it may be shown after a partial integration that $p'(\theta)$ is negative or positive depending on whether $\nu > \frac{1}{2}$ or $\nu < \frac{1}{2}$,

$$p'(\theta) = -\frac{2}{\pi} \cos\theta \int_0^\infty Q_{\lambda+\nu}(\cos^2\theta + \sin^2\theta \cosh 2t)$$

(38)

$$\times [2\nu \tanh 2\nu t - \tanh t] \cosh 2\nu t \tanh t \, dt \ .$$

The sign of the integrand is determined by the factor in brackets. Since $\tanh z$ is an increasing function of its argument, this factor is positive for $\nu > \frac{1}{2}$ and negative for $\nu < \frac{1}{2}$. Thus the function $p(\theta)$ decreases (increases) monotonically to its limiting value $p(\frac{\pi}{2})$ as θ increases from 0 to $\frac{\pi}{2}$ for $\nu > \frac{1}{2}$ ($\nu < \frac{1}{2}$) . This result is the precise analog for

366

Gegenbauer functions of the familiar result for Bessel functions, that $x[J_\nu^2(x) + Y_\nu^2(x)]$ is a decreasing or increasing function of x for $\nu > \frac{1}{2}$ or $\nu < \frac{1}{2}$.

The sign of $p''(\theta)$ is easily seen to be positive for $\nu > \frac{1}{2}$ and negative for $\nu < \frac{1}{2}$,

$$p''(\theta) = \frac{4}{\pi} \sin\theta \int_0^\infty [Q_{\lambda+\nu} - 2\cos^2\theta \,(\cosh 2t - 1)Q'_{\lambda+\nu}]$$

$$\times [2\nu \tanh 2\nu t - \tanh t] \cosh 2\nu t \tanh t \, dt , \qquad (39)$$

since $Q_{\lambda+\nu}$ and $-Q'_{\lambda+\nu}$ are both positive. Note that in the case of Bessel functions, there is no term analogous to the $Q_{\lambda+\nu}$ in Eq. (39) [3, Eq. 3.3]. This term is in fact of order λ^{-2} relative to that involving $Q'_{\lambda+\nu}$, and disappears in the limit $\lambda \to \infty$. The appearance of such terms seems to be a major complication as one goes to higher derivatives, and I do not at this point have any general result on the signs of higher derivatives.

The results established so far are sufficient to show that the first, second, and third differences of successive zeros θ_k of any linear combination of $c_\lambda^{(\nu+\frac{1}{2})}$ and $d_\lambda^{(\nu+\frac{1}{2})}$ (or $C_\lambda^{(\alpha)}$, $D_\lambda^{(\alpha)}$) form monotonic sequences with $\Delta\theta_k > 0$, $\Delta^2\theta_k < 0$, $\Delta^3\theta_k > 0$ provided $\nu > \frac{1}{2}$. In addition the first and second differences of the quantities

$$M_k = \int_{\theta_k}^{\theta_{k+1}} |y(\theta)|^\beta \, d\theta, \quad \beta > -1 , \qquad (40)$$

form monotonic sequences with $\Delta M_k < 0$, $\Delta^2 M_k > 0$ for $\nu > \frac{1}{2}$. Here y is any linear combination of $c_\lambda^{(\nu+\frac{1}{2})}$ and $d_\lambda^{(\nu+\frac{1}{2})}$. For generalizations, see [3] and [10]. These results extend, in particular, the classic results of Sturm by one step [3].

4. DEDUCTIONS FOR OTHER FUNCTIONS

a. Bessel Functions

The general results summarized earlier can be carried over to the
case of Bessel functions by taking the appropriate confluent limit of the
hypergeometric functions which define $C_\lambda^{(\alpha)}$ and $D_\lambda^{(\alpha)}$. Thus, if we
let $x = 1 - \dfrac{y^2}{2\lambda^2}$, $x < 1$, and $z = 1 + \dfrac{y^2}{2\lambda^2}$, $z > 1$, and consider the limit
$\lambda \to \infty$, we find that

$$C_\lambda^{(\alpha)}(x) \to \lambda^{2\alpha-1} \frac{\sqrt{\pi}}{\Gamma(\alpha)} (2y)^{-\alpha+\frac{1}{2}} J_{\alpha-\frac{1}{2}}(y), \quad x \to 1 - \frac{y^2}{2\lambda^2},$$

$$D_\lambda^{(\alpha)}(x) \to -\lambda^{2\alpha-1} \frac{\sqrt{\pi}}{\Gamma(\alpha)} (2y)^{-\alpha+\frac{1}{2}} Y_{\alpha-\frac{1}{2}}(y), \quad x \to 1 - \frac{y^2}{2\lambda^2},$$

$$e^{-i\pi\alpha} D_\lambda^{(\alpha)}(z) \to \lambda^{2\alpha-1} \frac{1}{\sqrt{\pi}\,\Gamma(\alpha)} (2y)^{-\alpha+\frac{1}{2}} K_{\alpha-\frac{1}{2}}(y), \quad z \to 1 + \frac{y^2}{2\lambda^2}. \quad (41)$$

If we use these relations in Eq. (15), let $\nu = \alpha - \frac{1}{2}$, and make the
substitution $z = \cosh 2t$, we obtain a generalization of Nicholson's
integral,

$$J_{n+\nu}^2(x) + Y_{n+\nu}^2(x) = \frac{4}{\pi^2} \frac{\Gamma(\nu)\Gamma(n+1)}{\Gamma(n+2\nu)} (4x)^\nu$$

$$\times \int_0^\infty K_\nu(2x\sinh t) C_n^{(\nu)}(\cosh 2t)(\cosh t)^{2\nu} (\sinh t)^\nu dt . \qquad (42)$$

Nicholson's result for $J_n^2 + Y_n^2$ is recovered for $\nu = 0$ using the limit
given in Eq. (32). The result above may be converted into an expression
involving derivatives of Bessel function through the use of the relation
(valid for a general bessel function $C_\nu(x)$)

$$x^{-\nu+n} C_{\nu+n}(x) = (-\frac{1}{x}\frac{d}{dx})^n (x^{-\nu} C_\nu(x)) . \qquad (43)$$

More generally, if we use the confluent limits in Eq. (13), with
$z_1 = 1 + \dfrac{z^2}{2\lambda^2}$, $z_2 = 1 + \dfrac{Z^2}{2\lambda^2}$, and the integration variable z in that
equation replaced by $\cosh\phi$, we obtain an expression for the product of
two hyperbolic Bessel functions analogous to that given by Gegenbauer
for ordinary Bessel functions [2, Sec. 11.42(17)],

$$2^{-\nu} \frac{\Gamma(n+2\nu)}{\Gamma(\nu)\Gamma(n+1)} \frac{K_{\nu+n}(z)}{z^{\nu}} \frac{K_{\nu+n}(Z)}{Z^{\nu}}$$

(44)

$$= \int_0^{\infty} [z^2+Z^2+2zZ\cosh\phi]^{-\frac{\nu}{2}} K_{\nu}([z^2+Z^2+2zZ\cosh\phi]^{\frac{1}{2}})C_n^{(\nu)}(\cosh\phi)(\sinh\phi)^{2\nu}d\phi .$$

This result is closely connected with generalizations of Gegenbauer's
addition formula [2, Secs. 11.4, 11.41].

b. Hermite Functions

The results which we obtain for Hermite functions are mostly new
and reasonably complete. We define Hermite functions $H_\lambda(x)$ and $G_\lambda(x)$
as independent solutions of the Hermite equation for arbitrary λ ,

$$\left[\frac{d^2}{dx^2} - 2x \frac{d}{dx} + 2\lambda\right] y(x) = 0 .$$

(45)

For $\lambda = n$, $H_n(x)$ will be the usual Hermite polynomial [8, Secs. 8.2-
8.4, 10.13]. We will adopt the following definitions for H_λ and G_λ in
terms of confluent hypergeometric functions,

$$H_\lambda(x) = 2^\lambda \Psi(-\frac{\lambda}{2}, \frac{1}{2}, x^2)$$

$$= \frac{2^\lambda}{\sqrt{\pi}}\left[\cos\frac{\lambda\pi}{2} \Gamma(\frac{\lambda}{2}+\frac{1}{2}) \Phi(-\frac{\lambda}{2}, \frac{1}{2}, x^2)\right.$$

$$\left. + 2x \sin\frac{\lambda\pi}{2} \Gamma(\frac{\lambda}{2}+1)\Phi(-\frac{\lambda}{2}+\frac{1}{2}, \frac{3}{2}, x^2)\right] ,$$

(46)

369

$$G_\lambda(x) = \frac{2^\lambda}{\sqrt{\pi}} \left[-\sin \frac{\lambda\pi}{2} \; \Gamma(\frac{\lambda}{2} + \frac{1}{2}) \Phi(-\frac{\lambda}{2}, \frac{1}{2}, x^2) \right.$$

$$\left. + 2x \cos \frac{\lambda\pi}{2} \; \Gamma(\frac{\lambda}{2} + 1) \Phi(-\frac{\lambda}{2} + \frac{1}{2}, \frac{3}{2}, x^2) \right] . \qquad (47)$$

Here $\Phi(a, b, z)$ and $\Psi(a, b, z)$ are the usual solutions of the confluent hypergeometric equation [8, Chapter 6]. In particular,

$$\Phi(a, b, z) = {}_1F_1(a, b, z) . \qquad (48)$$

For $x \to \infty$,

$$H_\lambda(x) \sim (2x)^\lambda, \quad x \to \infty ,$$

$$G_\lambda(x) \sim \frac{1}{\sqrt{\pi}} \; \Gamma(\lambda+1) x^{-\lambda-1} e^{x^2}, \quad x \to \infty . \qquad (49)$$

The functions as defined have a Wronskian

$$W_x(H_\lambda, G_\lambda) = \frac{2^\lambda}{\sqrt{\pi}} \; \Gamma(\lambda+1) e^{x^2} . \qquad (50)$$

We now use the fact that $H_\lambda(x)$ and $G_\lambda(x)$ can be obtained as confluent limits of the functions $C_\lambda^{(\alpha)}$ and $D_\lambda^{(\alpha)}$. The result for $H_n(x)$ is well-known [4, Sec. 5.6(3)]. In general,

$$H_\lambda(x) = \lim_{\alpha \to \infty} \alpha^{-\lambda/2} \; \Gamma(\lambda+1) \; C_\lambda^{(\alpha)}(x/\sqrt{\alpha})$$

$$G_\lambda(x) = \lim_{\alpha \to \infty} \alpha^{-\lambda/2} \; \Gamma(\lambda+1) \; D_\lambda^{(\alpha)}(x/\sqrt{\alpha}) . \qquad (51)$$

If we take the appropriate limits in Eq. (15), we obtain the following expression,

$$e^{-x^2}[H_\lambda^2(x) + G_\lambda^2(x)] = 2^{\lambda+1} \frac{1}{\pi} \Gamma(\lambda+1)$$

$$\times \int_0^\infty e^{-(2\lambda+1)t + x^2 \tanh t} (\cosh t \sinh t)^{-\frac{1}{2}} dt. \qquad (52)$$

It is easily checked that the right hand side is, in fact, a solution of the third order differential equation [6] satisfied by the product of two Hermite functions. The functions $e^{-\frac{1}{2}x^2} H_\lambda(x)$ and $e^{-\frac{1}{2}x^2} G_\lambda(x)$ which appear on the left are solutions of the modified Hermite equation

$$\left[\frac{d^2}{dx^2} + 2\lambda + 1 - x^2\right] y(x) = 0 . \qquad (53)$$

We can normalize these solutions to unit Wronskian by dividing by the numerical factor in Eq. (50). The left hand side of Eq. (52) then corresponds to the function $p(x)$ of Lorch and Szego [3],

$$p(x) = \frac{2^{-\lambda}\sqrt{\pi}}{\Gamma(\lambda+1)} e^{-x^2} [H_\lambda^2(x) + G_\lambda^2(x)]$$

$$\qquad (54)$$

$$= \frac{2}{\sqrt{\pi}} \int_0^\infty e^{-(2\lambda+1)t + x^2 \tanh t} (\cosh t \sinh t)^{-\frac{1}{2}} dt .$$

All derivatives of the right hand side with respect to x are positive for $x > 0$, so $p^{(n)}(x) > 0$, $n = 0, 1, \ldots$. The function $p(x)$ is therefore absolutely monotonic, and we conclude from the theorems of Lorch and Szego [3] and Lorch, Muldoon, and Szego [10] that the nth differences $\Delta^n x_k$ of the positive zeros of the Hermite functions and the nth differences $\Delta^n M_k$ of the quantities

$$M_k = \int_{x_k}^{x_{k+1}} W(x) |y(x)|^\beta dx, \quad \beta > -1, \ x_k \geq 0 , \qquad (55)$$

form absolutely monotonic sequences. Here y(x) is an arbitrary solution of Eq. (53) for real x, λ, and W(x) is an arbitrary absolutely monotonic function such that the integral converges. In particular, if we choose $W(x) = e^{x^2}$, we find that the theorems in [3] and [10] can be applied directly to the Hermite polynomials.

5. CONCLUSIONS

Although the problem of obtaining an analog of Nicholsons' formula for the Gegenbauer functions has been solved, the problem of exploiting the result fully remains. The calculations necessary to prove interesting monotonicity properties for the Gegenbauer functions are unfortunately much less tractable than those encountered for Bessel functions (also unsolved in general for order $\nu < \frac{1}{2}$) or Hermite functions. The case of Jacobi functions has not been considered at all. I know how to obtain two different analogs of Nicholson's formula for Jacobi functions by starting with Koornwinder's addition formula [11] or the angular momentum addition formula, but have so far not had a chance to work out the details. [It is clear that the methods used here can also be applied to other functions with suitable group and cut structure.]

It would also be interesting to derive the analog of Nicholson's expression for the product $K_\mu(x) K_\nu(x)$ [2, Sec. 13.73(1)], that is, expressions for $D_\lambda^{(\mu+\frac{1}{2})} D_\lambda^{(\nu+\frac{1}{2})}$, and the corresponding expressions for $D_\lambda^{(\alpha)} D_\sigma^{(\alpha)}$. It is probable that Glasser's expression [12] for $J_0(a)J_0(b) + Y_0(a)Y_0(b)$ would follow from the latter. Such expressions would also permit the study of the derivatives with respect to the order and degree (comp. [2, Sec. 13.73]).

I would like to thank Professor Richard Askey for bringing this problem to my attention, and for suppling a number of references on applications of Nicholson's formula to studies of higher monotonicity properties of Bessel functions.

Madison, March 30, 1975

FOOTNOTES

1. Nicholson's early papers are largely concerned with the theory of scattering of sound waves and electromagnetic waves from spherical and spheroidal systems, problems which lead naturally to the investigation of Bessel functions. The result in Eq. (1) was actually derived in Phil. Mag. $\underline{14}$, 697 (1907), without, however, being recognized explicitly as such. The modern scattering problems retain some features of the classical problems. There is presumably a moral in these two bits of history for those interested in sources for useful relations for special functions.

2. The polynomials $C_n^{(\alpha)}(z)$ obtained for $\lambda = n$ correspond to Szegö's polynomials $(\alpha + \frac{1}{2})_n P_n^{(\alpha)}(z) = (2\alpha)_n P_n^{(\alpha - \frac{1}{2}, \alpha - \frac{1}{2})}(z)$ [4]. The functions $D_n^{(\alpha)}(z)$ defined above differ from those of Robin [6, Sec. 170(93)] by a factor $\pi^{-1} e^{2\pi i (\alpha - \frac{1}{4})}$ which does not affect the recurrence relations, but simplifies the connections between the C's and D's and various other relations. Robin's definition preserves a symmetrical connection to the (unsymmetrically defined) Legendre functions. The definition in [8, p. 175, corrections on p. 2] is inappropriate even as corrected.

REFERENCES

1. J. W. Nicholson, Phil. Mag $\underline{19}$, 228 (1910); Quarterly J. of Pure and Applied Math. $\underline{42}$, 216 (1911).

2. G. N. Watson, A Treatise on the Theory of Bessel Functions, 2nd Edition, Cambridge University Press (1966), Secs. 13.73-13.75.

3. L. Lorch and P. Szego, Acta Math. $\underline{109}$, 55 (1963).

4. G. Szegö, Orthogonal Polynomials, American Mathematical Society Colloquium Publications, Vol. 23, 3rd Edition (1967).

5. L. Robin, Fonctions Sphérique de Legendre et Fonctions Sphéroïdales, Gauthiers-Villars, Paris (1959), Vol. 3, Chap. X.

6. E. T. Whittaker and G. N. Watson, A Course in Modern Analysis, Cambridge University Press (1952), p. 298.

7. L. Durand, P. M. Fishbane, and L. M. Simmons, Jr., to be published. P. Henrici, J. Rational Mech. Anal. 4, 983 (1955).

8. Higher Trancendental Functions, A. Erdélyi, editor, McGraw-Hill Book Company, New York (1953).

9. I. S. Gradshteyn and I. M. Ryzhik, Tables of Integrals, Series, and Products, Academic Press, New York (1965).

10. L. Lorch, M. E. Muldoon, and P. Szegö, Can. J. of Math. 22, 1238 (1970); 24, 349 (1972).

11. T. Koornwinder, Indag. Math. 34, 188 (1972); SIAM J. of Appl. Math. 25, 236 (1973).

12. M. L. Glasser, Math. of Computation 28, 613 (1974).

Department of Physics
University of Wisconsin
Madison, Wisconsin 53706

Supported in part by the U. S. Energy Research and Development Administration under contract AT(11-1)-881, C00-881-451.

Positivity and Special Functions
George Gasper

1. Introduction. The solutions to many problems (including the Riemann
hypothesis, see Pólya [98] and Titchmarsh [104, p. 45]) can be shown
to depend on the determination of when a specific function is positive or
nonnegative. In many cases it is possible to derive a series or integral
representation for the function in terms of more elementary functions, but
usually the positivity (or nonnegativity) is not at all obvious from the
first representations obtained. Thus one is forced to search for either a
positive representation or another applicable method.

Sometimes the problem can be reduced to a simpler one involving
fewer parameters or it can be transformed into another problem that is
easier to handle. For example, consider a two variable problem which
consists of proving

$$(1.1) \qquad \sum_n a_n p_n(x) p_n(y) \geq 0 \; ,$$

where $p_n(x)$ is a sequence of functions and x and y satisfy appropriate restrictions. If there is an integral representation of the form

$$(1.2) \qquad p_n(x) p_n(y) = \int p_n(z) d\mu_{x,y}(z), \qquad d\mu_{x,y}(z) \geq 0 \; ,$$

then the problem (1.1) can (at least formally) be reduced to the one variable problem

$$(1.3) \qquad \sum_n a_n p_n(z) \geq 0 \; ,$$

where z is in the support of the measures $d\mu_{x,y}(z)$. Also, if there is an integral representation of the form

(1.4)
$$p_n(z) = \int q_n(t)d\nu_z(t), \qquad d\nu_z(t) \geq 0 ,$$

where $q_n(t)$ is another sequence of functions, then the problem (1.3) can be transformed into the problem

(1.5)
$$\sum_n a_n q_n(t) \geq 0 ,$$

which might be much easier to prove than (1.3). Some examples of this will be given in this paper.

Similarly, it may be possible to simplify problems of the type

(1.6)
$$\int p_n(x)p_m(x)d\phi(x) \geq 0$$

by using formulas of the forms

(1.7)
$$p_n(x)p_m(x) = \sum_k a(k,m,n)p_k(x), \qquad a(k,m,n) \geq 0 ,$$

(1.8)
$$p_k(x) = \sum_j b(j,k)q_j(x) , \qquad b(j,k) \geq 0 ,$$

which are duals of (1.2) and (1.4).

Recently, many results have been obtained on when formulas of the types (1.2), (1.4), (1.7), (1.8) hold for the Hahn polynomials and their limit cases (Jacobi polynomials, Krawtchouk polynomials, etc.), and in this paper we shall concentrate on illustrating the main ideas behind the methods which have been used to prove or disprove the nonnegativity of the kernels (measures and coefficients) in these formulas. We shall also consider methods which have been used to prove the positivity of integrals of Bessel functions, Poisson kernels, Cesàro kernels, and of sums of Jacobi polynomials. Among the methods discussed are the use of computations, recurrence relations, maximum principles, estimates, absolutely monotonic and completely monotonic functions, and sums of squares. A general survey of orthogonal polynomials and their applications will appear in Askey's book [17].

2. Background material. In this section we present the notation and formulas which will be used in subsequent sections.

For $\alpha, \beta > -1$, N a nonnegative integer, and $n = 0, 1, \ldots, N$ we define the Hahn polynomials of degree n by

$$(2.1) \qquad Q_n(x; \alpha, \beta, N) = \sum_{k=0}^{n} \frac{(-n)_k (n+\alpha+\beta+1)_k (-x)_k}{k! \, (\alpha+1)_k (-N)_k}$$

$$= {}_3F_2 \left[\begin{array}{c} -n, \, n+\alpha+\beta+1, \, -x; 1 \\ \alpha+1, \, -N \end{array} \right]$$

where $(a)_0 = 1$, $(a)_k = a(a+1) \ldots (a+k-1) = \Gamma(k+a)/\Gamma(a)$ for $k = 1, 2, \ldots$. The Hahn polynomials satisfy the orthogonality relation

$$\sum_{x=0}^{N} \rho(x; \alpha, \beta, N) Q_m(x; \alpha, \beta, N) Q_n(x; \alpha, \beta, N) = \frac{\delta_{mn}}{\pi_n(\alpha, \beta, N)}$$

for $m, n = 0, 1, \ldots, N$ where

$$\rho(x; \alpha, \beta, N) = \frac{\dbinom{x+\alpha}{x} \dbinom{N-x+\beta}{N-x}}{\dbinom{N+\alpha+\beta+1}{N}},$$

$$\pi_n(\alpha, \beta, N) = \frac{(-1)^n (-N)_n (\alpha+1)_n (\alpha+\beta+1)_n}{n! \, (N+\alpha+\beta+2)_n (\beta+1)_n} \frac{2n+\alpha+\beta+1}{\alpha+\beta+1}.$$

These polynomials are a discrete analog of the Jacobi polynomials

$$(2.2) \qquad P_n^{(\alpha, \beta)}(x) = \binom{n+\alpha}{n} \lim_{N \to \infty} Q_n(N(1-x)/2; \alpha, \beta, N)$$

$$= \binom{n+\alpha}{n} {}_2F_1 \left[-n, n+\alpha+\beta+1; \alpha+1; \frac{1-x}{2} \right] \quad ,$$

which satisfy

$$\int_{-1}^{1} P_m^{(\alpha, \beta)}(x) \ P_n^{(\alpha, \beta)}(x)(1-x)^{\alpha}(1+x)^{\beta}dx = \frac{\delta_{mn}}{h_n^{(\alpha, \beta)}}$$

for $\alpha, \beta > -1$ with

$$h_n^{(\alpha, \beta)} = \frac{n! \ \Gamma(n+\alpha+\beta+1)(2n+\alpha+\beta+1)}{2^{\alpha+\beta+1} \ \Gamma(n+\alpha+1)\Gamma(n+\beta+1)} \quad .$$

Other limiting cases of the Hahn polynomials are the Meixner polynomials

$$(2.3) \qquad M_n(x; \beta, c) = \lim_{N \to \infty} Q_n(x; \beta-1, N(1-c)/c, N)$$

$$= {}_2F_1[-n, -x; \beta; 1-c^{-1}] \quad ,$$

the Krawtchouk polynomials

$$(2.4) \qquad K_n(x; p, N) = \lim_{t \to \infty} Q_n(x; pt, (1-p)t, N)$$

$$= {}_2F_1[-n, -x; -N; 1/p] \quad ,$$

and the Charlier polynomials

(2.5)
$$c_n(x;a) = \lim_{\beta \to \infty} M_n(x;\beta, a/\beta)$$

$$= \lim_{N \to \infty} K_n(x;a/N, N)$$

$$= {}_2F_0[-n, -x; \underline{\quad}; -1/a] \ .$$

These three polynomials are, respectively, orthogonal on the sets $\{0,1,2,\dots\}$, $\{0,1,\dots,N\}$, $\{0,1,2,\dots\}$ with respect to the jump functions

$$c^x(\beta)_x/x! , \qquad \beta > 0, \quad 0 < c < 1,$$

$$\binom{N}{x}p^x(1-p)^{N-x} , \qquad 0 < p < 1 ,$$

$$e^{-a}a^x/x! , \qquad a > 0 \ .$$

We shall also consider the Laguerre polynomials

(2.6)
$$L_n^\alpha(x) = \lim_{\beta \to \infty} P_n^{(\alpha, \beta)}(1 - 2x/\beta)$$

$$= \binom{n+\alpha}{n} {}_1F_1[-n;\alpha+1;x]$$

and the Bessel functions

(2.7)
$$J_\alpha(x) = \left(\frac{x}{2}\right)^\alpha \lim_{n \to \infty} n^{-\alpha} P_n^{(\alpha, \beta)}\left(1 - \frac{x^2}{2n^2}\right)$$

$$= \left(\frac{x}{2}\right)^\alpha \lim_{n \to \infty} n^{-\alpha} L_n^\alpha\left(\frac{x^2}{4n}\right)$$

$$= \left(\frac{x}{2}\right)^\alpha {}_0F_1[\alpha+1; -x^2/4]/\Gamma(\alpha+1) \ .$$

379

The Laguerre polynomials are orthogonal on $(0, \infty)$ with respect to the weight function $x^{\alpha} e^{-x}$, $\alpha > -1$. See Erdélyi et al. [53], Karlin and McGregor [81], Szegö [103], and Gasper [69], [70].

3. Computation. Here we consider the computational methods which have been applied to the projection formulas (1.8), (1.4) and to the product formulas (1.7), (1.2) for the Hahn polynomials and their limits.

Formula (1.8). For the Hahn polynomials we shall write formula (1.8) in the form

$$(3.1) \qquad Q_n(x;a, b, M) = \sum_{k=0}^{N} A(k, n) Q_k(x;\alpha, \beta, N)$$

where $A(k, n) = A(k, n;a, b, M, \alpha, \beta, N)$, $n = 0, 1, \ldots, M$, $x = 0, ., \ldots ,$ $\min(M, N)$, and it is assumed that $a, b, \alpha, \beta > -1$. By orthogonality

$$(3.2) \quad A(k, n) = \pi_k(\alpha, \beta, N) \sum_{x=0}^{N} \rho(x;\alpha, \beta, N) Q_n(x;a, b, M) Q_k(x;\alpha, \beta, N) .$$

Various hypergeometric series representations for the Hahn polynomials can be used in either (3.1) or (3.2) to write $A(k, n)$ as a sum of products of gamma functions. In particular, expanding the left hand side of (3.1) by means of the $_3F_2$ series in (2.1), using the expansion

$$(-x)_j = \sum_{k=0}^{j} \frac{(-1)^{j-k} N! \, j! \, (2k+\alpha+\beta+1) \, \Gamma(j+\alpha+1)}{(N-j)! \, k! \, (j-k)! \, \Gamma(\alpha+1)}$$

$$\times \frac{\Gamma(k+\alpha+\beta+1)}{\Gamma(j+k+\alpha+\beta+2)} Q_k(x;\alpha, \beta, N) ,$$

which is derived in Gasper [70], and then picking out the coefficient of $Q_k(x;\alpha, \beta, N)$ we obtain the formula

$$(3.3) \qquad A(k,n) = \frac{n!(-N)_k(\alpha+1)_k(n+a+b+1)_k}{k!(n-k)!(-M)_k(a+1)_k(k+\alpha+\beta+1)_k}$$

$$\times {}_4F_3 \left[\begin{array}{c} k-n,\ k-N,\ k+\alpha+1,\ n+k+a+b+1;1 \\ k-M,\ k+a+1,\ 2k+\alpha+\beta+2 \end{array} \right].$$

So far the only interesting nonnegativity results for $A(k,n)$ have been found for the case $M = N$, in which case we have

$$(3.4) \qquad Q_n(x;a,b,N) = \sum_{k=0}^{n} \frac{n!(\alpha+1)_k(n+a+b+1)_k}{k!(n-k)!(a+1)_k(k+\alpha+\beta+1)_k}$$

$$\times {}_3F_2 \left[\begin{array}{c} k-n,\ k+\alpha+1,\ n+k+a+b+1;1 \\ k+a+1,\ 2k+\alpha+\beta+2 \end{array} \right] Q_k(x;\alpha,\beta,N).$$

Since the coefficients are independent of N, application of the limit operation in (2.2) gives

$$(3.5) \qquad \frac{P_n^{(a,b)}(x)}{P_n^{(a,b)}(1)} = \sum_{k=0}^{n} \frac{n!(\alpha+1)_k(n+a+b+1)_k}{k!(n-k)!(a+1)_k(k+\alpha+\beta+1)_k}$$

$$\times {}_3F_2 \left[\begin{array}{c} k-n,\ k+\alpha+1,\ n+k+a+b+1;1 \\ k+a+1,\ 2k+\alpha+\beta+2 \end{array} \right] \frac{P_k^{(\alpha,\beta)}(x)}{P_k^{(\alpha,\beta)}(1)}.$$

For this and other proofs see Feldheim [57], Miller [97], Askey [2], [17], and Gasper [70]. The above ${}_3F_2$ can be summed in the three cases

(i) $\quad a = b,\ \alpha = \beta$ (ii) $\quad b = \beta$ (iii) $\quad a = \alpha$

by using the formulas of Watson [33, p. 16], Saalschütz [33, p. 9] and Gauss [33, p. 2], respectively, to give

$$(3.6) \qquad Q_n(x;a,a,N) = \sum_{k=0}^{[n/2]} \frac{n!(\alpha+1)_{n-2k}(n+2a+1)_{n-2k}}{(n-2k)!(2k)!(a+1)_{n-2k}(n-2k+2\alpha+1)_{n-2k}}$$

381

$$\times \frac{(1/2)_k (a-\alpha)_k}{(n-2k+a+1)_k (n-2k+\alpha+3/2)_k} \; Q_{n-2k}(x;\alpha, \alpha, N) \;,$$

(3.7)
$$Q_n(x;a,\beta,N) = \sum_{k=0}^{n} \frac{n!(\alpha+1)_k (n+a+\beta+1)}{k!(n-k)!(a+1)_k (k+\alpha+\beta+1)_k}$$

$$\times \frac{(a-\alpha)_{n-k}(k+\beta+1)_{n-k}}{(k+a+1)_{n-k}(2k+\alpha+\beta+2)_{n-k}} \; Q_k(x;\alpha,\beta,N) \;,$$

(3.8)
$$Q_n(x;\alpha,b,N) = \sum_{k=0}^{n} \frac{n!(n+a+b+1)_k}{k!(n-k)!(k+\alpha+\beta+1)_k}$$

$$\times \frac{(-1)^{n-k}(b-\beta)_{n-k}}{(2k+\alpha+\beta+2)_{n-k}} \; Q_k(x;\alpha,\beta,N) \;,$$

and similar formulas for Jacobi polynomials.

Clearly the coefficients in (3.6) and (3.7) are nonnegative when $a \geq \alpha$, the coefficients in (3.8) are nonnegative when $\beta - b = j$, $j = 0$, $1, \ldots,$ and some coefficients are negative if these conditions are not satisfied.

The analog of (3.6) for Jacobi polynomials was first derived by Gegenbauer in his work [75] on the ultraspherical polynomials

(3.9)
$$C_n^\lambda(x) = \frac{(2\lambda)_n}{(\lambda+1/2)_n} \; P_n^{(\lambda-1/2, \lambda-1/2)}(x) \;.$$

The coefficients in (3.7) for the limiting Jacobi polynomial case were computed by Szegö [103] who used them to prove the end point Cesàro summability theorem for Jacobi series [103, Theorem 9.1.3]. For some applications of these formulas to numerical analysis, transplantation theorems, positive definite functions on projective spaces, and positive polynomial sums see Rivlin and Wilson [100], Askey and Wainger [30], and Askey and Gasper [21], [22].

The transformation formulas for $_3F_2$ series in Bailey [33, Chapter 3] can be applied to the $_3F_2$ in (3.4) and the formulas (3.6) - (3.8) can be iterated to obtain other cases in which the $A(k, n)$ are nonnegative, but these procedures are not strong enough to yield the deeper results which have been obtained by using the method of recurrence relations described in the next section.

For Laguerre polynomials, application of the limit operation (2.6) to the Jacobi analog of (3.7) gives

$$(3.10) \qquad L_n^a(x) = \sum_{k=0}^n \frac{(a-\alpha)_{n-k}}{(n-k)!} L_k^\alpha(x)$$

which has nonnegative coefficients for all n if and only if $a \geq \alpha$.

For the Meixner, Krawtchouk, and the Charlier polynomials we can apply the limits (2.3)-(2.5) to (3.1) to obtain the formulas

$$(3.11) \qquad M_n(x;b, a) = \sum_{k=0}^n \frac{n!(\beta)_k}{k!(n-k)!(b)_k} \left[\frac{c(1-a)}{a(1-c)}\right]^k$$

$$\times {}_2F_1\left[k-n, k+\beta;k+b; \frac{c(1-a)}{a(1-c)}\right] M_k(x;\beta, c) \ ,$$

$$(3.12) \qquad K_n(x;q, M) = \sum_{k=0}^{\min(n, N)} \frac{n!(-N)_k}{k!(n-k)!(-M)_k} \left(\frac{p}{q}\right)^k$$

$$\times {}_2F_1[k-n, k-N;k-M;p/q]K_k(x;p, N) \ ,$$

$$(3.13) \qquad c_n(x;b) = \sum_{k=0}^n \binom{n}{k} a^k b^{-n}(b-a)^{n-k} c_k(x;a) \ ,$$

where it is assumed that x, n take appropriate integer values.

For $a, b > 0$ the coefficients in (3.13) are obviously nonnegative for all n if and only if $b \geq a$. However, since it is not obvious when the coefficients in (3.11) are nonnegative we use the transformation formula [53, 2.9(3)] to obtain

$$_2F_1\left[k-n,\,k+\beta;k+b;\ \frac{c(1-a)}{a(1-c)}\right]$$

$$=\left(\frac{a-c}{a(1-c)}\right)^{n-k}{}_2F_1\left[k-n,\,b-\beta,\,k+b;\ \frac{c(1-a)}{c-a}\right]\ ,$$

from which it is clear that the coefficients in (3.11) are nonnegative when $1 > a \geq c > 0$, $b \geq \beta > 0$; and these conditions are also necessary as is shown in [70, p. 192] by use of an asymptotic formula. Similarly, from

$$_2F_1[k-n,\,k-N;k-M;p/q]$$

$$=\left(\frac{q-p}{q}\right)^{n-k}{}_2F_1\left[k-n,\,N-M;k-M;\ \frac{p}{p-q}\right]$$

it is clear that the coefficients in (3.12) are nonnegative when $1 > q \geq p > 0$, $M \geq N$. For a study of projection formulas between different systems of orthogonal polynomials see Gasper [70], and for some other methods for proving (1.8) for general orthogonal polynomials see Wilson [114] and Askey [5].

Formula (1.7). One simple way to see that all of the coefficients in the product formulas

(3.14) $\quad Q_n(x;\alpha,\beta,N)Q_m(x;\alpha,\beta,N) = \sum_{k=0}^{N} B(k,\,m,\,n;\alpha,\beta,N)Q_k(x;\alpha,\beta,N)$,

where $x, m, n = 0, 1, \ldots, N$, cannot be nonnegative for $N = 0, 1, \ldots$, for some $\alpha, \beta > -1$ is to observe [24] that since

$$\sum_{k=0}^{N} B(k,\,m,\,n;\alpha,\beta,N) = 1$$

and

384

$$\nu(x; \alpha, \beta, N) \equiv \max_{0 \le n \le N} |Q_n(x; \alpha, \beta, N)| < \infty \ ,$$

if all the coefficients in (3.14) were nonnegative then it would follow that

$$\nu^2(x; \alpha, \beta, N) \le \nu(x; \alpha, \beta, N)$$

and hence $\nu(x; \alpha, \beta, N) \le 1$. But

$$Q_N(1; \alpha, \beta, N) = -\frac{N+\beta}{\alpha+1}$$

is not bounded for $N = 0, 1, \ldots$, and so for each (α, β) some coefficient in (3.14) must be negative when N is sufficiently large. Similarly, it can be shown that there is no (α, β) for which there is a constant C independent of m, n, N such that

$$\sum_{k=0}^{N} |B(k, m, n; \alpha, \beta, N)| \le C$$

for all m, n, N.

However, for the limiting Jacobi polynomial case of (3.14)

$$(3.15) \qquad P_n^{(\alpha, \beta)}(x)\, P_m^{(\alpha, \beta)}(x) = \sum_{k=|n-m|}^{n+m} B(k, m, n; \alpha, \beta)\, P_k^{(\alpha, \beta)}(x)$$

there are well-known cases in which all of the coefficients are non-negative. In particular, the nonnegativity of the coefficients in (3.15) for the two cases $\alpha = \beta = -\frac{1}{2}$ and $\alpha = \beta = \frac{1}{2}$ follow from the classical identities

$$(3.16) \qquad \cos n\theta \cos m\theta = \frac{1}{2} \cos(n+m)\theta + \frac{1}{2} \cos(n-m)\theta \ ,$$

$$(3.17) \qquad \frac{\sin(n+1)\theta}{\sin \theta} \frac{\sin(m+1)\theta}{\sin \theta} = \sum_{k=0}^{\min(m, n)} \frac{\sin(n+m-2k+1)\theta}{\sin \theta} \ ,$$

since

(3.18)
$$\frac{P_n^{(-\frac{1}{2}, -\frac{1}{2})}(\cos\theta)}{P_n^{(-\frac{1}{2}, -\frac{1}{2})}(1)} = \cos n\theta ,$$

(3.19)
$$\frac{P_n^{(\frac{1}{2}, \frac{1}{2})}(\cos\theta)}{P_n^{(\frac{1}{2}, \frac{1}{2})}(1)} = \frac{\sin(n+1)\theta}{(n+1)\sin\theta} .$$

The coefficients in (3.15) for the ultraspherical case $\alpha = \beta$ were computed by Dougall as products of gamma functions and from his formula (stated without proof in [48]) it is obvious that they are all nonnegative if and only if $\alpha \geq -\frac{1}{2}$. Proofs of Dougall's formula were later given by Hsü [79], Carlitz [44], Miller [96], Vilenkin [107], and Hylleraas [80] which also contains references to the Legendre polynomial case $\alpha = \beta = 0$. Dougall's formula has been used by Hirschman [77] to obtain a convolution structure, by Davis and Hirschman [47] to derive the strong Szegö limit theorem for Toeplitz matrices associated with ultraspherical polynomials, and by Askey and Wainger [30] to obtain a transplantation theorem for ultraspherical coefficients which then leads to an analog of the Marcinkiewicz multiplier theorem and an analog of a theorem of Hardy and Littlewood on Fourier coefficients of even functions.

By means of the quadratic transformation

(3.20)
$$\frac{P_n^{(\alpha, -\frac{1}{2})}(2x^2-1)}{P_n^{(\alpha, -\frac{1}{2})}(1)} = \frac{P_{2n}^{(\alpha, \alpha)}(x)}{P_{2n}^{(\alpha, \alpha)}(1)}$$

it follows from Dougall's result that the coefficients in (3.15) are also nonnegative when $\alpha \geq -\frac{1}{2}$, $\beta = -\frac{1}{2}$. Hylleraas computed a differential equation satisfied by the left side of (3.15) and used it to derive a three term recurrence relation in k for the coefficients in (3.15). For $\alpha = \beta$ his recurrence relation reduced to two terms which he could easily use to compute $B(k, m, n; \alpha, \alpha)$. The recurrence relation also simplified for $\alpha = \beta + 1$ so that he could compute the coefficients in this case, and from his formula it follows that the coefficients are nonnegative when $\alpha = \beta + 1 > 0$. In [97] Miller used the series (2.2) to write the left side of (3.15) as a polynomial in powers of $(1-x)/2$, expanded the powers by means of

$$(3.21) \quad \left(\frac{1-x}{2}\right)^j = j! \; \Gamma(j+\alpha+1) \sum_{k=0}^{j} \frac{(-1)^k (2k+\alpha+\beta+1)\Gamma(k+\alpha+\beta+1)}{(j-k)! \; \Gamma(k+\alpha+1) \; \Gamma(j+k+\alpha+\beta+2)} \; P_k^{(\alpha, \beta)}(x) \; ,$$

and then picked out the coefficient of $P_k^{(\alpha, \beta)}(x)$ to obtain it as a double sum of products of gamma functions, but no one has yet been able to use his formula to determine for which α, β the coefficients are nonnegative. In the next section we will describe the first and still the only known method which completely determines for which α, β the coefficients in (3.15) are nonnegative.

For the Meixner, Charlier, and the Laguerre polynomials one can compute the $k = m = n = 1$ coefficient in the corresponding limit cases of (3.14) and observe that it is negative. Also see Askey and Gasper [24] and Erdélyi [52]. Eagleson [51] used the generating function

$$\sum_{n=0}^{N} K_n(x; p, N) \frac{p^n (-N)_n}{n!} z^n = (1+z-pz)^x (1-pz)^{N-x}$$

to prove that the coefficients in

$$(3.22) \quad K_n(x; p, N) K_m(x; p, N) = \sum_{k=0}^{N} C(k, m, n; p, N) K_k(x; p, N)$$

are nonnegative when $\frac{1}{2} \le p < 1$, and then he gave some applications to statistics. This nonnegativity also follows from the explicit formula

387

$$(3.23) \qquad C(k, m, n; p, N) = \frac{m!(N-m)!n!(N-n)!}{p^{n+m}(1-p)^k N!}$$

$$\times \sum_{j \geq 0} \frac{p^j(1-p)^j(2p-1)^{k+m+n-2j}}{(j-k)!(j-m)!(j-n)!(k+m+n-2j)!(N-j)!}$$

given in [24]. An addition formula for Krawtchouk polynomials which gives (3.22) has been derived by Dunkl [50].

Formula (1.4). For the Hahn polynomials the integral (1.4) takes the form

$$(3.24) \qquad Q_n(x; a, b, M) = \sum_{y=0}^{N} D(x, y; a, b, M, \alpha, \beta, N) Q_n(y; \alpha, \beta, N)$$

where $x = 0, 1, \ldots, M$, $n = 0, 1, \ldots, \min(M, N)$, and

$$D(x, y; a, b, M, \alpha, \beta, N)$$

$$= \rho(y; \alpha, \beta, N) \sum_{n=0}^{N} \pi_n(\alpha, \beta, N) Q_n(x; a, b, M) Q_n(y; \alpha, \beta, N) \ .$$

This series can be summed only in some special cases. In Gasper [70] it is observed that from the easily proven formula

$$(3.25) \qquad {}_3F_2\left[\begin{array}{c} -x, a, b; 1 \\ c+\mu, d \end{array}\right] = \sum_{y=0}^{x} \binom{x}{y} \frac{(c)_y (\mu)_{x-y}}{(c+\mu)_x} \ {}_3F_2\left[\begin{array}{c} -y, a, b; 1 \\ c, d \end{array}\right]$$

where $x = 0, 1, \ldots$, we have

$$(3.26) \qquad Q_n(x; \alpha+\mu, \beta-\mu, N) = \sum_{y=0}^{x} \binom{x}{y} \frac{(\alpha+1)_y (\mu)_{x-y}}{(\alpha+\mu+1)_x} Q_n(y; \alpha, \beta, N) \ ,$$

$$(3.27) \qquad Q_n(x; \alpha, \beta, M) = \sum_{y=0}^{N} \binom{x}{y} \frac{(-N)_y (N-M)_{x-y}}{(-M)_x} Q_n(y; \alpha, \beta, N) \ .$$

The coefficients in (3.26) are nonnegative when $\mu \geq 0$, and those in (3.27) are nonnegative when $M \geq N$.

Formula (3.25) is a discrete analog of Bateman's [38] integral

$$(3.28) \quad {}_2F_1[a, b; c+\mu; x] =$$

$$= \frac{\Gamma(c+\mu)}{\Gamma(c)\,\Gamma(\mu)} \int_0^1 y^{c-1}(1-y)^{\mu-1}\, {}_2F_1[a, b; c; xy]dy \ ,$$

$$\operatorname{Re} c > 0, \quad \operatorname{Re} \mu > 0, \quad |x| < 1 \ ,$$

which is a limiting case of (3.25). Askey and Fitch [19] used (3.28) to obtain

$$(3.29) \quad \frac{P_n^{(\alpha+\mu, \beta-\mu)}(x)}{P_n^{(\alpha+\mu, \beta-\mu)}(1)} = \frac{\Gamma(\alpha+\mu+1)}{\Gamma(\alpha+1)\,\Gamma(\mu)}$$

$$\times \int_x^1 \frac{P_n^{(\alpha, \beta)}(y)}{P_n^{(\alpha, \beta)}(1)} \frac{(1-y)^\alpha}{(1-x)^{\alpha+\mu}} (y-x)^{\mu-1}dy, \qquad |x| < 1, \ \mu > 0 \ ,$$

and they derived the formula

$$(3.30) \quad \frac{P_n^{(\alpha+\mu, \beta)}(x)}{P_n^{(\alpha+\mu, \beta)}(1)} = \frac{2^\mu\, \Gamma(\alpha+\mu+1)}{\Gamma(\alpha+1)\,\Gamma(\mu)}$$

$$\times \int_x^1 \frac{P_n^{(\alpha, \beta)}(y)}{P_n^{(\alpha, \beta)}(1)} \frac{(1+x)^{n+\alpha+1}(1-y)^\alpha}{(1+y)^{n+\alpha+\mu+1}(1-x)^{\alpha+\mu}} (y-x)^{\mu-1}dy, \ |x|<1, \mu>0 \ ,$$

which can also be derived from (3.28) by using the formula [103, (4.3.2)]

$$(3.31) \quad \frac{P_n^{(\alpha, \beta)}(x)}{P_n^{(\alpha, \beta)}(1)} = \left(\frac{1+x}{2}\right)^n {}_2F_1\left[-n, -n-\beta; \alpha+1; \frac{x-1}{x+1}\right] \ .$$

If in (3.25) we use the formula [70, (2.6)]

(3.32) $$Q_n(x; \alpha, \beta, N) = \frac{(x-N)_n}{(-N)_n} \, {}_3F_2\left[\begin{array}{c} -n, \, -n-\beta, \, -x;1 \\ \alpha+1, \, 1+N-x-n \end{array}\right],$$

which has (3.31) as a limit case, we obtain

(3.33) $$Q_n(x; \alpha+\mu, \beta, N)$$

$$= \sum_{y=0}^{x} \binom{x}{y} \frac{(\alpha+1)_y (\mu)_{x-y}}{(\alpha+\mu+1)_x} \frac{(x-y-N)_n}{(-N)_n} Q_n(y; \alpha, \beta, N+y-x)$$

as a discrete analog of (3.30). Askey [4] used a simple argument based on the positivity of the Poisson kernel for Jacobi series to show that (3.30) implies

(3.34) $$\frac{P_n^{(\alpha+\mu, \beta)}(x)}{P_n^{(\alpha+\mu, \beta)}(1)} = \int_{-1}^{1} \frac{P_n^{(\alpha, \beta)}(y)}{P_n^{(\alpha, \beta)}(1)} \, d\nu_x(y), \qquad d\nu_x(y) \geq 0 \, ,$$

for $\alpha, \beta > -1$, $\mu > 0$, $-1 \leq x \leq 1$, with $d\nu_x(y)$ independent of n. This formula was not pointed out in Askey and Fitch [19] since at that time they did not know of any proof of the positivity of the Poisson kernel (the positivity did not follow from Watson's integral representation [109], [53, 19.12 (13)], and the product formula (3.44) was not yet known), and it was not until later that it was observed that the positivity follows immediately from Bailey's formula [34]

(3.35) $$\sum_{n=0}^{\infty} h_n^{(\alpha, \beta)} t^n P_n^{(\alpha, \beta)} (\cos 2\theta) P_n^{(\alpha, \beta)} (\cos 2\phi)$$

$$= \frac{\Gamma(\alpha+\beta+2) \, (1-t)}{2^{\alpha+\beta+1} \, \Gamma(\alpha+1) \, \Gamma(\beta+1) (1+t)^{\alpha+\beta+2}}$$

$$\times F_4[(\alpha+\beta+2)/2, (\alpha+\beta+3)/2; \alpha+1, \beta+1; a^2/k^2, b^2/k^2]$$

where F_4 is the fourth type of Appell function and

$$a = \sin \theta \sin \phi, \quad b = \cos \theta \cos \phi, \quad k = (t^{-\frac{1}{2}} + t^{\frac{1}{2}})/2 \; .$$

The crucial tool used by Bailey to derive (3.35) was Watson's formula [108] for the product of two terminating hypergeometric functions

(3.36) $\qquad {}_2F_1[-n, n+a;c;z] \; {}_2F_1[-n, n+a;c;Z]$

$$= \frac{(a-c+1)_n}{(-1)^n (c)_n} \; F_4[-n, n+a; c, a-c+1; zZ, (1-z)(1-Z)] \; .$$

In order to use (3.33) to prove the nonnegativity of the coefficients in (3.24) for $a = \alpha + \mu$, $\mu > 0$, $M = N$, the author considered the kernel

(3.37) $\quad S_z(x, y; \alpha, \beta, N, M)$

$$= \sum_{n=0}^{z} \frac{(-z)_n}{(-N)_n} \; \pi_n(\alpha, \beta, N) Q_n(x; \alpha, \beta, N) Q_n(y; \alpha, \beta, M), \quad z = 0, 1, \ldots ,$$

in [69], derived an extension of (3.36) to products of ${}_3F_2(1)$ series (also see [71]) and used it to derive the formula

(3.38) $\qquad S_z(x, y; \alpha, \beta, N, M)$

$$= \frac{(N-z)\,\Gamma(N+z)}{\Gamma(N+1)(N+\alpha+\beta+2)_z} \sum_{r=0}^{z} \sum_{s=0}^{z-r} \frac{(\alpha+\beta+2)_{2r+2s}}{(1-N-z)_{2r+2s}}$$

$$\times \frac{(-z)_{r+s}(-x)_r(-y)_r(x-N)_s(y-M)_s}{r!\,s!\,(-M)_{r+s}(\alpha+1)_s(\beta+1)_s} \; ,$$

for $z = 0, 1, \ldots$. From this it is clear that $S_z(x, y; \alpha, \beta, N, M) \geq 0$ for $\alpha, \beta > -1$ when x, y, z, N, M are nonnegative integers satisfying

$$0 \leq x \leq N, \quad 0 \leq y \leq M, \quad 0 \leq z \leq \min(N, M) \; .$$

A similar proof also gives the positivity of the kernel (3.37) with the factor $(-z)_n / (-N)_n$ replaced by t^n, $0 \le t < 1$. Since the coefficients in the expansion

$$\frac{(x-y-N)_n}{(-N)_n} Q_n(y;\alpha, \beta, N+y-x)$$

$$= \sum_{z=0}^{N} \rho(z;\alpha, \beta, N) S_{N+y-x}(z, y;\alpha, \beta, N, N+y-x) \, Q_n(z;\alpha, \beta, N)$$

are nonnegative, it then follows from (3.33) that the coefficients in (3.24) are nonnegative for $a = \alpha+\mu$, $\mu \ge 0$, $M = N$. Iterating this result with (3.26) and (3.27) we obtain

Theorem 1. Let $a, b, \alpha, \beta > -1$. If $a + b \ge \alpha + \beta$, $\beta \ge b$, and $M \ge N$ then the expansions (3.24) have nonnegative coefficients.

Consideration of the coefficients in (3.24) for $x = 1$ and $y = 0, 2$ shows that $a + b \ge \alpha + \beta$ and $(a+1)(\beta+1) \ge (\alpha+1)(b+1)$ are necessary conditions for the coefficients in (3.24) to be nonnegative for $0 \le x \le M = N$, $N = 0, 1, \ldots$.

For Jacobi polynomials it is known that

(3.39)
$$\frac{P_n^{(a, a)}(x)}{P_n^{(a, a)}(1)} = \int \frac{P_n^{(\alpha, \alpha)}(y)}{P_n^{(\alpha, \alpha)}(1)} \, d\nu_x(y), \qquad d\nu_x(y) \ge 0 \ ,$$

when $a \ge \alpha > -1$, $-1 \le x \le 1$, which can be obtained by applying (3.20) and the quadratic transformation

(3.40)
$$\frac{x P_n^{(\alpha, \frac{1}{2})}(2x^2 - 1)}{P_n^{(\alpha, \frac{1}{2})}(1)} = \frac{P_{2n+1}^{(\alpha, \alpha)}(x)}{P_{2n+1}^{(\alpha, \alpha)}(1)}$$

to (3.30) to obtain the Feldheim [58] and Vilenkin [106] formulas, and then using the positivity of the Poisson kernel for ultraspherical polynomials. The absolute continuity of the measures in (3.39) is con-sidered in Bingham [41]. However, the analogs of (3.20) and (3.40) for Hahn polynomials involve $_4F_3(1)$ series [70, (3.18) and (3.19)] and so the analog of (3.39) for Hahn polynomials does not follow from the case $b = \beta$ of Theorem 1.

Since the Meixner, Krawtchouk, and the Charlier polynomials are self dual (i.e. $M_n(x;\beta, c) = M_x(n;\beta, c)$, etc.) their projection formulas of the type (3.24) follow from the dual cases (3.11)-(3.13).

For Laguerre polynomials one can apply the limit (2.6) to formula (3.29) to derive

$$(3.41) \qquad \frac{L_n^a(x)}{L_n^a(0)} = \frac{\Gamma(a+1)x^{-a}}{\Gamma(\alpha+1)\Gamma(a-\alpha)} \int_0^x \frac{L_n^\alpha(y)}{L_n^\alpha(0)} y^\alpha (x-y)^{a-\alpha-1} dy, \quad a > \alpha > -1 ,$$

which is dual to (3.10) and has Sonine's first finite integral [111, 12.11(1)]

$$(3.42) \qquad J_a(x) = \frac{2^{\alpha-a+1}x^{-a}}{\Gamma(a-\alpha)} \int_0^x J_\alpha(y) y^{\alpha+1} (x^2-y^2)^{a-\alpha-1} dy, \quad a > \alpha > -1 ,$$

as a limit case. This formula is also a limit case of (3.10). It is used in Kingman [84] to obtain a projection operator which takes random walks in one space and projects them into another. For formulas of the Dirichlet-Mehler type see Gasper [72], and for the positivity and total positivity of some other kernels see Karlin and McGregor [82].

Formula (1.2). An argument similar to that given for the product formula (3.14) shows that for any $\alpha, \beta > -1$ the coefficients in

$$(3.43) \qquad Q_n(x;\alpha, \beta, N) Q_n(y;\alpha, \beta, N) = \sum_{z=0}^N E(x, y, z;\alpha, \beta, N) Q_n(z;\alpha, \beta, N)$$

where $x, y, n = 0, 1, \ldots, N,$ are not all nonnegative for $N = 0, 1, \ldots,$ and that there is no constant C independent of x, y, N such that

$$\sum_{z=0}^{N} |E(x, y, z; \alpha, \beta, N)| \le C$$

for all x, y, N. In addition, since the Meixner, Krawtchouk, and Charlier polynomials are self dual the above discussion of formulas of the type (1.7) also applies to formulas of the type (1.2) for these polynomials.

For Jacobi polynomials the author [66] recently proved the following:

Theorem 2. Let $\alpha, \beta > -1$ and $-1 < x, y < 1$. Then there is an integral representation of the form

$$(3.44) \qquad \frac{P_n^{(\alpha, \beta)}(x)}{P_n^{(\alpha, \beta)}(1)} \frac{P_n^{(\alpha, \beta)}(y)}{P_n^{(\alpha, \beta)}(1)} = \int_{-1}^{1} \frac{P_n^{(\alpha, \beta)}(z)}{P_n^{(\alpha, \beta)}(1)} d\mu_{x, y}(z), \quad n = 0, 1, \ldots ,$$

with the measures $d\mu_{x, y}(z) = d\mu_{x, y}^{(\alpha, \beta)}(z)$ satisfying

$$\int_{-1}^{1} |d\mu_{x, y}(z)| dz \le C$$

for some constant C independent of x, y, if and only if $\alpha \ge \beta$, $\alpha + \beta \ge -1$. Also $d\mu_{x, y}(z) \ge 0$ for $-1 < x, y < 1$ if and only if $\alpha \ge \beta$ and either $\beta \ge \frac{1}{2}$ or $\alpha + \beta \ge 0$.

This was then used to prove the positivity of a generalized translation operator, to extend Bochner's [42] results on homogeneous stochastic processes and heat and diffusion equations associated with ultraspherical series to Jacobi series, and to obtain a convolution structure for Jacobi series extending that previously obtained by Hirschman [75] for ultraspherical series and by Askey and Wainger [32] for Jacobi series when $\alpha \ge \beta \ge -\frac{1}{2}$. The convolution structure has been used [32] to obtain an analogue of the Hardy-Littlewood theorem on fractional integration, to obtain the uniform (C, γ), $\gamma > \max(\alpha + \frac{1}{2}, \beta + \frac{1}{2})$, summability for $-1 \leq x \leq 1$ of Jacobi series of continuous functions from the end point result in Szegö [103, Theorem 9.1.3.], and to interpolate certain norm inequalities for bounded linear operators which are not defined on a dense set. Other applications are given in Askey [9] and Bavinck [39].

The crucial observation used in proving Theorem 2 is that by means of appropriate [65] hypergeometric series representations for Jacobi polynomials it follows from Watson [111, 13.46 (7)] that

(3.45) $K(\cos 2\phi, \cos 2\psi, \cos 2\theta; \alpha, \beta)$

$$\equiv \sum_{n=0}^{\infty} h_n^{(\alpha, \beta)} P_n^{(\alpha, \beta)}(\cos 2\phi) P_n^{(\alpha, \beta)}(\cos 2\psi) P_n^{(\alpha, \beta)}(\cos 2\theta)/P_n^{(\alpha, \beta)} \quad (1)$$

$$= 2^{-\beta-2} \Gamma(\alpha+1) (\sin\phi \, \sin\psi \sin^2\theta)^{-\alpha} (\cos\phi \cos\psi \cos\theta)^{-\beta}$$

$$\times \int_0^{\infty} J_\alpha(t \sin\phi \sin\psi) J_\beta(t \cos\phi \cos\psi) J_\beta(t \cos\theta) t^{1-\alpha} \, dt$$

when

$$\alpha > -\frac{1}{2}, \quad \beta > -1, \quad \cos\theta \neq |\cos(\psi \pm \phi)|, \quad 0 < \phi, \psi, \theta < \frac{\pi}{2} \quad .$$

For $\alpha = \beta > -\frac{1}{2}$, the nonnegativity of (3.45) follows from Sonine's formula [111, 13.46 (3)]:

(3.46) $$\int_0^{\infty} J_\alpha(at) J_\alpha(bt) J_\alpha(ct) t^{1-\alpha} \, dt = \frac{2^{\alpha-1} \Delta^{2\alpha-1}}{(abc)^\alpha \Gamma(\alpha + \frac{1}{2}) \Gamma(\frac{1}{2})}$$

when a, b, c are the sides of a plane triangle of area Δ; otherwise the value of the integral is zero. For $\alpha \geq \beta > -\frac{1}{2}$ the nonnegativity of (3.45) can be shown by using the formula [111, 13.46 (1)] to write

395

(3.47) $\quad K(\cos 2\phi, \cos 2\psi, \cos 2\theta; \alpha, \beta)$

$$= \frac{\Gamma(\alpha+1)(\sin\phi \, \sin\psi \, \sin\theta)^{-2\alpha}}{2^{\alpha+\beta+1} \, \Gamma(\alpha-\beta)\Gamma(\beta+\frac{1}{2})\Gamma(\frac{1}{2})}$$

$$\times \int_0^A (1-\cos^2\phi-\cos^2\psi-\cos^2\theta+2\cos\phi\cos\psi\cos\theta\cos\gamma)^{\alpha-\beta-1} \, \sin^{2\beta}\gamma \, d\gamma$$

where ϕ, ψ, θ are as above and the value of A is

$$0, \quad \arccos \frac{\cos^2\phi+\cos^2\psi+\cos^2\theta-1}{2\cos\phi\cos\psi\cos\theta}, \quad \pi$$

according as $\sin^2\phi \, \sin^2\psi$ is less than, between, or greater than the two numbers $(\cos\phi\cos\psi \pm \cos\theta)^2$.

Macdonald's [111, 13.46] expressions for the integral in (3.47) in terms of hypergeometric functions were used in [66] to prove the non-negativity of (3.45) when $\alpha \geq \beta > -1$, $\alpha > -\frac{1}{2}$ and either $\beta \geq -\frac{1}{2}$ or $\alpha + \beta \geq 0$, and to prove that if $\alpha \geq \beta > -1$, $\alpha + \beta \geq -1$, $\alpha > -\frac{1}{2}$ then there is a constant C independent of x, y such that

$$\int_{-1}^1 |K(x, y, z; \alpha, \beta)| \, (1-z)^\alpha \, (1+z)^\beta \, dz \leq C .$$

Then Zygmund's [117] extension of the Riemannian theory of trigonometric series to Jacobi series and analytic continuation were used to show that if $\alpha \geq \beta > -1$, $\alpha + \beta > -1$ or if $\alpha > -\frac{1}{2}$, $\alpha + \beta = -1$, $x \neq y$, then the product formula (3.44) holds for $-1 < x, y < 1$ with

$$d\mu_{x,y}^{(\alpha,\beta)}(z) = K(x, y, z; \alpha, \beta) \, (1-z)^\alpha \, (1+z)^\beta \, dz .$$

In view of (3.18), the case $\alpha = \beta = -\frac{1}{2}$ of (3.44) follows from the identity

$$\cos n\phi \, \cos n\psi = \tfrac{1}{2} \cos n(\phi + \psi) + \tfrac{1}{2} \cos n(\phi - \psi) .$$

The case $\alpha = \beta > -\frac{1}{2}$ of (3.44) follows by a change of variables [77] from Gegenbauer's formula

$$(3.48) \quad C_n^\lambda(\cos \phi) \, C_n^\lambda(\cos \psi) = \frac{2^{1-2\lambda} \, \Gamma(n+2\lambda)}{n! \, \Gamma(\lambda) \, \Gamma(\lambda)}$$

$$\times \int_0^\pi C_n^\lambda(\cos \phi \cos \psi + \sin \phi \sin \psi \cos \theta) \, (\sin \theta)^{2\lambda-1} \, d\theta, \quad \lambda > 0,$$

which is a consequence of his addition formula [53, 10.9(34)] for ultraspherical polynomials. The limiting case of (3.48) for Bessel functions [111, 11.41 (17)] is used in Kingman [84] to give a necessary and sufficient condition for a distribution function to be associative.

Recently, Koornwinder [86], [87], [88] derived the addition formula for Jacobi polynomials and when $\alpha > \beta > -\frac{1}{2}$ it gives (3.44) in an equivalent double integral form. Koornwinder [89] has also Bateman's [37] addition formula for $L_n(x)$ to $L_n^\alpha(x)$; and Watson's [110] formula

$$(3.49) \quad L_n^\alpha(x) \, L_n^\alpha(y) = \frac{2^{\alpha-\frac{1}{2}} \, \Gamma(n+\alpha+1)}{n! \, \Gamma(\frac{1}{2})} \int_0^\pi L_n^\alpha(x+y+2(xy)^{\frac{1}{2}} \cos \theta)$$

$$\times \frac{e^{-(xy)^{\frac{1}{2}} \cos \theta} \, J_{\alpha-\frac{1}{2}}((xy)^{\frac{1}{2}} \sin \theta)}{(xy)^{\alpha/2-\frac{1}{4}} \, (\sin \theta)^{-\alpha-\frac{1}{2}}} \, d\theta, \quad \alpha > -\frac{1}{2},$$

follows from it by integration. The convolution structure which follows from (3.49) is considered in McCully [95] and Askey [4].

4. <u>Recurrence relations.</u> We shall first consider the proof given in Gasper [64] of the following result.

Theorem 3. Let $\alpha, \beta > -1$. Then the coefficients in the linearization of the product of Jacobi polynomials formula (3.15) are nonnegative for $m, n = 0, 1, \ldots$, if and only if $\alpha \geq \beta$ and

(4.1) $(\alpha+\beta+1)(\alpha+\beta+4)^2 (\alpha+\beta+6) \geq (\alpha-\beta)^2[(\alpha+\beta+1)^2 - 7(\alpha+\beta+1) - 24]$.

The curve (4.1) starts at $(\alpha, \beta) = ((-11 + (73)^{\frac{1}{2}})/8, -1)$ and approaches the line $\alpha + \beta = -1$ tangentially from the left, meeting it at the point $(-\frac{1}{2}, -\frac{1}{2})$. Thus the region in Theorem 3 includes the set $\alpha \geq \beta > -1$, $\alpha + \beta \geq -1$. In [64] this theorem is used to derive a convolution structure dual to that which follows from Theorem 2 and to derive a Wiener-Lévy theorem for Jacobi expansions. In order to have a convolution structure it suffices to have

(4.2) $\sum_k |B(k, m, n;\alpha, \beta)| \; P_k^{(\alpha, \beta)}(1) \leq C \; P_n^{(\alpha, \beta)}(1) \; P_m^{(\alpha, \beta)}(1)$

with C independent of m, n . Askey and Wainger [31] used asymptotic formulas to prove (4.2) for $\alpha \geq \beta \geq -\frac{1}{2}$, and their proof is extended in Askey and Gasper [20] to show that (4.2) holds for all m, n if and only if $\alpha \geq \beta > -1$, $\alpha \geq -\frac{1}{2}$.

The necessity of the conditions in Theorem 3 follow from the fact that if they are not satisfied then either $B(1, 1, 1;\alpha, \beta) < 0$ or $B(2, 2, 2;\alpha, \beta) < 0$. To simplify our discussion of the use of a recurrence relation to prove the nonnegativity of the coefficients in (3.15) we shall restrict ourselves to the case $\alpha \geq \beta > -1$, $\alpha + \beta > -1$. An explicit formula [63, (4)] gives the case m = 1, and so by symmetry of the left side of (3.15) in m and n we may assume that $n \geq m \geq 2$. Hylleraas' recurrence relation [80] for $B_k = B(k, m, n;\alpha, \beta)$ may be written in the form

(4.3) $a_j B_{j+n-m+1} = (\alpha - \beta)b_j \; B_{j+n-m} + c_j B_{j+n-m-1}$

where the coefficients also depend on m, n, α, β . From Hylleraas' explicit formulas for B_{n-m}, B_{n+m} and for the coefficients in (4.3), see [63], it is clear that $B_{n-m} > 0$, $B_{n+m} > 0$,

$$a_j > 0, \qquad j = 1, 2, \ldots, 2m-1 \ ,$$

$$c_j > 0, \qquad j = 1, 2, \ldots, 2m \ ,$$

and that b_j assumes both positive and negative values. However, computation of b_j for various α, β suggested that it is well behaved in the sense that it changes sign at most once as j increases from 0 to $2m$. This behavior was verified in [63] by writing b_j as a positive multiple of a fourth degree polynomial in $J = j - 1$,

(4.4) $$F(J) = aJ^4 + bJ^3 + cJ^2 + dJ + e \ ,$$

(where the coefficients depend on m, n, α, β), proving that the only pos-sible sign patterns for the coefficients of $F(J)$ are $-, -, +, +, +$ and $-, -, -, \pm, +$, and then applying Descartes' rule of signs. A computation gave $B_{n-m+1} \geq 0$ and then the nonnegativity of the B_k for $\alpha \geq \beta > -1$, $\alpha + \beta > -1$ followed by iterating (4.3) for $j = 1, 2, \ldots$ with the middle term on the right side when $b_j \geq 0$, and for $j = 2m, 2m-1, \ldots$, with the middle term on the left side when $b_j < 0$. A more complicated analysis was needed in [64] to handle the case $\alpha + \beta < -1$.

We now turn to the results which have been obtained by means of recurrence relations for the coefficients in the projection formulas (3.4) and (3.5). Let $a, b, \alpha, \beta > -1$ and let $N(\alpha, \beta)$ be the set of (a, b) for which the coefficients in (3.4), and hence in (3.5), are nonnegative for all n, N. Also, for any set $S(\alpha, \beta)$ of points in the (a, b)-plane and any integer j let

$$S(\alpha, \beta; j) = \{(a, b): (a, b + j) \in S(\alpha, \beta)\} \ .$$

Then the results in Askey and Gasper [21] may be stated as follows.

<u>Theorem 4.</u> If $\beta \geq \alpha$ and

$$R(\alpha, \beta) = \{(a, b): a + b \geq \alpha + \beta, \ \beta - \alpha \geq b - a\} \ ,$$

399

then

$$\bigcup_{j=0}^{\infty} R(\alpha, \beta; j) \subset N(\alpha, \beta)$$

$$\subset \{(a,b): \ a > \alpha, \ \beta - \alpha \geq b - a\} \cup [\bigcup_{j=0}^{\infty} \{(\alpha, \beta - j)\}] \ .$$

If $\alpha \geq \beta$ and

$$S(\alpha, \beta) = \{(a, b): \ a \geq \alpha, \ b = \beta\} \ ,$$

$$T(\alpha, \beta) = \{(a, b): \ a + b \geq \alpha + \beta, \ a - b \geq 2\alpha - 2\beta, \ a\beta - b\alpha \geq b - a + \alpha - \beta \ ,$$

$$(a+b)(a\beta - b\alpha + 3\beta - 3\alpha + 2a - 2b) \geq 2(2\alpha - 2\beta + b - a)\} \ ,$$

then

$$\bigcup_{j=0}^{\infty} [S(\alpha, \beta; j) \cup T(\alpha, \beta; j)] \subset N(\alpha, \beta)$$

$$\subset \{(a,b): \ a > \alpha, \ a - b \geq 2\alpha - 2\beta, \ a\beta - b\alpha \geq b - a + \alpha - \beta\} \cup [\bigcup_{j=0}^{\infty} S(\alpha, \beta; j)] \ .$$

In addition, if $-1 < \alpha \leq 0$, then

$$N(\alpha, \alpha) = \{(a, b): \ a \geq b, \ (\alpha + 2)(a^2 + b^2) - 2(\alpha + 1)ab - (\alpha - 2)(a + b) - 4\alpha \geq 0\} \ .$$

The set $R(\alpha, \beta)$ is a wedge to the right of the point (α, β) with vertex at (α, β) and sides of slope ± 1; and $R(\alpha, \beta; j)$, $j \geq 0$, is $R(\alpha, \beta)$ translated j units downward. For $\beta \geq \alpha$ the set $N(\alpha, \beta)$ is completely determined except for a small sawtooth set on the right of the line $a = \alpha$. When $\alpha \geq \beta$, $N(\alpha, \beta)$ contains the half-lines $S(\alpha, \beta; j)$, $j = 0, 1, \ldots$, and yet the restriction $a - b \geq 2\alpha - 2\beta$ is required when $\beta - b \neq j$, as is shown in [21] using an asymptotic expansion. Due to the restriction $a\beta - b\alpha \geq b - a + \alpha - \beta$, $N(\alpha, \beta)$ is below the straight line through $(-1, -1)$

400

and (α, β) . The necessity of the conditions in Theorem 4 were obtained by considering the $_3F_2(1)$ series in (3.4) for various values of k, n . The set inclusion

$$(4.5) \qquad\qquad R(\alpha, \beta) \subset N(\alpha, \beta), \qquad \beta \geq \alpha ,$$

was proved by deriving and analyzing a recurrence relation for $A_k = A(k, n; a, b, N, \alpha, \beta, N)$ of the form

$$(4.6) \qquad\qquad a_k A_k = b_k A_{k+1} + c_k A_{k+2}, \qquad k = 0, 1, \ldots ,$$

(where the coefficients also depend on n, a, b, α, β) which had the property that if $n \geq 1$, $\beta \geq \alpha$, $a + b \geq \alpha + \beta$ and $\beta - \alpha \geq b - a$, then

$$a_k > 0, \qquad k = 0, 1, \ldots, n\text{-}1 ,$$

$$b_k \geq 0, \qquad k = 0, 1, \ldots, n\text{-}1 ,$$

$$c_k \geq 0, \qquad k = 0, 1, \ldots, n\text{-}2 .$$

Since $A(k, n) = 0$ for $k > n$ and $A(n, n) > 0$, (4.5) followed by consecutive applications of (4.6) with $k = n - 1, n - 2, \ldots$. The set inclusions

$$R(\alpha, \beta; j) \subset N(\alpha, \beta), \qquad \beta \geq \alpha, \quad j \geq 0 ,$$

then follow by repeated applications of formula (3.8).

Formulas (3.7) and (3.8) give $S(\alpha, \beta; j) \subset N(\alpha, \beta)$, $j \geq 0$, and the set inclusions $T(\alpha, \beta; j) \subset N(\alpha, \beta)$, $j \geq 0$, are obtained by using a recurrence relation in n which is due to Bailey [36].

In addition to the applications of Theorem 4 mentioned in Askey and Gasper [21], Zaremba [116] has used it to extend certain bounds on $Q_n(x; 0, 0, N)$ to $Q_n(x; \alpha, \beta, N)$ for $(\alpha, \beta) \in N(0, 0)$. A recurrence relation is used in Askey and Gasper [24] to prove that

401

(4.7) $\quad e^{-x} L_n^\alpha(x) e^{-x} L_m^\alpha(x) = \sum_{k=0}^\infty a(k, m, n;\alpha) e^{-x} L_k^\alpha(x), \qquad x \geq 0 ,$

with nonnegative coefficients for $m, n = 0, 1, \ldots,$ if and only if $\alpha \geq (-5 + (17)^{\frac{1}{2}})/2;$ and this result is then used to obtain a convolution structure. Also, Askey [12] has used a recurrence relation for a hyper-geometric function to prove that if $-1 < \alpha < \gamma$ and the (C, γ) means of the series $\sum_{n=0}^\infty a_n$ are nonnegative, i.e.

$$\sum_{k=0}^n \frac{(\gamma+1)_{n-k}}{(n-k)!} a_k \geq 0, \qquad n = 0, 1, \ldots ,$$

then the (C, α) means of the series $\sum_{n=0}^\infty a_n t^n$ are nonnegative for the interval $0 \leq t \leq (\alpha+1)/(\gamma+1),$ which is best possible.

5. <u>Maximum principles.</u> One consequence of Theorem 2 is the fact that if $\alpha \geq \beta > -1$ and either $\beta \geq -\frac{1}{2}$ or $\alpha + \beta \geq 0,$ and if $\sum_{n=0}^\infty |a_n| < \infty$ and

(5.1) $$\sum_{n=0}^\infty a_n \frac{P_n^{(\alpha, \beta)}(x)}{P_n^{(\alpha, \beta)}(1)} \leq 0, \qquad -1 \leq x \leq 1 ,$$

then

(5.2) $$u(x, y) \equiv \sum_{n=0}^\infty a_n \frac{P_n^{(\alpha, \beta)}(x)}{P_n^{(\alpha, \beta)}(1)} \frac{P_n^{(\alpha, \beta)}(y)}{P_n^{(\alpha, \beta)}(1)} \leq 0, \qquad -1 \leq x, y \leq 1 .$$

For ultraspherical polynomials this was observed and used by Bochner in [42]. Since the Jacobi polynomials $y = P_n^{(\alpha, \beta)}(x)$ satisfy the differential equation

$$\frac{d}{dx} \left[(1-x)^{\alpha+1} (1+x)^{\beta+1} \frac{dy}{dx} \right] + n(n+\alpha+\beta+1)(1-x)^\alpha (1+x)^\beta y = 0 ,$$

the function $u(x, y)$ is at least formally the solution of the hyperbolic boundary value problem

(5.3) $\left[(1-x)^{\alpha+1} (1+x)^{\beta+1} (1-y)^{\alpha} \cdot (1+y)^{\beta} u_x \right]_x$

$- \left[(1-x)^{\alpha} (1+x)^{\beta} (1-y)^{\alpha+1} (1+y)^{\beta+1} u_y \right]_y = 0 \ ,$

$$u(x,1) = \sum_{n=0}^{\infty} a_n \frac{P_n^{(\alpha,\beta)}(x)}{P_n^{(\alpha,\beta)}(1)} \ ,$$

and so the inequality (5.2) is a maximum property of (5.3). In [112] Weinberger obtained a maximum principle for a class of hyperbolic equations and used it to prove the positivity of the Riemann function and to derive an extension of Bochner's result. From Weinberger's theorem, (5.1) and the fact that $u(x, y) = u(y, x)$ it follows that (5.2) holds for $\alpha \geq \beta > -1$, $\alpha + \beta + 1 \geq 0$ when $x + y \geq 0$. Weinberger observed that since $P_n^{(\alpha,\alpha)}(-x) = (-1)^n P_n^{(\alpha,\alpha)}(x)$ it then follows that (5.2) holds for the whole square $-1 \leq x, \ y \leq 1$ when $\alpha = \beta \geq -\frac{1}{2}$. Koornwinder [89] has used a maximum principle to show that (5.2) holds for $\alpha \geq \beta \geq -\frac{1}{2}$. Chebli [46] has also used a maximum principle to obtain similar results for Sturm-Liouville operators on $[0, \infty)$.

For difference equations Askey [3], [4] proved the following analog of Weinberger's maximum principle by using induction and observed that it implies Theorem 6.

Theorem 5.   Define the operator Δ_n by $\Delta_n k(n) = k(n+1) + \alpha_n k(n)$ $+ \beta_n k(n-1)$ and let $a(n, m)$ satisfy

$$\Delta_n a(n, m) = \Delta_m a(n, m)$$

with $a(n, 0) = a(0, n) = a(n)$ and $a(n, -1) = a(-1, n) \leq 0$. If $a(n) \geq 0$ and $\alpha_{n+1} \geq \alpha_n$, $\beta_{n+1} \geq \beta_n \geq 0$, $n = 0, 1, \ldots$, then

$$a(n, m) \geq 0, \qquad n, m = 0, 1, \ldots \ .$$

403

<u>Theorem 6.</u>   Let $p_0(x) = 1$, $p_1(x) = x + c$,   c real, $p_1(x)p_n(x) =$
$= p_{n+1}(x) + \alpha_n p_n(x) + \beta_n p_{n-1}(x)$. If $\alpha_n \geq 0$, $\beta_{n+1} > 0$, $\alpha_{n+1} \geq \alpha_n$,
$\beta_{n+2} \geq \beta_{n+1}$, $n = 0, 1, \ldots$,   then

$$p_n(x)\, p_m(x) = \sum_{k = |n-m|}^{n+m} a_{k, m, n}\, p_k(x)$$

with $a_{k, m, n} \geq 0$.

The hypotheses in Theorem 6 are satisfied for Jacobi polynomials
when $\alpha \geq \beta > -1$, $\alpha + \beta \geq 1$ and for some other (α, β); but not for
$-\frac{1}{2} < \alpha = \beta < \frac{1}{2}$. Thus, although Theorem 6 applies to a wide class of
orthogonal polynomials, it is unfortunately not strong enough to yield
Theorem 3 even for the $-\frac{1}{2} < \alpha = \beta < \frac{1}{2}$ case.

The method of induction is also used in Trebels [105, p. 27] to
prove the nonnegativity of a sum which has applications to approximation
theory. The maximum principle for elliptic equations in Protter and
Weinberger [99] has been used by Beckmann [40] to prove the nonnegativity of the Poisson kernel (3.35), which is then used to construct two-
dimensional probability density functions.

6. <u>Estimation.</u>   We have already referred to the two papers [31], [32]
of Askey and Wainger in which asymptotic estimates were used to obtain uniform bounds for the kernels in the product formulas (3.15) and
(3.44) for $\alpha \geq \beta \geq -\frac{1}{2}$. Due to the complicated nature of the kernels one
could not hope to prove their nonnegativity by extending the estimates in
[31], [32]. However, there are a number of interesting positivity results
which were first obtained by means of estimates and we shall point out
some of them here.

About fifty years ago Kogbetliantz [85] used asymptotic estimates
to prove that

$$(6.1) \quad \sum_{k=0}^{n} \frac{(\gamma+1)_{n-k}}{(n-k)!} \frac{(2k+\alpha+\beta+1)(\alpha+\beta+1)_k}{(\alpha+\beta+1)k!} \frac{P_k^{(\alpha, \beta)}(x)}{P_k^{(\beta, \alpha)}(1)} \geq 0, \quad -1 \leq x \leq 1 ,$$

when $\alpha = \beta > -\frac{1}{2}$ and $\gamma = 2\alpha + 2$. This is a special case of the non-negativity of the (C, γ) means of the Poisson kernel (3.35); and, in fact, by means of the product formula (3.44) and the positivity of (3.35) it follows from Kogbetliantz's result that the $(C, 2\alpha + 2)$ means of the Poisson kernel (3.35) are nonnegative for $0 \leq t \leq 1$ when $\alpha = \beta \geq -\frac{1}{2}$.

When $\alpha = \beta = -\frac{1}{2}$, $\gamma = 1$, the nonnegativity of (6.1) is equivalent to that of the Fejér [54] kernel

$$(6.2) \qquad K_n(\theta) = \frac{1}{2} + \sum_{k=1}^{n} \frac{n-k+1}{n+1} \cos k\theta \ ,$$

which Fejér used to prove that the $(C, 1)$ means of the Fourier series of a continuous function converge uniformly to the function. In connection with the formulas of Section 3 it is interesting to recall that Fejér's proof in [55] of the nonnegativity of (6.2) consists of first observing that the Dirichlet kernel satisfies

$$(6.3) \qquad D_n(\theta) = \frac{1}{2} + \sum_{k=1}^{n} \cos k\theta = \frac{\sin(k+1/2)\theta}{2 \sin \frac{\theta}{2}} \ ,$$

which is a limit case of the case $\alpha = \beta = -a = -\frac{1}{2}$ of (3.7), and then observing that

$$(6.4) \qquad K_n(\theta) = \frac{1}{n+1} \sum_{k=0}^{n} D_k(\theta)$$

$$= \frac{1}{n+1} \sum_{k=0}^{n} \frac{\sin(k+\frac{1}{2})\theta}{2 \sin\frac{\theta}{2}} = \frac{1}{2(n+1)} \left(\frac{\sin \frac{(n+1)\theta}{2}}{\sin \frac{\theta}{2}} \right)^2 \geq 0 \ ,$$

by the case $m = n$ of (3.17) . Fejér [55] also used (3.17) and the Dirichlet-Mehler formula for the Legendre polynomials $P_n(x) = P_n^{(0, 0)}(x)$,

$$(6.5) \qquad P_n(\cos \theta) = \frac{2^{\frac{1}{2}}}{\pi} \int_{\theta}^{\pi} \frac{\sin(k+\frac{1}{2})\phi}{(\cos \theta - \cos \phi)^{\frac{1}{2}}} d\phi \ ,$$

which is a special case of Bateman's integral (3.28), to prove that

(6.6) $$\sum_{k=0}^{n} P_k(x) > 0, \qquad -1 < x \leq 1 \ ,$$

and then used this inequality to prove (6.1) for $\alpha = \beta = 0$, $\gamma = 2$.

When $\alpha = \beta$ and $\gamma = 2\alpha + 2$ the sum (6.1) has the relatively simple generating function

(6.7) $$\frac{1 - t^2}{(1-t)^{2\alpha+3}(1-2xt+t^2)^{\alpha+3/2}}$$

$$= \sum_{n=0}^{\infty} t^n \sum_{k=0}^{n} \frac{(2\alpha+3)_{n-k}}{(n-k)!} \frac{(2k+2\alpha+1)(2\alpha+1)_k}{(2\alpha+1)k!} \frac{P_k^{(\alpha, \alpha)}(x)}{P_k^{(\alpha, \alpha)}(1)} \ ,$$

to which Kogbetliantz applied an asymptotic argument of Darboux type to prove his above-mentioned result.

Fejér [56] proved (6.1) for $\alpha = -\beta = \frac{1}{2}$, $\gamma = 2$ and later observed that the case $\alpha = \beta = 0$, $\gamma = 2$ of (6.1) also follows from this result by means of (6.5). Askey [9] extended this argument to obtain (6.1) for $\alpha + \beta = 0$, $0 \leq \alpha \leq \frac{1}{2}$, $\gamma = 2$ by rewriting (3.29) in the form

(6.8) $$\frac{P_n^{(\alpha-\mu, \beta+\mu)}(x)}{P_n^{(\beta+\mu, \alpha-\mu)}(1)} = \frac{\Gamma(\beta+\mu+1)}{\Gamma(\beta+1)\Gamma(\mu)}$$

$$\times \int_{-1}^{x} \frac{P_n^{(\alpha, \beta)}(y)}{P_n^{(\beta, \alpha)}(1)} \frac{(1+y)^\beta}{(1+x)^{\beta+\mu}} (x-y)^{\mu-1} dy, \qquad \mu > 0, \quad |x| < 1 \ ,$$

and then he used Fejér's and Kogbetliantz's results and some cases in which the sums (6.1) have simple generating functions to prove that (6.1) holds for $\gamma = \alpha + \beta + 2$ when $-\beta \leq \alpha \leq 1 + \beta$ and when $3 - \beta \leq \alpha \leq \beta + 2$. He also conjectured that (6.1) holds for $\gamma = \alpha + \beta + 2$ when $\alpha + \beta \geq -1$, $\beta \geq -\frac{1}{2}$, and that the limiting case

406

(6.9) $\int_0^x (x-t)^{\alpha+3/2} t^{\alpha+1} J_\alpha(t)dt \geq 0, \quad x > 0, \quad \alpha \geq -\frac{1}{2}$,

holds. Also see Askey [10].

Recently Fields and Ismail [59] proved (6.9) by using an asymptotic argument on an integral representation for a $_1F_2$ function and properties of completely monotonic functions discussed in the next section.

A different type of argument is given in Gasper [62] where the two identities (3.17) and (6.3), summation by parts, and estimates for trigonometric functions are used to determine for which α the cosine polynomials

$$\frac{1}{1+\alpha} + \frac{\cos\theta}{1+\alpha} + \frac{\cos 2\theta}{2+\alpha} + \dots + \frac{\cos n\theta}{n+\alpha}$$

of Young [115] and Rogosinski and Szegö [101] are nonnegative.

Estimates are used in Askey and Steinig [29] to prove that

(6.10) $\dfrac{d}{d\theta}\left\{\sum_{k=0}^n \dfrac{\sin(k+1)\theta}{(k+1)\sin(\theta/2)}\right\} < 0, \qquad 0 < \theta < \pi$,

which is equivalent to the case $\alpha = \frac{3}{2}$, $\beta = -\frac{1}{2}$ of the inequality

(6.11) $\sum_{k=0}^n \dfrac{P_k^{(\alpha,\beta)}(x)}{P_k^{(\beta,\alpha)}(1)} > 0, \qquad -1 < x < 1$,

and they are also used in Askey and Gasper [22] to prove (6.11) for $\alpha = \frac{5}{2}$, $\beta = -\frac{1}{2}$.

In Section 8 we will discuss a method which has led to a proof of (6.9), of Askey's conjecture concerning the case $\gamma = \alpha + \beta + 2$ of (6.1), and of (6.11) for $\alpha + \beta > 0$, $\beta \geq -\frac{1}{2}$.

7. <u>Absolutely monotonic and completely monotonic functions</u>. If $f(x)$, $x > 0$, is the Laplace transform of a nonnegative measure, i.e.

407

(7.1)
$$f(x) = \int_0^\infty e^{-xt} d\mu(t), \qquad d\mu(t) \geq 0 \ ,$$

then $f(x)$ is called completely monotonic (on $(0, \infty)$) .

For example, the function $x^{-\alpha}$, $\alpha > 0$, is completely monotonic since

$$x^{-\alpha} = \frac{1}{\Gamma(\alpha)} \int_0^\infty e^{-xt} t^{\alpha-1} dt \ , \qquad \alpha > 0 \ .$$

Since [111, p. 386]

$$x(x^2+1)^{-\alpha-3/2} = \frac{2^{-\alpha-1} \Gamma(1/2)}{\Gamma(\alpha+3/2)} \int_0^\infty e^{-xt} t^{\alpha+1} J_\alpha(t) dt \ , \qquad \alpha > -1 \ ,$$

we have

$$\int_0^\infty e^{-xt} \int_0^t (t-u)^{\alpha+3/2} u^{\alpha+1} J_\alpha(u) du$$

$$= \frac{2^{\alpha+1} \Gamma(\alpha+3/2) \Gamma(\alpha+5/2)}{\Gamma(1/2) x^{\alpha+3/2} (x^2+1)^{\alpha+3/2}} \ ,$$

and thus the inequality (6.9) is equivalent to the complete monotonicity of the function $x^{-\alpha-3/2} (x^2+1)^{-\alpha-3/2}$. For $\alpha = -1/2$ the complete monotonicity is a consequence of the formula

$$x^{-1} (x^2+1)^{-1} = \int_0^\infty e^{-xt} (1 - \cos t) dt \ .$$

Askey [9] used this and the fact that the product of completely monotonic functions is completely monotonic to prove that $x^{-\alpha-3/2} (x^2+1)^{-\alpha-3/2}$ is completely monotonic for $\alpha = -1/2, 1/2, 3/2, \ldots,$ and hence that (6.9) holds for these α . In [13] he used this result to give a sufficient condition for a function to be a radial characteristic function. In the Fields

408

and Ismail proof of (6.9) mentioned in Section 6, they first proved (6.9) for $-1/2 \le \alpha \le 1/2$ and then used the multiplicative property of completely monotonic functions to extend the result to $\alpha \ge -1/2$. For another application of completely monotonic functions see Askey and Pollard [27], where the Hausdorff-Bernstein-Widder theorem [113] on the equivalence of (7.1) with

$$(-1)^n \frac{d^n}{dx^n} f(x) \ge 0, \quad x > 0, \quad n = 0, 1, \ldots \; ,$$

is used to prove that $x^{-2\lambda}(x^2+1)^{-\lambda}$, $\lambda > 0$, is completely monotonic.

An excellent example of the use of absolutely monotonic functions [113, p. 114] is the following simple proof by Askey and Pollard [27] of Kogbetliantz's result that the function

$$f_\alpha(t) = \frac{1-t^2}{(1-t)^{2\alpha+3}(1-2xt+t^2)^{\alpha+3/2}}, \quad -1 \le x \le 1, \quad \alpha \ge -1/2 \; ,$$

has nonnegative power series coefficients and so is absolutely monotonic (on $[0,1)$). They first observed that since

(7.2) $$f_\alpha(t) = f_{-1/2}(t) \frac{1}{(1-t)^{2\alpha+1}(1-2xt+t^2)^{\alpha+1/2}}$$

and the function $f_{-1/2}(t)$ is absolutely monotonic by Fejér's result (6.4), it suffices to prove that

(7.3) $$\frac{1}{(1-t)^{2\alpha+1}(1-2xt+t^2)^{\alpha+\frac{1}{2}}} = \sum_{n=0}^{\infty} t^n \sum_{k=0}^{n} \frac{(2\alpha+1)_{n-k}}{(n-k)!} \frac{(2\alpha+1)_k}{k!} \frac{P_k^{(\alpha,\alpha)}(x)}{P_k^{(\alpha,\alpha)}(1)}$$

is absolutely monotonic for $\alpha > -1/2$. Letting $x = \cos\theta$ and

$$g(t) = \log[(1-t)^{-2}(1-2xt+t^2)^{-1}]$$

$$= -2\log(1-t) - \log(1-2t\cos\theta + t^2)$$

we have

$$g'(t) = \frac{2}{1-t} + \frac{2\cos\theta - 2t}{1 - 2t\cos\theta + t^2}$$

$$= \frac{2}{1-t} + \frac{e^{i\theta}}{1 - te^{i\theta}} + \frac{e^{-i\theta}}{1 - te^{-i\theta}}$$

$$= \sum_{n=0}^{\infty} (2 + e^{i(n+1)\theta} + e^{-i(n+1)\theta})t^n$$

$$= \sum_{n=0}^{\infty} 2(1 + \cos(n+1)\theta)t^n \quad ,$$

so $g'(t)$ is absolutely monotonic. Since $g(0) = 0$, $g(t)$ is also absolutely monotonic, and hence so is

$$\frac{1}{(1-t)^{2\alpha+1}(1-2xt+t^2)^{\alpha+\frac{1}{2}}} = e^{(\alpha+\frac{1}{2})g(t)} = \sum_{h=0}^{\infty} \frac{[(\alpha+\frac{1}{2})g(t)]^n}{n!} \quad , \quad \alpha \geq -\frac{1}{2} \quad ,$$

which concludes the proof. Absolutely monotonic functions are aslo used in Fields and Ismail [60], Askey [14], [15], [16], Askey and Gasper [22], Askey, Gasper and Ismail [26], and in Bustoz [43] which contains a simpler proof of Askey's result [12] mentioned at the end of Section 4.

8. <u>Sums of squares</u>. Fejér's proof (6.4) of the nonnegativity of the Fejér kernel and the following proof of the limiting Bessel function result

$$(8.1) \qquad \int_0^x (x-t)t^{\frac{1}{2}} J_{-\frac{1}{2}}(t)dt$$

$$= \int_0^x t^{\frac{1}{2}} J_{\frac{1}{2}}(t)dt = x(\frac{\pi}{2})^{\frac{1}{2}} J_{\frac{1}{2}}^2(x/2)$$

use only the square of one function. Nicholson's formula [111, p. 444]

$$J_\alpha^2(x) + Y_\alpha^2(x) = \frac{8}{\pi^2} \int_0^\infty K_0(2x \sinh t)\cosh 2\alpha t \, dt, \qquad \text{Re } \alpha > 0 \ ,$$

which has been extended to Legendre functions by Durand in this volume, uses the sum of two squares. One of the simplest proofs of the Turán type inequality

$$J_\alpha^2(x) - J_{\alpha-1}(x) J_{\alpha+1}(x) > 0, \qquad \alpha \geq -1, \ x > 0 \ ,$$

is by means of the expansion [111, p. 152]

$$(8.2) \qquad J_\alpha^2(x) - J_{\alpha-1}(x) J_{\alpha+1}(x)$$

$$= 4x^{-2} \sum_{n=0}^\infty (2n+\alpha+1) J_{2n+\alpha+1}^2(x) \ .$$

Formulas (8.1) and (8.2) suggest that it might be possible to prove the nonnegativity of the integral (6.9) and of other integrals of Bessel functions by using a sum of squares of Bessel functions with nonnegative coefficients. From the $_0F_1$ formula for $J_\alpha(x)$ in (2.7) we have

$$(8.3) \qquad \int_0^x (x-t)^\lambda t^\mu J_\alpha(t) dt = \frac{\Gamma(\lambda+1)\Gamma(\alpha+\mu+1)}{\Gamma(\alpha+1)\Gamma(\alpha+\lambda+\mu+2)} 2^{-\alpha} x^{\alpha+\lambda+\mu+1}$$

$$\times {}_2F_3\left[\begin{array}{cc} (\alpha+\mu+1)/2, \ (\alpha+\mu+2)/2 & ; \ -x^2/4 \\ \alpha+1, \ (\alpha+\lambda+\mu+2)/2, \ (\alpha+\lambda+\mu+3)/2 & \end{array}\right]$$

for $\alpha + \mu > -1$, $\lambda > -1$. Thus to write the integral (8.3) as a sum of squares of Bessel functions it suffices to use the expansion [111, 5.5 (1)]

$$(8.4) \qquad z^{2\nu} = \frac{\Gamma^2(\nu+1)2^{2\nu+1}}{\Gamma(2\nu+1)} \sum_{n=0}^\infty \frac{(n+\nu)\,\Gamma(n+2\nu)}{n!} J_{n+\nu}^2(z)$$

411

in the $_2F_3$ series in (8.3). The expansions in Gasper [73] were obtained by using (8.4) with $z = x/2$ to write

$$(8.5) \int_0^x (x-t)^\lambda t^\mu J_\alpha(t)dt = \frac{\Gamma(\lambda+1)\Gamma(\alpha+\mu+1)\Gamma^2(\nu+1)}{\Gamma(\alpha+1)\Gamma(\alpha+\lambda+\mu+2)} 2^{4\nu-\alpha} x^{\alpha+\lambda+\mu+1-2\nu}$$

$$\times \sum_{n=0}^\infty {}_5F_4 \left[\begin{array}{c} -n,\ n+2\nu,\ \nu+1,\ (\alpha+\mu+1)/2,\ (\alpha+\mu+2)/2;1 \\ \nu+\frac{1}{2},\ \alpha+1,\ (\alpha+\lambda+\mu+2)/2,\ (\alpha+\lambda+\mu+3)/2 \end{array} \right]$$

$$\times \frac{(2\nu+1)_n}{n!} \frac{2n+2\nu}{n+2\nu} J^2_{n+\nu}(x/2) ,$$

where $\alpha + \mu > -1$, $\lambda > -1$, $2\nu \ne -1, -2, \ldots$, and the factor $(2n+2\nu)/(n+2\nu)$ must be replaced by 1 when $n = 0$.

When $\lambda = \alpha + 3/2$ and $\mu = \nu = \alpha + 1 > 0$ the $_5F_4(1)$ reduces to a $_3F_2(1)$ which can be summed by Saalschütz's formula to give

$$(8.6) \int_0^x (x-t)^{\alpha+3/2} t^{\alpha+1} J_\alpha(t)dt$$

$$= \frac{\Gamma(\alpha+5/2)\Gamma(2\alpha+3)\Gamma(\alpha+2)}{\Gamma(3\alpha+9/2)} 2^{3\alpha+3} x^{\alpha+3/2}$$

$$\times \sum_{n=0}^\infty \frac{((2\alpha+1)/4)_n\ ((2\alpha+3)/4)_n}{((6\alpha+9)/4)_n\ ((6\alpha+11)/4)_n} \frac{(2\alpha+3)_n}{n!} \frac{2n+2\alpha+2}{n+2\alpha+2} J^2_{n+\alpha+1}(x/2) ,$$

which is clearly nonnegative for $\alpha \ge -1/2$, $x > 0$ and positive for $\alpha > -1/2$, $x > 0$, since the positive zeros of $J_\nu(x)$ are interlaced [111, p. 479] with those of $J_{\nu+1}(x)$.

To use (8.5) to prove

$$(8.7) \int_0^x (x-t)^\lambda t^{\lambda+\frac{1}{2}} J_\alpha(t)dt > 0 , \qquad 0 \le \lambda \le \alpha-1/2,\ \alpha > 1/2,\ x > 0 ,$$

we set $\mu = \lambda + 1/2$, $\nu = (\alpha+\lambda+1/2)/2$, so that the $_5F_4(1)$ reduces to the Saalschützian $_4F_3(1)$ series

412

(8.8) $\quad _4F_3\left[\begin{array}{c} -n, \quad n+\alpha+\lambda+1/2, \quad (\alpha+\lambda+5/2)/2, \quad (\alpha+\lambda+5/2)/2;1 \\ \alpha+1, \quad (\alpha+2\lambda+5/2)/2, \quad (\alpha+2\lambda+7/2)/2 \end{array}\right]$

and then use Whipple's transformation formula [33, 4.3 (4)] to show that this $_4F_3(1)$ equals

(8.9) $\quad \dfrac{(\lambda+1/2)_n \; ((2\alpha-2\lambda-1)/4)_n}{(\alpha+1)_n \; ((2\alpha+3\lambda+7)/4)_n}$

$\times \; _7F_6\left[\begin{array}{c} (2\alpha+6\lambda+3)/4, \; (2\alpha+6\lambda+11)/8, \; (\lambda+1)/2, \; \lambda/2, \; (2\alpha+2\lambda+5)/4, \; n+\alpha+\lambda+1/2, \; -n;1 \\ (2\alpha+6\lambda+3)/8, \lambda+(2\alpha+5)/4, \lambda+(2\alpha+7)/4, \lambda+1/2, (5+2\lambda-2\alpha)/4 -n, n+(2\alpha+6\lambda+7)/4 \end{array}\right],$

which is clearly positive when $0 < \lambda < \alpha -1/2$, since each of its $n+1$ terms are positive. When $\lambda = 0$ or $\lambda = \alpha -1/2$ the series in (8.8) reduces to a Saalschützian $_3F_2(1)$ which can be summed to give the positivity of (8.8) for $\alpha > 1/2$ and so (8.7) follows from (8.5). For other consequences of (8.5) and references to applications (e.g., to univalent functions, Lommel functions, and to stability of least square smoothing) see [73].

In Askey and Gasper [22] formula (2.2) is used to obtain

(8.10) $\quad \displaystyle\sum_{k=0}^{n} P_k^{(\alpha,\,0)}(x) = \dfrac{(\alpha+2)_n}{n!} \; _3F_2\left[\begin{array}{c} -n, \, n+\alpha+2, \, (\alpha+1)/2; \, (1-x)/2 \\ \alpha+1, \quad (\alpha+3)/2 \end{array}\right]$

and then a fractional integral representation for this $_3F_2$ in terms of ultraspherical polynomials, Gegenbauer's limiting case of (3.6), and the fact that

(8.11) $\quad \left\{ C_n^{\lambda}((\tfrac{1+x}{2})^{\frac{1}{2}}) \right\}^2 = \left(\dfrac{(2\lambda)_n}{n!}\right)^2 \; _3F_2\left[\begin{array}{c} -n, \, n+2\lambda, \; \lambda; \, (1-x)/2 \\ 2\lambda, \; \lambda+1/2 \end{array}\right]$

are used to derive the expansion

(8.12)
$$\sum_{k=0}^{n} P_k^{(\alpha,\,0)}(x)$$

$$= \sum_{j=0}^{[n/2]} \frac{(n-2j)!\,(\frac{\alpha+2}{2})_{n-j}\,(\frac{\alpha+3}{2})_{n-2j}\,(\frac{1}{2})_j}{j!\,(\alpha+1)_{n-2j}\,(\frac{\alpha+3}{2})_{n-j}\,(\frac{\alpha+1}{2})_{n-2j}} \left\{ C_{n-2j}^{(\alpha+1)/2}\left((\frac{1+x}{2})^{\frac{1}{2}}\right) \right\}^2 \,,$$

which is clearly nonnegative for $-1 \le x \le 1$ when $\alpha \ge -2$. Application
of the integral (6.8) then gives

(8.13)
$$\sum_{k=0}^{n} \frac{P_k^{(\alpha,\,\beta)}(x)}{P_k^{(\beta,\,\alpha)}(1)} \ge 0\,, \qquad\qquad -1 \le x \le 1\,,$$

for $\alpha + \beta \ge -2$, $\beta \ge 0$. The restriction $\alpha + \beta \ge -2$ is necessary since
(8.13) is negative for $n = x = 1$ when $\alpha + \beta < -2$. The positivity results which follow from (8.12) are discussed in [22].

Bailey's formula [35]

(8.14) $\displaystyle \int_0^x J_\alpha(t)dt = x \int_0^{\pi/2} J_{\alpha/2}^2 (\frac{x}{2} \sin\theta) \sin\theta\, d\theta, \quad \alpha > -1 \,,$

is a limit case of (8.12). For the integral on the left of (8.14), formula
(8.5) gives the expansion [73]

(8.15)
$$\int_0^x J_\alpha(t)dt = \frac{\Gamma^2(\frac{\alpha+3}{2})2^{\alpha+2}}{\Gamma(\alpha+2)}$$

$$\times \sum_{n=0}^{\infty} \frac{(\alpha+2)_{2n}(\frac{1}{2})_n(\frac{1}{2})_n}{(2n)!\,(\frac{\alpha+2}{2})_n(\frac{\alpha+2}{2})_n} \frac{4n+\alpha+1}{2n+\alpha+1} J_{2n+(\alpha+1)/2}^2(x/2), \quad \alpha > -1\,,$$

which suggests that there should be another expansion for the left side
of (8.12) as a sum of squares which has (8.15) as a limit case. To

derive it, first observe that as an extension of (8.4), which is a special case of the addition formula for Bessel functions, we have

$$(8.16) \quad (\frac{1-x}{2})^j = \sum_{k=0}^{n-j} \frac{2^{2k}(n-j)!\,(n-j-k)!\,(2j+2a-1)_k \left\{(j+a)_k\right\}^2}{k!\,(2j+2a)_{n-j}(2j+2a)_{n-j+k}}$$

$$\times \frac{2j+2k+2a-1}{2j+2a-1} (\frac{1-x}{2})^{j+k} \left\{ C_{n-j-k}^{j+k+a} ((\frac{1+x}{2})^{\frac{1}{2}}) \right\}^2 \,,$$

which is a special case of the addition formula for ultraspherical poly-nomials [53, 10.9 (34)]. Then using (8.16) with $a = (\alpha+2)/2$ in the right side of (8.10), changing the order of summation, and using Watson's formula [33, 3.3 (1)] for the sum of a $_3F_2(1)$, we obtain the desired expansion

$$(8.17) \quad \sum_{k=0}^{n} P_k^{(\alpha, 0)}(x) = \sum_{j=0}^{[n/2]} \frac{2^{2j}(n-2j)!\,(\alpha+2)_{2j}(\frac{1}{2})_j}{j!\,(\alpha+2)_{n+2j}}$$

$$\times \frac{(\frac{\alpha+2}{2})_{2j}(\frac{\alpha+2}{2})_{2j}}{(\frac{\alpha+2}{2})_j(\frac{\alpha+2}{2})_j} \frac{4j+\alpha+1}{2j+\alpha+1} (\frac{1-x}{2})^{2j} \left\{ C_{n-2j}^{2j+\frac{\alpha+2}{2}} ((\frac{1+x}{2})^{\frac{1}{2}}) \right\}^2 \,.$$

After (8.12) and (8.17) were derived the author observed that (8.1), (8.11) and a general expansion formula of Fields and Wimp [61] (also in Luke [92, 9.1 (13)]) could be used to obtain

$$(8.18) \quad \sum_{k=0}^{n} P_k^{(\alpha, 0)}(1-2(1-x^2)(1-y^2))$$

$$= \sum_{j=0}^{n} \frac{(\alpha+2)_n (n+\alpha+2)_j (\frac{\alpha+2}{2})_j}{j!\,(n-j)!\,(\frac{\alpha+3}{2})_j\,(j+\alpha+1)_j} (1-y^2)^j$$

$$\times \left\{ \frac{j!\,(n-j)!}{(\alpha+1)_j (2j+\alpha+2)_{n-j}} C_j^{(\alpha+1)/2}(x)\, C_{n-j}^{j+(\alpha+2)/2}(y) \right\}^2 \,,$$

which combines (8.12) and (8.17).

A simple proof of Kogbetliantz's result can be given by using (2.2) and then (8.16) or [92, 9.1 (13)] to derive

$$(8.19) \quad \sum_{k=0}^{n} \frac{(2\alpha+3)_{n-k}}{(n-k)!} \frac{(2k+2\alpha+1)(2\alpha+1)_k}{(2\alpha+1)k!} \frac{P_k^{(\alpha,\alpha)}(x)}{P_k^{(\alpha,\alpha)}(1)}$$

$$= \frac{(4\alpha+4)_n (2\alpha+3)_n}{n!(2\alpha+2)_n} \, {}_3F_2\left[\begin{array}{c} -n,\ n+4\alpha+4,\ \alpha+3/2;\ (1-x)/2 \\ 2\alpha+3,\ 2\alpha+5/2 \end{array}\right]$$

$$= \frac{n+2\alpha+2}{2\alpha+2} \sum_{j=0}^{[n/2]} \frac{2^{2j}(n-2j)!(2\alpha+2)_{2j}(2\alpha+2)_{2j}(4\alpha+4)_{2j}(\alpha+\frac{1}{2})_j}{j!(\alpha+2)_j(2\alpha+2)_j(n+4\alpha+4)_{2j}(4\alpha+4)_n}$$

$$\times \frac{4j+4\alpha+3}{2j+4\alpha+3}\left(\frac{1-x}{2}\right)^{2j}\left\{C_{n-2j}^{2j+2\alpha+2}\left(\left(\frac{1+x}{2}\right)^{\frac{1}{2}}\right)\right\}^2 \quad,$$

which is obviously positive for $\alpha > -1/2$, $x > -1$.

Similarly, the positivity of the power series coefficients in (7.3) for $\alpha > -1/2$, $x > -1$ follows directly from

$$(8.20) \quad \sum_{k=0}^{n} \frac{(2\alpha+1)_{n-k}}{(n-k)!} \frac{(2\alpha+1)_k}{k!} \frac{P_k^{(\alpha,\alpha)}(x)}{P_k^{(\alpha,\alpha)}(1)}$$

$$= \frac{(4\alpha+2)_n}{n!} \, {}_3F_2\left[\begin{array}{c} -n,\ n+4\alpha+2,\ \alpha+1/2;\ (1-x)/2 \\ 2\alpha+1,\ 2\alpha+3/2 \end{array}\right]$$

$$= \sum_{j=0}^{[n/2]} \frac{2^{2j}(n-2j)!(2\alpha+1)_{2j}(2\alpha+1)_{2j}(4\alpha+2)_{2j}(\alpha+\frac{1}{2})_j}{j!(\alpha+1)_j(2\alpha+1)_j(n+4\alpha+2)_{2j}(4\alpha+2)_n}$$

$$\times \frac{4j+4\alpha+1}{2j+4\alpha+1}\left(\frac{1-x}{2}\right)^{2j}\left\{C_{n-2j}^{2j+2\alpha+1}\left(\left(\frac{1+x}{2}\right)^{\frac{1}{2}}\right)\right\}^2 \quad.$$

One can also use a sum of squares of ultraspherical polynomials to prove that the partial sums of the expansion

$$(8.21) \quad (1-x)^{-(2\alpha+3)/4} \sim A_\alpha \sum_{k=0}^{\infty} \frac{(\alpha+\frac{1}{2})_k \, (\frac{2\alpha+5}{4})_k}{(\alpha+1)_k \, (\frac{2\alpha+1}{4})_k} P_k^{(\alpha, \, -\frac{1}{2})}(x)$$

are positive for $\alpha > \frac{1}{2}$, $-1 \leq x \leq 1$, where A_α is a positive constant. For, proceeding as above, we have

$$(8.22) \quad \sum_{k=0}^{n} \frac{(\alpha+\frac{1}{2})_k (\frac{2\alpha+5}{4})_k}{(\alpha+1)_k (\frac{2\alpha+1}{4})_k} P_k^{(\alpha, \, -\frac{1}{2})}(x)$$

$$= \frac{(4n+2\alpha+3)(\alpha+3/2)_n}{(2\alpha+3)n!} \, {}_3F_2 \left[\begin{matrix} -n, \; n+\alpha+3/2, \; (2\alpha+3)/4; \; (1-x)/2 \\ \alpha+1, \; (2\alpha+7)/4 \end{matrix} \right]$$

$$= \frac{4n+2\alpha+3}{2\alpha+3} \sum_{j=0}^{n} \frac{(\alpha+3/2)_{n+j} (\frac{1}{2})_j (\frac{2\alpha+3}{4})_j (\frac{2\alpha-1}{4})_j}{j!(n-j)!(j+\alpha+\frac{1}{2})_j (\alpha+1)_j (\frac{2\alpha+5}{4})_j (\frac{2\alpha+7}{4})_j}$$

$$\times (\frac{1-x}{2})^j \left\{ \frac{(n-j)!}{(2j+\alpha+3/2)_{n-j}} C_{n-j}^{j+(2\alpha+3)/4} ((\frac{1+x}{2})^{\frac{1}{2}}) \right\}^2 > 0 \, ,$$

$$\alpha > 1/2, \quad -1 \leq x \leq 1 \; .$$

Askey and Fitch [18] proved the positivity of the partial sums of (8.21) for the case $\alpha = 3/2$ by using estimates and Carslaw's investigation [45, p. 297-303] of the graph of the series

$$\sum_{k=0}^{n} \frac{\sin(2k+1)\theta}{2k+1} \; .$$

This result was then used to prove the positivity of Cotes' numbers for ultraspherical abscissas. In [6] Askey used the results in [7] on when (8.13) holds to prove the positivity of Cotes' numbers for some Jacobi abscissas. The values of the n-th partial sum of the series

$$(8.23) \qquad (1-x)^{-a}(1+x)^{-b} \sim \sum_{k=0}^{\infty} c_k P_k^{(\alpha,\beta)}(x)$$

at the zeros of $P_n^{(\alpha,\beta)}(x)$ are the generalized Cotes' numbers when integrating with respect to the weight function $(1-x)^{\alpha-a}(1+x)^{\beta-b}$, and so (8.22) partially answers a question posed in Horton [78] on when these Cotes' numbers are positive.

One of the main reasons that we were able to handle the integrals (8.6), (8.7), (8.15) and the sums (8.10), (8.19), (8.20), (8.22) so easily is that they could be expressed as generalized hypergeometric functions. However except for some special cases, the sums (6.1), (8.13) and

$$(8.24) \qquad \sum_{k=0}^{n} \frac{(\lambda+1)_{n-k}}{(n-k)!} \frac{(\lambda+1)_k}{k!} \frac{P_k^{(\alpha,\beta)}(x)}{P_k^{(\beta,\alpha)}(1)}$$

cannot be expressed in terms of a single generalized hypergeometric function, and so one has to work with a double sum to which, in particular, the Fields and Wimp expansion [61] is not applicable.

The sum (8.24) is a common generalization of (8.13) and the case $\gamma = \alpha+\beta+2$ of (6.1) since, by a summation by parts,

$$(8.25) \qquad \sum_{k=0}^{n} \frac{(\alpha+\beta+3)_{n-k}}{(n-k)!} \frac{(2k+\alpha+\beta+1)(\alpha+\beta+1)_k}{(\alpha+\beta+1)k!} \frac{P_k^{(\alpha,\beta)}(x)}{P_k^{(\beta,\alpha)}(1)}$$

$$= \sum_{k=0}^{n} \frac{(\alpha+\beta+2)_{n-k}}{(n-k)!} \frac{(\alpha+\beta+2)_k}{k!} \frac{P_k^{(\alpha+1,\beta)}(x)}{P_k^{(\beta,\alpha+1)}(1)} .$$

Hence Askey's conjecture concerning the first sum in (8.25) is equivalent to the nonnegativity of the sum (8.24) for $\lambda = \alpha+\beta \geq 0$, $\beta \geq -1/2$, $-1 \leq x \leq 1$; which is a special case of the conjecture in Askey and Gasper [22] that if $0 \leq \lambda \leq \alpha+\beta$, $\beta \geq -1/2$, $-1 < x \leq 1$, then the sum (8.24) is positive, except when $\lambda = 0$, $\alpha = -\beta = 1/2$, $n \geq 2$, when the sum is non-negative and has zeros.

This conjecture will be proven by the author in [74] by showing that there is an expansion of the form

$$(8.26) \qquad \sum_{k=0}^{n} \frac{(\lambda+1)_{n-k}}{(n-k)!} \frac{(\lambda+1)_k}{k!} \frac{P_k^{(\alpha,\beta)}(x)}{P_k^{(\beta,\alpha)}(1)}$$

$$= \sum_{j=0}^{[n/2]} a_{n,j} (1+x)^{n-2j} \left\{ P_j^{((\lambda+\alpha-\beta)/2,\,\beta+n-2j)}(x) \right\}^2$$

with positive or nonnegative coefficients when $0 \leq \lambda \leq \alpha + \beta$, $\beta \geq -1/2$. When $\lambda = 0$, the $a_{n,j}$ are multiples of Saalschützian ${}_4F_3(1)$ series, and their nonnegativity is proven for $\alpha + \beta \geq 0$, $\beta \geq -1/2$ and for $\alpha + \beta \geq -2$, $\beta \geq 0$ by using the transformation formulas for Saalschützian ${}_4F_3(1)$ series in Bailey [33]. A similar proof gives the nonnegativity of $a_{n,j}$ for $\lambda = \alpha + \beta$ when $\alpha + \beta \geq 0$, $\beta \geq -1/2$ and when $\alpha + \beta \geq -1$, $\beta \geq \alpha$. This result is then used to prove that the $(C, \alpha+\beta+2)$ means of the Poisson kernel for Jacobi series (3.35) are nonnegative for $0 \leq t \leq 1$ if and only if $\alpha, \beta \geq -1/2$. When $0 < \lambda < \alpha + \beta$, $\beta \geq -1/2$, the coefficients in (8.26) are multiples of Saalschützian ${}_5F_4(1)$ series, and their positivity is proven by using a projection formula to write them as sums of Saalschützian ${}_4F_3(1)$ series to which transformation formulas are applied.

9. <u>Additional comments.</u> Another powerful method for proving positivity depends on the identities

$$(9.1) \qquad \frac{p_n'}{p_n} = \sum_{k=1}^{n} \frac{1}{x-x_k}, \qquad -\left(\frac{p_n'}{p_n}\right)' = \frac{p_n'^2 - p_n p_n''}{p_n^2} = \sum_{k=1}^{n} \frac{1}{(x-x_k)^2},$$

where $p_n = p_n(x)$ is a polynomial of degree n with simple zeros

$x_1, \ldots, x_n$. These identities have been used to derive bounds for the zeros of the classical orthogonal polynomials [91], [103, p. 120], and they are the main tool used in [67] to prove that if $\beta \geq \alpha > -1$ then the Jacobi polynomials satisfy the Turán type inequalities

$$(9.2) \quad \left(\frac{P_n^{(\alpha, \beta)}(x)}{P_n^{(\alpha, \beta)}(1)} \right)^2 - \frac{P_{n-1}^{(\alpha, \beta)}(x)}{P_{n-1}^{(\alpha, \beta)}(1)} \frac{P_{n+1}^{(\alpha, \beta)}(x)}{P_{n+1}^{(\alpha, \beta)}(1)} > 0, \quad -1 < x < 1, \quad n \geq 1 \; .$$

The analog of (9.2) for $Q_n(x; \alpha, \beta, N)$ has been proven by Karlin and Szegö [83] for $\alpha = \beta > -1$ by using a generating function, but it is still open for $\beta > \alpha > -1$. The identities (9.1) are also used in [68] to derive inequalities for determinants containing orthogonal polynomials, and in Askey, Gasper and Harris [25] to prove that if $\beta \geq \alpha > -1$, $1 \leq r \leq x \leq y \leq s$ and $rs = xy$, then

$$(9.3) \quad P_n^{(\alpha, \beta)}(r) \, P_n^{(\alpha, \beta)}(s) \leq P_n^{(\alpha, \beta)}(x) \, P_n^{(\alpha, \beta)}(y) \; .$$

For $\alpha = \beta = -1/2$, $r = 1$, this is proved in [25] by using an extremal property of the Tchebycheff polynomials. Also, Gegenbauer's integral (3.48) is used to prove that if $\lambda > 0$ and $x, y \geq 1$, then

$$(9.4) \quad \frac{C_n^\lambda(x)}{C_n^\lambda(1)} \frac{C_n^\lambda(y)}{C_n^\lambda(1)} \leq \frac{C_n^\lambda(xy + (x^2-1)^{\frac{1}{2}} (y^2-1)^{\frac{1}{2}})}{2 C_n^\lambda(1)}$$

$$+ \frac{C_n^\lambda(xy - (x^2-1)^{\frac{1}{2}} (y^2-1)^{\frac{1}{2}})}{2 C_n^\lambda(1)} \; ,$$

with equality only when $x = 1$ or $y = 1$. The Legendre polynomial case $\lambda = 1/2$, $x = y$ of (9.4) was found by Malkov [94] is his work on interaction operators.

The nonnegativity results for the kernels in the projection formulas

420

(3.1), (3.5), and in (3.24) and its limiting Jacobi polynomial case

(9.5) $$\frac{P_n^{(a,\,b)}(x)}{P_n^{(a,\,b)}(1)} = \int_{-1}^{1} \frac{P_n^{(\alpha,\,\beta)}(y)}{P_n^{(\alpha,\,\beta)}(1)} \, d\nu_x(y), \quad -1 \le x \le 1, \quad n = 0,\, 1,\, \cdots ,$$

are still not complete, and one interesting open case is the conjecture [70] that if $M = N$, $\beta \ge \alpha > -1$, $a = \alpha + \delta$, $b = \beta + \delta$, $\delta > 0$, then (3.24) has nonnegative coefficients. The limiting Jacobi polynomial case of this conjecture is only known for $\alpha = \beta$ (see Section 3) and for $\beta = \alpha + 1$, [4, (4.13)].

Reasons for considering the problem of when the expansions

(9.6) $$(1-x)^{\gamma}(1+x)^{\delta} P_n^{(a,\,b)}(x) = \sum_{k=0}^{\infty} G(k,\,n;a,\,b,\,\alpha,\,\beta,\,\gamma,\,\delta) P_k^{(\alpha,\,\beta)}(x)$$

have nonnegative coefficients are given in [21]. In addition to applying to the case $\gamma = \delta = 0$, Theorem 4 also applies to the case $\gamma = a - \alpha$, $\delta = b - \beta$ of (9.6) since, for $a, b, \alpha, \beta > 1$, $G(k,\,n;a,\,b,\,\alpha,\,\beta,\,a-\alpha,\,b-\beta)$ is a positive multiple of

$$_3F_2 \left[\begin{array}{c} n-k, \ n+a+1, \ n+k+\alpha+\beta+1;1 \\ n+\alpha+1, \ 2n+a+b+2 \end{array} \right]$$

when $k \ge n$, and is zero when $k < n$. Also see Dunkl [49] and Askey [1], [11].

So far the only known proof of Theorem 3 is the recurrence relation proof in [64]. It would be of interest to have additional proofs, possibly by explicit formulas or by an extension of Askey's maximum principle, even for the case $\alpha \ge \beta > -1$, $\alpha + \beta \ge -1$. A maximum principle proof of the nonnegativity of the kernel (3.37), analogous to Beckmann's [40] for the nonnegativity of (3.35), would also be of interest.

The main tool that Askey needed in his work [8], [9] on the L^p convergence of Lagrange interpolation polynomials at the zeros of Jacobi polynomials is the nonnegativity of the Cesàro means of some order of the Poisson kernel (3.35). The nonnegativity of some Cesàro mean is proven in [9] for a region slightly larger than $\alpha, \beta \geq -1/2$, and Askey has conjectured that it also holds for $\alpha + \beta \geq 0$ when $-1 < \beta < -1/2$.

The coefficient $a(k, m, n; \alpha)$ in (4.7) is a positive multiple of

$$(9.7) \qquad \int_0^\infty L_k^\alpha(x) \, L_m^\alpha(x) \, L_n^\alpha(x) \, x^\alpha \, e^{-\rho x} \, dx$$

for $\rho = 2$. Askey and Gasper [24] proved that the integrals (9.7) are nonnegative for all k, m, n, if $\alpha \geq -1/2$ and $\rho \geq 3$ or if $\alpha \geq (-5 + (17)^{\frac{1}{2}})/2$ and $\rho \geq 2$. We also proved that at least one integral is negative if $\alpha < -1/2$ or $\rho < 3/2$ or if $(\alpha+1)^2(\rho-1)^3 + 3(\alpha+1)(\rho-1) < 2$. The case $\rho = 3$ is considered in Szegö [102] and Askey and Gasper [23], and the case $\rho = 3/2$ is considered in Gillis [76].

Kuttner's paper [90] on the Riesz (R, n^λ, κ) means suggests the problem of when

$$(9.8) \qquad \int_0^x (x^\nu - t^\nu)^\lambda \, t^\mu \, J_\alpha(t) dt \geq 0, \qquad x > 0 .$$

Several cases of (9.8) for $\nu = 1, 2$ are proved in [73] by the method of sums of squares, and additional methods will be needed to handle some other cases. The case $\lambda = 0$ has been completely solved by Makai [93] and Askey and Steinig [28] by using an oscillation theorem for second order differential equations.

One of the problems pointed out in Askey and Gasper [22] is that of determining for which values of $a, b, \alpha, \beta, \lambda$ we have

$$(9.9) \qquad \sum_{k=0}^n \frac{(\lambda+1)_{n-k}}{(n-k)!} \, c_k P_k^{(\alpha, \beta)}(x) \geq 0 , \qquad -1 \leq x \leq 1, \quad n = 0, 1, \ldots ,$$

where the c_k are the coefficients in the expansion (8.23). The connection of the case $\lambda = 0$ of this problem with the problem of the nonnegativity of the generalized Cotes' numbers was discussed in Section 8.

Among Jensen's necessary and sufficient conditions for the Riemann hypothesis given in Pólya [98] is the condition that

$$(9.10) \qquad \int_{-\infty}^{\infty} \int_{-\infty}^{\infty} \Phi(\alpha) \, \Phi(\beta) \, e^{i(\alpha+\beta)x} (\alpha-\beta)^{2n} \, d\alpha d\beta \geq 0$$

for all real x when $n = 0, 1, \ldots,$ where

$$\Phi(t) = 4 \sum_{n=1}^{\infty} (2n^4 \pi^2 e^{9t} - 3n^2 \pi e^{5t}) e^{-n^2 \pi e^{4t}} .$$

$\Phi(t)$ is an even function of t, which is positive for all real t. Since the above integral is a square when $n = 0$, the method of sums of squares is suggested for proving (9.10). A computer analysis of (9.10) and of the other necessary and sufficient conditions in [98] might lead to some interesting observations.

Added in proof. The editor has drawn my attention to the formula

$$(9.11) \qquad P_i(u) \, P_j(u) = \sum_{k=0}^{n} P_{ijk} \, P_k(u), \qquad P_{ijk} \geq 0, \qquad u = 0, 1, \ldots, n ,$$

in Sloane's paper in this volume. Since the Eberlein polynomials $E_k(u; V, n)$ are related to the Hahn polynomials by the relation

$$E_k(u; V, n) = (-1)^k \binom{n}{k} Q_{n-u}(k; 0, V-2n, n) ,$$

we find by using (9.11) with $P_k(u) = E_k(u;V, n)$ that if $\alpha = 0$ and $\beta = 0, 1,$ $\ldots,$ then the Hahn polynomials have a product formula of the form

$$(9.12) \qquad \frac{Q_n(x;\alpha, \beta, N)}{Q_N(x;\alpha, \beta, N)} \frac{Q_n(y;\alpha, \beta, N)}{Q_N(y;\alpha, \beta, N)} = \sum_{z=0}^{N} g(x, y, z) \frac{Q_n(z;\alpha, \beta, N)}{Q_N(z;\alpha, \beta, N)}$$

with $g(x, y, z) = g(x, y, z;\alpha, \beta, N) \geq 0$ and thus $\sum_z |g(x, y, z)| = \sum_z g(x, y, z) = 1$, where $x, y, z, n = 0, 1, \ldots, N$.

Similarly, from Lemma 2.4 and formulas (2.26) and (4.34) in Delsarte [An algebraic approach to the association schemes of coding theory,

423

Philips Research Reports Supplements, No. 10, 1973] it follows that if $\alpha = 0$ and $\beta = 0, 1, \ldots,$ then

$$(9.13) \quad \frac{Q_n(x;\alpha, \beta, N)}{Q_N(x;\alpha, \beta, N)} \frac{Q_m(x;\alpha, \beta, N)}{Q_N(x;\alpha, \beta, N)} = \sum_{k=0}^{N} G(k, m, n) \frac{Q_k(x;\alpha, \beta, N)}{Q_N(x;\alpha, \beta, N)}$$

with $G(k, m, n) = G(k, m, n;\alpha, \beta, N) \geq 0$ and thus $\sum_k |G(k, m, n)| = 1$, where $k, m, n, x = 0, 1, \ldots, N$. Both of these product formulas yield the inequality

$$(9.14) \quad \left| \frac{Q_n(x;\alpha, \beta, N)}{Q_N(x;\alpha, \beta, N)} \right| \leq 1, \quad n, x = 0, 1, \ldots, N ,$$

for $\alpha = 0$, $\beta = 0, 1, \ldots$ (this also follows from (2.29) in Delsarte's paper) and, by proceeding as in [23, §7], they can be used to construct convolution algebras. Thus it is of interest to determine for which α, β these formulas hold. Note that since the left side of (9.14) equals $(\alpha+1)_N / (\beta+1)_N$ when $n = 0$, $x = N$, a necessary condition for (9.14) and hence (9.12) and (9.13) (with nonnegative coefficients) to hold is that $\beta \geq \alpha$.

REFERENCES

1. R. Askey, Orthogonal expansions with positive coefficients, Proc. Amer. Math. Soc. 16 (1965), 1191-1194.

2. R. Askey, Jacobi polynomial expansions with positive coefficients and imbeddings of projective spaces, Bull. Amer. Math. Soc. 74 (1968), 301-304.

3. R. Askey, Linearization of the product of orthogonal polynomials, in Problems in Analysis, edited by R. Gunning, Princeton Univ. Press, Princeton, N. J., 1970, 223-228.

4. R. Askey, Orthogonal polynomials and positivity, in Studies in Applied Mathematics, 6, Special Functions and Wave Propagation, edited by D. Ludwig and F. W. J. Olver, SIAM, Philadelphia, Pa. (1970), 64-85.

5. R. Askey, Orthogonal expansions with positive coefficients. II, SIAM Jour. Math. Anal. 2 (1971), 340-346.

6. R. Askey, Positivity of the Cotes numbers for some Jacobi absciscas, Numer. Math. 19 (1972), 46-48.

7. R. Askey, Positive Jacobi polynomial sums, Tôhoku Math. Jour. 24 (1972), 109-119.

8. R. Askey, Mean convergence of orthogonal series and Lagrange interpolation, Acta Math. Acad. Sci. Hung. 23 (1972), 71-85.

9. R. Askey, Summability of Jacobi series, Trans. Amer. Math. Soc. 179 (1973), 71-84.

10. R. Askey, Refinements of Abel summability for Jacobi series, in Proc. Symp. Pure Math. vol. 26, Harmonic Analysis on Homogeneous Spaces, ed. by C. Moore, Amer. Math. Soc., Providence, 1973, 335-338.

11. R. Askey, Certain rational functions whose power series have positive coefficients. II, SIAM Jour. Math. Anal. 5 (1974), 53-57.

12. R. Askey, Positive Cesàro means of numerical series, Proc. Amer. Math. Soc. 45 (1974), 63-68.

13. R. Askey, Radial characteristic functions, Math. Research Center, Madison, Technical Report #1262, 1973.

14. R. Askey, Some characteristic functions of unimodal distributions, J. Math. Anal. Appl., to appear.

15. R. Askey, Positive Jacobi polynomial sums III, ISNM 25 Birkhäuser verlag, Basel and Stuttgart, 1974, pp. 305-312.

16. R. Askey, Some absolutely monotonic functions, Studia Sci. Math. Hung., to appear.

17. R. Askey, Orthogonal Polynomials and Special Functions, Regional Conference Series in Applied Mathematics, volume 21, SIAM, Philadelphia, 1975.

18. R. Askey and J. Fitch, Positivity of the Cotes numbers for some ultraspherical abscissas, SIAM J. Numer. Anal. 5 (1968), 199-201.

19. R. Askey and J. Fitch, Integral representations for Jacobi polynomials and some applications, Jour. Math. Anal. Appl. 26 (1969), 411-437

20.   R. Askey and G. Gasper, Linearization of the product of Jacobi polynomials. III, Can. J. Math. 23 (1971), 332-338.

21.   R. Askey and G. Gasper, Jacobi polynomial expansions of Jacobi polynomials with non-negative coefficients, Proc. Camb. Phil. Soc. 70 (1971), 243-255.

22.   R. Askey and G. Gasper, Positive Jacobi polynomial sums. II, Amer. J. Math., to appear.

23.   R. Askey and G. Gasper, Certain rational functions whose power series have positive coefficients, Amer. Math. Monthly 79 (1972), 327-341.

24.   R. Askey and G. Gasper, Convolution structures for Laguerre polynomials, Jour. d'Analyse Math., to appear.

25.   R. Askey, G. Gasper, and L. A. Harris, An inequality for Tchebycheff polynomials and extensions, J. Approx. Theory 13 (1975), to appear.

26.   R. Askey, G. Gasper, and M. E.-H. Ismail, A positive sum from summability theory, J. Approx. Theory 13 (1975), to appear.

27.   R. Askey and H. Pollard, Some absolutely monotonic and completely monotonic functions, SIAM Jour. Math. Anal. 5 (1974), 58-63.

28.   R. Askey and J. Steinig, Some positive trigonometric sums, Trans. Amer. Math. Soc. 187 (1974), 295-307.

29.   R. Askey and J. Steinig, A monotonic trigonometric sum, Amer. Jour. Math., to appear.

30.   R. Askey and S. Wainger, A transplantation theorem for ultra-spherical coefficients, Pacific J. Math. 16 (1966), 393-405.

31.   R. Askey and S. Wainger, A dual convolution structure for Jacobi polynomials, in Orthogonal Expansions and their Continuous Analogues, edited by D. Haimo, Southern Illinois Univ. Press, Carbondale, Illinois, 1968, 25-36.

32.   R. Askey and S. Wainger, A convolution structure for Jacobi series, Amer. J. Math. 91 (1969), 463-485.

33. W. N. Bailey, Generalized Hypergeometric Series, Cambridge Univ. Press, Cambridge, 1935.

34. W. N. Bailey, The generating function of Jacobi polynomials, Journal London Math. Soc. 13 (1938), 8-12.

35. W. N. Bailey, Some integrals involving Bessel functions, Quart. J. Math. (Oxford) 9 (1938), 141-147.

36. W. N. Bailey, Contiguous hypergeometric functions of the type $_3F_2(1)$, Proc. Glasgow Math. Assoc. 2 (1954), 62-65.

37. H. Bateman, Partial Differential Equations of Mathematical Physics, Cambridge Univ. Press, Cambridge, 1932.

38. H. Bateman, The solution of linear differential equations by means of definite integrals, Trans. Camb. Phil. Soc. 21 (1909), 171-196.

39. H. Bavinck, On positive convolution operators for Jacobi series, Tôhoku Math. J. 24 (1972), 55-69.

40. P. Beckmann, Orthogonal Polynomials for Engineers and Physicists, Golem Press, Boulder, Colorado, 1973.

41. N. H. Bingham, Integral representations for ultraspherical polynomials, Jour. London Math. Soc. (2) 6 (1973), 1-11.

42. S. Bochner, Sturm-Liouville and heat equations whose eigenfunctions are ultraspherical polynomials or associated Bessel functions, Proc. Conf. on Diff. Eq., Univ. of Maryland, 1955, 23-48.

43. J. Bustoz, Note on "Positive Cesaro means of numerical series", Proc. Amer. Math. Soc. 45 (1974), 69.

44. L. Carlitz, The product of two ultraspherical polynomials, Proc. Glasgow Math. Assoc. 5 (1961/2), 76-79.

45. H. S. Carslaw, Introduction to the Theory of Fourier Series and Integrals, 3rd. ed., Dover, New York, 1931.

46. H. Chébli, Sur la positivité des opérateurs de "translation géneralisée" associés à un opérateur de Sturm-Liouville sur $[0,\infty]$, C. R. Acad. Sci. Paris, Sér A 275 (1972), 601-604.

47. J. Davis and I. I. Hirschman, Jr., Toeplitz forms and ultraspherical polynomials, Pacif. Jour. Math. 18 (1966), 73-95.

48. J. Dougall, A theorem of Sonine in Bessel functions, with two extensions to spherical harmonics, Proc. Edin. Math. Soc. 37 (1919), 33-47.

49. C. F. Dunkl, An expansion in ultraspherical polynomials with non-negative coefficients, SIAM Jour. Math. Anal. 5 (1974), 51-52.

50. C. F. Dunkl, A Krawtchouk polynomial addition theorem and wreath products of symmetric groups, to appear.

51. G. K. Eagleson, A characterization theorem for positive definite sequences on the Krawtchouk polynomials, Austral. J. Statist. 11 (1969), 29-38.

52. A. Erdélyi, On some expansions in Laguerre polynomials, J. London Math. Soc. 13 (1938), 154-156.

53. A. Erdélyi, W. Magnus, F. Oberhettinger and F. G. Tricomi, Higher Transcendental Functions, Vols. I, II, III, McGraw-Hill, New York, 1953.

54. L. Fejér, Untersuchungen über Fouriersche Reihen, Math. Annalen 58 (1904), 51-69. Gesammelte Arbeiten I, 142-160.

55. L. Fejér, Über die Laplacesche Reihe, Math. Ann. 67 (1909), 76-109. Gesammelte Arbeiten I, 503-537.

56. L. Fejér, Neue Eigenschaften der Mittelwerte bei den Fourier-reihen, Jour. London Math. Soc. 8 (1933), 53-62. Gesammelte Arbeiten II, 493-501.

57. E. Feldheim, Contributions à la theorie des polynomes de Jacobi, Mat. Fiz. Lapok 48°(1941), 453-504 (Hungarian, French summary).

58. E. Feldheim, On the positivity of certain sums of ultraspherical polynomials, Jour. d'Anal. Math. 11 (1963), 275-284.

59. J. Fields and M. E. -H. Ismail, On the positivity of some $_1F_2$'s SIAM J. Math. Anal., 6 (1975), 551-559.

60. J. L. Fields and M. Ismail, On some conjectures of Askey concerning completely monotonic functions, in Spline Functions and Approximation Theory, edited by A. Meir and A. Sharma, ISNM Vol. 21, Birkhäuser Verlag, Basel, 1973, pp. 101-111.

61. J. L. Fields and J. Wimp, Expansions of hypergeometric functions in hypergeometric functions, Math. of Computation 15 (1961), 390-395.

62. G. Gasper, Nonnegative sums of cosine, ultraspherical and Jacobi polynomials, J. Math. Anal. Appl. 26 (1969), 60-68.

63. G. Gasper, Linearization of the product of Jacobi polynomials. I, Canad. J. Math. 22 (1970), 171-175.

64. G. Gasper, Linearization of the product of Jacobi polynomials. II, Can. J. Math. 22 (1970), 582-593.

65. G. Gasper, Positivity and the convolution structure for Jacobi series, Annals of Math. 93 (1971), 112-118.

66. G. Gasper, Banach algebras for Jacobi series and positivity of a kernel, Annals of Math. 95 (1972), 261-280.

67. G. Gasper, An inequality of Turán type for Jacobi polynomials, Proc. Amer. Math. Soc. 32 (1972), 435-439.

68. G. Gasper, On two conjectures of Askey concerning normalized Hankel determinants for the classical polynomials, SIAM J. Math. Anal. 4 (1973), 508-513.

69. G. Gasper, Non-negativity of a discrete Poisson kernel for the Hahn polynomials, J. Math. Anal. Appl. 42 (1973), 438-451.

70. G. Gasper, Projection formulas for orthogonal polynomials of a discrete variable, J. Math. Anal. Appl. 45 (1974), 176-198.

71. G. Gasper, Products of terminating $_3F_2(1)$ series, Pacific J. Math., to appear.

72. G. Gasper, Formulas of the Dirichlet-Mehler type, to appear in Proc. Conf. on Fractional Calculus and its Applications to the Mathematical Sciences.

73. G. Gasper, Positive integrals of Bessel functions, SIAM Jour. Math. Anal., to appear.

74. G. Gasper, Positive sums of the classical orthogonal polynomials, in preparation.

75. L. Gegenbauer, Zur Theorie der Functionen $C_n^v(x)$, Denkschriften der Akademie der Wiss. in Wien, Math. Naturwiss. Kl. 48 (1884), 293-316.

76. J. Gillis, Integrals of products of Laguerre polynomials, SIAM J. Math. Anal. 6 (1975), 318-339

77. I. I. Hirschman, Jr., Harmonic analysis and ultraspherical polynomials, Symposium on Harmonic Analysis and Related Integral Transforms, Cornell University, 1956.

78. R. L. Horton, Expansions using orthogonal polynomials, Ph. D Thesis, Univ. Of Wisconsin, Madison, 1973.

79. H. Y. Hsü, Certain integrals and infinite series involving ultraspherical polynomials and Bessel functions, Duke Math. Jour. 4 (1938), 374-383.

80. E. Hylleraas, Linearization of products of Jacobi polynomials, Math. Scand. 10 (1962), 189-200.

81. S. Karlin and J. McGregor, The Hahn polynomials, formulas and an application, Scripta Math. 26 (1961), 33-46.

82. S. Karlin and J. McGregor, Classical diffusion processes and total positivity, Jour. Math. Anal. Appl. 1 (1960), 163-183.

83. S. Karlin and G. Szegö, On certain determinants whose elements are orthogonal polynomials, Jour. d'Anal. Math. 8 (1961), 1-157.

84. J. F. C. Kingman, Random walks with spherical symmetry, Acta Math. 109 (1963), 11-53.

85. E. Kogbetliantz, Recherches sur la sommabilité des séries ultrasphériques par la méthode des moyennes arithmétiques, J. Math. Pures Appl. (9) 3 (1924), 107-187.

86. T. H. Koornwinder, The addition formula for Jacobi polynomials. III, Completion of the proof, Math. Centrum Amsterdam, Rep. TW 135 (1972).

87. T. H. Koornwinder, The addition formula for Jacobi polynomials and spherical harmonics, SIAM Jour. Appl. Math. 25 (1973), 236-246.

88. T. H. Koornwinder, Jacobi polynomials. III, An analytic proof of the addition formula, SIAM Jour. Math. Anal., 6 (1975), 533-543.

89. T. H. Koornwinder, Three notes on classical orthogonal polynomials; I. Yet another proof of the addition formula for Jacobi polynomials; II. The addition formula for Laguerre polynomials; III. New proof of the positivity of generalized translation for Jacobi series, Math. Centrum Amsterdam Rep., to appear.

90. B. Kuttner, On the Riesz means of a Fourier series (II), J. London. Math. Soc. 19 (1944), 77-84.

91. E. N. Laguerre, Sur les équations algébriques dont le premier membre satisfait à une équation linéaire du second ordre, Comptes Rendus de l'Académie des Sciences, Paris, 90 (1880), 809-812. Oeuvres, Vol. 1, 126-132.

92. Y. L. Luke, The Special Functions and their Approximations, Vols. I, II, Academic Press, New York, 1969.

93. E. Makai, An integral inequality satisfied by Bessel functions, Acta Math. Acad. Sci. Hungar. 25 (1974), 387-390.

94. E. I. Malkov, Positive definiteness of the absorptive part of the interaction operator, Soviet Jour. of Nuclear Physics 10 (1970), 491-494, or Yad. Fiz. 10 (Oct. 1969), 849-855.

95. J. McCully, The Laguerre transform, SIAM Rev. 2 (1960), 185-191.

96. W. Miller, Special functions and the complex Euclidean group in 3-space. I, J. Math. Physics 9 (1968), 1163-1175.

97. W. Miller, Special functions and the complex Euclidean group in 3-space. II, J. Math. Physics 9 (1968), 1175-1187.

98. G. Pólya, Über die algebraisch-funktionentheoretischen Untersuchungen von J. L. W. V. Jensen, Kgl. Danske Videnskabernes Selskab. 7 (1927), No. 17.

99. M. H. Protter and H. F. Weinbergerger, Maximum Principles in Differential Equations, Prentice-Hall, Englewood Cliffs, New Jersey, 1967.

100. T. J. Rivlin and M. W. Wilson, An optimal property of Chebyshev expansions, J. Approx. Theory 2 (1969), 312-317.

101. W. Rogosinski and G. Szegö, Über die Abschnitte von Potenz-
      reihen, die in einem Kreise beschränkt bleiben, Math. Zeit. 28
      (1928), 73-94.

102. G. Szegö, Über gewisse Potenzreihen mit lauter positiven Koef-
      fizienten, Math. Zeit. 37 (1933), 674-688.

103. G. Szegö, Orthogonal Polynomials, Amer. Math. Soc. Colloq. Pub.,
      Vol. 23, Amer. Math. Soc., Providence, R. I., Third Edition, 1967.

104. E. C. Titchmarsh, The Zeta-Function of Riemann, Cambridge Univ.
      Press, 1930.

105. W. Trebels, Multipliers for (C, α)-bounded Fourier expansions in
      Banach spaces and approximation theory, Lecture Notes in Mathe-
      matics, Springer-Verlag, New York, 1973.

106. N. Ja. Vilenkin, Some relations for Gegenbauer functions, Uspehi
      Math. Nauk (N. S.) 13 (1958), No. 3 (81), 167-172.

107. N. Ja. Vilenkin, Special Functions and the Theory of Group Repre-
      sentations, Vol. 22, Translations of Math. Monographs, Amer.
      Math. Soc., Providence, 1968.

108. G. N. Watson, The product of two hypergeometric functions, Proc.
      London Math. Soc., (2) 20 (1922), 189-195.

109. G. N. Watson, Notes on generating functions of polynomials: (4)
      Jacobi polynomials, J. London Math. Soc. 9 (1934), 22-28.

110. G. N. Watson, Another note on Laguerre polynomials, Jour. London
      Math. Soc. 14 (1939), 19-22.

111. G. N. Watson, A Treatise on the Theory of Bessel Functions,
      Cambridge Univ. Press, Cambridge, 1966.

112. H. Weinberger, A maximum property of Cauchy's problem, Ann. of
      Math. (2) 64 (1956), 505-513.

113. D. V. Widder, The Laplace Transform, Princeton Univ. Press,
      Princeton, 1941.

114. M. W. Wilson, Non-negative expansions of polynomials, Proc.
      Amer. Math. Soc. 24 (1970), 100-102.

115. W. H. Young, On a certain series of Fourier, Proc. London Math. Soc. (2) 11 (1912), 357-366.

116. S. K. Zaremba, Some properties of polynomials orthogonal over the set $\langle 1, 2, \ldots, N \rangle$, Math. Research Center, Madison, Technical Report #1342.

117. A. Zygmund, Sur la théorie riemannienne de certains systèmes orthogonaux. II, Prace Matematyczno Fizyczne 39 (1932), 73-117.

Supported in part by NSF Grant GP-32116. The author is an Alfred P. Sloan Fellow.

Two-Variable Analogues of the Classical Orthogonal Polynomials

Tom Koornwinder

Contents

1. Introduction

By now the theory of Jacobi polynomials is rather well settled cf.
Askey [5] and Gasper [24]. The surprising richness of this theory is
closely tied up with the group theoretic interpretation of Jacobi polynomials for certain values of the parameters. Hardly any comparable results
have been obtained for orthogonal polynomials in two or more variables,
probably because people did not study the right classes of polynomials.
Most emphasis was laid on certain biorthogonal systems, while some remarkable orthogonal systems of polynomials in two variables, scattered
over the literature, did not get as much attention as they deserved.

Analogues in severable variables of the Jacobi polynomials seem
to be highly nontrivial generalizations of the one-variable case. Therefore, it is a good approach to restrict oneself first to the case of two variables. In this paper a number of distinct classes of orthogonal polynomials in two variables will be introduced for which many properties hold
which are analogous to properties of Jacobi polynomials. The polynomials
belonging to these orthogonal systems are eigenfunctions of two algebraically independent partial differential operators. In the Chebyshev cases

436

(i. e. if all parameters are equal to $\pm\frac{1}{2}$) these polynomials can be interpreted as quotients of two eigenfunctions of the Laplacian on a two-dimensional torus or sphere, which satisfy symmetry relations with respect
to certain reflections. For two of the classes which will be considered
and for certain values of the parameters the polynomials can be interpreted
as spherical functions on certain compact symmetric Riemannian spaces
of rank 2.

This paper is an extended version of the survey given in the
author's thesis [47, sections 8, 9, 10, 11]. It also includes some new
material, see in particular sections 4. 4 and 4. 5. Most results are given
without proofs.

The Proceedings of this Advanced Seminar include three other papers
dealing with orthogonal polynomials in several variables, cf. James [35]
and Karlin and McGregor [40], [41]. The present paper has some relationship with [40], cf. section 3. 7. 3, and an important relationship
with the zonal polynomials and the generalized Jacobi polynomials considered by James, cf. sections 4. 4 and 4. 5.

Notation.   Throughout this paper c denotes a constant factor, which is
usually nonzero. The symbol $\dfrac{\partial^k}{\partial x_1 \partial x_2 \ldots \partial x_k}$ $x_1 x_2 \ldots x_k$ denotes the partial derivative

2. Jacobi polynomials
    The classical orthogonal polynomials in one variable are the Jacobi
polynomials, the Laguerre and the Hermite polynomials. Since Laguerre
and Hermite polynomials are limit cases of the Jacobi polynomials, we
shall restrict ourselves to Jacobi polynomials and their two-variable
analogues, i. e., to the case of a bounded orthogonality region.

Two-variable analogues of the Jacobi polynomials may be related
to Jacobi polynomials in several different ways:

(a) It may be possible to express them in terms of Jacobi polynomials.

(b) They may occur in certain formulas for Jacobi polynomials.

437

(c) Their properties will be analogous to the properties of Jacobi polynomials.

(d) The methods of proving these properties may be analogous to the proofs in the Jacobi case.

We shall briefly consider those properties of the Jacobi polynomials which we want to generalize. For standard results about Jacobi polynomials the reader is referred to Szegö [68, Chap. 4] and Erdélyi [17, Chap. 10].

2.1. Simple analytic properties

Let $\alpha, \beta > -1$. Jacobi polynomials $P_n^{(\alpha, \beta)}(x)$, $n = 0, 1, 2, \ldots$, are orthogonal polynomials with respect to the weight function $(1-x)^{\alpha}(1+x)^{\beta}$ on the interval $(-1, 1)$. They are normalized such that $P_n^{(\alpha, \beta)}(1) = (\alpha+1)_n/n!$. We shall often use the notation $R_n^{(\alpha, \beta)}(x) = P_n^{(\alpha, \beta)}(x)/P_n^{(\alpha, \beta)}(1)$

The pair of differential recurrence relations

$$(2.1) \qquad \frac{d}{dx} P_n^{(\alpha, \beta)}(x) = \frac{1}{2}(n+\alpha+\beta+1) P_{n-1}^{(\alpha+1, \beta+1)}(x) \ ,$$

$$(2.2) \qquad (1-x)^{-\alpha}(1+x)^{-\beta} \frac{d}{dx} [(1-x)^{\alpha+1}(1+x)^{\beta+1} P_{n-1}^{(\alpha+1, \beta+1)}(x)] = -2n P_n^{(\alpha, \beta)}(x)$$

can immediately be derived from the definition. There are three important corollaries of (2.1) and (2.2). Combination of (2.1) and (2.2) gives the second order differential equation

$$(2.3) \qquad [(1-x^2) \frac{d^2}{dx^2} + (\beta-\alpha-(\alpha+\beta+2)x)\frac{d}{dx}] P_n^{(\alpha, \beta)}(x) = -n(n+\alpha+\beta+1) P_n^{(\alpha, \beta)}(x) \ .$$

Iteration of (2.2) leads to the Rodrigues formula

$$(2.4) \qquad P_n^{(\alpha, \beta)}(x) = \frac{(-1)^n}{2^n n!} (1-x)^{-\alpha}(1+x)^{-\beta} \frac{d^n}{dx^n} [(1-x)^{\alpha+n}(1+x)^{\beta+n}] \ .$$

Finally, repeated application of (2.1) together with the value of $P_n^{(\alpha, \beta)}(1)$ gives the power series expansion

$$(2.5) \qquad R_n^{(\alpha, \beta)}(x) = \sum_{k=0}^{n} \frac{(-n)_k (n+\alpha+\beta+1)_k}{(\alpha+1)_k \; k!} \left(\frac{1-x}{2}\right)^k =$$

$$= {}_2F_1(-n, n+\alpha+\beta+1; \alpha+1; \tfrac{1}{2}(1-x)) .$$

Orthogonal polynomials $p_n(x)$ with respect to a weight function $w(x)$ are called classical if they satisfy one of the following three equivalent conditions:

(a) The polynomials $p_n(x)$ are eigenfunctions of a second order linear differential operator.

(b) The system of polynomials $dp_{n+1}(x)/dx$, $n = 0, 1, 2, \ldots$, is an orthogonal system.

(c) There is a polynomial $\rho(x)$ such that $p_n(x)$ is given by the Rodrigues formula

$$p_n(x) = c \cdot (w(x))^{-1} \frac{d^n}{dx^n} [(\rho(x))^n \, w(x)] .$$

For Jacobi polynomials these three properties are contained in formulas (2.3), (2.1) and (2.4), respectively.

2.2. Chebyshev polynomials

Consider the unit circle parametrized by the angle θ, let $\sigma = \pm 1$, and let the functions $f_n^\sigma(\theta)$, $n = 0, 1, 2, \ldots$, be the successive eigenfunctions of $d^2/d\theta^2$ which satisfy the symmetry relation

$$f_n^\sigma(-\theta) = \sigma \, f_n^\sigma(\theta) .$$

Then $f_n^1(\theta) = \cos n\theta$, $f_n^{-1}(\theta) = \sin(n+1)\theta$, and $f_n^\sigma(\theta)/f_0^\sigma(\theta)$ is a polynomial of degree n in $\cos \theta$. In this way we obtain the special Jacobi polynomials

$$(2.6) \qquad T_n(\cos\theta) = R_n^{(-\frac{1}{2}, -\frac{1}{2})}(\cos\theta) = \cos n\theta \ ,$$

$$(2.7) \qquad U_n(\cos\theta) = (n+1)\, R_n^{(\frac{1}{2}, \frac{1}{2})}(\cos\theta) = \frac{\sin(n+1)\theta}{\sin\theta} \ ,$$

which are called Chebyshev polynomials of the first and second kind, respectively.

In a similar way we can consider eigenfunctions $f_n^{\sigma,\,\tau}(\theta)$ $(\sigma, \tau = \pm 1,$ $n = 0, 1, 2, \ldots)$ of $d^2/d\theta^2$ which satisfy symmetry relations

$$f_n^{\sigma,\,\tau}(-\theta) = \sigma\, f_n^{\sigma,\,\tau}(\theta), \quad f_n^{\sigma,\,\tau}(\pi-\theta) = \tau\, f_n^{\sigma,\,\tau}(\theta)$$

with respect to the reflections in $\theta = 0$ and $\theta = \pi/2$. Then $f_n^{\sigma,\,\tau}(\theta)/$ $f_0^{\sigma,\,\tau}(\theta)$ is a polynomial in $\cos 2\theta$. Thus we obtain (2.6) and (2.7) with θ replaced by 2θ, and, furthermore, the special Jacobi polynomials

$$(2.8) \qquad (2n+1)\, R_n^{(\frac{1}{2}, -\frac{1}{2})}(\cos 2\theta) = \frac{\sin(2n+1)\theta}{\sin\theta} \ ,$$

$$(2.9) \qquad R_n^{(-\frac{1}{2}, \frac{1}{2})}(\cos 2\theta) = \frac{\cos(2n+1)\theta}{\cos\theta} \ .$$

This interpretation of Jacobi polynomials of order $(\pm\frac{1}{2}, \pm\frac{1}{2})$ is a motivation for the study of Jacobi polynomials of general order (α, β) . Many simple formulas for Jacobi polynomials of general order (α, β) (for instance the differential equation (2.3)) are "analytic continuations" of the cases $\alpha, \beta = \pm\frac{1}{2}$. So we may first derive these formulas in the Chebyshev cases, then predict the formulas in the general case, and finally give a formal proof.

2.3. <u>Jacobi polynomials as spherical functions and deeper analytic properties</u>

First we give some definitions and results about homogeneous spaces and spherical functions. For further details we refer to

Helgason [26, Chap. 10] and Coifman and Weiss [10, Chap. 1 and 2].

Let G be a compact group acting transitively on a compact Hausdorff space M, fix e ϵ M and let K be the subgroup of G which leaves e fixed. Then M = G/K is called a homogeneous space of G . Let a function f on M be called zonal if f is invariant with respect to K . It is possible to decompose $L^2(M)$ as the orthogonal direct sum of finite dimensional subspaces which are invariant and irreducible with respect to G . The functions belonging to these irreducible subspaces are continuous on M . Each G-invariant subspace of $L^2(M)$ of nonzero dimension contains zonal functions which are not identically zero. The class of zonal L^1-functions on M can be considered as a subalgebra of the convolution algebra of L^1-functions on G .

Theorem 2.1. The following four statements are equivalent:

(a) In each subspace of $L^2(M)$ which is irreducible with respect to G, the class of zonal functions has dimension 1 .

(b) The decomposition of $L^2(M)$ into irreducible subspaces with respect to G is unique.

(c) The representation of G in $L^2(M)$ contains each irreducible representation of G at most once.

(d) The convolution algebra of zonal L^1-functions on M is commutative. (Helgason 2 4.1 p. 408)

Definition 2.2. Let M = G/K have the equivalent properties of Theorem 2.1.

(a) A function f on M is called a harmonic if f belongs to an irreducible subspace of $L^2(M)$ with respect to G .

(b) A function f on M is called a spherical function if (i) f is a harmonic, (ii) f is zonal, (iii) f(e) = 1 . also called spherical harmonics

Unless otherwise stated, the homogeneous spaces considered in this paper will satisfy the equivalent properties of Theorem 2.1.

Sometimes it is possible to parametrize the set of K-orbits Kx in M such that the spherical functions expressed in terms of these parameters become well-known special functions. On the other hand, new special functions may be obtained in this way. If special functions can be interpreted as spherical functions then certain deeper analytic results immediately follow from the group theoretic interpretation. Among these results are an integral representation, a product formula, a positive convolution structure and, sometimes, an addition formula. A related, but more elementary result is the inequality

(2.10) $|f(x)| \le f(e) = 1, \quad x \in M$,

for spherical functions f on M .

For certain discrete values of (α, β) (see Figure 1) Jacobi polynomials $R_n^{(\alpha, \beta)}(x)$ can be interpreted as spherical functions on two-point homogeneous spaces or, equivalently, symmetric spaces of rank 1, cf. Helgason [27] and Gangolli [21]. The most elementary case is the unit circle $S^1 = O(2)/O(1)$ with spherical functions $\cos n\theta = R_n^{(-\frac{1}{2}, -\frac{1}{2})}(\cos \theta)$. Gegenbauer polynomials $R_n^{(q/2-3/2, q/2-3/2)}(x)$, $q = 3, 4, \ldots$, can be interpreted as spherical functions on the sphere $S^{q-1} = O(q)/O(q-1)$. The harmonics on this homogeneous space are the well-known spherical harmonics, cf. for instance Müller [59]. The other cases are the real projective space $SO(q)/O(q-1)$ $(\alpha = q/2-3/2, \beta = -\frac{1}{2})$, the complex projective space $SU(q)/U(q-1)$ $(\alpha = q-2, \beta = 0)$, the quaternionic projective space $Sp(q)/Sp(q-1) \times Sp(1)$ $(\alpha = 2q - 3, \beta = 1)$ and the Cayley projective plane $(\alpha = 7, \beta = 3)$. Jacobi polynomials of order $(\frac{1}{2}, \frac{1}{2})$ can also be interpreted as the characters on the group $SU(2)$.

Among the deeper analytic properties of Jacobi polynomials are the inequality

(2.11) $|R_n^{(\alpha, \beta)}(x)| \le 1, \quad |x| \le 1, \quad \alpha \ge \beta > -1, \quad \alpha \ge -\frac{1}{2}$,

(cf. Szegö [68, §7. 32]), the Laplace type integral representation, the

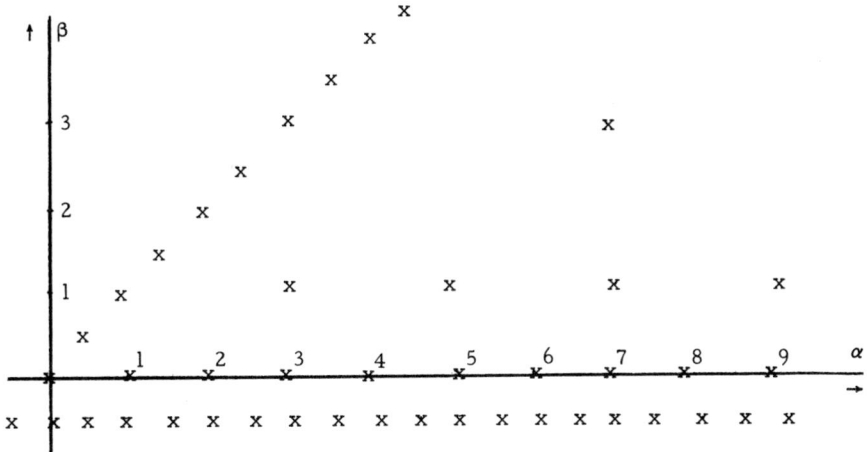

Figure 1

product formula and the addition formula for Jacobi polynomials (cf.
Koornwinder [42]), and a positive convolution structure for Jacobi series
(cf. Gasper [22], [23]). For the values of (α, β) given in Figure 1 these
results follow from the group theoretic interpretation. If it is known that
a certain result for Jacobi polynomials holds for special values of (α, β)
then it is easier to obtain such a result in the general case, either by
analytic manipulation of the known cases or by first predicting the general result as an "analytic continuation" of the known cases and next
giving some new analytic proof.

In section 2.2 we pointed out that the analytic structure of elementary formulas for Jacobi polynomials can already be predicted from the
Chebyshev cases. This is no longer true for the deeper results, since
these formulas may become degenerate if $\alpha = \beta$ or $\beta = \frac{1}{2}$. For instance,
the formula expressing the product $R_n^{(\alpha, \beta)}(x)\, R_n^{(\alpha, \beta)}(y)$ as an integral
of a Jacobi polynomial of the same degree and order, is a sum of two
terms if $\alpha = \beta = -\frac{1}{2}$, a single integral if $\alpha = \beta$ or $\beta = -\frac{1}{2}$ and a double
integral if $\alpha > \beta > -\frac{1}{2}$. Hence, $SU(3)/U(2)$ is the least complicated
homogeneous space on which the product formula is nondegenerate.

443

3. Methods of constructing two-variable analogues of the Jacobi polynomials

3.1. General orthogonal polynomials in two variables

For the few things which are known about general orthogonal polynomials in two variables the reader is referred to Jackson [32], Erdélyi [17, Chap. 12] and Bertran [8]. Some further results have been obtained in connection with numerical cubature problems, cf. Stroud [67].

Let Ω be an open subset of $\mathbb{R}^2$ and let $w(x, y)$ be a nonnegative L^1-function on Ω. For convenience we shall always suppose that the region Ω is bounded and that the function $w(x, y)$ is strictly positive and continuous on Ω. The set $\mathscr{V}_n$, $n = 0, 1, \ldots$, of orthogonal polynomials of degree n with respect to the weight function $w(x, y)$ on the region Ω is defined as the $(n+1)$-dimensional set of all polynomials $p(x, y)$ of degree n with complex coefficients such that

$$\iint\limits_{\Omega} p(x, y)\, q(x, y)\, w(x, y)\, dx\, dy = 0$$

if $q(x, y)$ is a polynomial of degree less than n.

For a special region Ω and a special weight function $w(x, y)$ it may be possible to find an explicit basis of polynomials $p_{n,k}(x, y)$, $k = 0, 1, \ldots, n$, for $\mathscr{V}_n$. In particular, it is of interest to find an orthogonal basis for $\mathscr{V}_n$. There are infinitely many possibilities to choose an orthogonal basis for $\mathscr{V}_n$. One evident method for constructing such an orthogonal basis is to define $p_{n,k}(x, y)$ as the polynomial with highest term $c \cdot x^{n-k} y^k$ obtained by applying the Gram-Schmidt orthogonalization process to the sequence of monomials

$$(3.1) \qquad 1, x, y, x^2, xy, y^2, x^3, x^2 y, \ldots, x^n, x^{n-1} y, \ldots, x^{n-k} y^k .$$

However, this is not a very canonical method, since the transformation $(x, y) \to (y, x)$ generally does not map the thus obtained orthogonal basis $\{p_{n,0}(x, y), p_{n,1}(x, y), \ldots, p_{n,n}(x, y)\}$ of $\mathscr{V}_n$ onto itself, even if Ω and $w(x, y)$ are invariant under this transformation.

Sometimes it is possible to preserve the symmetry in x and y by choosing a nonorthogonal basis $p_{n,k}(x,y)$, $k = 0, 1, \ldots, n$, for $\mathscr{V}_n$, and another nonorthogonal basis $q_{n,k}(x,y)$, $k = 0, 1, \ldots, n$, for $\mathscr{V}_n$, such that $p_{n,k}(x,y)$ is orthogonal to $q_{n,\ell}(x,y)$ if $k \neq \ell$. Then the systems $\{p_{n,k}(x,y)\}$ and $\{q_{n,k}(x,y)\}$ together are called a biorthogonal system (cf. Erdélyi [17, Chap. 12] and Schleusner [65]).

It is the author's opinion that if a general theory of sufficient depth is possible for orthogonal polynomials in two variables then it probably will be given for the classes $\mathscr{V}_n$ rather than for some basis of $\mathscr{V}_n$. However, for certain special regions and weight functions there are quite interesting properties for polynomials belonging to some explicit basis of $\mathscr{V}_n$.

Let us return to the polynomials $p_{n,k}(x,y)$ which are obtained by orthogonalizing the sequence of monomials (3.1). Let us arrange the pairs of integers (n, k), $n \geq k \geq 0$, by lexicographic ordering, i.e. $(m, \ell) <$ (n, k) if either $m < n$ or $m = n$ and $\ell < k$. Then $p_{n,k}(x,y)$ has a power series expansion

$$(3.2) \qquad p_{n,k}(x,y) = \sum_{(m,\ell) \leq (n,k)} c(m, \ell; n, k)\, x^{m-\ell}\, y^{\ell} ,$$

and there are recurrence relations

$$(3.3) \qquad x p_{n,k}(x,y) = \sum_{(m,\ell)=(n-1,k)}^{(n+1,k)} a(m, \ell; n, k)\, p_{m,\ell}(x,y)$$

and

$$(3.4) \qquad y\, p_{n,k}(x,y) = \sum_{(m,\ell)=(n-1,k-1)}^{(n+1,k+1)} b(m, \ell; n, k)\, p_{m,\ell}(x,y) .$$

Note that the number of terms in the recurrence relations becomes arbitrarily large if $n \to \infty$. This is in striking contrast with the one-variable case, where the recurrence relation contains at most three terms, independent of n.

445

Next consider the trivial example of a direct product of two systems of orthogonal polynomials in one variable. Let $\{p_n(x)\}$ and $\{q_n(y)\}$ be orthogonal systems with respect to the weight functions $w_1(x)$ and $w_2(y)$, respectively. Define

$$(3.5) \qquad p_{n,k}(x,y) = p_{n-k}(x)\, q_k(y) \ .$$

Then $p_{n,k}(x,y)$ can be obtained by orthogonalizing the sequence (3.1) with respect to the weight function $w_1(x)\, w_2(y)$. Now the formulas (3.2), (3.3) and (3.4) can be simplified, since $c(m,\ell;n,k)$ can only be non-zero if $m-\ell \le n-k$ and $\ell \le k$, $a(m,\ell;n,k)$ vanishes except for $(m,\ell) = (n+1,k)$, (n,k) or $(n-1,k)$, and $b(m,\ell;n,k)$ vanishes except for $(m,\ell) = (n+1,k+1)$, (n,k) or $(n-1,k-1)$.

We shall meet less trivial examples of orthogonal polynomials $p_{n,k}(x,y)$ for which certain coefficients in (3.1), (3.2) and (3.3) vanish, such that the number of nonvanishing coefficients in the recurrence relations remains bounded for $n \to \infty$. It is worthwhile to consider systems having these properties, even if the symmetry in x and y is destroyed.

In some examples which will be considered in this paper, there is a partial ordering $<$ of the pairs of integers (n,k), $n \ge k \ge 0$, such that $(m,\ell) < (n,k)$ implies that $(m,\ell) \le (n,k)$ and such that

$$(3.6) \qquad p_{n,k}(x,y) = \sum_{(m,\ell) < (n,k)} c(m,\ell;n,k)\, x^{m-\ell}\, y^{\ell} \ .$$

The following useful lemma gives some consequences of this property.

Lemma 3.1. Let $\phi_1, \phi_2, \phi_3, \ldots$ be a linearly independent sequence of polynomials in x and y and let p_n be a nonzero polynomial with "highest" term $c \cdot \phi_n$ which is obtained by orthogonalizing the sequence $\phi_1, \phi_2, \ldots, \phi_n$ with respect to a certain weight function on a certain region. Let $<$ be a partial ordering of the set $\mathbb{N}$ of natural numbers such

446

that $i < j$ implies $i \le j$ and such that for each $n \in \mathbb{N}$ p_n is a linear combination of polynomials ϕ_m, $m < n$. Then:

(a) For each n, ϕ_n is a linear combination of polynomials p_m, $m < n$.

(b) If q is a linear combination of polynomials ϕ_m, $m < n$, and if q is orthogonal to ϕ_m for $m < n$ and $m \ne n$, then $q = c \cdot p_n$.

(c) If not $m > n$ then p_n is orthogonal to ϕ_m.

The easy proof is left to the reader.

3.2. The definition of classical orthogonal polynomials in two variables

It seems natural to look for some two-variable analogue of the one-variable criterium for classical orthogonal polynomials that the polynomials must be eigenfunctions of a second order differential operator. Krall and Sheffer [53] and independently Engelis [16] considered the case that the classes $\mathcal{U}_n$, $n = 0, 1, 2, \ldots$, of orthogonal polynomials of degree n with respect to a certain weight function $w(x, y)$ on a certain region Ω are eigenspaces of a second order linear partial differential operator. They classified all partial differential operators having this property. The only bounded regions occurring in their classification are the unit disk with weight function $(1-x^2-y^2)^\alpha$ and the triangular region $\{(x, y) \mid 0 < y < < x < 1\}$ with weight function $(1-x)^\alpha (x-y)^\beta y^\gamma$.

In the present paper we shall consider examples of orthogonal systems $\{p_{n,k}(x, y)\}$, such that the polynomials $p_{n,k}(x, y)$ are the joint eigenfunctions of two commuting partial differential operators D_1 and D_2, where D_1 has order two and D_2 may have any arbitrary order, and where D_1 and D_2 are algebraically independent, i.e., if Q is a polynomial in two variables and if $Q(D_1, D_2)$ is the zero operator then Q is the zero polynomial. If the eigenvalue of $p_{n,k}(x, y)$ with respect to D_1 only depends on the degree n then we are back in the situation studied by Krall and Sheffer [53]. In that case the operator D_2 provides us a canonical method to choose an orthogonal basis for $\mathcal{U}_n$, i.e. by taking the eigenfunctions of D_2.

447

It does not yet seem to be the time to make a final decision, which systems should be called two-variable analogues of the classical orthogonal polynomials. Rather than trying to classify all orthogonal systems which are eigenfunctions of differential operators, we shall emphasize the methods by which such systems can be constructed. These methods, which are suggested by the results for Jacobi polynomials stated in section 2, are the following:

(a) Consider orthogonal polynomials in two variables which can be expressed in terms of Jacobi polynomials in some elementary way.

(b) Consider orthogonal polynomials in two variables which are analogous to Chebyshev polynomials, i. e., which can be expressed as elementary trigonometric functions in two variables or as spherical harmonics on the sphere S^2 satisfying symmetry relations with respect to certain reflections.

(c) Consider orthogonal polynomials in two variables which can be interpreted as spherical functions on homogeneous spaces of rank 2. Informally stated, a homogeneous space $M = G/K$ has rank r if G and K are Lie groups and the set of K-orbits on M is a manifold of dimension r (except possibly for a set of measure zero).

(d) Construct new orthogonal polynomials in two variables by performing quadratic transformations on known ones.

(e) Construct new orthogonal polynomials in two variables from known ones by doing "analytic continuation" with respect to some parameter.

3. 3. <u>Examples of two-variable analogues of the Jacobi polynomials</u>

Below we introduce seven different classes of orthogonal polynomials in two variables.

<u>Class I.</u> For $\alpha > -1$, $z = x + iy$, $\bar{z} = x - iy$, the polynomials

$$(3.7) \quad {}_1P^\alpha_{m,n}(z, \bar{z}) = \begin{cases} P^{(\alpha, m-n)}_n(2z\bar{z} - 1)\, z^{m-n} & \text{if} \quad m \geq n \ , \\ P^{(\alpha, n-m)}_m(2z\bar{z} - 1)\, \bar{z}^{n-m} & \text{if} \quad m < n \ , \end{cases}$$

448

are orthogonal with respect to the weight function $(1 - x^2 - y^2)^\alpha$ on the unit disk.

<u>Class II.</u> For $\alpha > -1$ the polynomials

$$(3.8) \qquad {}_2P^\alpha_{n,\,k}(x, y) = P^{(\alpha+k+\frac{1}{2},\,\alpha+k+\frac{1}{2})}_{n-k}(x)\ (1 - x^2)^{\frac{1}{2}k} \cdot$$

$$\cdot\ P^{(\alpha,\,\alpha)}_{k}((1-x^2)^{-\frac{1}{2}}y), \qquad n \geq k \geq 0 \ ,$$

are also orthogonal with respect to the weight function $(1-x^2-y^2)^\alpha$ on the unit disk.

<u>Class III.</u> For $\alpha, \beta > -1$ the polynomials

$$(3.9) \qquad {}_3P^{\alpha,\,\beta}_{n,\,k}(x, y) = P^{(\alpha,\,\beta+k+\frac{1}{2})}_{n-k}(2x-1)x^{\frac{1}{2}k}\ P^{(\beta,\,\beta)}_{k}(x^{-\frac{1}{2}}y), \qquad n \geq k \geq 0 \ ,$$

are orthogonal with respect to the weight function $(1 - x)^\alpha\ (x-y^2)^\beta$ on the region $\{(x, y)\,|\,y^2 < x < 1\}$, which is bounded by a straight line and a parabola.

<u>Class IV.</u> For $\alpha, \beta, \gamma > -1$ the polynomials

$$(3.10) \qquad {}_4P^{\alpha,\,\beta,\,\gamma}_{n,\,k}(x, y) = P^{(\alpha,\,\beta+\gamma+2k+1)}_{n-k}(2x-1)x^{k}$$

$$\cdot\ P^{(\beta,\,\gamma)}_{k}(2x^{-1}y-1), \qquad n \geq k \geq 0 \ ,$$

are orthogonal with respect to the weight function $(1-x)^\alpha(x-y)^\beta\ y^\gamma$ on the triangular region $\{(x, y)\,|\,0 < y < x < 1\}$.

<u>Class V.</u> For $\alpha, \beta, \gamma, \delta > -1$ the polynomials

$$(3.11) \qquad {}_5P^{\alpha,\,\beta,\,\gamma,\,\delta}_{n,\,k}(x, y) = P^{(\alpha,\,\beta)}_{n-k}(x)\ P^{(\gamma,\,\delta)}_{k}(y), \qquad n \geq k \geq 0 \ ,$$

are orthogonal with respect to the weight function $(1-x)^{\alpha}(1+x)^{\beta}(1-y)^{\gamma}(1+y)^{\delta}$. on the square $\{(x, y) \mid -1 < x < 1, \ -1 < y < 1\}$.

__Class VI.__ Let $\alpha, \beta, \gamma > -1$, $\alpha + \gamma + 3/2 > 0$, $\beta + \gamma + 3/2 > 0$. Let $_6P^{\alpha, \beta, \gamma}_{n, k}(u, v)$, $n \geq k \geq 0$, be a polynomial with highest term $c \cdot u^{n-k} v^k$ obtained by orthogonalization of the sequence $1, u, v, u^2, uv, \dots$ with respect to the weight function

$$(1-u+v)^{\alpha} (1+u+v)^{\beta} (u^2 - 4v)^{\gamma}$$

on the region $\{(u, v) \mid |u| < v + 1, \ u^2 - 4v > 0\}$, which is bounded by two straight lines and a parabola touching these lines. In particular we have

(3.12) $\quad _6P^{\alpha, \beta, -\frac{1}{2}}_{n, k}(x+y, xy) =$

$$= c \cdot [P^{(\alpha, \beta)}_n(x) P^{(\alpha, \beta)}_k(y) + P^{(\alpha, \beta)}_k(x) P^{(\alpha, \beta)}_n(y)]$$

and

(3.13) $\quad _6P^{\alpha, \beta, \frac{1}{2}}_{n, k}(x+y, xy) =$

$$= c \cdot (x-y)^{-1}[P^{(\alpha, \beta)}_{n+1}(x) P^{(\alpha, \beta)}_k(y) - P^{(\alpha, \beta)}_k(x) P^{(\alpha, \beta)}_{n+1}(y)] .$$

__Class VII.__ Let $\alpha > -5/6$, $z = x + iy$, $\bar{z} = x - iy$, $m, n \geq 0$. Let $_7P^{\alpha}_{m, n}(z, \bar{z}) = c \cdot z^m \bar{z}^n +$ polynomial in $z, \bar{z}$ of degree less than $m + n$, such that $_7P^{\alpha}_{m, n}(z, \bar{z})$ is orthogonal to all polynomials $q(z, \bar{z})$ of degree less than $m + n$ with respect to the weight function

$$[-(x^2 + y^2 + 9)^2 + 8(x^3 - 3xy^2) + 108]^{\alpha}$$

on the region bounded by the three-cusped deltoid (or Steiner's hypocycloid)

$$-(x^2 + y^2 + 9)^2 + 8(x^3 - 3xy^2) + 108 = 0 .$$

450

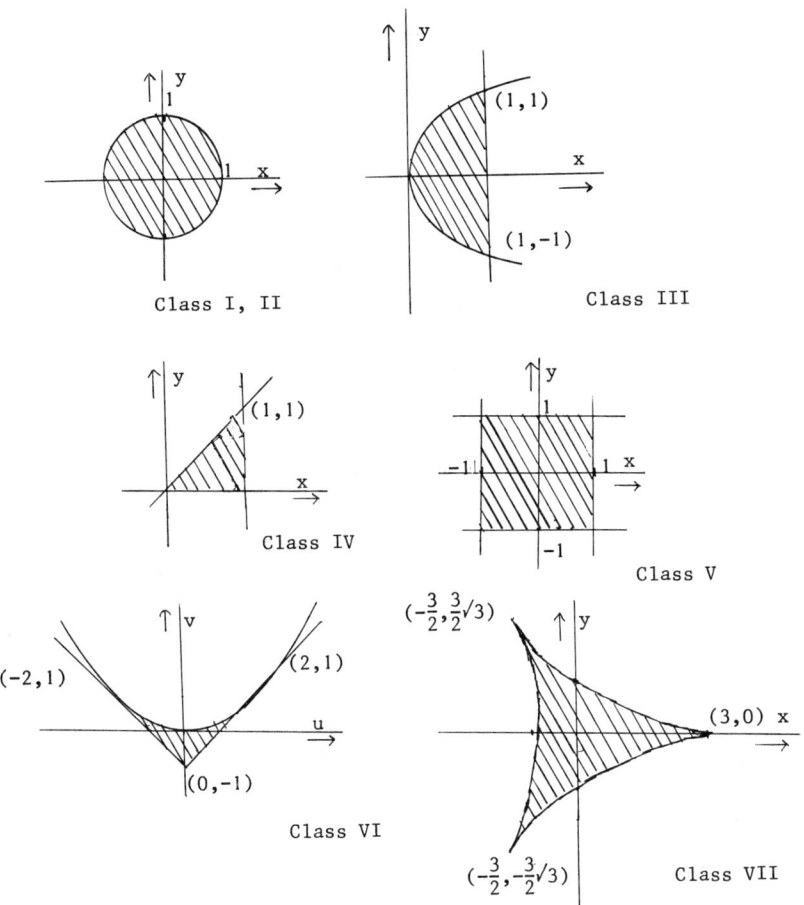

Figure 2

It can be proved that the polynomials $_7P^{\alpha}_{n-k,\,k}(z,\bar{z})$, $k = 0, 1, \ldots, n$, form an orthogonal basis for the class $\mathcal{V}_n$ with respect to this weight function and region. For $\alpha = \pm\frac{1}{2}$ we have an explicit expression. Let

$$(3.14) \qquad e^{\pm}_{m,\,n}(\sigma,\tau) = e^{i(m\sigma+n\tau)} \pm e^{i((m+n)\sigma-n\tau)} +$$

$$+ e^{i(-(m+n)\sigma+m\tau)} \pm e^{i(-n\sigma-m\tau)} +$$

$$+ e^{i(n\sigma-(m+n)\tau)} \pm e^{i(-m\sigma+(m+n)\tau)}$$

and

$$(3.15) \qquad z = e^{i\sigma} + e^{-i\tau} + e^{i(-\sigma+\tau)} \quad .$$

Then

$$(3.16) \qquad _7P^{-\frac{1}{2}}_{m,\,n}(z,\bar{z}) = c \cdot e^{+}_{m,\,n}(\sigma,\tau)$$

and

$$(3.17) \qquad _7P^{\frac{1}{2}}_{m,\,n}(z,\bar{z}) = c \cdot e^{-}_{m+1,\,n+1}(\sigma,\tau)/e^{-}_{1,\,1}(\sigma,\tau) \quad .$$

In Figure 2 we give pictures of the orthogonality regions for these seven classes.

At this stage it is useful to make a remark about the notation chosen in this paper. The author believes that the theory of special orthogonal polynomials in several variables is not yet developed far enough in order to make a final decision about the notation. The notation introduced above is therefore only intended for the present paper. We shall write $p_{n,k}$ instead of $P_{n,k}$ if the coefficient of the highest term of the orthogonal polynomial equals 1. The notation $R_{n,k}$ instead of $P_{n,k}$ will be used if the orthogonal polynomial equals 1 at a vertex of the orthogonality region. The left under index denoting the class will be deleted if no confusion is possible.

In the classes I and VII the polynomials have the form $c \cdot z^m \bar{z}^n +$ + a polynomial of degree less than $m + n$. This is an important difference with the polynomials of the other classes, which are obtained by orthogonalizing the sequence $1, x, y, x^2, xy, \ldots$. The polynomials of the classes I-V have a simple explicit expression in terms of Jacobi polynomials. However, in the classes VI and VII such elementary expressions only exist for certain special values of the parameters. It can therefore be expected that the study of these last two classes is much more difficult than the study of the other classes. On the other hand, the theory of the classes I-IV is by no means a trivial corollary of the theory of Jacobi polynomials. It is helpful to consider the classes VI and VII as more difficult analogues of the classes IV and I, respectively.

3.4. Some references and applications

Orthogonal polynomials on the disk of class I were introduced by Zernike and Brinkman [74]. Zernike [73] used the case $\alpha = 0$ for the study of diffraction problems. For further applications in optics see the references in Myrick [62], cf. also Marr [58]. If $\alpha = q - 2$, $q = 2, 3, 4,$ ..., then these polynomials are the spherical functions on the sphere S^{2q-1} considered as homogeneous space $U(q)/U(q-1)$, cf. Vilenkin and Šapiro [70], [64], Ikeda and Kayama [30], [31], Koornwinder [43], Boyd [9], Folland [19], [20], Dunkl [13], and Annabi [2]. From this group theoretic interpretation there follows an addition formula for the polynomials of class I (cf. Šapiro [64]), which implies the addition formulas for Jacobi polynomials and for Laguerre polynomials (cf. Koornwinder [44] and [50], respectively). A related result is a positive convolution structure for polynomials of class I, cf. Annabi and Trimèche [3], Trimèche [69] and Kanjin [37].

Orthogonal polynomials on the disk of class II were introduced by Didon [12] in the case $\alpha = 0$ and by Koschmieder [51] in the general case. A generating function is given by Koschmieder [52]. Using this class of polynomials Koornwinder [49] obtained a new proof of the addition formula for Gegenbauer polynomials. Hermite and Didon introduced a well-

known biorthogonal system of polynomials on the unit disk with respect
to the weight function $(1-x^2-y^2)^\alpha$, cf. Erdélyi [17, sections 12.5 and
12.6].

Orthogonal polynomials of class III are implicitly contained in
Agahanov [1]. The addition formula for Jacobi polynomials can be con-
sidered as an orthogonal expansion in terms of the polynomials of class
III, cf. Koornwinder [48, §3].

Orthogonal polynomials of class IV on a triangular region were introduced by Proriol [63]. They were applied to the problem of solving the
Schrödinger equation for the Helium atom, cf. Munschy and Pluvinage
[61], [60]. The same class was independently obtained by Karlin and
McGregor [39] in view of applications to genetics. Appell's polynomials
on the triangle (cf. Erdélyi [17, §12.4]) provide a nonorthogonal basis for
the class $\mathcal{V}_n$ with respect to the weight function and region considered
in class IV. In the case of order $(\alpha, \beta, 0)$ Appell extended this basis to
a biorthogonal system. Engelis [15] and independently Fackerell and
Littler [18] obtained a biorthogonal system in the case of general order.

The polynomials of class VI were introduced by Koornwinder [45].
The motivation was that for special values of the parameters and in terms
of suitable coordinates these polynomials are eigenfunctions of the
Laplace-Beltrami operator on certain compact symmetric spaces of rank 2,
cf. section 5. A further analysis was given by Sprinkhuizen [66], see
also section 4.3. For $\gamma = 0$ and $n = k$ these are the generalized
Jacobi polynomials of 2×2 matrix argument introduced by Herz [29],
cf. section 4.4. For the interpretation of polynomials of class VI as
spherical functions the reader is referred to section 5.

The author [46] also introduced the polynomials of class VII, motivated in a similar way as in the case of class VI. These polynomials
were independently considered by Eier and Lidl [14], [54] for $n = 0$. The
symmetric (or antisymmetric) polynomials generated by the monomial
$x_1^{n_1} x_2^{n_2} x_3^{n_3}$ can be expressed in terms of the elementary symmetric polynomials in three variables by using the polynomials of class VII of order

454

$\pm\frac{1}{2}$. Zonal polynomials of 3×3 matrix argument (cf. James [35]) can be expressed in terms of polynomials of class VII of order 0 . For all these results we refer to section 4.5. Section 5 contains references about the interpretation by means of spherical functions.

3.5. Two-variable analogues of the Chebyshev polynomials

The best way to bring more systematics in the examples of section 3.3 is to look at the Chebyshev cases, i.e. when the parameters are equal to $+\frac{1}{2}$ or $-\frac{1}{2}$. For these cases there exist interpretations analogous to the interpretation of Chebyshev polynomials given in section 2.2. It turns out that the classes I-IV are related to symmetries on the sphere S^2 and the classes V-VII to symmetries on certain two-dimensional tori. The results below are just observations which can be verified for our examples. However, these observations suggest the existence of a general theory.

On the sphere S^2 we use spherical coordinates θ $(0 \le \theta \le \pi)$ and ϕ (mod 2π) such that the mapping $(\theta, \phi) \rightarrow (\sin\theta\cos\phi, \sin\theta\sin\phi, \cos\theta)$ gives the natural embedding of S^2 as a subspace of $\mathbb{R}^3$, cf. Fig. 3.

A two-dimensional torus is obtained by choosing two linearly independent vectors e_1 and e_2 in $\mathbb{R}^2$ and by identifying points in $\mathbb{R}^2$ whose difference is $ke_1 + \ell e_2$, k, ℓ integers. All such tori are topologically equivalent with $S^1 \times S^1$, but they are different as Riemannian spaces. We shall consider two particular tori for which the group of all isometries is sufficiently rich for our purposes. The first one is the square torus denoted by T^2, where $e_1 = (2\pi, 0)$, $e_2 = (0, 2\pi)$ and each point in $\mathbb{R}^2$ has one and only one representative in the square region $\{(s,t) \mid -\pi < s \le \pi, \ -\pi < t \le \pi\}$, cf. Figure 4. The other one is the hexagonal torus denoted by H, where $e_1 = (\pi, \pi\sqrt{3})$, $e_2 = (\pi, -\pi\sqrt{3})$ and each point in $\mathbb{R}^2$ has one and only one representative in the hexagonal region

$$\{(s,t) \mid -\pi < s \le \pi, \ -\pi < \tfrac{1}{2}s + \tfrac{1}{2}\sqrt{3}\,t \le \pi, \ -\pi < \tfrac{1}{2}s - \tfrac{1}{2}\sqrt{3}\,t \le \pi\} ,$$

cf. Figure 5.

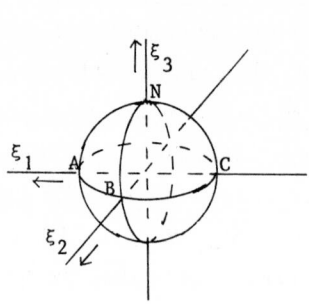

Figure 3

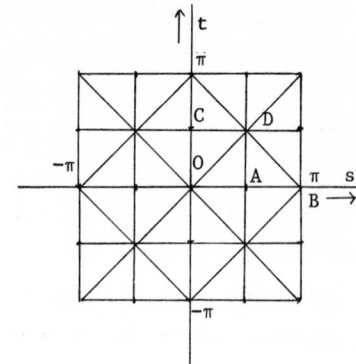

Figure 4

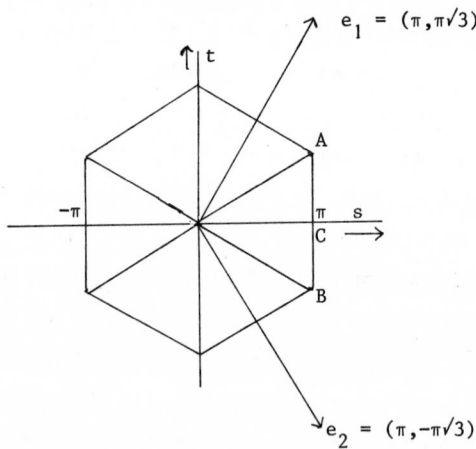

Figure 5

456

Corresponding with each of the seven classes introduced in sec-tion 3. 3 we shall choose a Riemannian manifold M (one of the spaces S^2, T^2 or H), a group G of isometries acting transitively on M (not necessarily the group of all isometries), a polygonal region R in M which is bounded by geodesics such that the reflections in these geodes-ics generate a discrete subgroup Γ of G having R as a fundamental region, and finally a C^∞-mapping F from M onto the orthogonality re-gion Ω of the polynomials such that $F(T\xi) = F(\xi)$ for all $T \in \Gamma$ and F restricted to R is a diffeomorphism. By using the mapping F any orthog-onal system of polynomials belonging to one of the seven classes can be considered as an orthogonal system of functions on R with respect to a certain measure $\rho(\xi) d\xi$, where $d\xi$ is the measure on M induced by the Riemannian metric ($d\xi = \sin\theta \, d\theta \, d\phi$ on S^2, $d\xi = ds \, dt$ on T^2 and H).

The choices of M, G, R and F corresponding to the different classes, and the resulting weight function ρ are listed in Table 1. The Table contains references to the Figures 3, 4 and 5.

After having made these choices the following results can be ob-tained by inspection.

(a) The manifold M as a homogeneous space of the group G satisfies the equivalent statements of Theorem 2.1. Hence the harmonics on M with respect to G are well-defined by Definition 2. 2.

(b) The two coordinates of the mapping F are harmonics which are invariant with respect to Γ .

(c) Let σ be a one-dimensional representation of the group Γ . Then $\sigma(T) = \pm 1$ for each $T \in \Gamma$. Call a function f on M of type σ if $f(T\xi) = \sigma(T) f(\xi)$ for each $T \in \Gamma$.

(d) If an orthogonal system of polynomials as given in Table 1 has parameters $\pm\frac{1}{2}$ then the corresponding weight function ρ on R is the square of a harmonic on M of certain type σ which is strictly positive on R . If all parameters are equal to $-\frac{1}{2}$ then ρ is the constant func-tion on R and if all parameters are equal to $+\frac{1}{2}$ then $\rho(\xi)$ is the square

Table 1

system of polynomials	manifold M	(generators of) G	region R	mapping F: M→Ω̄	weight function ρ with respect to invariant measure on M
$_1 P^\alpha_{m,n}(z,\bar z)$	S^2	O(3)	$0 < \theta < \pi/2$	$z = \sin\theta\, e^{i\phi}$, $\bar z = \sin\theta\, e^{-i\phi}$	$(\cos\theta)^{2\alpha+1}$
$_2 P^\alpha_{n,k}(x,y)$	S^2	O(3)	$0 < \phi < \pi$	$x = \cos\theta$, $y = \sin\theta\cos\phi$	$(\sin\theta\sin\phi)^{2\alpha+1}$
$_3 P^{\alpha,\beta}_{n,k}(x,y)$	S^2	O(3)	ACN	$x = \sin^2\theta$, $y = \sin\theta\cos\phi$	$(\cos\theta)^{2\alpha+1}(\sin\theta\sin\phi)^{2\beta+1}$
$_4 P^{\alpha,\beta,\gamma}_{n,k}(x,y)$	S^2	O(3)	ABN	$x = \sin^2\theta$, $y = \sin^2\theta\cos^2\phi$	$(\cos\theta)^{2\alpha+1}(\sin\theta\sin\phi)^{2\beta+1}$ $\cdot\,(\sin\theta\cos\phi)^{2\gamma+1}$
$_5 P^{\alpha,\beta,\gamma,\delta}_{n,k}(x,y)$	T^2	O(2) × O(2)	OADC	$x = \cos 2s$, $y = \cos 2t$	$(\sin s)^{2\alpha+1}(\cos s)^{2\beta+1}$ $\cdot\,(\sin t)^{2\gamma+1}(\cos t)^{2\delta+1}$
$_6 P^{\alpha,\alpha,\gamma}_{n,k}(u,v)$	T^2	O(2) × O(2), $(s,t)\to(t,s)$	OBE	$u = \cos s + \cos t$, $v = \cos s \cos t$	$(\sin s\sin t)^{2\alpha+1}$ $\cdot\,(\sin(\frac{s+t}{2})\,\sin(\frac{s-t}{2}))^{2\gamma+1}$.
$_6 P^{\alpha,\beta,\gamma}_{n,k}(u,v)$	T^2	O(2) × O(2), $(s,t)\to(t,s)$	OAD	$u = \cos 2s + \cos 2t$, $v = \cos 2s \cos 2t$	$(\sin s\sin t)^{2\alpha+1}(\cos s\cos t)^{2\beta+1}$ $\cdot\,(\sin(s+t)\sin(s-t))^{2\gamma+1}$.
$_7 P^\alpha_{m,n}(z,\bar z)$	H	SO(2) × SO(2), reflections in OA and OB	OAB	$z = e^{i(s+t/\sqrt3)}$ $+ e^{i(-s+t/\sqrt3)} +$ $+ e^{-2it/\sqrt3}$	$(\sin s)^{2\alpha+1}$ $\cdot\,(\sin(\frac12 s + \frac12\sqrt3\, t))^{2\alpha+1}$ $\cdot\,(\sin(\frac12 s - \frac12\sqrt3\, t))^{2\alpha+1}$

Table 1

458

of a harmonic which equals the Jacobian $\partial F(\xi)/\partial \xi$ and which changes sign with respect to any reflection in a boundary of R .

(e) There is a one-to-one correspondence between the one-dimensional representations σ of Γ and the Chebyshev cases of the class of orthogonal polynomials under consideration. In fact, for fixed values $\pm \frac{1}{2}$ of the parameters, the polynomials considered as functions of ξ can be written as a quotient of two harmonics of type σ, where the denominator equals $(\rho(\xi))^{\frac{1}{2}}$.

Let us illustrate these rather abstract statements by means of a few examples. The harmonics on $S^2 = O(3)/O(2)$ are the well-known spherical harmonics of degree $n = 0, 1, 2, \ldots$. If the class of spherical harmonics of degree n is decomposed into irreducible subclasses with respect so $O(2)$ then we obtain the functions

$$(3.18) \qquad P_{n-k}^{(k, k)}(\cos \theta) (\sin \theta)^k (A \cos k\phi + B \sin k\phi) ,$$

$k = 0, 1, 2, \ldots, n$, where A and B are constants. A further decomposition into irreducible subclasses with respect to $S\,O(2)$ gives the functions

$$(3.19) \qquad P_{n-k}^{(k, k)}(\cos \theta)(\sin \theta)^k e^{\pm ik\phi}, \qquad k = 0, 1, \ldots, n .$$

Examples of Chebyshev cases for the classes I-IV are, for instance:

$$(3.20) \qquad {}_1 P_{m, n}^{-\frac{1}{2}}(\sin \theta \, e^{i\phi}, \, \sin \theta \, e^{-i\phi}) =$$

$$= c \cdot P_{2n}^{(m-n, m-n)}(\cos \theta) (\sin \theta)^{m-n} e^{i(m-n)\phi} ,$$

$$m \geq n ,$$

$$(3.21) \qquad {}_1 P_{m, n}^{\frac{1}{2}}(\sin \theta \, e^{i\phi}, \, \sin \theta \, e^{-i\phi}) =$$

$$= \frac{P_{2n+1}^{(m-n, m-n)}(\cos \theta)(\sin \theta)^{m-n} e^{i(m-n)\phi}}{\cos \theta} ,$$

$$m \geq n ,$$

459

(3.22)
$$_2P_{n,k}^{\frac{1}{2}}(\cos\theta,\ \sin\theta\cos\phi)=$$

$$=c\cdot\frac{P_{n-k}^{(k+1,\,k+1)}(\cos\theta)(\sin\theta)^{k+1}\sin(k+1)\phi}{\sin\theta\ \sin\phi}\ ,$$

(3.23)
$$_3P_{n,k}^{\frac{1}{2},\,\frac{1}{2}}(\sin^2\theta,\ \sin\theta\cos\phi)=$$

$$=c\cdot\frac{P_{2n-2k+1}^{(k+1,\,k+1)}(\cos\theta)\,(\sin\theta)^{k+1}\sin(k+1)\phi}{\cos\theta\ \sin\theta\ \sin\phi}\ ,$$

(3.24)
$$_4P_{n,k}^{-\frac{1}{2},\,-\frac{1}{2},\,\frac{1}{2}}(\sin^2\theta,\ \sin^2\theta\cos^2\phi)=$$

$$=c\cdot\frac{P_{2n-2k}^{(2k+1,\,2k+1)}(\cos\theta)\,(\sin\theta)^{2k+1}\cos(2k+1)\phi}{\sin\theta\ \cos\phi}\ .$$

The harmonics on the square torus T^2 with respect to the group generated by $O(2)\times O(2)$ and the reflection $(s,t)\to(t,s)$ are the linear combinations of $\begin{Bmatrix}\cos ns\\\sin ns\end{Bmatrix}\cdot\begin{Bmatrix}\cos kt\\\sin kt\end{Bmatrix}$ and $\begin{Bmatrix}\cos ks\\\sin ks\end{Bmatrix}\cdot\begin{Bmatrix}\cos nt\\\sin nt\end{Bmatrix}$, where (n,k), $n\ge k\ge 0$, is fixed. We have, for instance

(3.25)
$$_6P_{n,k}^{-\frac{1}{2},\,-\frac{1}{2},\,\frac{1}{2}}(\cos s+\cos t,\ \cos s\cos t)=$$

$$=c\cdot\frac{\cos(n+1)s\ \cos kt-\cos ks\ \cos(n+1)t}{\cos t-\cos s}\ ,$$

(3.26)
$$_6P_{n,k}^{\frac{1}{2},\,-\frac{1}{2},\,\frac{1}{2}}(\cos 2s+\cos 2t,\ \cos 2s\cos 2t)=$$

$$=c\cdot\frac{\sin(2n+3)s\ \sin(2k+1)t-\sin(2k+1)s\ \sin(2n+3)t}{\sin 3s\ \sin t-\sin s\ \sin 3t}\ .$$

Finally let us consider the hexagonal torus H as a homogeneous space of the group generated by $SO(2)\times SO(2)$ and the reflections in OA and OB. In terms of new coordinates

$$\sigma = s + t/\sqrt{3}, \quad \tau = s - t/\sqrt{3}$$

the harmonics on H are the linear combinations of

$$e^{i(m\sigma + n\tau)}, \quad e^{i((m+n)\sigma - n\tau)}, \quad e^{i(-(m+n)\sigma + m\tau)},$$

$$e^{i(-n\sigma - m\tau)}, \quad e^{i(n\sigma - (m+n)\tau)} \text{ and } e^{i(-m\sigma + (m+n)\tau)},$$

where m and n are integers ≥ 0 . Now the formulas (3.14), (3.15), (3.16) and (3.17) give an interpretation of the Chebyshev cases of class VII.

Observe that the groups Γ are abelian in the cases I-V but non-abelian in the cases VI and VII. If the classes of harmonics on M are decomposed into irreducible subclasses with respect to nonabelian Γ then some of these subclasses do not correspond with one-dimensional representations σ of Γ . The harmonics corresponding with higher dimensional irreducible representations of Γ do not fit in our picture. It is not clear how to express such harmonics as orthogonal polynomials.

In the cases V-VII each one-dimensional representation σ of Γ occurs at most once in each class of harmonics on M . This is no longer true for the cases I-IV. There we first have to split up the spherical harmonics with respect to O(2) or SO(2) in order to get uniqueness with respect to the one-dimensional representations σ of Γ .

The reader may observe some missing cases in Table 1. If M is the hexagonal torus H, if G is the group of all isometries of H (i.e. the reflection in OC included) and if R is the region OCA (cf. Figure 5), then probably the Chebyshev cases of another interesting class of orthogonal polynomials can be obtained. On the sphere S^2 we may take for the region R a spherical triangle with angles $\{\pi/2, \pi/3, \pi/3\}$ or $\{\pi/2, \pi/3, \pi/4\}$ or $\{\pi/2, \pi/3, \pi/5\}$. It is not clear at all whether the approach of this subsection applied to these regions will give new classes of orthogonal polynomials.

3.6. Partial differential operators for which the orthogonal polynomials are eigenfunctions

It follows easily from the results of section 3.5 that for the Chebyshev cases of the classes I-VII the polynomials are eigenfunctions of certain partial differential operators. In this subsection it will be discussed how these differential operators can be generalized for other values of the parameters.

The harmonics on the sphere S^2 are eigenfunctions of the Laplace-Beltrami operator

$$(3.27) \qquad \partial_{\theta\theta} + \cot\theta \; \partial_\theta + (\sin\theta)^{-2} \partial_{\phi\phi} \; .$$

The harmonics on S^2 belonging to irreducible subclasses with respect to $SO(2)$ are also eigenfunctions of ∂_ϕ (cf. (3.19)), and the harmonics belonging to irreducible subclasses with respect to $O(2)$ are also eigenfunctions of $\partial_{\phi\phi}$ (cf. (3.18)).

The harmonics on the tori T^2 and H are eigenfunctions of the Laplace operator $\partial_{ss} + \partial_{tt}$. In case V the harmonics on T^2 with respect to the symmetry group $O(2) \times O(2)$ are also eigenfunctions of ∂_{ss}. In case VI the harmonics on T^2 with respect to the group of all isometries of T^2 are also eigenfunctions of ∂_{sstt}. In case VII the harmonics on H with respect to the group generated by $SO(2) \times SO(2)$ and the reflections in OA and OB are also eigenfunctions of the operator

$$(3.28) \qquad \partial_t(\sqrt{3}\; \partial_s + \partial_t)(-\sqrt{3}\; \partial_s + \partial_t) \; .$$

It follows that in each case of Table 1 the harmonics are eigenfunctions of two algebraically independent partial differential operators, one of which is the Laplace-Beltrami operator on M.

Fix one of the classes I-VII and let D be a partial differential operator for which the harmonics are eigenfunctions. Fix the parameters of the orthogonal system of polynomials equal to $\pm\frac{1}{2}$. By using result

(e) of section 3.5 it follows that each polynomial of the system considered as a function of ξ is an eigenfunction of the operator

$$(\rho(\xi))^{-\frac{1}{2}} D \circ (\rho(\xi))^{\frac{1}{2}} .$$

Since $(\rho(\xi))^{\frac{1}{2}}$ is a harmonic (cf. result (d) in section 3.5), the polynomials of the system are also eigenfunctions of the operator

(3.29) $(\rho(\xi))^{-\frac{1}{2}} D \circ (\rho(\xi))^{\frac{1}{2}} - (\rho(\xi))^{-\frac{1}{2}} (D(\rho(\xi))^{\frac{1}{2}}) ,$

which is a partial differential operator without zero order part.

Let us next consider an orthogonal system of polynomials of arbitrary order belonging to one of the classes I-VII and let ρ be the corresponding weight function on the region R as given in Table 1. Then, if the partial differential operator D has order 1 or 2, the polynomials of the system considered as functions of ξ are still eigenfunctions of the operator (3.29). This can be proved by first observing that the operator (3.29) is selfadjoint on the region R with respect to the measure $\rho(\xi)d\xi$. Next, for each individual class one has to rewrite the operator (3.29) in terms of coordinates x, y (or $z, \bar{z}$) and one has to verify that this operator maps $x^{n-k} y^k$ to a polynomial with "highest" term $\lambda_{n,k} x^{n-k} y^k$ (or $z^m \bar{z}^n$ to a polynomial with highest term $\lambda_{m,n} z^m \bar{z}^n$). Then $\lambda_{n,k}$ is the eigenvalue of the eigenfunction $P_{n,k}$. For the classes VI and VII (with $D = \partial_{ss} + \partial_{tt}$) the detailed proof has been given in Koornwinder [45, §4] and [46, §5], respectively. Note that the operator (3.29) is the unique analytic continuation of the Chebyshev cases such that it depends linearly on the parameters.

In the cases VI and VII there are partial differential operators of order 4 and 3, respectively, for which the harmonics are eigenfunctions. In these cases it is much more difficult to find the generalization of (3.29) for arbitrary values of the parameters. It can be proved that $_6P_{n,k}^{\alpha,\beta,\gamma}$ ($\cos 2s + \cos 2t, \cos 2s \cos 2t$) is an eigenfunction of the fourth order operator

(3. 30)
$$\rho^{-1}(\partial_s \circ \rho_2 \partial_t + \partial_t \circ \rho_2 \partial_s) \circ$$

$$\circ \rho_1 \rho_2^{-1}(\partial_s \circ \rho_2 \partial_t + \partial_t \circ \rho_2 \partial_s) \ ,$$

where $\rho_1(s, t) = (\sin s \ \sin t)^{2\alpha+1}(\cos s \ \cos t)^{2\beta+1}$, $\rho_2(s, t) = (\sin(s+t) \cdot$
$\cdot \ \sin(s-t))^{2\gamma+1}$ and $\rho = \rho_1 \rho_2$, cf. Koornwinder [45, (5.15)]. It can also
be proved that $_7 P_{m, n}^\alpha (F(s, t))$ is an eigenfunction of the third order operator

(3. 31) $\quad X_1 X_2 X_3 + (\alpha + \tfrac{1}{2}) \ \rho^{-1}[(X_1 \rho)X_2 X_3 + (X_2 \rho)X_3 X_1 + (X_3 \rho)X_1 X_2] \ +$

$$+ \ (\alpha+\tfrac{1}{2})^2 \ \rho^{-1}[(X_1 X_2 \ \rho)X_3 + (X_2 X_3 \ \rho)X_1 + (X_3 X_1 \ \rho)X_2] \ ,$$

where $F(s, t)$ and $\rho(s, t)$ are given in Table 1 and

$$X_1 = i(\tfrac{3}{2} \ \partial_s - \tfrac{1}{2}\sqrt{3} \ \partial_t), \ X_2 = -i(\tfrac{3}{2} \ \partial_s + \tfrac{1}{2}\sqrt{3} \ \partial_t) \ ,$$

$$X_3 = i\sqrt{3} \ \partial_t \ ,$$

cf. Koornwinder [46, §6].

Summarizing our results we have for each of the classes I-VII and
for each choice of the parameters two partial differential operators (say
D_1 and D_2) for which the polynomials of the system are eigenfunctions.
The operator D_1 has order 2 in all cases and the operator D_2 has
order 1, 2, 2, 2, 2, 4 or 3, respectively. It can be proved that the operators D_1 and D_2 commute, that they are algebraically independent and
that they generate the algebra of all differential operators which have the
polynomials of the system as eigenfunctions (cf. Koornwinder [45, §6]
and [46, §7] for the classes VI and VII, respectively). It can also be
proved that two distinct polynomials belonging to the same orthogonal
system cannot have the same pair of eigenvalues with respect to D_1 and
D_2.

3.7. General methods of constructing orthogonal polynomials in two variables from orthogonal polynomials in one variable

There are known some rather general classes of orthogonal polynomials in two variables which can be expressed in terms of orthogonal polynomials in one variable in some elementary way. Many of the special classes considered in section 3.3 are included in these general classes. We shall discuss three general classes. For each of these classes certain coefficients in the expansions (3.2), (3.3) and (3.4) will vanish.

3.7.1. Rotation invariant weight functions

Let $w(x)$ be a weight function on the interval $(0,1)$. For each integer $k \geq 0$ consider polynomials $p_n^k(x)$ of degree n which are orthogonal with respect to the weight function $x^k w(x)$ on the interval $(0,1)$. Define for integers $m, n \geq 0$ and for $z = x + iy$, $\bar{z} = x - iy$, $x, y \in \mathbb{R}$,

(3.32)
$$p_{m,n}(z, \bar{z}) = \begin{cases} p_n^{m-n}(z\bar{z})\, z^{m-n} & \text{if } m \geq n \ , \\[2mm] p_m^{n-m}(z\bar{z})\, \bar{z}^{n-m} & \text{if } m < n \ . \end{cases}$$

Then

(3.33)
$$\iint_{x^2+y^2<1} p_{m,n}(x+iy, x-iy) \ \overline{p_{k,\ell}(x+iy, x-iy)} \cdot$$

$$\cdot \ w(x^2+y^2)\,dx\,dy = 0 \quad \text{if } (m,n) \neq (k,\ell) \ ,$$

and $p_{m,n}(z, \bar{z}) - c \cdot z^m \bar{z}^n$ is a polynomial of degree less than $m + n$. Hence, the polynomials $p_{n-k,k}(z, \bar{z})$, $k = 0, 1, \ldots, n$, form an orthogonal basis of the class $\mathscr{V}_n$ with respect to the rotation invariant weight function $w(x^2 + y^2)$ on the unit disk. This method of constructing orthogonal polynomials in two variables is due to Maldonado [57]. Clearly, the polynomials of class I (cf. section 3.3) have the form (3.32). Since

(3.34)
$$p_{m,n}(z, \bar{z}) = p_{n,m}(\bar{z}, z) \ ,$$

(3.35)
$$p_{m,n}(e^{i\phi}z, \overline{e^{i\phi}z}) = e^{i(m-n)\phi} p_{m,n}(z, \bar{z}) \ ,$$

the orthogonal basis $\{p_{n,0}, p_{n-1,1}, \ldots, p_{0,n}\}$ of $\mathcal{N}_n$ is essentially invariant with respect to orthogonal transformations of the (x, y)-plane.

The power series expansion of $p_{m,n}(z, \bar{z})$ only contains terms $c \cdot z^{m-j} \bar{z}^{n-j}$, $j = 0, 1, \ldots, \min(m, n)$. The recurrence relations express $z\, p_{m,n}(z, \bar{z})$ as a linear combination of $p_{m+1,n}$ and $p_{m,n-1}$, and $\bar{z}\, p_{m,n}(z, \bar{z})$ as a linear combination of $p_{m,n+1}$ and $p_{m-1,n}$.

3.7.2. <u>Weight functions of the form $w_1(x)\, w_2((\rho(x))^{-1}y)$</u>

Let $w_1(x)$ be a weight function on the interval (a, b). Let $\rho(x)$ be a positive function on (a, b) which is either a polynomial of degree r $(r = 0, 1, 2, \ldots)$ or the square root of a polynomial of degree $2r$ $(r = \frac{1}{2}, 1, \frac{3}{2}, \ldots)$. For each integer $k \geq 0$ let the polynomials $p_n^k(x)$, $n = 0, 1, 2, \ldots$, be orthogonal with respect to the weight function $(\rho(x))^{2k+1} w_1(x)$ on (a, b). Let $w_2(y)$ be a weight function on the interval (c, d). If $\rho(x)$ is not a polynomial then suppose that $c = -d$ and that $w_2(y)$ is an even function on $(-d, d)$. Let the polynomials $q_n(y)$ be orthogonal with respect to the weight function $w_2(y)$ on (c, d). Define

(3.36)
$$P_{n,k}(x, y) = p_{n-k}^k(x)\, (\rho(x))^k\, q_k(\tfrac{y}{\rho(x)}), \quad n \geq k \geq 0 \ .$$

Then $P_{n,k}(x, y)$ is a linear combination of monomials $x^{m-\ell} y^\ell$ such that $\ell \leq k$ and $m + (r-1)\ell \leq n + (r-1)k$, and for $(n, k) \neq (m, \ell)$ we have

(3.37)
$$\int_{x=a}^{b} \int_{y=c\rho(x)}^{d\rho(x)} P_{n,k}(x, y)\, P_{m,\ell}(x, y) \ .$$

$$\cdot \ w_1(x)\, w_2((\rho(x))^{-1}y)\, dx\, dy = 0 \ .$$

If $\rho(x) \equiv 1$ then (3.36) reduces to (3.5). If $r = 0$, $\frac{1}{2}$ or 1 then the polynomial $p_{n,k}(x,y)$ has degree n and it can be obtained by orthogonalizing the sequence $1, x, y, x^2, xy, \ldots$, but for $r > 1$ this is no longer true.

In the special cases that $\rho(x) = (1-x^2)^{\frac{1}{2}}$, $(a,b) = (-1,1)$, $(c,d) = (-1,1)$, $r = 1$, or $\rho(x) = x^{\frac{1}{2}}$, $(a,b) = (0,1)$, $(c,d) = (-1,1)$, $r = \frac{1}{2}$, or $\rho(x) = x$, $(a,b) = (0,1)$, $(c,d) = (0,1)$, $r = 1$, this method of constructing orthogonal polynomials in two variables has been described by Agahanov [1]. Further specialization of the weight functions leads to the classes II, III and IV, respectively (cf. section 3.3).

Let us make some further remarks about the case $r = 1$. Introduce a partial ordering $\prec$ such that $(m, \ell) \prec (n, k)$ if $m \leq n$ and $\ell \leq k$. For this partial ordering and for $p_{n,k}(x,y)$ defined by (3.36) with $r = 1$, the power series expansion (3.6) is valid, cf. Figure 6. By using (for instance) Lemma 3.1 it can be proved that $x\, p_{n,k}(x,y)$ is a linear combination of the three polynomials $p_{n-1,k}$, $p_{n,k}$, $p_{n+1,k}$ (cf. Figure 7) and that $y\, p_{n,k}(x,y)$ is a linear combination of the nine polynomials $p_{m,\ell}$ such that $(n-1, k-1) \prec (m, \ell) \prec (n+1, k+1)$ (cf. Figure 8).

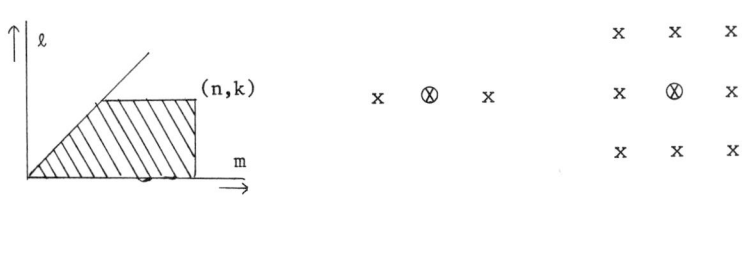

Figure 6   Figure 7   Figure 8

3.7.3. Symmetric and antisymmetric products

Let the polynomials $p_n(x)$ be orthogonal with respect to the weight function $w(x)$ on some interval. Then both the system of symmetric polynomials

467

(3. 38) $\qquad p_n(x) \, p_k(y) + p_k(x) \, p_n(y), \qquad n \geq k \geq 0 \; ,$

and the system of antisymmetric polynomials

(3. 39) $\qquad p_{n+1}(x) \, p_k(y) - p_k(x) \, p_{n+1}(y), \qquad n \geq k \geq 0 \; ,$

are orthogonal systems with respect to the weight function $w(x) \, w(y)$, $y < x$. The system (3. 39) is a special case of the systems considered by Karlin and McGregor [38], [40].

Let $u = x + y$, $v = xy$ and define

(3. 40) $\qquad P_{n,k}^{-\frac{1}{2}}(u, v) = p_n(x) \, p_k(y) + p_k(x) \, p_n(y), \quad n \geq k \geq 0 \; ,$

(3. 41) $\qquad P_{n,k}^{\frac{1}{2}}(u, v) = \dfrac{p_{n+1}(x) \, p_k(y) - p_k(x) \, p_{n+1}(y)}{x - y} \; , \quad n \geq k \geq 0 \; ,$

(3. 42) $\qquad W(u, v) = w(x) \, w(y), \quad x > y \; .$

Note that $P_{n,n}^{\frac{1}{2}}(x + y, \, xy)$ denotes the Christoffel-Darboux kernel (up to a constant factor) for the orthogonal polynomials $p_k(x)$, $k = 0, 1, \ldots, n$.

Since any symmetric polynomial in x and y is a polynomial in the elementary symmetric polynomials $u = x + y$, $v = xy$, it follows that the systems $\{P_{n,k}^{-\frac{1}{2}}(u, v)\}$ and $\{P_{n,k}^{\frac{1}{2}}(u, v)\}$ are orthogonal systems with respect to the weight function $(u^2 - 4v)^{-\frac{1}{2}} \, W(u, v)$ and $(u^2 - 4v)^{\frac{1}{2}} \, W(u, v)$, respectively.

Note that $P_{n,k}^{-\frac{1}{2}}(x+y, \, xy)$ is a linear combination of terms $(x^m y^\ell + x^\ell y^m)$, $m \geq \ell \geq 0$, $m \leq n$, $\ell \leq k$, and that $P_{n,k}^{\frac{1}{2}}(x+y, \, xy)$ is a linear combination of terms $(x-y)^{-1} (x^{m+1} y^\ell - x^\ell y^{m+1})$, $m \geq \ell \geq 0$, $m \leq n$, $\ell \leq k$. Let us define the polynomials $Z_{n,k}^{\pm\frac{1}{2}}(u, v)$, $n \geq k \geq 0$, by

(3. 43) $\qquad Z_{n,k}^{-\frac{1}{2}}(x+y, \, xy) = \dfrac{x^n y^k + x^k y^n}{1 + \delta_{n,k}} \; ,$

$$(3.44) \qquad Z^{\frac{1}{2}}_{n,\,k}(x+y,\,xy) = \frac{x^{n+1}\,y^k - x^k\,y^{n+1}}{x-y} \quad .$$

Since $\frac{1}{2}(t^n + t^{-n}) = T_n(\frac{1}{2}(t+t^{-1}))$ and $\dfrac{t^{n+1} - t^{-n-1}}{t - t^{-1}} = U_n(\frac{1}{2}(t+t^{-1}))$, cf. (2.6) and (2.7), it follows that

$$(3.45) \qquad Z^{-\frac{1}{2}}_{n,\,k}(u,\,v) = \frac{2}{1+\delta_{n,\,k}}\; v^{\frac{1}{2}(n+k)}\; T_{n-k}(\tfrac{1}{2}v^{-\frac{1}{2}}u)$$

and

$$(3.46) \qquad Z^{\frac{1}{2}}_{n,\,k}(u,\,v) = v^{\frac{1}{2}(n+k)}\; U_{n-k}(\tfrac{1}{2}\,v^{-\frac{1}{2}}u) \quad .$$

Hence, $Z^{-\frac{1}{2}}_{n,\,k}(u,\,v)$ and $Z^{\frac{1}{2}}_{n,\,k}(u,\,v)$ are both linear combinations of monomials $u^{n-k-2i}v^{k+i}$, $i = 0,1,\ldots,[\frac{1}{2}(n-k)]$, cf. Figure 9, and $P^{-\frac{1}{2}}_{n,\,k}(u,\,v)$ and $P^{\frac{1}{2}}_{n,\,k}(u,\,v)$ have expansions of the form

$$(3.47) \qquad P^{\pm\frac{1}{2}}_{n,\,k}(u,\,v) = \sum_{\ell=0}^{k}\; \sum_{m=\ell}^{n}\; c^{\pm\frac{1}{2}}_{m,\,\ell\,;n,\,k}\; Z^{\pm\frac{1}{2}}_{m,\,\ell}(u,\,v) \quad ,$$

cf. Figure 10. It follows that both $P^{-\frac{1}{2}}_{n,\,k}(u,\,v)$ and $P^{\frac{1}{2}}_{n,\,k}(u,\,v)$ are linear combinations of monomials $u^{m-\ell}v^{\ell}$, $m \leq n$, $m + \ell \leq n+k$, cf. Figure 11.

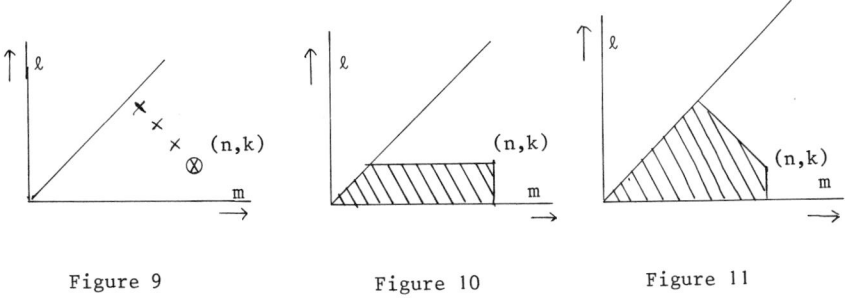

Figure 9 Figure 10 Figure 11

The above results imply that the polynomials $P^{-\frac{1}{2}}_{n,\,k}(u,\,v)$ and $P^{\frac{1}{2}}_{n,\,k}(u,\,v)$ can be obtained by orthogonalizing the sequence $1,\,u,\,v,\,u^2,\,uv,\,\ldots$.

For special choices of the weight function $w(x)$ we obtain the polynomials of class VI of order $(\alpha, \beta, \pm\frac{1}{2})$, cf. section 3.3.

Regarding the recurrence relations it can easily be proved that $u\,P_{n,k}(u, v)$ is a linear combination of the five polynomials $P_{n+1,k}$, $P_{n,k+1}$, $P_{n,k}$, $P_{n,k-1}$, $P_{n-1,k}$, cf. Figure 12, and that $v\,P_{n,k}(u, v)$ is a linear combination of the nine polynomials $P_{m,\ell}$ such $n-1 \leq m \leq n+1$, $k-1 \leq \ell \leq k+1$, cf. Figure 13.

$$
\begin{array}{ccc}
 & x & \\
x & \otimes & x \\
 & x &
\end{array}
\qquad\qquad
\begin{array}{ccc}
x & x & x \\
x & \otimes & x \\
x & x & x
\end{array}
$$

Figure 12   Figure 13

4. Differential recurrence relations and other analytic properties

In this section we shall discuss some analytic properties of the classes I, IV, VI and VII. An important tool for the analysis is certain differential recurrence relations analogous to (2.1) and (2.2), which have been obtained for most of the classes I-VII. Instead of one pair of ordinary differential recurrence relations we now have two pairs of partial differential recurrence relations such that all values of (n, k) (or of (m, n) in the case of Class I) are connected with each other by the recurrence relations. The partial differential operators may have order more than one and they may depend on the parameters.

By using the differential recurrence relations we obtain properties analogous to the properties characterizing classical orthogonal polynomials in one variable (cf. section 2.1):

(a) There exist partial differential operators which map the orthogonal system onto another orthogonal system of polynomials in two variables.

(b) Iteration of differential recurrence relations gives a Rodrigues type formula.

(c) Composition of two differential operators belonging to the same pair of differential recurrence relations leads to partial differential equations as considered in section 3.6.

470

On applying the differential recurrence relations which raise the degree, it is possible to prove certain properties of the orthogonal polynomials by complete induction with respect to the degree, cf. Sprinkhuizen [66].

In subsections 4.1, 4.2 and 4.3 we shall give the differential recurrence relations for the classes I, IV and VI, respectively. The results for class III are quite analogous to those of class IV. For class II probably no satisfactory results exist. In the case of class VII the research has not yet been done. Finally, in subsections 4.4 and 4.5 we shall discuss the relation of the polynomials of class VI and VII with the polynomials discussed by James [35].

4.1. Polynomials on the disk

For the polynomials $P^\alpha_{m,n}(z,\bar{z})$ of class I there is a pair of first order differential recurrence relations

(4.1)
$$\partial_z P^\alpha_{m,n}(z,\bar{z}) = c \cdot P^{\alpha+1}_{m-1,n}(z,\bar{z}) \ ,$$

(4.2)
$$(1-z\bar{z})^{-\alpha} \partial_{\bar{z}} [(1-z\bar{z})^{\alpha+1} P^{\alpha+1}_{m-1,n}(z,\bar{z})] =$$
$$= c \cdot P^\alpha_{m,n}(z,\bar{z}) \ ,$$

and a similar pair connecting $P^\alpha_{m,n}$ and $P^{\alpha+1}_{m,n-1}$, which can be obtained from (4.1) and (4.2) by complex conjugation. Iteration of (4.2) and its complex conjugate gives the Rodrigues type formula

(4.3)
$$P^\alpha_{m,n}(z,\bar{z}) = c \cdot (1-z\bar{z})^{-\alpha} (\partial_z)^n (\partial_{\bar{z}})^m [(1-z\bar{z})^{\alpha+m+n}] \ .$$

It is of interest to compare (4.3) with the Rodrigues type formula

$$U^{2\alpha+1}_{m,n}(x,y) = c \cdot (1-x^2-y^2)^{-\alpha} (\partial_x)^m (\partial_y)^n [(1-x^2-y^2)^{\alpha+m+n}]$$

(cf. Erdélyi [17, 12.6(11)]) for polynomials $U_{m,n}^{2\alpha+1}(x, y)$ belonging to a biorthogonal system of polynomials on the unit disk with respect to the weight function $(1-x^2-y^2)^\alpha$.

4.2. Polynomials on the triangle

Consider the polynomials $P_{n,k}^{\alpha,\beta,\gamma}(x, y)$ of class IV and let

$$w_{\alpha,\beta,\gamma}(x, y) = (1-x)^\alpha (x-y)^\beta y^\gamma, \qquad 0 < y < x < 1 \ ,$$

denote the weight function. There is a pair of first order differential recurrence relations given by

$$(4.4) \qquad \partial_y P_{n,k}^{\alpha,\beta,\gamma} = c \cdot P_{n-1,k-1}^{\alpha,\beta+1,\gamma+1} \ ,$$

$$(4.5) \qquad (w_{\alpha,\beta,\gamma})^{-1} \partial_y (w_{\alpha,\beta+1,\gamma+1} P_{n-1,k-1}^{\alpha,\beta+1,\gamma+1}) = c \cdot P_{n,k}^{\alpha,\beta,\gamma} \ .$$

Define the second order partial differential operator $\underline{E}^{\beta,\gamma}$ by

$$(4.6) \qquad \underline{E}^{\beta,\gamma} = x\,\partial_{xx} + 2y\,\partial_{xy} + y\,\partial_{yy} + (\beta + \gamma + 2)\,\partial_x + (\gamma + 1)\,\partial_y \ .$$

Let D^* denote the formal adjoint of a partial differential operator D . A calculation shows that $(\underline{E}^{\beta,\gamma})^* = \underline{E}^{-\beta,-\gamma}$. Now we have the pair of second order differential recurrence relations

$$(4.7) \qquad \underline{E}^{\beta,\gamma} P_{n,k}^{\alpha,\beta,\gamma} = c \cdot P_{n-1,k}^{\alpha+2,\beta,\gamma} \ ,$$

$$(4.8) \qquad (w_{\alpha,\beta,\gamma})^{-1} \underline{E}^{-\beta,-\gamma}(w_{\alpha+2,\beta,\gamma} P_{n-1,k}^{\alpha+2,\beta,\gamma}) =$$

$$= c \cdot P_{n,k}^{\alpha,\beta,\gamma} \ .$$

472

These formulas can be proved by using the second order differential re-currence relations for Jacobi polynomials which are derived in Koornwinder [48, (2.4) and (2.5)].

Iteration of (4.8) gives the Rodrigues type formula

$$
(4.9) \qquad P^{\alpha, \beta, \gamma}_{n, k} = c \cdot (w_{\alpha, \beta, \gamma})^{-1} (E_{-}^{-\beta, -\gamma})^{n-k} (\partial_y)^k \cdot
$$

$$
\cdot [w_{\alpha+2n-2k, \beta+k, \gamma+k}] .
$$

It is of interest to compare (4.9) with the Rodrigues type formula for Appell's polynomials on the triangle, cf. Erdélyi [17, 12.4(4)] .

4.3. **Polynomials on a region bounded by two straight lines and a parabola**

Consider polynomials $P^{\alpha, \beta, \gamma}_{n, k}(u, v)$ of class VI. Let

$$
w_{\alpha, \beta, \gamma}(u, v) = (1-u+v)^{\alpha} (1+u+v)^{\beta} (u^2 - 4v)^{\gamma}
$$

denote the weight function. The differential recurrence relations involve the two second order partial differential operators

$$
(4.10) \qquad D_{-}^{\gamma} = \partial_{uu} + u \partial_{uv} + v \partial_{vv} + (\gamma + \tfrac{3}{2}) \partial_v ,
$$

$$
(4.11) \qquad E_{-}^{\alpha, \beta} = u \partial_{uu} + 2(v+1) \partial_{uv} + u \partial_{vv} + (\alpha+\beta+2) \partial_u + (\beta-\alpha) \partial_v .
$$

A calculation shows that $(D_{-}^{\gamma})^* = D_{-}^{-\gamma}$ and $(E_{-}^{\alpha, \beta})^* = E_{-}^{-\alpha, -\beta}$.

The first pair of partial differential recurrence relations is given by

$$
(4.12) \qquad D_{-}^{\gamma} P^{\alpha, \beta, \gamma}_{n, k} = c \cdot P^{\alpha+1, \beta+1, \gamma}_{n-1, k-1} ,
$$

$$
(4.13) \quad (w_{\alpha, \beta, \gamma})^{-1} D_{-}^{-\gamma} (w_{\alpha+1, \beta+1, \gamma} P^{\alpha+1, \beta+1, \gamma}_{n-1, k-1}) = c \cdot P^{\alpha, \beta, \gamma}_{n, k} ,
$$

cf. Koornwinder [45, §5]. If $\gamma = \pm\frac{1}{2}$ then these formulas immediately follow from (2.1), (2.2), (3.12) and (3.13).

The second pair of differential recurrence relations is given by

$$(4.14) \qquad E_-^{\alpha,\beta} \, P_{n,k}^{\alpha,\beta,\gamma} = c \cdot P_{n-1,k}^{\alpha,\beta,\gamma+1} \quad ,$$

$$(4.15) \qquad (w_{\alpha,\beta,\gamma})^{-1} E_-^{-\alpha,-\beta} (w_{\alpha,\beta,\gamma+1} \, P_{n-1,k}^{\alpha,\beta,\gamma+1}) = c \cdot P_{n,k}^{\alpha,\beta,\gamma} \quad ,$$

cf. Sprinkhuizen [66, §4]. If $\gamma = -\frac{1}{2}$ then these formulas are corollaries of (2.3), (3.12) and (3.13).

In the remainder of this subsection we summarize some interesting consequences of (4.12), (4.13), (4.14) and (4.15), which were obtained by Sprinkhuizen [66].

Iteration of (4.13) and (4.15) gives the Rodrigues type formula

$$(4.16) \qquad P_{n,k}^{\alpha,\beta,\gamma} = c \cdot (w_{\alpha,\beta,\gamma})^{-1} (D_-^{-\gamma})^k (E_-^{-\alpha-k,-\beta-k})^{n-k} \cdot$$

$$\cdot [w_{\alpha+k,\beta+k,\gamma+n-k}] \cdot$$

Let $(m,\ell) < (n,k)$ if $m \le n$ and $m + \ell \le n + k$. This defines a partial ordering.

<u>Theorem 4.1.</u> $P_{n,k}^{\zeta,\ell,\gamma}(u,v)$ is a linear combination of monomials $u^{m-\ell} v^{\ell}$ such that $(m,\ell) < (n,k)$ (cf. Figure 11).

For $\gamma = \pm\frac{1}{2}$ this result was already obtained in section 3.7.3. In the case of general γ the theorem is evident if $n = k$. Complete induction with respect to $n - k$ and use of (4.15) gives the general theorem.

Denote the second order operator (3.29) and the fourth order operator (3.30), expressed in terms of u and v, by $D_1^{\alpha,\beta,\gamma}$ and $D_2^{\alpha,\beta,\gamma}$, respectively (see Koornwinder [45] for the explicit expressions). Then the polynomials $P_{n,k}^{\alpha,\beta,\gamma}(u,v)$ are eigenfunctions of both differential operators with eigenvalues

(4.17) $\lambda_{n,k}^{\alpha,\beta,\gamma} = -n(n+\alpha+\beta+2\gamma+2) - k(k+\alpha+\beta+1)$

and

(4.18) $\mu_{n,k}^{\alpha,\beta,\gamma} = k(k+\alpha+\beta+1)(n+\gamma+\frac{1}{2})(n+\alpha+\beta+\gamma+\frac{3}{2})$,

respectively. If $(n,k) \neq (m,\ell)$ then $(\lambda_{n,k}^{\alpha,\beta,\gamma}, \mu_{n,k}^{\alpha,\beta,\gamma}) \neq (\lambda_{m,\ell}^{\alpha,\beta,\gamma}, \mu_{m,\ell}^{\alpha,\beta,\gamma})$.
However, if we only consider $\lambda^{\alpha,\beta,\gamma}$ then degeneracies may occur. For
instance, $\lambda_{2,2}^{3/2,-1/2,0} = \lambda_{3,0}^{3/2,-1/2,0}$. But if $(m,\ell) \neq (n,k)$ and $(m,\ell) <$
(n,k) then $\lambda_{m,\ell}^{\alpha,\beta,\gamma} > \lambda_{n,k}^{\alpha,\beta,\gamma}$, i.e., $\lambda_{n,k}^{\alpha,\beta,\gamma}$ is a monotonic function of
(n,k) with respect to the partial ordering. This result together with
Theorem 4.1 implies:

<u>Theorem 4.2.</u>   If a function $f(u,v)$ is an eigenfunction of $D_1^{\alpha,\beta,\gamma}$ and if

$$f(u,v) = \sum_{(m,\ell)\,<\,(n,k)} c_{m,\ell}\, u^{m-\ell}\, v^{\ell}$$

for certain coefficients $c_{m,\ell}$ with $c_{n,k} \neq 0$, then $f(u,v) = c \cdot P_{n,k}^{\alpha}(u,v)$.
By applying Theorem 4.1 there follow quadratic transformation form-
ulas

(4.19) $$\frac{P_{n+k,n-k}^{\alpha,\alpha,\gamma}(u,v)}{P_{n+k,n-k}^{\alpha,\alpha,\gamma}(2,1)} = \frac{P_{n,k}^{\gamma,-\frac{1}{2},\alpha}(2v, u^2-2v-1)}{P_{n,k}^{\gamma,-\frac{1}{2},\alpha}(2,1)} \quad ,$$

(4.20) $$\frac{P_{n+k+1,n-k}^{\alpha,\alpha,\gamma}(u,v)}{P_{n+k+1,n-k}^{\alpha,\alpha,\gamma}(2,1)} = \frac{u\, P_{n,k}^{\gamma,\frac{1}{2},\alpha}(2v, u^2-2v-1)}{2\, P_{n,k}^{\gamma,\frac{1}{2},\alpha}(2,1)} \quad .$$

Application of Theorem 4.1 and Lemma 3.1 gives the recurrence relations

(4.21) $$u\, P_{n,k}^{\alpha,\beta,\gamma}(u,v) = \sum_{(n-1,k)<\,(m,\ell)<\,(n+1,k)} a(m,\ell;n,k)\, P_{m,\ell}^{\alpha,\beta,\gamma}(u,v)$$

475

and

$$(4.22) \quad v \, P_{n,k}^{\alpha, \beta, \gamma}(u, v) = \sum_{(n-1, k-1) < (m, \ell) < (n+1, k+1)} b(m, \ell; n, k) \, P_{m, \ell}^{\alpha, \beta, \gamma}(u, v) .$$

These recurrence relations involve nine terms (cf. Figure 14) and fifteen terms (cf. Figure 15), respectively. However, a calculation shows that four of the nine coefficients in (4.21) vanish and that six of the fifteen coefficients in (4.22) vanish, as is indicated in Figures 14 and 15. Hence, the recurrence relations have the same structure as in the cases $\gamma = \pm \frac{1}{2}$, cf. Figures 12 and 13 in section 3.7.3.

Figure 14   Figure 15

4.4. Expansions in terms of James type zonal polynomials

It seems difficult to derive some explicit power series expansion analogous to (2.5) for the polynomials of class VI. However, some results can be obtained for expansions in terms of the so-called James type zonal polynomials, which are explicitly known in the case of two variables. In particular, the polynomials $_6P_{n,n}^{\alpha, \beta, 0}(u, v)$ can be identified with certain hypergeometric functions of 2×2 matrix argument. The results announced in this subsection will be published elsewhere in more details.

It is convenient to introduce new variables $\xi = 1 - \frac{1}{2}u$, $\eta = \frac{1}{4}(1-u+v)$, and to define

(4.23)
$$R_{n,k}^{\alpha, \beta, \gamma}(\xi, \eta) = \frac{{}_6P_{n,k}^{\alpha, \beta, \gamma}(2-2\xi, 1-2\xi+4\eta)}{{}_6P_{n,k}^{\alpha, \beta, \gamma}(2,1)} .$$

Then the polynomials $R_{n,k}^{\alpha, \beta, \gamma}(\xi, \eta)$ are polynomials with highest term $c \cdot \xi^{n-k} \eta^k$ obtained by orthogonalization of the sequence $1, \xi, \eta, \xi^2, \xi\eta$, $\eta^2, \xi^3, \xi^2\eta, \ldots$ with respect to the weight function $\eta^{\alpha}(1-\xi+\eta)^{\beta}(\xi^2 - 4\eta)^{\gamma}$ on the region $\{(\xi, \eta) \mid \eta > 0, 1 - \xi + \eta > 0, \xi^2 - 4\eta > 0\}$, cf. Figure 16. They are normalized such that $R_{n,k}^{\alpha, \beta, \gamma}(0, 0) = 1$.

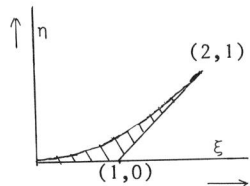

Figure 16

It follows from Theorem 4.1 that $R_{n,k}^{\alpha, \beta, \gamma}(\xi, \eta)$ is a linear combination of monomials $\xi^{m-\ell}\eta^{\ell}$ such that $(m, \ell) < (n, k)$. The recurrence relations for $\xi R_{n,k}^{\alpha, \beta, \gamma}(\xi, \eta)$ and $\eta R_{n,k}^{\alpha, \beta, \gamma}(\xi, \eta)$ have the structure of Figures 12 and 13, respectively.

For $\gamma = \pm\frac{1}{2}$ we have

(4.24)
$$R_{n,k}^{\alpha, \beta, -\frac{1}{2}}(x+y, xy) =$$

$$= \frac{1}{2}[R_n^{(\alpha, \beta)}(1-2x) R_k^{(\alpha, \beta)}(1-2y) + R_k^{(\alpha, \beta)}(1-2x) R_n^{(\alpha, \beta)}(1-2y)]$$

$$= \sum_{\ell=0}^{k} \sum_{m=\ell}^{n} \frac{1}{2}\left[\frac{(-n)_m(-k)_\ell(n+\alpha+\beta+1)_m(k+\alpha+\beta+1)_\ell}{(\alpha+1)_m(\alpha+1)_\ell \, m! \, \ell!} + \right.$$

$$\left. + \frac{(-k)_m(-n)_\ell(k+\alpha+\beta+1)_m(n+\alpha+\beta+1)_\ell}{(\alpha+1)_m(\alpha+1)_\ell \, m! \, \ell!}\right] \cdot \frac{x^m y^\ell + x^\ell y^m}{1+\delta_{m,\ell}}$$

477

and

(4.25) $R_{n,k}^{\alpha,\beta,\frac{1}{2}}(x+y, xy) =$

$$= -\frac{(\alpha+1)(R_{n+1}^{(\alpha,\beta)}(1-2x) R_k^{(\alpha,\beta)}(1-2y) - R_k^{(\alpha,\beta)}(1-2x) R_{n+1}^{(\alpha,\beta)}(1-2y))}{(n-k+1)(n+k+\alpha+\beta+2)(x-y)}$$

$$= \sum_{\ell=0}^{k} \sum_{m=\ell}^{n} \left[\frac{(-n-1)_{m+1}(-k)_\ell(n+\alpha+\beta+2)_{m+1}(k+\alpha+\beta+1)_\ell}{(\alpha+1)_{m+1}(\alpha+1)_\ell(m+1)!\,\ell!} \right.$$

$$\left. - \frac{(-k)_{m+1}(-n-1)_\ell(k+\alpha+\beta+1)_{m+1}(n+\alpha+\beta+2)_\ell}{(\alpha+1)_{m+1}(\alpha+1)_\ell(m+1)!\,\ell!} \right]$$

$$\cdot \frac{\alpha+1}{(-n+k-1)(n+k+\alpha+\beta+2)} \cdot \frac{x^{m+1}y^\ell - x^\ell y^{m+1}}{x-y}.$$

Hence, in view of (3.43) and (3.44), we have explicit expansions of $R_{n,k}^{\alpha,\beta,\gamma}(\xi,\eta)$ in terms of $Z_{m,\ell}^{\gamma}(\xi,\eta)$, $\gamma = \pm\frac{1}{2}$.

As a generalization of (3.45) and (3.46) let us define

(4.26) $Z_{n,k}^{\gamma}(\xi,\eta) = \frac{2^{2n-2k}(n-k)!}{(2\gamma+n-k+1)_{n-k}} \eta^{\frac{1}{2}(n+k)} P_{n-k}^{(\gamma,\gamma)}(\frac{1}{2}\eta^{-\frac{1}{2}}\xi)$

$$= \sum_{i=0}^{[\frac{1}{2}(n-k)]} \frac{(-n+k)_{2i}}{(-n+k-\gamma+\frac{1}{2})_i\,i!} \xi^{n-k-2i}\eta^{k+i}, \quad n \geq k \geq 0.$$

Note that the coefficient of $\xi^{n-k}\eta^k$ equals 1.

Then the coefficients in the expansion

(4.27) $R_{n,k}^{\alpha,\beta,\gamma}(\xi,\eta) = \sum_{(m,\ell)<(n,k)} c_{m,\ell;n,k}^{\alpha,\beta,\gamma} Z_{m,\ell}^{\gamma}(\xi,\eta)$

are uniquely determined. It is our purpose to find explicit expressions for these coefficients. The expansion (4.27) is motivated by formulas (4.24) and (4.25) and also by the following facts.

478

Comparing formula (4.16) with a similar Rodrigues type formula in Herz [29, (6.4')] we can express $R_{n,n}^{\alpha,\beta,0}(x+y, xy)$ as a hypergeometric function of matrix argument:

(4.28)
$$R_{n,n}^{\alpha,\beta,0}(x+y, xy) =$$

$$= {}_2F_1\left(-n, n+\alpha+\beta+\frac{3}{2}; \alpha + \frac{3}{2}; \begin{bmatrix} x & 0 \\ 0 & y \end{bmatrix}\right) \quad .$$

Herz [29] defined hypergeometric functions of matrix argument by means of generalized Laplace transforms and their inverses. However, Constantine [11, (25)] obtained the series expansion

(4.29)
$$ {}_2F_1(a, b; c; X) = \sum_{m=0}^{\infty} \sum_{\ell=0}^{m} \frac{(a)_{m,\ell} (b)_{m,\ell}}{(c)_{m,\ell} (m+\ell)!} C_{m,\ell}(X) \quad , $$

where $(a)_{m,\ell} = (a)_m (a-\frac{1}{2})_\ell$ and the functions $C_{m,\ell}(X)$ are the so-called zonal polynomials depending on the eigenvalues of the 2×2 symmetric matrix X. In fact, the theory was given for hypergeometric functions depending on the eigenvalues of $m \times m$ symmetric matrices. The zonal polynomials were introduced by James [33]. The polynomial $C_{m,\ell}(X)$ is the spherical function on $GL(2, \mathbb{R})/O(2)$ belonging to the irreducible representation $\{2m, 2\ell\}$ of $GL(2, \mathbb{R})$. James [34, (7.9)] pointed out that

$$C_{n,k}\left(\begin{bmatrix} x & 0 \\ 0 & y \end{bmatrix}\right) = C_{n,k}(I_2)(xy)^{\frac{1}{2}(n+k)} P_{n-k}\left(\frac{x+y}{2(xy)^{\frac{1}{2}}}\right) \quad .$$

It follows that

(4.30)
$$C_{n,k}\left(\begin{bmatrix} x & 0 \\ 0 & y \end{bmatrix}\right) = \frac{(n+k)!(\frac{3}{2})_{n-k}}{(n-k)! \, k! (\frac{3}{2})_n} Z_{n,k}^0(x+y, xy) \quad .$$

Combination of (4.28), (4.29) and (4.30) gives

(4.31) $R_{n,n}^{\alpha,\beta,0}(\xi,\eta) =$

$$= \sum_{\ell=0}^{k} \sum_{m=\ell}^{n} \frac{(-n)_m (-n-\frac{1}{2})_\ell (n+\alpha+\beta+\frac{3}{2})_m (n+\alpha+\beta+1)_\ell (\frac{3}{2})_{m-\ell}}{(\alpha+\frac{3}{2})_m (\alpha+1)_\ell (\frac{3}{2})_m \, \ell! \, (m-\ell)!} \cdot$$

$$\cdot Z_{m,\ell}^{0}(\xi,\eta) \ ,$$

which gives the coefficients in (4.27) if $\gamma = 0$ and $n = k$.

Let us now consider the general case of (4.27). The operator D_{-}^{γ} (cf (4.10)), expressed in terms of the coordinates ξ and η, acts on $R_{n,k}^{\alpha,\beta,\gamma}(\xi,\eta)$ and $Z_{n,k}^{\gamma}(\xi,\eta)$ in a similar way. We have

(4.32) $D_{-}^{\gamma} R_{n,k}^{\alpha,\beta,\gamma}(\xi,\eta) =$

$$\frac{k(k+\alpha+\beta+1)(n+\gamma+\frac{1}{2})(n+\alpha+\beta+\gamma+\frac{3}{2})}{4(\alpha+1)(\alpha+\gamma+\frac{3}{2})} R_{n-1,k-1}^{\alpha+1,\beta+1,\gamma}(\xi,\eta) \ ,$$

(4.33) $D_{-}^{\gamma} Z_{n,k}^{\gamma}(\xi,\eta) = \frac{1}{4} k(n+\gamma+\frac{1}{2}) Z_{n-1,k-1}^{\gamma}(\xi,\eta)$.

If $k = 0$ then (4.32) and (4.33) have to be interpreted such that the right hand sides become zero. By complete induction with respect to k formulas (4.32) and (4.33) imply that the coefficients in (4.27) can only be nonzero if $\ell \leq k$ and $m \leq n$, cf. Figure 10. Using this result and Lemma 3.1 we can now immediately prove that the recurrence relation for $\eta \, R_{n,k}^{\alpha,\beta,\gamma}(\xi,\eta)$ has the structure given by Figure 13. It also follows from (4.32), (4.33) and (4.27) that

(4.34) $c_{m,\ell;n,k}^{\alpha,\beta,\gamma} = \dfrac{k(k+\alpha+\beta+1)(n+\gamma+\frac{1}{2})(n+\alpha+\beta+\gamma+\frac{3}{2})}{(\alpha+1)(\alpha+\gamma+\frac{3}{2})\ell(m+\gamma+\frac{1}{2})} \cdot$

$$\cdot c_{m-1,\ell-1;n-1,k-1}^{\alpha+1,\beta+1,\gamma} \ .$$

Hence, it is sufficient to calculate the coefficients $c_{m,\ell;n,k}^{\alpha,\beta,\gamma}$ if $\ell = 0$. These are also the coefficients in the power series of the boundary value $R_{n,k}^{\alpha,\beta,\gamma}(\xi, 0)$:

$$(4.35) \qquad \overset{*}{R}_{n,k}^{\alpha,\beta,\gamma}(\xi, 0) = \sum_{m=0}^{n} c_{m,0;n,k}^{\alpha,\beta,\gamma} \, \xi^m .$$

Using the action of the differential operator $E_{-}^{\alpha,\beta}$ (cf. (4.11)) on $R_{n,k}^{\alpha,\beta,\gamma}(\xi,\eta)$ and $Z_{n,k}^{\gamma}(\xi,\eta)$ we can obtain a four-term recurrence relation expressing $c_{m+1,0;n,k}^{\alpha,\beta,\gamma}$ as a linear combination of $c_{m,0;n,k}^{\alpha,\beta,\gamma}$, $c_{m-1,0;n-1,k-1}^{\alpha+1,\beta+1,\gamma}$ and $c_{m-1,0;n-1,k}^{\alpha,\beta,\gamma+1}$. If either $n = k$ or $k = 0$ then this reduces to a three-term recurrence relation and then we are able to calculate the coefficients. The results are:

$$(4.36) \qquad c_{m,\ell;n,n}^{\alpha,\beta,\gamma} = \frac{(-n)_m (-n-\gamma-\frac{1}{2})_\ell (n+\alpha+\beta+\gamma+\frac{3}{2})_m (n+\alpha+\beta+1)_\ell (\gamma+\frac{3}{2})_{m-\ell}}{(\alpha+\gamma+\frac{3}{2})_m (\alpha+1)_\ell (\gamma+\frac{3}{2})_m \, \ell! \, (m-\ell)!} ,$$

$$(4.37) \qquad c_{m,0;n,0}^{\alpha,\beta,\gamma} = \frac{(-n)_m (n+\alpha+\beta+2\gamma+2)_m (\gamma+\frac{1}{2})_m}{(\alpha+\gamma+\frac{3}{2})_m (2\gamma+1)_m \, m!} .$$

Formula (4.36) was first proved by Sprinkhuizen (unpublished). For $\gamma = 0$ formula (4.36) is the Herz-Constantine-James result (4.31). For $\gamma = \pm\frac{1}{2}$ formulas (4.36) and (4.37) imply (4.24) and (4.25) ($n = k$ or $k = 0$). Finally formulas (4.36) and (4.37) give nice explicit boundary values:

$$(4.38) \qquad R_{n,n}^{\alpha,\beta,\gamma}(\xi, 0) = R_n^{(\alpha+\gamma+\frac{1}{2},\beta)}(1-2\xi) ,$$

$$(4.39) \qquad R_{n,0}^{\alpha,\beta,\gamma}(\xi, 0) = {}_3F_2 \left(\begin{array}{c} -n, \; n+\alpha+\beta+2\gamma+2, \; \gamma+\frac{1}{2} \\ \alpha+\gamma+\frac{3}{2}, \quad 2\gamma+1 \end{array} \middle| \, \xi \right) .$$

4.5. Polynomials on the deltoid

Consider the polynomials $P_{m,n}^{\alpha}(z,\bar{z})$ of class VII. They have power series expansion

$$(4.40) \qquad P^{\alpha}_{m, n}(z, \bar{z}) = c_{m, n} z^m \bar{z}^n + \sum_{k+\ell < m+n} c_{k, \ell} z^k \bar{z}^\ell .$$

Because of the symmetries of the orthogonality region it follows that $c_{k, \ell} = 0$ if $k - \ell \neq m - n \pmod 3$.

Let us denote the second order operator (3.29) and the third order operator (3.31), expressed in terms of z and $\bar{z}$, by D^{α}_1 and D^{α}_2, respectively (cf. Koornwinder [46] for the explicit expressions). Both differential operators have the polynomials $P^{\alpha}_{m, n}(z, \bar{z})$ as eigenfunctions. The eigenvalues are

$$(4.41) \qquad \lambda^{\alpha}_{m, n} = -\frac{4}{3}(m^2 + mn + n^2 + 3(\alpha + \frac{1}{2})(m + n))$$

and

$$(4.42) \qquad \mu^{\alpha}_{m, n} = (m-n)(2m + n + 3\alpha + \frac{3}{2})(m + 2n + 3\alpha + \frac{3}{2}) ,$$

respectively. If $(k, \ell) \neq (m, n)$ then $(\lambda^{\alpha}_{k, \ell}, \mu^{\alpha}_{k, \ell}) \neq (\lambda^{\alpha}_{m, n}, \mu^{\alpha}_{m, n})$. The two differential equations for $P^{\alpha}_{m, n}(z, \bar{z})$ give recurrence relations for the coefficients in (4.40). From these recurrence relations it can be derived that $c_{k, \ell} \neq 0$ only if $k + 2\ell \leq m + 2n$ and $2k + \ell \leq 2m + n$, cf. Figure 17.

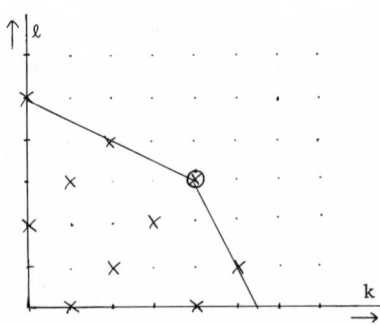

Figure 17

Observe that $\lambda^{\alpha}_{k,\ell} > \lambda^{\alpha}_{m,n}$ if $(k, \ell) \neq (m, n)$, $k + 2\ell \leq m + 2n$, $2k + \ell \leq 2m + n$. This implies:

Theorem 4.3. Let $P(z, \bar{z})$ be a linear combination of monomials $z^k \bar{z}^\ell$ such that $2k + \ell \leq 2m + n$, $k + 2\ell \leq m + 2n$, $k - \ell = m - n$ (mod 3) and the coefficient of $z^m \bar{z}^n$ is nonzero. If $P(z, \bar{z})$ is also an eigenfunction of D^{α}_1 then $P(z, \bar{z}) = c \cdot P^{\alpha}_{m,n}(z, \bar{z})$.

Lidl [54, §3.c.β] defines generalized Gegenbauer polynomials $C^a_k(x, y)$ by the generating function

(4.43)
$$(1 - xz + yz^2 - z^3)^{-a} = \sum_{k=0}^{\infty} C^a_k(x, y) z^k .$$

By using Theorem 4.3 it can be proved that $C^a_k(x, y)$ is a constant multiple of $_7P^{a-\frac{1}{2}}_{k, 0}(x, y)$.

Let us define

(4.44)
$$Z^{\alpha}_{n_1, n_2, n_3}(\xi, \eta, \zeta) =$$
$$= c \cdot \zeta^{(n_1 + n_2 + n_3)/3} {}_7P^{\alpha}_{n_1 - n_2, n_2 - n_3}(\zeta^{-\frac{1}{3}} \xi, \zeta^{-\frac{2}{3}} \eta) ,$$

where $n_1 \geq n_2 \geq n_3 \geq 0$ and the coefficient of $\xi^{n_1 - n_2} \eta^{n_2 - n_3} \zeta^{n_3}$ equals 1. These functions are polynomials in ξ, η, ζ and they can be considered as three-variable analogues of the polynomials defined by (4.26). If $\alpha = -\frac{1}{2}, \frac{1}{2}$ or 0 and if

$$\xi = x_1 + x_2 + x_3, \quad \eta = x_1 x_2 + x_2 x_3 + x_3 x_1, \quad \zeta = x_1 x_2 x_3$$

are the elementary symmetric polynomials in x_1, x_2, x_3, then the polynomials defined by (4.44) have special interpretations.

It follows from (3.14), (3.15), (3.16) and (3.17) that

(4.45)
$$Z^{-\frac{1}{2}}_{n_1, n_2, n_3}(\xi, \eta, \zeta) = c \cdot \sum x^{n_1}_{i_1} x^{n_2}_{i_2} x^{n_3}_{i_3},$$

where the sum is taken over all permutations (i_1, i_2, i_3) of $(1, 2, 3)$, and that

$$(4.46) \quad Z^{\frac{1}{2}}_{n_1, n_2, n_3}(\xi, \eta, \zeta) =$$

$$= \begin{vmatrix} x_1^{n_1+2} & x_2^{n_1+2} & x_3^{n_1+2} \\ x_1^{n_2+1} & x_2^{n_2+1} & x_3^{n_2+1} \\ x_1^{n_3} & x_2^{n_3} & x_3^{n_3} \end{vmatrix} \Bigg/ \begin{vmatrix} x_1^2 & x_2^2 & x_3^2 \\ x_1 & x_2 & x_3 \\ 1 & 1 & 1 \end{vmatrix}.$$

The polynomials satisfying (4.45) and (4.46) in the special case that $n_2 = n_3 = 0$, have been studied by Eier and Lidl [14] and Lidl [54], respectively.

The polynomial $Z^{\alpha}_{n_1, n_2, n_3}(\xi, \eta, \zeta)$ can be characterized up to a constant factor as a symmetric polynomial in x_1, x_2, x_3, which is a linear combination of monomials $x_1^{k_1} x_2^{k_2} x_3^{k_3}$ such that $k_1 + k_2 + k_3 = n_1 + n_2 + n_3$, $k_1 \leq n_1$, $k_3 \geq n_3$ and the coefficient of $x_1^{n_1} x_2^{n_2} x_3^{n_3}$ is nonzero, and which is an eigenfunction of a certain second order differential operator. By using this characterization and by using the results in James [34], the case $\alpha = 0$ of these polynomials can be identified with the zonal polynomials of 3×3 matrix argument considered by James.

It seems probable that, analogous to the results of section 4.5, the three-variable analogues of the polynomials of class VI will have nice expansions in terms of the polynomials $Z_{n_1, n_2, n_3}(\xi, \eta, \zeta)$.

5. <u>Orthogonal polynomials in two variables as spherical functions</u>

In this final section we briefly discuss the known cases in which polynomials of the classes I-VII can be interpreted as spherical functions. The importance of such a group theoretic interpretation should be clear from section 2.3.

The polynomials, $_1P_{m,n}^{q-2}(z,\bar{z})$, $q = 2, 3, 4, \ldots$, are the spherical functions on the sphere S^{2q-1} considered as the homogeneous space $U(q)/U(q-1)$, cf. the references given in section 3.4. A related result is the inequality

$$(5.1) \qquad |_1P_{m,n}^{\alpha}(z,\bar{z})| \le P_{m,n}^{\alpha}(1,1), \quad |z| \le 1 \ ,$$

which holds for $\alpha \ge 0$. The author proved (unpublished) analogous inequalities

$$(5.2) \qquad |_3P_{n,k}^{\alpha,\beta}(x,y)| \le {}_3P_{n,k}^{\alpha,\beta}(1,1), \quad y^2 \le x \le 1, \quad \alpha \ge |\beta + \tfrac{1}{2}| \ ,$$

and

$$(5.3) \qquad |_4P_{n,k}^{\alpha,\beta,\gamma}(x,y)| \le {}_4P_{n,k}^{\alpha,\beta,\gamma}(1,1), \quad 0 \le y \le x \le 1 \ ,$$

$$\alpha \ge |\beta + \gamma + 1|, \quad \beta \ge \max(\gamma, -\tfrac{1}{2}) \ .$$

A spherical function interpretation and a positive convolution structure are unknown for the classes III and IV, but formulas (5.2) and (5.3) suggest that further research in this direction may be worthwhile.

In the case of the classes VI and VII and for certain values of the parameters the second order differential operator (3.29) (with $D = \partial_{ss} + \partial_{tt}$ and $\rho(s,t)$ as given in Table 1) is the radial part of the Laplace-Beltrami operator on certain compact symmetric Riemannian spaces of rank 2, cf. Harish-Chandra [25, §7], or Helgason [28, (3.3)] together with the volume element ratio given in Helgason [26, Chap. 10, §1.5]. The spherical functions on these symmetric spaces are eigenfunctions of the Laplace-Beltrami operator. Since the Laplace-Beltrami operator may have degenerate eigenvalues, this does not prove that the polynomials of class VI and VII can be interpreted as spherical functions. Still it strongly motivated the author to introduce the classes VI and VII.

In the case of class VI and of order (α, β, γ) the restricted root vectors of the corresponding symmetric space have a Dynkin diagram $0 \Rightarrow 0$ and a vector diagram as given by Figure 18, where the restricted root vectors λ_1, $2\lambda_1$ and λ_2 have multiplicities $2\alpha - 2\beta$, $2\beta + 1$ and $2\gamma + 1$, respectively.

In the case of class VII and of order α the restricted root vectors have Dynkin diagram $0 - 0$ and multiplicity $2\alpha + 1$. The vector diagram for this case is given by Figure 19.

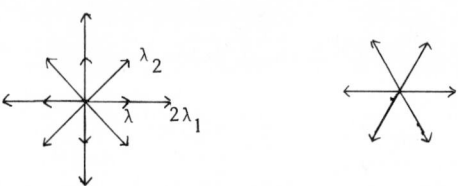

Figure 18   Figure 19

A list of all compact symmetric spaces of rank 2 is included in Helgason [26, p. 354, Table II]. The corresponding Dynkin diagrams and multiplicities of the restricted root vectors are given by Araki [4, pp. 32, 33] and by Loos [55, pp. 119, 146]. For our classes VI and VII we give the results in Tables 2 and 3, respectively.

In these tables the tori T^2 and H are considered as homogeneous spaces of the groups given in Table 1. The groups $SO(5)$ and $SU(3)$ are considered as homogeneous spaces of $SO(5) \times SO(5)$ and $SU(3) \times SU(3)$, respectively.

On a symmetric space M of rank r the class of all invariant differential operators is a commutative algebra with r generators, one of which is the Laplace-Beltrami operator (cf. Helgason [26, Chap. 10, §2]). The spherical functions on M can be characterized as the zonal functions which are eigenfunctions of all invariant differential operators on

homogeneous space	α	β	γ
square torus T^2	$-\frac{1}{2}$	$-\frac{1}{2}$	$-\frac{1}{2}$
SO(5)	$\frac{1}{2}$	$-\frac{1}{2}$	$\frac{1}{2}$
$O(q)/SO(2) \times O(q-2)$	0	0	$\frac{1}{2}q - \frac{5}{2}$
$O(q)/O(2) \times O(q-2)$	$\frac{1}{2}q - \frac{5}{2}$	$-\frac{1}{2}$	0
$U(q)/U(2) \times U(q-2)$	$q - 4$	0	$\frac{1}{2}$
$Sp(q)/Sp(2) \times Sp(q-2)$	$2q - 7$	1	$\frac{3}{2}$
$SO(10)/U(5)$	2	0	$\frac{3}{2}$
E III	4	0	$\frac{5}{2}$

Table 2

homogeneous space	α
hexagonal torus H	$-\frac{1}{2}$
SU(3)	$\frac{1}{2}$
SU(3)/SO(3)	0
SU(6)/Sp(3)	$\frac{3}{2}$
E IV	$\frac{7}{2}$

Table 3

M. Except for the Laplace-Beltrami operator, the radial parts of the invariant differential operators are not known in general. However, it follows from our results that for each of the spaces M listed in Table 2 (or Table 3) the following two statements are equivalent:

(a) The polynomials $_6P^{\alpha,\beta,\gamma}_{m,n}(u,v)$ (or $_7P^{\alpha}_{m,n}(z,\bar{z})$) as functions of s and t are the spherical functions on M considered as functions of the radial coordinates s and t .

(b) The differential operator (3.30) (or (3.31)) is the radial part of an invariant differential operator on M .

In several special cases it has been proved that statements (a) and (b) are true. For the two tori T^2 and H it is evident. In the group cases SO(5) and SU(3) the spherical functions are the characters on the group. Then Weyl's character formula (cf. Weyl [71, (37)], [72,(29)]) together with formulas (3.26) and (3.17) proves statement (a), and statement (b) follows from Berezin [6]. For complex Grassmann manifolds U(q)/U(2) × U(q-2) the spherical functions and the invariant differential operators have been given by Berezin and Karpelevic [7].

In the case of real Grassmann manifolds O(q)/O(2) × O(q-2) Maass [56] first derived the invariant differential operators and then obtained the spherical functions as orthogonal polynomials. Another approach was followed by James and Constantine [36]. They obtained the spherical functions by using zonal polynomials of 2 × 2 matrix argument. More generally, it follows from their results that our polynomials of class VI and of order ($\frac{1}{2}$(q-p-3), $\frac{1}{2}$(p-3), 0) can be interpreted as functions on O(q), right invariant with respect to O(2) × O(q-2), left invariant with respect to O(p) × O(q-p), and belonging to some irreducible representation of O(q) . This also gives a group theoretic explanation for the results of section 4.4 if $\gamma = 0$ and α and β are integers or half integers. Recently, the author found yet another approach to the analysis on Grassmann manifolds of rank two. In this approach the harmonics on the Grassmann manifold are obtained as restrictions of certain doubly homogeneous polynomials which satisfy certain orthogonality properties. These results will be soon available in preprint form.

Deeper analytic properties of polynomials of class VI and VII may be first derived in the cases, where a group theoretic interpretation is known, and next be generalized to other values of the parameters. A study of the most simple case, where all parameters are equal to $-\frac{1}{2}$, already gives an indication of the difficulties which can be expected in the general case.

Note Added in Proof

Recently I. Sprinkhuizen together with the author proved that

$$R^{\alpha,\beta,\gamma}_{n,k}(\xi,0) = \sum_{m=k}^{n} \frac{(\gamma+\frac{1}{2})_{m-k}(\gamma+\frac{1}{2})_{n-m}(n-k)!}{(m-k)!(n-m)!(2\gamma+1)_{n-k}} \cdot$$

$$\cdot \frac{(n+k+\alpha+\beta+2\gamma+2)_{m-k}(m+k+\alpha+\beta+2)_{n-m}}{(m+k+\alpha+\beta+\gamma+\frac{3}{2})_{m-k}(2m+\alpha+\beta+\gamma+\frac{5}{2})_{n-m}} R^{(\alpha+\gamma+\frac{1}{2},\beta)}_{m}(1-2\xi) ,$$

where $R^{\alpha,\beta,\gamma}_{n,k}(\xi,0)$ is defined by (4.23).

REFERENCES

[1]   C. A. Agahanov, A method of constructing orthogonal polynomials of two variables for a certain class of weight functions (Russian), Vestnik Leningrad Univ. 20(1965), no. 19, pp. 5-10.

[2]   H. Annabi, Analyse harmonique sur $U(q)/U(q-1)$, Thèse, Université de Tunis, 1974.

[3]   H. Annabi and K. Trimèche, Convolution généralisée sur le disque unité, C. R. Acad. Sci. Paris Sér. A278 (1974), 21-24.

[4]   S. Araki, On root systems and an infinitesimal classification of irreducible symmetric spaces, J. Math. Osaka City Univ. 13(1962), 1-34.

[5] R. Askey, Orthogonal polynomials and special functions, Lectures delivered at a Regional Conference held at V. P. I. and S. U. from June 10 to June 14, 1974.

[6] F. A. Berezin, Laplace operators on semisimple Lie groups, Trudy Moskov. Mat. Obšč 6(1957), 371-463 (Russian) , Amer. Math. Soc. Transl. (2) 21 (1962), 239-339.

[7] F. A. Berezin and F. I. Karpelevič, Zonal spherical functions and Laplace operators on some symmetric spaces (Russian), Dokl. Akad. Nauk SSSR (N. S.) 118(1958), 9-12.

[8] M. Bertran, Note on orthogonal polynomials in v-variables, SIAM J. Math. Anal. 6(1975), 250-257.

[9] J. N. Boyd, Orthogonal polynomials on the disc, M. A. Thesis, University of Virginia, 1972.

[10] R. R. Coifman and G. Weiss, Analyse harmonique non-commutative sur certains espaces homogènes, Lecture Notes in Math. 242, Springer-Verlag, Berlin, 1971.

[11] A. G. Constantine, Some non-central distribution problems in multivariate analysis, Ann. Math. Statistics 34 (1963), 1270-1285.

[12] F. Didon, Développements sur certaines séries de polynomes à un nombre quelconque de variables, Ann. Sci. Ecole Norm. Sup. (1) 7(1870), 247-268.

[13] C. F. Dunkl, An expansion in ultraspherical polynomials with non-negative coefficients, SIAM J. Math. Anal. 5(1974), 51-52.

[14] R. Eier and R. Lidl, Tschebyscheffpolynome in einer und zwei Variabelen, Abh. Math. Sem. Univ. Hamburg 41(1974), 17-27.

[15] G. K. Engelis, On polynomials orthogonal on a triangle (Russian), Latvijas Valsts Univ. Zinātn. Raksti 58(1964), vyp. 2, pp. 43-48.

[16] G. K. Engelis, On some two-dimensional analogues of the classical orthogonal polynomials (Russian), Latviĭskiĭ Matematičeskiĭ Ežegodnik 15(1974), 169-202.

490

[17]   A. Erdélyi, W. Magnus, F. Oberhettinger and F. G. Tricomi, Higher Transcendental Functions, vol. II, McGraw-Hill, New York, 1953.

[18]   E. D. Fackerell and R. A. Littler, Polynomials biorthogonal to Appell's polynomials, Bull. Austr. Math. Soc. 11(1974), 181-195.

[19]   G. G. Folland, The tangential Cauchy-Riemann complex on spheres, Trans. Amer. Math. Soc. 171(1972), 83-133.

[20]   G. B. Folland, Spherical harmonic expansion of the Poisson-Szegö kernel for the ball, Proc. Amer. Math. Soc. 47(1975), 401-408.

[21]   R. Gangolli, Positive definite kernels on homogeneous spaces and certain stochastic processes related to Levy's Brownian motion of several parameters, Ann. Inst. H. Poincaré, sect. B (N.S.) 3(1967), 121-226.

[22]   G. Gasper, Positivity and the convolution structure for Jacobi series, Ann. of Math. 93(1971), 112-118.

[23]   G. Gasper, Banach algebras for Jacobi series and positivity of a kernel, Ann. of Math. 95(1972), 261-280.

[24]   G. Gasper, Positivity and special functions, These Proceedings.

[25]   Harish-Chandra, Spherical functions on a semi-simple Lie group, I, Amer. J. Math. 80(1958), 241-310.

[26]   S. Helgason, Differential geometry and symmetric spaces, Academic Press, New York, 1962.

[27]   S. Helgason, The Radon transform on Euclidean spaces, compact two-point homogeneous spaces and Grassmann manifolds, Acta Math. 113(1965), 153-180.

[28]   S. Helgason, A formula for the radial part of the Laplace-Beltrami operator, J. Diff. Geometry 6(1972), 411-419.

[29]   C. S. Herz, Bessel functions of matrix argument, Ann. of Math 61(1955), 474-523.

[30]   M. Ikeda, On spherical functions for the unitary group, I, II, III, Mem. Fac. Engrg. Hiroshima Univ. 3(1967), 17-75.

[31] M. Ikeda and T. Kayama, On spherical functions for the unitary group, IV, Mem. Fac. Engrg. Hiroshima Univ. 3(1967), 77-100.

[32] D. Jackson, Formal properties of orthogonal polynomials in two variables, Duke Math. J. 2(1936), 423-434.

[33] A. T. James, Zonal polynomials of the real positive definite symmetric matrices, Ann. of Math. 74(1961), 456-469.

[34] A. T. James, Calculation of zonal polynomial coefficients by use of the Laplace-Beltrami operator, Ann. Math. Statistics 39(1968), 1711-1718.

[35] A. T. James, Special functions of matrix and single argument in statistics, These Proceedings.

[36] A. T. James and A. G. Constantine, Generalized Jacobi polynomials as spherical functions of the Grassmann manifold, Proc. London Math. Soc. (3) 29(1974), 174-192.

[37] Y. Kanjin, A convolution measure algebra on the unit disc, Tôhoku Math. J., to appear.

[38] S. Karlin and J. McGregor, Determinants of orthogonal polynomials, Bull. Amer. Math. Soc. 68(1962), 204-209.

[39] S. Karlin and J. McGregor, On some stochastic models in genetics, pp. 245-271 in "Stochastic models in medicine and biology" (J. Gurland, ed.), University of Wisconsin Press, Madison, 1964.

[40] S. Karlin and J. McGregor, Some properties of determinants of orthogonal polynomials, These Proceedings.

[41] S. Karlin and J. McGregor, Linear growth processes and Moran models with many types, These Proceedings.

[42] T. H. Koornwinder, The addition formula for Jacobi polynomials, I, Summary of results, Nederl. Akad. Wetensch. Proc. Ser. A75 = Jndag. Math. 34(1972), 188-191.

[43] T. H. Koornwinder, The addition formula for Jacobi polynomials, II, The Laplace type integral representation and the product formula, Math. Centrum Amsterdam Report TW133(1972).

[44] T. H. Koornwinder, The addition formula for Jacobi polynomials, III, Completion of the proof, Math. Centrum Amsterdam Report TW135(1972).

[45] T. H. Koornwinder, Orthogonal polynomials in two variables which are eigenfunctions of two algebraically independent partial differential operators, I, II, Nederl. Akad. Wetensch. Proc. Ser. A77 = Jndag. Math. 36(1974), 48-66.

[46] T. H. Koornwinder, Orthogonal polynomials in two variables which are eigenfunctions of two algebraically independent partial differential operators, III, IV, Nederl. Akad. Wetensch. Proc. Ser. A77 = Jndag. Math. 36(1974), 357-381.

[47] T. H. Koornwinder, The addition formula for Jacobi polynomials and the theory of orthogonal polynomials in two variables, a survey, Math. Centrum Amsterdam Report TW145(1974).

[48] T. H. Koornwinder, Jacobi polynomials, III, An analytic proof of the addition formula, SIAM J. Math. Analysis 6(1975), 533-543.

[49] T. H. Koornwinder, Yet another proof of the addition formula for Jacobi polynomials, to appear.

[50] T. H. Koornwinder, The addition formula for Laguerre polynomials, to appear.

[51] L. Koschmieder, Orthogonal polynomials on certain simple domains in the plane and in space (Spanish), Tecnica Rev. Fac. Ci. Ex. Tec. Univ. Nac. Tucuman 1(1951), 173-181.

[52] L. Koschmieder, A generator of orthogonal polynomials in the circle and in the triangle (Spanish), Rev. Mat. Hisp. -Amer. (4) 17(1957), 291-298.

[53] H. L. Krall and I. M. Sheffer, Orthogonal polynomials in two variables, Ann. Mat. Pura Appl. (4) 76(1967), 325-376.

[54] R. Lidl, Tschebyscheffpolynome in mehreren Variabelen, J. Reine Angew. Math., 273(1975), 178-198.

[55] O. Loos, Symmetric spaces, vol. II: Compact spaces and classification, Benjamin, New York, 1969.

[56] H. Maass, Zur Theorie der Kugelfunktionen einer Matrixvariabelen, Math. Ann. 135(1958), 391-416.

[57] C. D. Maldonado, Note on orthogonal polynomials which are "invariant in form" to rotations of axes, J. Math. Physics 6(1965), 1935-1938.

[58] R. B. Marr, On the reconstruction of a function on a circular domain from a sampling of its line integrals, J. Math. Anal. Appl. 45(1974), 357-374.

[59] C. Müller, Spherical harmonics, Lecture Notes in Math. 17, Springer-Verlag, Berlin, 1966.

[60] G. Munschy, Résolution de l'équation de Schrödinger des atomes à deux électrons, III, J. Phys. Radium (8) 18(1957), 552-558.

[61] G. Munschy and P. Pluvininage, Résolution de l'équation de Schrödinger des atomes à deux électrons, II, J. Phys. Radium (8) 18(1957), 157-160.

[62] D. R. Myrick, A generalization of the radial polynomials of F. Zernike, SIAM J. Appl. Math. 14(1966), 476-489.

[63] J. Proriol, Sur une famille de polynomes à deux variables orthogonaux dans un triangle, C. R. Acad. Sci. Paris 245 (1957), 2459-2461.

[64] R. L. Šapiro, Special functions related to representations of the group SU(n), of class I with respect to SU(n-1) (n ≥ 3) (Russian), Izv. Vysš Učeb. Zaved. Matematika (1968), no. 4 (71), pp. 97-107.

[65] J. W. Schleusner, A note on biorthogonal polynomials in two variables, SIAM J. Math. Anal. 5(1974), 11-18.

[66] I. G. Sprinkhuizen, Orthogonal polynomials in two variables. A further analysis of the polynomials orthogonal on a region bounded by two lines and a parabola, Math. Centrum Amsterdam Report TW 144(1974), to appear in SIAM J. Math. Anal.

[67] A. H. Stroud, Approximate calculation of multiple integrals,

Prentice Hall, Englewood Cliffs, New Jersey, 1971.

[68] G. Szegö, Orthogonal polynomials, Amer. Math. Soc. Colloquium Publications, vol. 23, Providence (R. I.), Third ed. , 1967.

[69] K. Trimèche, Convolution generalisée sur le disque unité, Thèse, Université de Tunis, 1974.

[70] N. Ja. Vilenkin and R. L. Šapiro, Irreducible representations of the group SU(n) of class I relative to SU(n-1) (Russian), Izv. Vysš. Učebn. Zaved. Matematika (1967), no. 7(62), pp. 9-20.

[71] H. Weyl, Theorie der Darstellung kontinuierlicher halbeinfacher Gruppen durch lineare transformationen, I, Math. Z. 23(1925), 271-309.

[72] H. Weyl, Theorie der Darstellung kontinuierlicher halbeinfacher Gruppen durch lineare transformationen, II, Math. Z. 24(1926), 328-376.

[73] F. Zernike, Beugungstheorie des Schneidenverfahrens und seiner verbesserten form, der Phasenkontrast-Methode, Physica 1(1934), 689-704.

[74] F. Zernike and H. C. Brinkman, Hypersphärishe Funktionen und die in sphärischen Bereichen orthogonalen Polynome, Proc. Kon. Akad. v. Wet. , Amsterdam, 38(1935), 161-170.

Mathematisch Centrum
Amsterdam

Special Functions of Matrix and Single Argument in Statistics

Alan T. James

FOREWORD

It is well-known that many of the 'ordinary' special functions of mathematical physics, and most of their properties, can be derived from the theory of group representations; Miller [1968]. The zonal spherical functions in particular are basically related to certain symmetric spaces, and thus to the representation theory of Lie groups. Although the theory for such functions appears quite general, the cases treated in the literature [Vilenkin [1965]] are invariably one-dimensional (i. e. functions of one variable). However, by applying group representation theory to more complicated symmetric spaces, namely $Gl(n, R)/0(n)$, the positive definite symmetric matrices, and $0(n)/0(p) \times 0(n-p)$, the Grassmann manifold, their respective zonal spherical functions for the finite dimensional irreducible subspaces have been derived. The spherical functions, $C_{\kappa}(X)$, of $Gl(n, R)/0(n)$ have been called zonal polynomials, since they turned out being homogeneous polynomials in the latent roots of the argument matrix, X. These also generate the spherical functions for $0(n)/0(p) \times 0(n-p)$, and other related functions. More importantly, they form a basis for a meaningful generalization of the whole family of hypergeometric functions to functions of matrix argument.

These generalizations are of importance in statistics, where they form the non-null or non-central parts of all multivariate distribution functions.

INTRODUCTION

Some detailed results of spherical functions on certain symmetric spaces will be outlined. Helgason [1962] in Chapter 10 deals with the abstract general theory. When he comes to examples of explicit formulae for functions he limits himself to spaces of rank 1, leading to classical functions of a single variable, but bringing out, however, the group theoretic significance. We seek explicit results for general rank.

The spherical functions on the Grassmann manifold of unoriented m subspaces of n space, $G_{m, n-m}$, will be the <u>even</u> spherical functions on the symmetric space, $SO(n)/(SO(m) \times SO(n-m))$ of oriented subspaces.

The spaces have an invariant measure and an invariant metric, the dual of which is an invariant differential operator called the Laplace Beltrami operator. A function on the space belonging to an irreducible representation of the transformation group is called a spherical function, and must be an eigenfunction of the Laplace Beltrami operator. The subgroup of the transformation group which leaves fixed a point of the space taken as origin, is called the isotropy, stationary or stability subgroup. Spherical functions invariant under the isotropy group are called zonal spherical functions.

The spherical functions open the way to a definition of scalar valued hypergeometric functions of matrix argument which supply integrals over matrix spaces which are needed in multivariate statistics.

The Lie algebra of differential operators which arise from the Laplace Beltrami operators supplies the differential operators which occur in the differential equations for the hypergeometric functions.

I. Zonal Polynomials of the Real Positive Definite Symmetric Matrices
1. The Symmetric Space $Gl(m, R)/O(m)$

The $m \times m$ positive definite real symmetric matrices X form a homogeneous space, $Gl(m, R)/O(m)$, with respect to congruence transformations by the linear group $Gl(m, R)$

$$X \rightarrow LXL' \qquad L \in Gl(m, R) \qquad\qquad (1.1)$$

The invariant measure is

$$\det X^{-\frac{1}{2}(m+1)}(dX) \qquad\qquad (1.2)$$

and the Hessian of the logarithm of the density function of the invariant measure namely,

$$dd \log(\det X^{-\frac{1}{2}(m+1)})$$
$$= -\tfrac{1}{2}(m+1) \, dd \log \det X$$
$$= -\tfrac{1}{2}(m+1) \, d \, tr(X^{-1}dX)$$
$$= +\tfrac{1}{2}(m+1) \, tr(X^{-1}dXX^{-1}dX)$$

is an invariant metric. Dropping the constant, we put

$$(ds)^2 = tr(X^{-1}dXX^{-1}dX) \qquad\qquad (1.3)$$

$$\text{Maass [1956]}$$

If we put

$$X = HYH'$$

where Y is the diagonal matrix of the latent roots $y_1, \ldots, y_m$ of X, then see James [1968], [1973]

$$(ds)^2 = \sum_{i=1}^{m} y_i^{-2}(dy_i)^2 + *(dh_{ij}) \ . \qquad\qquad (1.4)$$

The invariant measure is

$$\prod_{i=1}^{m} y_i^{-\frac{1}{2}(m+1)} \prod_{i<j}^{m} (y_i - y_j) \prod_{i=1}^{m} dy_i (dH) \qquad\qquad (1.5)$$

where (dH) is the invariant measure on the orthogonal group O(m). From these we shall later derive the Laplace Beltrami operator.

2. The Zonal Polynomials and Hypergeometric Functions

The zonal polynomial $C_\kappa(X)$ for the partition $\kappa = (k_1, \ldots, k_m)$ of k, is a homogeneous symmetric polynomial of degree k in the latent roots of X, which belongs to the irreducible representation $\{2\kappa\}$ of $Gl(m, R)$ in the space $R_k[X]$ of homogeneous polynomials $\varphi(X)$ in the elements of X transformed by

$$\varphi(X) \to (T_L \varphi)(X) = \varphi(L^{-1} X L^{-1'}) \qquad L \in Gl(m, R) \ . \qquad (1.6)$$

The polynomial $C_\kappa(X)$ is normalized such that

$$(\text{tr } X)^k = \sum_\kappa C_\kappa(X) \qquad (1.7)$$

where the summation is over all partitions κ of k into not more than m parts. Hence the exponential function of the trace has a power series (Constantine [1963])

$$\text{etr}(X) = \sum_{k=0}^{\infty} \sum_\kappa C_\kappa(X)/k! \equiv {}_0F_0(X) \ . \qquad (1.8)$$

A generalization of the negative binomial series is

$$\det(I_m - X)^{-a} = \sum_{k=0}^{\infty} \sum_\kappa (a)_\kappa C_\kappa(X)/k! \equiv {}_1F_0(a;X) \qquad (1.9)$$

where

$$(a)_\kappa = \prod_{i=1}^{m} (a - \tfrac{1}{2}(i-1))_{k_i} \qquad (1.10)$$

and

$$(a)_k = a(a+1)\ldots(a+k-1) \ . \qquad (1.11)$$

Since $C_\kappa(X)$ is a generalization of x^k and $(a)_\kappa$ a generalization of $(a)_k$, it seems natural to define

$$_pF_q(a_1, \ldots, a_p; b_1, \ldots, b_q; X) = \sum_{k=0}^{\infty} \sum_{\kappa} \frac{(a_1)_\kappa \cdots (a_p)_\kappa}{(b_1)_\kappa \cdots (b_q)_\kappa} \frac{C_\kappa(X)}{k!} \qquad (1.12)$$

In statistics, functions of two matrix variables are also required viz.

$$\int_{0(m)} {}_pF_q(a_1, \ldots, a_p; b_1, \ldots, b_q; XHYH') \, (dH) =$$

$$\sum_{k=0}^{\infty} \sum_{\kappa} \frac{(a_1)_\kappa \cdots (a_p)_\kappa}{(b_1)_\kappa \cdots (b_q)_\kappa} \frac{C_\kappa(X) C_\kappa(Y)}{C_\kappa(I_m) k!}$$

$$\overset{\text{def.}}{\equiv} {}_pF_q^{(m)}(a_1, \ldots, a_p; b_1, \ldots, b_q; X, Y) \qquad (1.13)$$

3. Is there an explicit formula for zonal polynomial coefficients and products of zonal polynomials?

Hua [1959] has considered zonal polynomials for the space $SU(n)/SO(n)$ which yields the same zonal polynomials. He calculates them in terms of characters of the linear group, but the coefficients involve integrals which seem difficult to evaluate. Bhanu Murti [1960] gives an integral representation.

By means of the Schur tensor representation theory which relates the finite dimensional representations of the linear group to those of the symmetric group, it can be shown (see James [1961]) that the coefficients of the zonal polynomials are the same as the coefficients of certain primitive idempotents of the group algebra of the symmetric group. Now since the characters of the symmetric group are the coefficients of principal idempotents, it seems that the calculation of zonal polynomial coefficients poses a combinatorial problem similar to the calculation of characters of the symmetric group. The latter problem has been under attack for more than half a century and although a general algorithm has been found, no explicit formula has been produced. An explicit workable

formula for zonal polynomial coefficients is presumably just about as remote.

Calculation of products of zonal polynomials as linear combinations of other zonal polynomials, seems likewise, rather intractable.
The complex or hermitian case is usually simpler than the real case. In
the hermitian case, we have an explicit formula as characters of the
linear group which are given by Van der Monde determinants.

Multiplication of characters of the linear group corresponds to decomposition of tensor products of the representations, a problem of interest to physicists for 50 years or so. The most explicit algorithm for
this, of which I know, is the so-called "raising operator" of G. de B.
Robinson [1961] in the equivalent problem of outer products of representations of the symmetric group. If a really explicit formula is lacking in
the complex case, what hope is there for the more difficult real case!

Consequently, in the study of functions of X we have to develop
a theory of power series with no explicit formula for the terms in them,
or the product of two such terms.

A related algebraic problem is that of the coefficients in the "binomial expansion"

$$C_\kappa^*(I_m + X) = \sum_{\nu \leq \kappa} \binom{\kappa}{\nu} C_\nu^*(X) \tag{1.14}$$

where
$$C_\kappa^*(X) = C_\kappa(X)/C_\kappa(I_m) . \tag{1.15}$$

The "binomial coefficients" $\binom{\kappa}{\nu}$ are positive rational numbers but
I know of no general formula for them, except in terms of the coefficients
g for products

$$C_\nu(X) C_\mu(X) = \sum_\kappa g_{\nu\mu}^\kappa C_\kappa(X) . \tag{1.16}$$

Leach[†] [1969] gives the result

[†] Quoted with his permission.

$$\binom{\kappa}{\nu} = \binom{k}{n} \sum_{\mu} g^{\kappa}_{\nu\mu} \; . \tag{1.17}$$

However the real zonal polynomials have been tabulated up to order 12 by Parkhurst and James [1974], and McLaren [1975] has published a computer program.

Tables of binomial coefficients have been given by Constantine [1966] and Pillai and Jouris [1969] and they have been studied by Muirhead [1974] and Bingham [1974].

In contrast to the algebraic side which quickly runs into combinatorial complexities, the analytic side develops very smoothly.

From the invariant metric and measure, we obtain the radial part of the Laplace Beltrami operator as (see James [1968])

$$\Delta = \sum_{i=1}^{m} \frac{\partial}{\partial y_i} \log \left\{ \prod_{i=1}^{m} y_i^{-\frac{1}{2}(m+1)} \prod_{j<k}^{m} (y_j - y_k) \right\} y_i^2 \frac{\partial}{\partial y_i} \tag{1.18}$$

$$= D^* + \frac{1}{2}(m-3)E$$

where

$$D^* = \sum_{i=1}^{m} y_i^2 \frac{\partial^2}{\partial y^2} + \sum_{i \neq j}^{m} y_i^2 (y_i - y_j)^{-1} \frac{\partial}{\partial y_i} \tag{1.19}$$

and the Euler operator

$$E = \sum_{i=1}^{m} y_i \frac{\partial}{\partial y_i} \tag{1.20}$$

simply multiplies homogeneous polynomials of degree k by k .

Since the zonal polynomial $C_\kappa(X)$ belongs to an irreducible representation it must be an eigenfunction of Δ, and, being homogeneous of degree k, an eigenfunction of E and therefore D^* ;

$$D^* C_\kappa(X) = (\rho_\kappa + k(m-1)) C_\kappa(X) \tag{1.21}$$

where

$$\rho_{\kappa} = \sum_i k_i(k_i - i) \tag{1.22}$$

To find $C_{\kappa}^{*}(X) = C_{\kappa}(X)/C_{\kappa}(I_m)$, we need the formula

$$C_{\kappa}(I_m) = 2^{2k} k! [(2k)!]^{-1} (\tfrac{1}{2}m)_{\kappa} \chi_{[2\kappa]}(1) \tag{1.23}$$

and $\chi_{[2\kappa]}(1)$ is the dimension of the representation $[\kappa]$ of the symmetric group given by

$$\chi_{[\kappa]}(1) = k! \prod_{i<j}^{p} (k_i - k_j - i+j) / \prod_{i=1}^{p} (k_i + p - i)! \tag{1.24}$$

where p is the number of nonzero parts of κ .

Since $C_{\kappa}^{*}(X)$ belongs to an irreducible representation and is invariant under the isotropy group $0(m)$ we have, corresponding to

$$(xz)^k = x^k z^k$$

that

$$\int_{0(m)} C_{\kappa}^{*}(XHZH') (dH) = C_{\kappa}^{*}(X) C_{\kappa}^{*}(Z) , \tag{1.25}$$

a formula that can also be proved by means of the Laplace Beltrami operator.

From this Constantine [1963] proved that

$$(\Gamma_m(a))^{-1} \int_{X>0} \mathrm{etr}(-X) \det X^{a-\frac{1}{2}(m+1)} C_{\kappa}(XZ) (dX) = (a)_{\kappa} C_{\kappa}(Z) , \tag{1.26}$$

where

$$\Gamma_m(a) = \int_{X>0} \mathrm{etr}(-X) \det X^{a-\frac{1}{2}(m+1)} (dX) \tag{1.27}$$

$$= \pi^{\frac{1}{4}m(m-1)} \prod_{i=1}^{m} \Gamma(a-\tfrac{1}{2}(i-1)) \tag{1.28}$$

and $(a)_{\kappa}$ is given by (1.10).

The two integrals suffice, together with the integral

$$\int_{0(n)} \text{etr}(LH)(dH) = {}_0F_1(\tfrac{1}{2}n; \tfrac{1}{4}LL') \ , \tag{1.29}$$

to yield power series for many multivariate statistical distributions, see Constantine [1963],[1968] James [1964], [1960], [1961] Khatri [1972] Sugiura [1972], [1973].

II. Orthogonal Polynomials of Matrix Argument

1. Generalized Jacobi Polynomials

They are symmetric polynomials in the latent roots $y_1, \cdots, y_m$ of X orthogonal relative to the beta density

$$\det X^{a-\frac{1}{2}(m+1)} \det(I_m - X)^{c-a-\frac{1}{2}(m+1)}(dX) \qquad 0 < X < I_m \tag{2.1}$$

or equivalently

$$\prod_{i=1}^{m} \left[y_i^{a-\frac{1}{2}(m+1)}(1-y_i)^{c-a-\frac{1}{2}(m+1)}(1-y_i)^{c-a-\frac{1}{2}(m+1)} \right] \prod_{i<j}^{m} (y_i - y_j) dy_1 \cdots dy_m$$

$$\tag{2.2}$$

$$0 < y_m < \ldots < y_1 < 1 \ .$$

In fact they are a Gramm Schmidt orthogonalization relative to the density, of the monomials

$$1, \ \sum_{i=1}^{m} y_i, \ \sum_{i<j}^{m} y_i y_j, \ \sum_{i=1}^{m} y_i^2, \ \sum_{i<j<k} y_i y_j y_k, \ \cdots, \ \sum y_1^{k_1} y_2^{k_2} \cdots y_m^{k_m}$$

of ascending weights given by the partitions

$$0, \ 1, \ 1^2, \ 2, \ 1^3, \ \ldots, \ \kappa = (k_1 k_2 \cdots k_m)$$

$P_\kappa(X)$ is an eigenfunction of the Laplace Beltrami operator

505

$$\Delta = D^* + (c-m+1)E - \delta^* - (a-\tfrac{1}{2}(m-1))\varepsilon$$

where D^* and E are given in (1.19) and (1.20)

$$\varepsilon = \sum_i \frac{\partial}{\partial y_i} \tag{2.3}$$

$$\delta^* = \sum_i y_i \frac{\partial^2}{\partial y_i^2} + \sum_{i,j}^m y_i(y_i-y_j)^{-1} \frac{\partial}{\partial y_i} = \frac{1}{2}(\varepsilon D^* - D^*\varepsilon) \equiv \frac{1}{2}[\varepsilon, D^*] \tag{2.4}$$

with eigenvalue

$$\rho_\kappa + kc \tag{2.5}$$

where ρ_κ is given by (1.22).

If we substitute

$$P_\kappa = \sum_{\sigma \le \kappa} \frac{(-1)^s \binom{\kappa}{\sigma} (c)_{(\kappa,\sigma)}}{(a)_\sigma} C^*_\sigma \tag{2.6}$$

in the differential equations, then from the relations (1.21), $E C^*_\kappa = kC^*_\kappa$

$$\varepsilon C^*_\kappa = \sum_i \binom{\kappa}{\kappa^{(i)}} C^*_{\kappa^{(i)}} \tag{2.7}$$

$$\delta^* C^*_\kappa = \sum_i (k_i + \tfrac{1}{2}(m-i)) \binom{\kappa}{\kappa^{(i)}} C^*_{\kappa^{(i)}} \tag{2.8}$$

where $\kappa^{(i)} = (k_1, \ldots, k_i-1, \ldots, k_m)$ and the summation is over those indices i for which $k_i > k_{i+1}$, we derive the recurrence relation

$$(c)_{(\kappa,\sigma)} = \sum_i \frac{\binom{\kappa}{\sigma_i} \binom{\sigma_i}{\sigma} (c)_{(\kappa,\sigma_i)}}{(k-s) \binom{\kappa}{\sigma} \left(c + \frac{\rho_\kappa - \rho_\sigma}{k-s}\right)} \tag{2.9}$$

where $\sigma_i = (s_1, \ldots, s_i+1, \ldots, s_m)$ and the summation is over indices i such that $s_{i-1} > s_i$.

If we put $a = \frac{1}{2}q$, and $c = \frac{1}{2}n$ for $m \leq q \leq \frac{1}{2}n$, then $P_\kappa(X)$ is the intertwining function which maps the representations $[2\kappa]$ of $0(n)$ from the representation of $0(n)$ given by the functions on the Grassmann manifold $G_{m,n-m}$ i.e. the m-subspaces of R^n, to the functions on $G_{q,n-q}$ see James and Constantine [1974].

2. The Generalized Gegenbauer Polynomial as a Zonal Function

A special case of the generalized Jacobi polynomials are the generalized Gegenbauer polynomials which are zonal spherical functions instead of intertwining spherical functions i.e. they produce idempotent operators which project on the spherical functions which form the representation space of $[2\kappa]$ in the space of functions on the Grassmann manifold, instead of mapping the spherical functions of one Grassmann manifold to those of another.

In this case, we take $q = m$ and put $a = \frac{1}{2}m$, $c = \frac{1}{2}n$ and substitute for $C_\kappa(I_m)$ in $C_\kappa^*(X) = C_\kappa(X)/C_\kappa(I_m)$. When X is possibly singular, $C_\kappa(X)$ is a symmetric function of its nonzero latent roots irrespective of its size. The same will be true of the generalized Gegenbauer polynomial which we define as

$$P_\kappa^{(n,m)}(X) = \sum_{\sigma \leq \kappa} \frac{(-1)^s \binom{\kappa}{\sigma} (\frac{1}{2}n)_{(\kappa,\sigma)} (2s)!}{((\frac{1}{2}m)_\sigma)^2 \, 2^s \, s! \, X_{[2\sigma]}^{(1)}} C_\sigma(X) . \qquad (2.10)$$

For arbitrary size of X, we define

$$P_\kappa^{(n,m)*}(X) = P_\kappa^{(n,m)}(X)/P_\kappa^{(n,m)}(I_m) \qquad (2.11)$$

For $m = 1$, the polynomials agree with ordinary Gegenbauer polynomials of even order viz.

$$P_k^{(n,1)}(x^2) = {}_2F_1(-k, \tfrac{1}{2}n-1+k; \tfrac{1}{2}; x^2) . \qquad (2.12)$$

507

According to the basic property of zonal functions

$$\int_{H_1 \epsilon \, 0(m)} \int_{H_2 \epsilon \, 0(n-m)} P_{\kappa}^{(n,\,m)*} (EH \, FH') \, (dH_1) \, (dH_2) =$$

$$P_{\kappa}^{(n,\,m)*} (E_{11}) P_{\kappa}^{(n,\,m)*} (F_{11}) \qquad (2.13)$$

where $H = \mathrm{diag}(H_1, H_2)$, E and F are symmetric $n \times n$ idempotent matrices of rank m, and E_{11} and F_{11} are their top left $m \times m$ submatrices.

3. Generalized Laguerre Polynomials

If we replace X by $c^{-1}X$ and let $c \to \infty$ the density (2.1) tends to

$$(\Gamma_m(a))^{-1} \, \mathrm{etr}(-X) \det X^{a-\frac{1}{2}(m+1)}(dX) \qquad (2.14)$$

and

$$P_{\kappa}(c^{-1}X) \to \sum_{\sigma \le \kappa} \frac{(-1)^s \binom{\kappa}{\sigma}}{(a)_{\sigma}} C_{\sigma}^{*}(X) \qquad (2.15)$$

which are the generalized Laguerre polynomials orthogonal with respect to the gamma density (see Constantine [1963], [1966] and James and Constantine [1974]). Herz [1955] derived polynomials orthogonal for different orders.

The integral of the square of the Laguerre polynomial with respect to the gamma density is given by

$$h_{\kappa} = k!/(C_{\kappa}(I_m)(a)_{\kappa}) \quad . \qquad (2.16)$$

If we replace y_i by $c^{-1}y_i$ and divide the differential equation by $(-c)$, it becomes a generalization of the Laguerre equation

$$(\delta^{*} - E + (a-\tfrac{1}{2}(m+1))\varepsilon + k)f(\chi) = 0 \quad .$$

508

Although the differential equation does not depend upon the partition κ of k specifically but only upon k, it yields a recurrence relation which determines the Laguerre polynomial uniquely given the top term as a multiple of $C_\kappa^*(X)$. Having verified the form of the denominator $(a)_\sigma$, the differential equation must imply a recurrence relation between the binomial coefficients $\begin{pmatrix} \kappa \\ \sigma \end{pmatrix}$ which can be obtained from (2.9) by taking the limit as $c \to \infty$ to yield

$$(k-s)\begin{pmatrix} \kappa \\ \sigma \end{pmatrix} = \sum_i \begin{pmatrix} \kappa \\ \sigma_i \end{pmatrix}\begin{pmatrix} \sigma_i \\ \sigma \end{pmatrix} , \qquad (2.18)$$

from which all binomial coefficients can be calculated from a knowledge of contiguous coefficients $\begin{pmatrix} \sigma_i \\ \sigma \end{pmatrix}$.

4. Hermite Polynomials

If we replace X by $aI_m + a^{\frac{1}{2}}X$ in the density (2.14) and let $a \to \infty$, then by taking the logarithm of the density, one can readily verify that the limiting density is

$$\text{etr}(-\tfrac{1}{2}X^2) \ (dX) \qquad -\infty < X < \infty . \qquad (2.19)$$

It is natural to define the limit of the Laguerre polynomials as the Hermite polynomials of a symmetric matrix (Hayakawa [1969] has considered a different case of Hermite polynomials of a rectangular matrix).

The limit of the Laguerre differential equation is the Hermite equation

$$(\delta^{**} - E + k) f(\chi) = 0 , \qquad (2.20)$$

where

$$\delta^{**} = \sum_{i=1}^{m} \frac{\partial^2}{\partial y_1^2} + \sum_{\substack{i,j \\ i \neq j}}^{m} (y_i - y_j)^{-1} \frac{\partial}{\partial y_i}$$

$$= \tfrac{1}{2}[\varepsilon, [\varepsilon, D^*]] . \qquad (2.21)$$

Substitution of a series with undetermined coefficients

$$H_\kappa(X) = \sum_{\lambda \leq \kappa} b_\lambda \, C_\lambda^*(X) \tag{2.22}$$

leads to a recurrence relation

$$b_\lambda = (k-\ell)^{-1} \left\{ \sum_i \binom{\lambda_{ii}}{\lambda_i}\binom{\lambda_i}{\lambda} b_{\lambda_{ii}} \right.$$

$$\left. + \sum_{i<j} (k_i - k_j - \tfrac{1}{2}(i-j)) \left[\binom{\lambda_{ij}}{\lambda_j}\binom{\lambda_j}{\lambda} - \binom{\lambda_{ij}}{\lambda_i}\binom{\lambda_i}{\lambda} \right] b_{\lambda_{ij}} \right\} \tag{2.23}$$

where λ_{ii} is the partition obtained from λ by increasing the ith element by 2, and λ_{ij} have the ith and jth elements increased by 1. The summations with respect to i and with respect to i, j are over values such that λ_{ii} and λ_{ij} are admissible.

From the recurrence relation, the coefficients b_λ of the Hermite polynomial $H_\kappa(X)$ can be successively calculated.

III. Differential Equations for Hypergeometric Functions of Matrix Argument.

Muirhead [1970] has derived differential equations for the hypergeometric functions:

Function	Differential Equation
$_0F_1(c;X)$	$\delta^* F + (c-\tfrac{1}{2}(m-1))\varepsilon F - mF = 0$
$_1F_1(a;c;X)$	$\delta^* F + [(c-\tfrac{1}{2}(m-1))\varepsilon - E]F - amF = 0$
$_2F_1(a;b;c;X)$	$(\delta^* - D^*)F + [(c-\tfrac{1}{2}(m-1))\varepsilon - (a+b+1-\tfrac{1}{2}(m-1))E]F - mabF = 0$

and proved that they are the unique solutions subject to the conditions

(a) they are symmetric functions of the latent roots of X

(b) they are analytic at X = 0, with value 1.

When $m = 1$, they reduce to the ordinary Bessel and confluent and Gaussian hypergeometric equations.

If one replaces X by $b^{-1}X$ in the Gaussian d. e. and lets $c \to \infty$, one obtains the confluent equation and likewise if one then goes on to replace X by $a^{-1}X$ and lets $a \to \infty$ one comes down to the Bessel equation.

To establish the differential equations, Muirhead [1970a] found the identities

$$\sum_i \binom{\kappa \ i}{\kappa} C_{\kappa_i}(I) = m(k+1) \, C_\kappa(I)$$

$$\sum_i \binom{\kappa \ i}{\kappa} (k_i - \tfrac{1}{2}(i-1)) C_{\kappa_i}(I) = k(k+1) C_\kappa(I)$$

$$\sum_i \binom{\kappa_i}{\kappa} (k_i - \tfrac{1}{2}(i-1))^2 C_{\kappa_i}(I) = (k+1)(\rho_\kappa + \tfrac{1}{2}k(m+1)) C_\kappa(I) \ .$$

They may be derived from a formula involving a p-fold Lie product

$$2^{-p}[[\dots[\varepsilon, D^*], D^*]\dots, D^*]C_\kappa^*(Y) = \sum_i (k_i + \tfrac{1}{2}(m-i))^p \, C^*_{\kappa_i}(Y) \ ,$$

which is readily established by induction.

Applying the p-fold Lie operator for $p = 0, 1, 2$ to the identity etr $Y = \sum_\kappa C_\kappa(Y)/k!$, using the formula to evaluate the R. H. S. and equating coefficients of $C_\kappa^*(Y)$ yields the successive identities.

Muirhead [1970b] has shown that as a generalization of equation 6. 8. 1 of Erdelyi et al. [1953],

$$\lim_{c \to \infty} {}_2F_1(a, b; c; 1 - cX^{-1}) = \det X^b \, \psi(b, b - a + \tfrac{1}{2}(m+1); X)$$

where

$$\psi(a, c; X) = (\Gamma_m(a))^{-1} \int_{Z > 0} \text{etr}(-XZ) \det Z^{a - \frac{1}{2}(m+1)} \cdot$$

$$\cdot \det(I_m + Z)^{c - a - \frac{1}{2}(m+1)} dZ$$

and ψ is a solution of the confluent hypergeometric equation. He uses it to derive asymptotic expansions of probability integrals.

Constantine and Muirhead [1972] have derived the following d.e.'s for the function of two variables, where now $X = \text{diag}(x_i)$ and $Y = \text{diag}(y_i)$.

Function	Differential Equation
$_0F_0(X, Y)$	$\delta_X^* F - \gamma_Y F = \frac{1}{2}(m-1)\text{tr}(Y)F$
$_1F_0(X, Y)$	$\delta_X^* F - a\gamma_Y F - \eta_Y F = \frac{1}{2}a(m-1)\,\text{tr}(Y)F$
$_0F_1(X, Y)$	$\delta_X^* F + [c - \frac{1}{2}(m-1)]\epsilon_X F = \text{tr}(Y)F$
$_1F_1(X, Y)$	$\delta_X^* F + [c - \frac{1}{2}(m-1)]\epsilon_X F - \gamma_Y F = a\,\text{tr}(Y)F$
$_2F_1(X, Y)$	$\delta_X^* F + [c - \frac{1}{2}(m-1)]\epsilon_X F - [a+b - \frac{1}{2}(m-1)]\gamma_Y F$

$$-\eta_Y F = ab\,\text{tr}(Y)F$$

where

$$\gamma_X = \sum_{i=1}^m x_i^2 \frac{\partial}{\partial x_i}$$

$$\eta_X = \sum_{i=1}^m x_i^3 \cdot \frac{\partial^2}{\partial x_i^2} + \sum_{i \neq j} x_i^3(x_i - x_j)^{-1} \frac{\partial}{\partial x_i} + (1 - \frac{1}{2}(m-1))\gamma_X$$

$$= \frac{1}{2}[D_X^*, \gamma_X] \ .$$

Fujikoshi [1975] has given differential equations for $_3F_2$ and $_2F_2$ functions of a single matrix variable.

IV. Probability Functions of Ordered Roots.

1. Differential Equations

I would like to announce some interesting as yet unpublished results of S. R. Eckert, having obtained his permission.

They concern the integral

$$F^{(m)}(x) = h_m \int_{-\infty < X < xI_m} \mathrm{etr}(-\tfrac{1}{4}X^2)\,(dX) \qquad (4.1)$$

where X is an $m \times m$ real symmetric matrix,

$$\mathrm{etr}(A) = \exp(\mathrm{tr}\,A), \quad (dX) = \prod_{i \le j}^{m} dx_{ij} \quad \text{and}$$

$$h_m = \left[2^{\frac{1}{4}m(m+3)} \pi^{\frac{1}{4}m(m+1)}\right]^{-1}. \qquad (4.2)$$

$F^m(x)$ is the distribution function of the largest latent root of a random symmetric matrix X of order m whose components are normally distributed.

If $y_1, \ldots, y_m$ are the latent roots of X in descending order, then

$$F^m(x) = k_m \int_{y_m=-\infty}^{x} \int_{y_{m-1}=-\infty}^{y_m} \cdots \int_{y_1=-\infty}^{y_2} \exp\left(-\tfrac{1}{4}\sum_{i=1}^{m} y_i^2\right) \cdot$$

$$\prod_{i<j}^{m} (y_i-y_j)dy_1\cdots dy_m \qquad (4.3)$$

where $k_m = c_m h_m$ \qquad (4.4)

and

$$c_m = \frac{\pi^{\frac{1}{2}m^2}}{\Gamma_m(\frac{1}{2}m)} \qquad (4.5)$$

and the multivariate gamma function is given by

$$\Gamma_m(a) = \int_{X>0} \mathrm{etr}(-X)\det X^{a-\frac{1}{2}(m+1)}\,(dX)$$

$$= \pi^{\frac{1}{4}m(m-1)} \prod_{i=1}^{m} \Gamma(a-\tfrac{1}{2}(i-1)). \qquad (4.6)$$

513

The derivative is the density function

$$f^{(m)}(x) = \frac{d}{dx} F^{(m)}(x)$$

$$= k_m \left(\exp\left(-\tfrac{1}{4}x^2\right) \right) \int_{y_m=-\infty}^{x} \cdots \int_{y_2=-\infty}^{y_3} \exp\left(-\tfrac{1}{4}\sum_{i=2}^{m} y_i^2\right)$$

$$\prod_{i>j\geq 2}^{m} (y_i-y_j) \prod_{i=2}^{m} (x-y_i)dy_2 \cdots dy_m \; . \tag{4.7}$$

If we put $z_i = x-y_{i+1}$ for $i = 1, \ldots, m-1$ we have

$$f^{(m)}(x) = k_m \exp\left(-\frac{m}{4} x^2\right) \int_{z_{m-1}=0}^{\infty} \cdots \int_{z_2=0}^{z_3} \int_{z_1=0}^{z_2} \exp\left(\tfrac{1}{2}x\sum_{i=1}^{m-1} z_i\right)$$

$$\prod_{i>j}^{m-1} (z_i-z_j) \exp\left(-\tfrac{1}{4}\sum_{i=1}^{m-1} z_i^2\right) \prod_{i=1}^{m-1} z_i dz_1 \cdots dz_{m-1} \tag{4.8}$$

To obtain differential equations, we introduce auxiliary functions similar to those given by Davis [1972]

$$E_r^{(m)}(x) = \exp\left(-\frac{m-1}{4}x^2\right) \int_{z_{m-1}=0}^{\infty} \cdots \int_{z_1=0}^{z_2} \exp\left(\tfrac{1}{2}x\sum_{i=1}^{m-1} z_i\right) \exp\left(-\tfrac{1}{4}\sum_{i=1}^{m-1} z_i^2\right)$$

$$\prod_{i>j}^{m-1} (z_i-z_j) \, \sigma_{m-r-1}(z) dz_1 \cdots dz_{m-1} \tag{4.9}$$

where $\sigma_p(z)$ is the pth elementary symmetric function of $z_1, \ldots, z_{m-1}$. Then

$$F^{(m-1)}(x) = k_{m-1} E_{m-1}^{(m)}(x) \tag{4.10}$$

and

514

$$f^{(m)}(x) = k_m \exp(-\tfrac{1}{4}x^2) \, E_0^{(m)}(x) \, . \qquad (4.11)$$

By differentiation with respect to x followed by partial integration with respect to the z_i to eliminate quadratic factors in $\sum z_i \, \sigma_{m-r-1}(\underset{\sim}{z})$, one obtains the differential equations

$$\frac{d}{dx} \, E_r^{(m)}(x) = -\tfrac{1}{2}rxE_r^{(m)}(x) + \tfrac{1}{2}(m-r) \, E_{r-1}^{(m)}(x)$$

$$+\tfrac{1}{2}(r+1) \, (r+2) \, E_{r+1}^{(m)}(x)$$

$$E_{-1} \equiv 0 \equiv E_m \qquad\qquad r = 0, 1, \ldots, m-1 \, . \qquad (4.12)$$

They are a limit of those obtained by Davis [1972].

The differential equations can be solved recursively starting at

$$E_0^{(1)} = 1 \, .$$

Since the distribution function $F^{(m)}(x)$ is the integral of the density function $f^{(m)}(x)$ we have

$$E_{m-1}^{(m)}(x) = k_m^{-1} \, k_{m-1} \int_{-\infty}^{x} \varphi(\xi) \, E_0^{(m-1)}(\xi) d\xi \qquad (4.13)$$

$$(4.14)$$

where $\varphi(x) = \exp(-\tfrac{1}{4}x^2) \, .$

The differential equation can be used to calculate $E_{r-1}^{(m)}(x)$ in terms of $E_{r+1}^{(m)}(x)$ and $E_r^{(m)}(x)$ and the derivative of the latter, for $r = m-1, m-2, \ldots, 1$. Hence by iteration, $E_0^{(m)}(x)$ can be obtained from $E_{m-1}^{(m)}(x)$. Multiplication by $\varphi(x)$ and integration carries us on to $E_m^{(m+1)}(x)$ etc.

$$E_0^{(m-1)} \;\rightarrow\; E_{m-1}^{(m)} \;\rightarrow\; E_0^{(m)} \;\rightarrow\; E_m^{(m+1)} \;\rightarrow\; \text{etc.} \qquad (4.15)$$

$\times \varphi(x)$ and | d. e. s. | $\times \varphi(x)$ and
integration | r times | integration

2. The Transcendental Basis

An interesting feature of the solutions of the differential equations is that they are not transcendent with respect to classical functions but only involve products of powers of error integrals with polynomial coefficients.

Let

$$\varphi = \exp(-x^2/4)$$

$$\Phi_1 = \int_{-\infty}^{x} \varphi(\xi)d\xi \tag{4.16}$$

$$\Phi_2 = \int_{-\infty}^{x} (\varphi(\xi))^2 d\xi \ , \tag{4.17}$$

and put

$$\Omega_0 = 1$$

$$\Omega_1 = \Phi_1$$

$$\Omega_{2k} = \Phi_2^k$$

$$\Omega_{2\,k+1} = \Phi_1 \Phi_2^k \tag{4.18}$$

Then the functions $E_r^{(m)}(x)$, $r = 0, 1, \ldots, m-1$ are linear combinations, with polynomials, π_i, in x as coefficients, of the transcendental basis

$$\varphi^{m-1}, \ \varphi^{m-2} \Omega_1, \ldots, \Omega_{m-1} \ . \tag{4.19}$$

3. The Mapping Theorem

For each $p = 1, \ldots, m-1$, there is an independent solution $\tilde{E}_{r,\,p}^{(m)}(x)$ of the differential equation which has the transcendental basis

$$\varphi^{m-p-1}, \ \varphi^{m-p-2} \Omega_1, \ldots, \Omega_{m-p-1} \tag{4.20}$$

Writing $\tilde{E}_{r,\,1}^{(m)}(x) \equiv E_r^{(m)}(x)$, $r = 0, \ldots, m-1$ as the auxiliary functions of the first solution, then we obtain the pth solution $\tilde{E}_{r,\,p}^{(m)}(x)$ by the mapping

$$\tilde{E}_{r,\,p}^{(m)}(x) = T_p \tilde{E}_{r,\,1}^{(m)}(x) \qquad (4.21)$$

where, if

$$v = \sum_{j=0}^{m-1} \pi_j \, \varphi^{m-j-1} \, \Omega_j \qquad (4.22)$$

then

$$T_p(v) = \sum_{j=p}^{m-1} t_{pj} \, \pi_j \, \varphi^{m-j-1} \, \Omega_{j-p} \qquad (4.23)$$

and

$$t_{pj} = \begin{cases} 0 \ \text{if} \ p > j \ \text{or if} \ p \ \text{is odd and} \ j \ \text{is even} \\[2mm] \dbinom{[\frac{1}{2}j]}{[\frac{1}{2}p]} \ \text{otherwise} \end{cases} \qquad (4.24)$$

The distribution of the sth ordered root has similar E functions which satisfy the same differential equation, and which are therefore linear combinations of the $\tilde{E}$ functions.

The corresponding functions derived from the gamma density

$$(\Gamma_m(a))^{-1} \ \text{etr}\,(-X) \ \det X^{a-\frac{1}{2}(m+1)}(dX) \qquad (4.25)$$

and the beta density

$$\Gamma_m(c)\,(\Gamma_m(a))^{-1}\,(\Gamma_m(c-a))^{-1} \det X^{a-\frac{1}{2}(m+1)} \det(I_m-X)^{c-a-\frac{1}{2}(m+1)}(dX) \ (4.26)$$

satisfy differential equations given by Davis [1972] and can be solved recursively. Solutions for m up to 5 have finite transcendental bases.

Dr. Eckert's results have an interesting structure which must probably occur in many other mathematical situations yet to be discovered.

517

REFERENCES

BHANU MURTI, T. S. [1960]. Plancherel's measure for the factor space SL(n, R)/S0(n, R). Dokl. Akad. Nauk. S. S. S. R., 133, 503-506.

BINGHAM, C. [1974]. An identity involving partitional generalized binomial coefficients. J. Multivariate Analysis 4, 210-223.

CONSTANTINE, A. G. [1963]. Some noncentral distribution problems in multivariate analysis. Ann. Math. Statist. 34, 1270-85.

CONSTANTINE, A. G. [1966]. The distribution of Hotelling's generalized T_0^2, Ann. Math. Statist. 37, 215-25.

CONSTANTINE, A. G. and MUIRHEAD, R. J. [1972], Partial differential equations for hypergeometric functions of two argument matrices. J. Multivariate Anal. 3, 332-8.

DAVIS, A. W. [1972]. On the marginal distributions of the latent roots of the multivariate beta matrix. Ann. Math. Statist., 43, 1664-1670.

ERDÉLYI, A. et al. [1953]. Higher transcendental functions, vol. 1 McGraw-Hill, New York.

FUJIKOSHI, Y. [1975]. Partial differential equations for hypergeometric functions $_3F_2$ of matrix argument. J. Multivariate Anal. (to appear).

HELGASON, S. [1962]. Differential Geometry and Symmetric Spaces, Academic Press, New York.

HERZ, C. S. [1955]. Bessel functions of Matrix Argument. Ann. of Math. 61, 474-523.

HUA, LOO-KENG [1959]. Harmonic Analysis of Functions of Several Complex Variables in the Classical Domains. Moscow.

JAMES, A. T. [1960]. The distribution of the latent roots of the covariance matrix. Ann. Math. Statist., 31, 151-8.

JAMES, A. T. [1961a]. Zonal polynomials of the real positive definite symmetric matrices. Ann. of Math. 74, 456-69.

JAMES, A. T. [1961b]. The distribution of noncentral means with known covariance. Ann. Math. Statist., 32, 874-82.

JAMES, A. T. [1964]. Distributions of matrix variates and latent roots derived from normal samples. Ann. Math. Statist., 35, 475-501.

JAMES, A. T. [1968]. Calculation of zonal polynomial coefficients by use of the Laplace Beltrami operator. Ann. of Math. Statist. 39, 1711-1718.

JAMES, A. T. [1973]. The variance information manifold and the functions on it. In Multivariate Analysis (P. R. Krishnaiah. ed.) Academic Press, New York.

JAMES, A. T. and CONSTANTINE, A. G. [1974]. Generalized Jacobi polynomials as spherical functions of the Grassmann Manifold, Proc. Lond. Math. Soc. 28, 174-192.

KHATRI, C. G. [1972]. On the exact finite series distribution of the smallest or the largest root of matrices in three situations. J. Multivariate Anal. 2, 201-207.

LEACH, Brian G. [1969]. Bessel functions of matrix argument with statistical applications. Ph. D. thesis Adelaide.

MAASS, H. [1956]. Spherical functions and quadratic forms. J. Indian Math. Soc. 20, 117-162.

McLAREN, M. L. [1975]. Coefficients of the zonal polynomials. Applied Statistics. (JRSS Series C) to appear.

MILLER, W. [1968]. Lie Theory and Special Functions. Academic Press, New York.

MUIRHEAD, R. J. [1970a]. Systems of partial differential equations for hypergeometric functions of matrix argument. Ann. Math. Statist. 41, 991-1001.

MUIRHEAD, R. J. [1970b]. Asymptotic distributions of some multivariate tests. Ann. Math. Statist. 41, 1002-1010.

MUIRHEAD, R. J. [1974]. On the calculation of generalized binomial coefficients. J. Mult. Anal. 4. 3, 341-346.

PARKHURST, A. M. and JAMES, A. T. [1974]. Zonal polynomials of order 1 through 12, Selected tables in mathematical statistics, vol. 2, Ed. by H. L. Harter and D. B. Owen, American Mathematical Society, Providence, Rhode Island, 199-388.

PILLAI, K. C. S. and JOURIS, G. M. [1969]. On the moments of elementary symmetric functions of the roots of two matrices. Ann. Inst. Statist. Math. 21, 309-20.

ROBINSON, G. de B. [1961]. Representation Theory of the Symmetric Group Toronto.

SUGIURA, N. [1972]. Asymptotic solutions of the hypergeometric function $_1F_1$ of matrix argument, useful in multivariate analysis. Ann. Inst. Statist. Math. 24, 517-524.

SUGIURA, N. [1973]. Derivatives of the characteristic root of a symmetric or a Hermitian matrix with two applications in multivariate analysis. Communication in Statist. 1, 393-417.

VILENKIN, N. Y. [1965]. Special Functions and the Theory of Group Representations. Izd. Nauka., Moscow (in Russian); AMS Transl., Providence, Rhode Island 1968 (English Transl.).

The University of Adelaide

520

Some Properties of Determinants of Orthogonal Polynomials
Samuel Karlin and James McGregor

Introduction

In the study of coincidence problems for birth and death stochastic processes [3;4;5] the authors encountered systems of orthogonal polynomials in several variables constructed as follows. Let $\psi(x)$ be a distribution function on the semi-axis $[0, \infty)$ with infinitely many points of increase, and with finite moments of all orders. Let $Q_n(x)$, $n = 0, 1, 2, \ldots$ be the orthogonal polynomials for the distribution ψ, and taking integers $0 \leq i_1 < i_2 < \ldots < i_n$ we form the determinant

$$(1) \qquad Q \begin{pmatrix} i_1, i_2, \ldots, i_n \\ x_1, x_2, \ldots, x_n \end{pmatrix} = \det \| Q_{i_\alpha}(x_\beta) \| \ .$$

This is a polynomial in the n variables $x_1, \ldots, x_n$ and the collection of all such determinants $(0 \leq i_1 < \ldots < i_n)$ constitutes an orthogonal system on the simplex $S = \{(x_1, \ldots, x_n; \ 0 \leq x_1 < x_2 < \ldots < x_n\}$. In fact if $\int_0^\infty Q_i(x)Q_j(x)d\psi(x) = \delta_{ij}/\pi_j$, then (see [4])

$$(2) \qquad \int \ldots \int_S Q \begin{pmatrix} i_1, \ldots, i_n \\ x_1, \ldots, x_n \end{pmatrix} Q \begin{pmatrix} j_1, \ldots, j_n \\ x_1, \ldots, x_n \end{pmatrix} d\psi(x_1) \ldots d\psi(x_n)$$

$$= \delta_{i_1 j_1} \delta_{i_2 j_2} \ldots \delta_{i_n j_n} / (\pi_{j_1} \pi_{j_2} \ldots \pi_{j_n}) \ .$$

The importance of these determinantal systems for the analysis of multi-particle birth death processes is set forth in [4]. The collection (l) is also a good prototype case of the theory of multi parameter eigenvalue problems studied to an extent in a broader algebraic setting by Atkinson [1]. There are many properties of orthogonal polynomials in one variable that possess analogues for the determinantal systems (l). These include recurrence formulas, a version of the Christoffel Darboux formula and Wronskian identities, the completeness property, quadrature formulas, generating functions in the classical cases, etc. Here we elaborate a number of their detailed developments.

We assume throughout that the $Q_n(x)$ are normalized such that $Q_n(0) = 1$. This normalization is possible since all zeros of Q_n lie in the open interval $(0, \infty)$. The recurrence relation then has the form

(3) $\qquad -xQ_n = \mu_n Q_{n-1} - (\lambda_n + \mu_n)Q_n + \lambda_n Q_{n+1}$

$\qquad\qquad$ where $\mu_0 = 0$, $\mu_{n+1} > 0$, $\lambda_n > 0$ for $n \geq 0$.

The constants π_j are

(3a) $\qquad \pi_0 = 1, \quad \pi_j = \dfrac{\lambda_0 \lambda_1 \cdots \lambda_{j-1}}{\mu_1 \mu_2 \cdots \mu_j}, \quad j \geq 1$.

1. Recurrence formulas

The determinantal polynomials $Q\begin{pmatrix} i_1, \ldots, i_n \\ x_1, \ldots, x_n \end{pmatrix}$ of order n satisfy n different recurrence formulas which may be derived from the basic recurrence formula $-xQ(x) = AQ(x)$ written explicitly in (3). Here $A = \|a_{ij}\|$ is the infinitesimal Jacobi matrix, and $Q(x) = (Q_0(x), Q_1(x), Q_2(x), \ldots)$. Of these recurrence formulas the one most directly related to the probabilistic applications is

$$-(x_1 + \ldots + x_n) \, Q\begin{pmatrix} i_1, \ldots, i_n \\ x_1, \ldots, x_n \end{pmatrix} = \sum_{k=1}^{n} \left\{ \mu_{i_k} \, Q\begin{pmatrix} i_1, \ldots, i_{k-1}, i_k-1, i_{k+1}, \ldots, i_n \\ x_1, \ldots\ldots\ldots\ldots\ldots\ldots\ldots, x_n \end{pmatrix} \right.$$

$$\left. - (\lambda_{i_k} + \mu_{i_k}) \, Q\begin{pmatrix} i_1, \ldots, i_n \\ x_1, \ldots, x_n \end{pmatrix} + \lambda_{i_k} \, Q\begin{pmatrix} i_1, \ldots, i_{k-1}, i_k+1, i_{k+1}, \ldots, i_n \\ x_1, \ldots\ldots\ldots\ldots\ldots\ldots\ldots, x_n \end{pmatrix} \right\}$$

which may be written in the form

$$(4) \quad -(x_1 + \ldots + x_n) \, Q\begin{pmatrix} i_1, \ldots, i_n \\ x_1, \ldots, x_n \end{pmatrix} = \sum_{0 \le j_1 < \ldots < j_n} D_{n,1}\begin{bmatrix} i_1, \ldots, i_n \\ j_1, \ldots, j_n \end{bmatrix} Q\begin{pmatrix} j_1, \ldots, j_n \\ x_1, \ldots, x_n \end{pmatrix}$$

where

$$D_{n,1}\begin{bmatrix} i_1, \ldots, i_n \\ j_1, \ldots, j_n \end{bmatrix} = 0 \quad \text{if} \quad \sum_{\nu=1}^{n} |i_\nu - j_\nu| > 1 \, ,$$

$$D_{n,1}\begin{bmatrix} i_1, \ldots, i_n \\ i_1, \ldots, i_n \end{bmatrix} = -\sum_{\nu=1}^{n} (\lambda_{i_\nu} + \mu_{i_\nu})$$

$$D_{n,1}\begin{bmatrix} i_1, \ldots, i_{k-1}, i_k, i_{k+1}, \ldots, i_n \\ i_1, \ldots, i_{k-1}, j, i_{k+1}, \ldots, i_n \end{bmatrix} = \begin{cases} \lambda_{i_k} & \text{if} \quad j = i_k + 1 \, , \\ \mu_{i_k} & \text{if} \quad j = i_k - 1 \, . \end{cases}$$

In these and in subsequent formulas we adhere to the natural convention that $Q\begin{pmatrix} i_1, \ldots, i_n \\ x_1, \ldots, x_n \end{pmatrix}$ is zero if two of the i_ν are equal. The above formula is most easily derived if in the identity

$$-(x_1 + \ldots + x_n) \, Q\begin{pmatrix} i_1, \ldots, i_n \\ x_1, \ldots, x_n \end{pmatrix} = \sum_{\sigma} (\text{sign } \sigma) \sum_{r=1}^{n}$$

$$Q_{i_1}(x_{\sigma_1}) \ldots Q_{i_{r-1}}(x_{\sigma_{r-1}})[-x_{\sigma_r} Q_{i_r}(x_{\sigma_r})] Q_{i_{r+1}}(x_{\sigma_{r+1}}) \ldots Q_{i_n}(x_{\sigma_n})$$

we apply the basic recurrence formula to the factors in square brackets on the right and then recombine terms. The notation here is as follows:

$\sum\limits_{\sigma}$ indicates a summation over all permutations σ of the n letters $1, \ldots, n$ and σ_r is the image of r under the permutation σ.

Another recurrence formula, obtained by applying the composition formula of [2, (p. 98)] to the equation $-x\underline{Q}(x) = A\underline{Q}(x)$ is

$$(5) \quad (-1)^n x_1 x_2 \cdots x_n \; Q\begin{pmatrix} i_1, \cdots, i_n \\ x_1, \cdots, x_n \end{pmatrix} = \sum_{0 \le j_1 < \cdots < j_n} D_{n,n}\begin{bmatrix} i_1, \cdots, i_n \\ j_1, \cdots, j_n \end{bmatrix} Q\begin{pmatrix} j_1, \cdots, j_n \\ x_1, \cdots, x_n \end{pmatrix}$$

where

$$D_{n,n}\begin{bmatrix} i_1, \cdots, i_n \\ j_1, \cdots, j_n \end{bmatrix} = A\begin{pmatrix} i_1, \cdots, i_n \\ j_1, \cdots, j_n \end{pmatrix} \quad .$$

The following facts concerning these minors of A are useful in studying the general recurrence formula. $A\begin{pmatrix} i_1, \cdots, i_n \\ j_1, \cdots, j_n \end{pmatrix}$ is zero if $\max|i_\nu - j_\nu| > 1$. However if $\max\limits_{\nu}(|i_\nu - j_\nu|) \le 1$ then

$$(-1)^n (-1)^{\sum(i_\nu + j_\nu)} A\begin{pmatrix} i_1, \cdots, i_n \\ j_1, \cdots, j_n \end{pmatrix} > 0 \quad .$$

If k is an integer $1 \le k \le n$ the elementary symmetric function of degree k of the n variables $x_1, \ldots, x_n$ is

$$E_k(x_1, \ldots, x_n) = \sum_{1 \le \alpha_1 < \cdots < \alpha_k \le n} x_{\alpha_1} x_{\alpha_2} \cdots x_{\alpha_k} \quad .$$

The general recurrence formula is

(6)
$$(-1)^k E_k(x_1, \ldots, x_n) Q\begin{pmatrix} i_1, \ldots, i_n \\ x_1, \ldots, x_n \end{pmatrix} =$$

$$\sum_{0 \le j_1 < \ldots < j_n} D_{n,k}\begin{bmatrix} i_1, \ldots, i_n \\ j_1, \ldots, j_n \end{bmatrix} Q\begin{pmatrix} j_1, \ldots, j_n \\ x_1, \ldots, x_n \end{pmatrix}$$

$$(1 \le k \le n)$$

and formulas (4) and (5) are the special cases $k = 1$ and $k = n$, respectively. The left side of (6), invoking Laplace's expansion by minors, is equal to

$$\sum_{1 \le \alpha_1 < \ldots < \alpha_k \le n} (-1)^k x_{\alpha_1} \ldots x_{\alpha_k} \sum_{1 \le \beta_1 < \ldots < \beta_k \le n}$$

$$(-1)^{\alpha_1 + \ldots + \alpha_k + \beta_1 + \ldots + \beta_k} Q\begin{pmatrix} i_{\beta_1}, \ldots, i_{\beta_k} \\ x_{\alpha_1}, \ldots, x_{\alpha_k} \end{pmatrix} Q^c\begin{pmatrix} i_{\beta_1}, \ldots, i_{\beta_k} \\ x_{\alpha_1}, \ldots, x_{\alpha_k} \end{pmatrix}$$

where $Q^c\begin{pmatrix} i_{\beta_1}, \ldots, i_{\beta_k} \\ x_{\alpha_1}, \ldots, x_{\alpha_k} \end{pmatrix}$ is the minor of $Q\begin{pmatrix} i_1, \ldots, i_n \\ x_1, \ldots, x_n \end{pmatrix}$ complementary

to $Q\begin{pmatrix} i_{\beta_1}, \ldots, i_{\beta_k} \\ x_{\alpha_1}, \ldots, x_{\alpha_k} \end{pmatrix}$. Using (5) for $n = k$ and interchanging order of summation this becomes

(7)
$$\sum_{0 \le r_1 < \ldots < r_k} \sum_{1 \le \beta_1 < \ldots < \beta_k \le n} A\begin{pmatrix} i_{\beta_1}, \ldots, i_{\beta_k} \\ r_1, \ldots, r_k \end{pmatrix} \sum_{1 \le \alpha_1 < \ldots < \alpha_k \le n}$$

$$(-1)^{\sum \alpha_\nu + \sum \beta_\nu} Q\begin{pmatrix} r_1, \ldots, r_k \\ x_{\alpha_1}, \ldots, x_{\alpha_k} \end{pmatrix} Q^c\begin{pmatrix} i_{\beta_1}, \ldots, i_{\beta_k} \\ x_{\alpha_1}, \ldots, x_{\alpha_k} \end{pmatrix}.$$

Let $Q\begin{pmatrix} (r_1,\ldots,r_k),\ (i_{\beta_1},\ldots,i_{\beta_k})^c \\ x_1,\ldots\ldots\ldots\ldots\ldots,x_n \end{pmatrix}$ denote the determinant $Q\begin{pmatrix} j_1,\ldots,j_n \\ x_1,\ldots,x_n \end{pmatrix}$

where $(j_1,\ldots,j_n)$ consists of the indices $(r_1,\ldots,r_k)$ together with the n-k indices remaining when $i_{\beta_1},\ldots,i_{\beta_k}$ are deleted from $(i_1,\ldots,i_n)$, and then arranged in increasing order $j_1 < \ldots < j_n$. Then the inner sum in (7) is identified to be, by virtue of the Laplace expansion,

$\pm Q\begin{pmatrix} (r_1,\ldots,r_k),\ (i_{\beta_1},\ldots,i_{\beta_k})^c \\ x_1,\ldots\ldots\ldots\ldots\ldots,x_n \end{pmatrix}$ where the sign depends on $r_1,\ldots,r_k$

and is for the moment ambiguous. Now $A\begin{pmatrix} i_{\beta_1},\ldots,i_{\beta_k} \\ r_1,\ldots,r_k \end{pmatrix}$ is zero if

$\max|i_{\beta_\nu} - r_\nu| > 1$ and the determinant $Q\begin{pmatrix} (r_1,\ldots,r_k),\ (i_{\beta_1},\ldots,i_{\beta_k})^c \\ x_1,\ldots\ldots\ldots\ldots\ldots,x_n \end{pmatrix}$ is

zero if any two of its rows are the same. Hence for the non-zero terms in the sum

$$\sum_{(0 \le r_1 < \ldots < r_k)} \ \sum_{1 \le \beta_1 < \ldots < \beta_k \le n} (\pm)\, A\begin{pmatrix} i_{\beta_1},\ldots,i_{\beta_k} \\ r_1,\ldots,r_k \end{pmatrix} Q\begin{pmatrix} (r_1\ldots r_k),\ (i_{\beta_1}\ldots i_{\beta_k})^c \\ x_1,\ldots\ldots\ldots\ldots,x_n \end{pmatrix}$$

we find that the ambiguous sign is always $+$. Thus (6) is valid with

$$D_{n,k}\begin{bmatrix} i_1,\ldots,i_n \\ j_1,\ldots,j_n \end{bmatrix} = 0 \quad \text{if} \quad \begin{cases} \max|i_\nu - j_\nu| > 1, \\ \text{or} \\ \sum |i_\nu - j_\nu| > k. \end{cases}$$

Moreover, if $\sum |i_\nu - j_\nu| = p \le k$ and if

$$|i_{\alpha_1} - j_{\alpha_1}| = \ldots = |i_{\alpha_p} - j_{\alpha_p}| = 1$$

then $D_{n,k}\begin{bmatrix} i_1,\ldots,i_n \\ j_1,\ldots,j_n \end{bmatrix}$ is the sum of all the kth order minors of A

formed from k of the rows $i_1, \ldots, i_n$ including the particular rows $i_{\alpha_1}, \ldots, i_{\alpha_p}$ and from k of the columns of $j_1, \ldots, j_n$ including the particular columns $j_{\alpha_1}, \ldots, j_{\alpha_p}$. (For the special case $p = 0$ this means that $D_{n,k} \begin{bmatrix} i_1, \ldots, i_n \\ i_1, \ldots, i_n \end{bmatrix}$ is the sum of all the kth order principal minors of $A \begin{pmatrix} i_1, \ldots, i_n \\ i_1, \ldots, i_n \end{pmatrix}$.) Since all of these minors have the sign $(-1)^{k+p}$ we conclude that $(-1)^{k+p} D \begin{bmatrix} i_1, \ldots, i_n \\ j_1, \ldots, j_n \end{bmatrix} > 0$ in this

case. Furthermore, each of these minors $A \begin{pmatrix} \beta_1, \ldots, \beta_k \\ \gamma_1, \ldots, \gamma_k \end{pmatrix}$ has the property

$$A \begin{pmatrix} \beta_1, \ldots, \beta_k \\ \gamma_1, \ldots, \gamma_k \end{pmatrix} \pi_{i_1} \cdots \pi_{i_n} = A \begin{pmatrix} \gamma_1, \ldots, \gamma_k \\ \beta_1, \ldots, \beta_k \end{pmatrix} \pi_{j_1} \cdots \pi_{j_n}$$

and hence we conclude

$$(8) \quad D_{n,k} \begin{bmatrix} i_1, \ldots, i_n \\ j_1, \ldots, j_n \end{bmatrix} \pi_{i_1} \pi_{i_2} \cdots \pi_{i_n} = D_{n,k} \begin{bmatrix} j_1, \ldots, j_n \\ i_1, \ldots, i_n \end{bmatrix} \pi_{j_1} \pi_{j_2} \cdots \pi_{j_n} .$$

We next show that for given n the complete set of formulas (6), $k = 1, 2, \ldots, n$ determine all the polynomials $Q \begin{pmatrix} i_1, \ldots, i_n \\ x_1, \ldots, x_n \end{pmatrix}$ when $Q \begin{pmatrix} 0, \ldots, n-1 \\ x_1, \ldots, x_n \end{pmatrix}$ is prescribed. This is expressed in

Theorem 1. Let $x_1, \ldots, x_n$ and γ be given complex numbers. Then there is a unique solution $\{\phi(i_1, \ldots, i_n)\}; \ 0 \le i_1 < \ldots < i_n,$ of the equations

$$(-1)^k E_k(x_1, \ldots, x_n) \phi(i_1, \ldots, i_n) =$$

$$\sum_{0 \le j_1 < \ldots < j_n} D_{n,k} \begin{bmatrix} i_1, \ldots, i_n \\ j_1, \ldots, j_n \end{bmatrix} \phi(j_1, \ldots, j_n)$$

$$(1 \le k \le n)$$

<u>with</u> $\phi(0, 1, \ldots, n-1) = \gamma$. <u>However if ϕ is merely required to satisfy</u> <u>the initial condition and the equations for fewer than n values of k</u> <u>then there is more than one solution.</u>

<u>Proof.</u> The first statement is proved by induction. Assume that the init-ial condition and recurrence formulas determine $\phi(i_1, \ldots, i_n)$ for $0 \leq i_1$ $< \ldots < i_n \leq m$ where $m \geq n-1$. If $0 \leq i_1 < \ldots < i_{n-1} \leq m-1$ then in the formula

$$-E_1(x_1, \ldots, x_n)\, \phi(i_1, \ldots, i_{n-1}, m) =$$

$$\sum D_{n,1}\begin{bmatrix} i_1, \ldots, i_{n-1}, m \\ j_1, \ldots, j_{n-1}, j_n \end{bmatrix} \phi(j_1, \ldots, j_n)$$

the only undetermined quantity on the right which has a non-zero co-efficient is $\phi(i_1, \ldots, i_{n-1}, m+1)$. This quantity can therefore be computed and ϕ is determined for $0 \leq i_1 < \ldots < i_{n-1} \leq m-1,\ i_n \leq m+1$.
Now if $0 \leq i_1 < \ldots < i_{n-2} \leq m-2$ then in the formula

$$E_2(x_1, \ldots, x_n)\, \phi(i_1, \ldots, i_{n-2}, m-1, m) =$$

$$\sum D_{n,2}\begin{bmatrix} i_1, \ldots, i_{n-2}, m-1, m \\ j_1, \ldots\ldots\ldots\ldots, j_n \end{bmatrix} \phi(j_1, \ldots, j_n)$$

the only undetermined quantity on the right which has a non-zero coef-ficient is $\phi(i_1, \ldots, i_{n-2}, m, m+1)$. This quantity can therefore be computed and ϕ is determined for $0 \leq i_1 < \ldots < i_{n-2} \leq m-2, \ i_{n-1} < i_n \leq m+1$. Proceeding in this way we use the formula with E_k to determine ϕ for $0 \leq i_1 < \ldots < i_{n-k} \leq m-k, \ i_{n-k+1} < \ldots < i_n \leq m+1$ and eventually find that ϕ is determined for $0 \leq i_1 < \ldots < i_n \leq m+1$. This completes the proof.

(It is easily seen that if γ is fixed and $\neq 0$ then as a function of $x_1, \ldots, x_n, \ \phi(i_1, \ldots, i_n)$ is a symmetric polynomial of degree exactly $\sum_{v=1}^{n} i_v - \frac{1}{2} n(n-1)$.)

To prove the second statement let $\phi \begin{pmatrix} i_1, \ldots, i_n \\ x_1, \ldots, x_n \end{pmatrix}$ denote the unique solution of all the formulas (6) with $1 \leq k \leq n$ and $\phi \begin{pmatrix} 0, \ldots, n-1 \\ x_1, \ldots, x_n \end{pmatrix} = 1$.

The formulas

$$E_k(x_1, \ldots, x_n) = (-1)^k \sum_{0 \leq j_1 < \ldots < j_n} D_{n,k} \begin{bmatrix} 0, \ldots, n-1 \\ j_1, \ldots, j_n \end{bmatrix} \phi \begin{pmatrix} j_1, \ldots, j_n \\ x_1, \ldots, x_n \end{pmatrix}$$

show that $x_1, \ldots, x_n$ are uniquely determined by the values $\phi \begin{pmatrix} i_1, \ldots, i_n \\ x_1, \ldots, x_n \end{pmatrix}$ for $0 \leq i_1 < \ldots < i_n \leq n$. If $k_1, \ldots, k_p$ are $p < n$ values of k then for given $x_1, \ldots, x_n$ there are infinitely many different choices $y_1, \ldots, y_n$ such that

$$E_{k_v}(x_1, \ldots, x_n) = E_{k_v}(y_1, \ldots, y_n), \quad v = 1, \ldots, p$$

and hence $\gamma \phi \begin{pmatrix} i_1, \ldots, i_n \\ x_1, \ldots, x_n \end{pmatrix}, \ \gamma \phi \begin{pmatrix} i_1, \ldots, i_n \\ y_1, \ldots, y_n \end{pmatrix}$ are both solutions of (6) for $k = k_1, \ldots, k_p$ with the initial value γ. If $\gamma \neq 0$ they are distinct, in fact they cannot coincide for all indices $0 \leq i_1 < \ldots < i_n \leq n$. This proves the second statement when $\gamma \neq 0$. If $\gamma = 0$ then the p equations have the non-trivial solution $\phi \begin{pmatrix} i_1, \ldots, i_n \\ x_1, \ldots, x_n \end{pmatrix} - \phi \begin{pmatrix} i_1, \ldots, i_n \\ y_1, \ldots, y_n \end{pmatrix}$ with initial value $\gamma = 0$. The theorem is now complete.

2. CHRISTOFFEL-DARBOUX FORMULA AND WRONSKIAN IDENTITIES.

The polynomials of the second kind associated with the orthogonal polynomial $\{Q_n\}$ can be defined by the equation

$$(9) \qquad Q_n^{(0)}(x) = \int_0^\infty \frac{Q_n(x)-Q_n(y)}{x-y} \, d\psi(y), \qquad n \geq 1$$

and these satisfy the recurrence relation (3) for $n \geq 1$ with $Q_0^{(0)} = 0$, by definition.

We will obtain generalizations of the formulas

$$(10a) \qquad (x-y) \sum_{k=0}^{n} Q_k(x) Q_k(y) \pi_k = \lambda_n \pi_n Q\begin{pmatrix} n, & n+1 \\ x, & y \end{pmatrix},$$

$$(10b) \qquad (x-y) \sum_{k=0}^{n} Q_k(x) Q_k^{(0)}(y)\pi_k = \lambda_n \pi_n \begin{vmatrix} Q_n(x), & Q_n^{(0)}(y) \\ Q_{n+1}(x), & Q_{n+1}^{(0)}(y) \end{vmatrix} + 1$$

$$(10c) \qquad (x-y) \sum_{k=0}^{n} Q_k^{(0)}(x) Q_k^{(0)}(y)\pi_k = \lambda_n \pi_n Q^{(0)}\begin{pmatrix} n, & n+1 \\ x, & y \end{pmatrix},$$

valid for $n \geq 0$ and the formula

$$(10d) \qquad \lambda_n \pi_n \begin{vmatrix} Q_n(x), & Q_n^{(0)}(x) \\ Q_{n+1}(x), & Q_{n+1}^{(0)}(x) \end{vmatrix} = -1,$$

which is obtained by setting $y = x$ in (10.b). The first three formulas may be regarded as special cases of Green's formula, and formula (10.d) asserts essentially that the Wronskian of two solutions of $A\underline{u} = x\underline{u}$ is a constant (independent of n). Formula (10.a) is known in the theory of orthogonal polynomials as the Christoffel Darboux formula.

We present for completeness a proof of (10.b) and the reader will have no difficulty in proving (10.a) and (10.c) by a similar method. Since

$Q_0^{(0)} \equiv 0$ the term $k = 0$ gives no contribution to the sum in (10.b) and the left side of (10.b) with the aid of the recurrence formula becomes

$$= - \sum_{k=1}^{n} \{ (\lambda_k \pi_k [Q_{k+1}(x) - Q_k(x)] - \lambda_{k-1} \pi_{k-1} [Q_k(x) - Q_{k-1}(x)]) Q_k^{(0)}(y)$$

$$- Q_k(x)(\lambda_k \pi_k [Q_{k+1}^{(0)}(y) - Q_k^{(0)}(y)] - \lambda_{k-1} \pi_{k-1} [Q_k^{(0)}(y) - Q_{k-1}^{(0)}(y)]) \}$$

$$= \sum_{k=1}^{n} \left\{ \lambda_k \pi_k \begin{vmatrix} Q_k(x), & Q_k^{(0)}(y) \\ Q_{k+1}(x), & Q_{k+1}^{(0)}(y) \end{vmatrix} - \lambda_{k-1} \pi_{k-1} \begin{vmatrix} Q_{k-1}(x), & Q_{k-1}^{(0)}(y) \\ Q_k(x), & Q_k^{(0)}(y) \end{vmatrix} \right\}$$

$$= \lambda_n \pi_n \begin{vmatrix} Q_n(x), & Q_n^{(0)}(y) \\ Q_{n+1}(x), & Q_{n+1}^{(0)}(y) \end{vmatrix} + 1 .$$

Other formulas of this kind involving the higher order polynomials $Q_i^{(j)}(x)$, $j > 0$ can be found. We record only

(11)
$$\begin{vmatrix} Q_n(x) & , Q_n^{(0)}(x) & , Q_n^{(1)}(x) & , \ldots, Q_n^{(r)}(x) \\ Q_{n+1}(x), Q_{n+1}^{(0)}(x) & , Q_{n+1}^{(1)}(x) & , \ldots, Q_{n+1}^{(r)}(x) \\ \vdots & & & \vdots \\ Q_{n+r+1}(x), Q_{n+r+1}^{(0)}(x), Q_{n+r+1}^{(1)}(x), \ldots, Q_{n+r+1}^{(r)}(x) \end{vmatrix} = 0, \ r \geq 1, \ n \geq 1 ,$$

which is an extension of (10.d).

For later purposes it is best to write (10a) in the form

(12)
$$\frac{\lambda_n \pi_n}{\lambda_0 \pi_0} \frac{Q \begin{pmatrix} n, n+1 \\ x, y \end{pmatrix}}{Q \begin{pmatrix} 0, 1 \\ x, y \end{pmatrix}} = \sum_{k=0}^{n} Q_k(x) Q_k(y) \pi_k$$

The determinental Christoffel Darboux formula has the generalization

$$\left(\prod_{r=1}^{p} \prod_{s=1}^{m} \frac{\pi_{n+r-1} \lambda_{n+r+s-2}}{\pi_{r-1} \lambda_{r+s-2}}\right) \frac{Q\left(\begin{array}{c} n, n+1, \ldots, n+p-1, n+p, n+p+1, \ldots, n+p+m-1 \\ x_1, x_2, \ldots, x_p, y_1, y_2, \ldots, y_m \end{array}\right)}{Q\left(\begin{array}{c} 0, 1, \ldots, p-1, p, p+1, \ldots, p+m-1 \\ x_1, x_2, \ldots, x_p, y_1, y_2, \ldots, y_m \end{array}\right)}$$

(13)

$$\sum_{0 \le k_1 < \ldots < k_p \le n+p-1} \frac{Q\left(\begin{array}{c} k_1, \ldots, k_p \\ x_1, \ldots, x_p \end{array}\right)}{Q\left(\begin{array}{c} 0, \ldots, p-1 \\ x_1, \ldots, x_p \end{array}\right)}$$

$$\times \frac{Q\left(\begin{array}{c} k_1, \ldots, k_p, n+p, n+p+1, \ldots, n+m-1 \\ y_1, \ldots, y_m \end{array}\right)}{Q\left(\begin{array}{c} 0, \ldots, m-1 \\ y_1, \ldots, y_m \end{array}\right)} \pi_{k_1} \cdots \pi_{k_p}$$

valid for $2 \le p \le m$.

The proof of (13) is rather lengthy, derived by a series of inductive arguments based on Sylvester's determinant identity now stated. The identity relates subdeterminants of order $m-2, m-1$ and m (where $m \ge 2$) for a kernel $B(m, x)$ as follows:

$$B\left(\begin{array}{c} n+1, n+2, \ldots, n+m-2 \\ x_1, x_2, \ldots, x_{m-2} \end{array}\right) B\left(\begin{array}{c} n, n+1, \ldots, n+m-1 \\ x_1, x_2, \ldots, x_m \end{array}\right) =$$

$$\left| \begin{array}{cc} B\left(\begin{array}{c} n, n+1, \ldots, n+m-3, n+m-2 \\ x_1, x_2, \ldots, x_{m-2}, x_{m-1} \end{array}\right) & , B\left(\begin{array}{c} n, n+1, \ldots, n+m-3, n+m-2 \\ x_1, x_2, \ldots, x_{m-2}, x_m \end{array}\right) \\ B\left(\begin{array}{c} n+1, n+2, \ldots, n+m-2, n+m-1 \\ x_1, x_2, \ldots, x_{m-2}, x_{m-1} \end{array}\right) & , B\left(\begin{array}{c} n+1, n+2, \ldots, n+m-2, n+m-1 \\ x_1, x_2, \ldots, x_{m-2}, x_m \end{array}\right) \end{array} \right| .$$

[See 2, p. 3]. In addition to the above identity we will use those obtained from it by permuting rows and columns. The above identity, for
example, will be referred to as "Sylvester's identity for $B\begin{pmatrix} n, \ldots, n+m-1 \\ x_1, \ldots, x_m \end{pmatrix}$
with the pivot minor $B\begin{pmatrix} n+1, \ldots, n+m-2 \\ x_1, \ldots, x_{m-2} \end{pmatrix}$".

Consider also the composed kernel

$$(14) \qquad K(\alpha, \gamma) = \int_B L(\alpha, \beta)\, M(\beta, \gamma) d\nu(\beta)$$

where the variables α, β, γ range over linear sets A, B, C on the real
axis respectively and ν is a measure on B . Recall now the continuous Cauchy Binet formula where the determinants of the kernel K are expressed in terms of those of L and M, viz.

$$(15) \qquad K\begin{pmatrix} \alpha_1, \ldots, \alpha_n \\ \gamma_1, \ldots, \gamma_n \end{pmatrix} = \int \ldots \int_{\beta_1 < \ldots < \beta_n} L\begin{pmatrix} \alpha_1, \ldots, \alpha_n \\ \beta_1, \ldots, \beta_n \end{pmatrix}$$

each $\beta_k \in B$

$$\times M\begin{pmatrix} \beta_1, \ldots, \beta_n \\ \gamma_1, \ldots, \gamma_n \end{pmatrix} d\nu(\beta_1) \ldots d\nu(\beta_n)$$

The process of forming (15) from (14) will be referred to as the basic
composition formula (cf. Karlin [2, p. 17].

We will have occasion to deal with the kernel $Y(\alpha, \beta)$ defined once
and for all by

$$Y(\alpha, \beta) = \begin{cases} 1 & \text{if } \beta \le \alpha \\ 0 & \text{if } \beta > \alpha \end{cases} .$$

The Christoffel-Darboux formula (10a) with x regarded as fixed may be
written in the form

(16)
$$K(n, y) = \sum_{k=0}^{\infty} Y(n, k) Q_k(y) \nu_k$$

where $\nu_k = Q_k(x)\pi_k$

and
$$K(n, y) = \frac{\lambda_n \pi_n Q\begin{pmatrix} n, n+1 \\ x, y \end{pmatrix}}{\lambda_0 \pi_0 Q\begin{pmatrix} 0, 1 \\ x, y \end{pmatrix}} .$$

Invoking the basic composition formula (15) on (16) we obtain

(17)
$$K\begin{pmatrix} n, n+1, \ldots, n+m-1 \\ y_1, y_2, \ldots, y_m \end{pmatrix} =$$

$$\sum_{0 < k_1 < \ldots < k_m} Y\begin{pmatrix} n, n+1, \ldots, n+m-1 \\ k_1, k_2, \ldots, k_m \end{pmatrix} \phi\begin{pmatrix} k_1, \ldots, k_m \\ y_1, \ldots, y_m \end{pmatrix} \nu_{k_1} \cdots \nu_{k_m}$$

Now $Y\begin{pmatrix} n, \ldots, n+m-1 \\ k_1, \ldots, k_m \end{pmatrix}$ with $k_1 < \ldots < k_m$ is zero except if $k_1 \leq n$ and

$k_r = n+r-1$ for $r = 2, 3, \ldots, m$ in which case the value of the determinant is 1. Thus (17) is equivalent to

(18)
$$K\begin{pmatrix} n, n+1, \ldots, n+m-1 \\ y_1, y_2, \ldots, y_m \end{pmatrix} =$$

$$\left[\prod_{r=1}^{m-1} Q_{n+r}(x)\pi_{n+r}\right] \sum_{k=0}^{n} Q\begin{pmatrix} k, n+1, n+2, \ldots, n+m-1 \\ y_1, y_2, y_3, \ldots, y_m \end{pmatrix} Q_k(x)\pi_k .$$

The left member of (18) is equal to

$$\left[\prod_{r=1}^{m} \left(\frac{\lambda_{n+r-1} \pi_{n+r-1}}{\lambda_0 \pi_0 \, Q\binom{0,\,1}{x,\,y_r}} \right) B\binom{n,\,n+1,\,\ldots,\,n+m-1}{y_1,\,y_2,\,\ldots,\,y_m} \right]$$

where $B(n, x) = Q\binom{n,\,n+1}{x,\,y}$. We next prove that

$$(19) \qquad B\binom{n,\,n+1,\,\ldots,\,n+m-1}{y_1,\,y_2,\,\ldots,\,y_m} = Q\binom{n,\,n+1,\,\ldots,\,n+m}{x,\,\,y_1,\,\ldots,\,y_m} \prod_{r=1}^{m-1} Q_{m+r}(x) .$$

If $m = 2$ this formula is Sylvester's identity for the determinant $Q\binom{n,\,n+1,\,n+2}{x,\,\,y_1,\,y_2}$ with "pivot minor $Q_{n+1}(x)$" . The general result is now proved by induction on m . We write Sylvesters identity for the determinant $B\binom{n,\,n+1,\,\ldots,\,n+m-1}{y_1,\,y_2,\,\ldots,\,y_m}$ with the "pivot minor $B\binom{n+1,\,n+2,\,\ldots,\,n+m-2}{y_2,\,\,y_3,\,\ldots,\,y_{m-1}}$

yielding

$$(20) \quad B\binom{n,\,n+1,\,\ldots,\,n+m-1}{y_1,\,y_2,\,\ldots,\,y_m} \; B\binom{n+1,\,\ldots,\,n+m-2}{y_2,\,\ldots,\,y_{m-1}}$$

$$= \begin{vmatrix} B\binom{n,\,n+1,\,\ldots,\,n+m-2}{y_1,\,y_2,\,\ldots,\,y_{m-1}}, & B\binom{n+1,\,n+2,\,\ldots,\,n+m-1}{y_1,\,\,y_2,\,\ldots,\,y_{m-1}} \\[2em] B\binom{n,\,n+1,\,\ldots,\,n+m-2}{y_2,\,y_3,\,\ldots,\,y_m}, & B\binom{n+1,\,n+2,\,\ldots,\,n+m-1}{y_2,\,y_3,\,\ldots,\,y_m} \end{vmatrix}$$

We now use the result, assumed valid for the $(m-1)^{st}$ order determinants occurring on the right of (20), so this side becomes

$$
= \prod_{r=1}^{m-3} Q_{n+r}(x) \prod_{s=2}^{m-1} Q_{n+s}(x) \begin{vmatrix} Q\begin{pmatrix} n, n+1, \ldots, n+m-1 \\ x, \; y_1, \ldots, y_{m-1} \end{pmatrix}, & Q\begin{pmatrix} n+1, n+2, \ldots, n+m \\ x \; , \; y_1, \ldots, y_{m-1} \end{pmatrix} \\ Q\begin{pmatrix} n, n+1, \ldots, n+m-1 \\ x, \; y_2, \ldots, y_m \end{pmatrix}, & Q\begin{pmatrix} n+1, n+2, \ldots, n+m \\ x, \; y_2, \ldots, y_m \end{pmatrix} \end{vmatrix}
$$

$$
= \left(\prod_{r=2}^{m-3} Q_{n+r}(x) \right)\left(\prod_{s=1}^{m-1} Q_{n+s}(x) \right) Q\begin{pmatrix} n, n+1, \ldots, n+m \\ x, \; y_1, \ldots, y_m \end{pmatrix} Q\begin{pmatrix} n+1, n+2, \ldots, n+m-1 \\ x, \; y_2, \ldots, y_{m-1} \end{pmatrix}
$$

where Sylvester's identity has been used again. Now the left side of (20) can be transformed by using the result assumed valid for the $(m-2)^{nd}$ order determinant

$$
B\begin{pmatrix} n+1, \ldots, n+m-2 \\ y_2, \ldots, y_{m-1} \end{pmatrix} = \left(\prod_{r=2}^{m-3} Q_{n+r}(x) \right) Q\begin{pmatrix} n+1, n+2, \ldots, n+m-1 \\ x, \; y_2, \ldots, y_{m-1} \end{pmatrix}
$$

and we see that this quantity is a factor of both sides of (20). After cancelling it the induction is complete.

Using (19) the left member of (18) can now be written as

$$
\left[\prod_{r=1}^{m} \frac{\lambda_{n+r-1} \pi_{n+r-1}}{\lambda_0 \pi_0} Q\begin{pmatrix} 0, 1 \\ x, y_r \end{pmatrix} \right]\left[\prod_{s=1}^{m-1} Q_{n+s}(x) \right] Q\begin{pmatrix} n, n+1, \ldots, n+m \\ x, \; y_1, \ldots, y_m \end{pmatrix}
$$

and hence after obvious cancellations (19) becomes

$$
(21) \qquad \left[\pi_n \prod_{r=1}^{m} \frac{\lambda_{n+r-1}}{\lambda_0 \pi_0} Q\begin{pmatrix} 0, 1 \\ x, y_r \end{pmatrix} \right] Q\begin{pmatrix} n, n+1, \ldots, n+m \\ x, \; y_1, \ldots, y_m \end{pmatrix}
$$

$$
= \sum_{k=0}^{n} Q\begin{pmatrix} k, n+1, n+2, \ldots, n+m-1 \\ y_1, y_2, y_3, \ldots, \; y_m \end{pmatrix} Q_k(x) \pi_k \; .
$$

Setting $n = 0$ in (21) gives

(21a)
$$\left[\prod_{r=1}^{m} \frac{\lambda_{r-1}}{\lambda_0 \pi_0 \, Q\!\begin{pmatrix} 0 \, , 1 \\ x \, , y_r \end{pmatrix}}\right] Q\!\begin{pmatrix} 0, & 1, & \ldots, & m \\ x, & y_1, & \ldots, & y_m \end{pmatrix} = Q\!\begin{pmatrix} 0, & 1, & \ldots, & m-1 \\ y_1, & y_2, & \ldots, & y_m \end{pmatrix}.$$

Hence if both sides of (21) are divided by $Q\!\begin{pmatrix} 0, & 1, & \ldots, & m-1 \\ y_1, & y_2, & \ldots, & y_m \end{pmatrix}$ the result is

(22)
$$\frac{\pi_n}{\pi_0} \left(\prod_{r=1}^{m} \frac{\lambda_{n+r-1}}{\lambda_{r-1}}\right) \frac{Q\!\begin{pmatrix} n, & n+1, & \ldots, & n+m \\ x, & y_1, & \ldots, & y_m \end{pmatrix}}{Q\!\begin{pmatrix} 0, & 1, & \ldots, & m \\ x, & y_1, & \ldots, & y_m \end{pmatrix}}$$

$$= \sum_{k=0}^{n} \frac{Q\!\begin{pmatrix} k, & n+1, & \ldots, & n+m-1 \\ y_1, & y_2, & \ldots, & y_m \end{pmatrix}}{Q\!\begin{pmatrix} 0, & 1, & \ldots, & m-1 \\ y_1, & y_2, & \ldots, & y_m \end{pmatrix}} Q_k(x)\pi_k .$$

With m and $y_1, \ldots, y_m$ fixed let $R(n, x)$ denote the left member of (22) and let

(23)
$$Z(n, k) = Y(n, k) \frac{Q\!\begin{pmatrix} k, & n+1, & \ldots, & n+m-1 \\ y_1, & y_2, & \ldots, & y_m \end{pmatrix}}{Q\!\begin{pmatrix} 0, & 1, & \ldots, & m-1 \\ y_1, & y_2, & \ldots, & y_m \end{pmatrix}} .$$

Then (22) can be written as

$$R(n, x) = \sum_{k=0}^{\infty} Z(n, k) Q_k(x)\pi_k$$

and to this we apply the basic composition formula to determinants of order p obtaining

(24) $R\begin{pmatrix} n\,,n+1,\,\ldots\,,n+p-1\\ x_1,x_2\,,\ldots\,,x_p\end{pmatrix} = \sum_{0 \le k_1 <\ldots< k_p} Z\begin{pmatrix} n\,,n+1,\,\ldots\,,n+p-1\\ k_1,k_2\,,\ldots\,,k_p\end{pmatrix}$

$Q\begin{pmatrix} k_1,\,\ldots\,,k_p\\ x_1,\,\ldots\,,x_p\end{pmatrix} \pi_{k_1} \pi_{k_2}\;\cdots\;\pi_{k_p}\;.$

In the determinant

$Z\begin{pmatrix} n,n+1,\,\ldots\,,n+p-1\\ k_1,k_2,\,\ldots\,,k_p\end{pmatrix} = \begin{vmatrix} Z(n,k_1) & ,Z(n,k_2) & ,\,\ldots\\ Z(n+1,k_1) & ,Z(n+1,k_2) & ,\,\ldots\\ \vdots & & \\ Z(n+p-1,k_1), & Z(n+p-1,k_2), & \ldots \end{vmatrix},\;k_1 <\ldots< k_p\,,$

if for some $r \le p$ we have $k_r > n+r-1$ then the first r rows of the determinant are vectors whose last $p-r+1$ components are all zero, so the determinant is zero. Hence the summation in (24) can be restricted to the range where

(25) $$0 \le k_1 <\ldots< k_p \le n+p-1\;.$$

We will <u>assume that</u> $2 \le p \le m$ and show that if $r \le p$ and if $k_1,\,\ldots\,,k_r$ satisfy (25) with p replaced by r then

(26) $Z\begin{pmatrix} n,n+1,\,\ldots\,,n+r-1\\ k_1,k_2,\,\ldots\,,k_r\end{pmatrix} =$

$= \dfrac{Q\begin{pmatrix} k_1,\,\ldots\,,k_r,n+r,n+r+1,\,\ldots\,,n+m-1\\ y_1,\ldots\ldots\ldots\ldots\ldots\ldots\ldots\,,y_m\end{pmatrix}\displaystyle\prod_{\alpha=1}^{r-1} Q\begin{pmatrix} n+\alpha,n+\alpha+1,\,\ldots\,,n+\alpha+m-1\\ y_1\,,\;\;y_2,\ldots,y_m\end{pmatrix}}{\left[Q\begin{pmatrix} 0,1\,,\,\ldots\,,m-1\\ y_1,y_2,\,\ldots\,,y_m\end{pmatrix}\right]^r}$

This is trivial if $r = 1$ and the proof will be by induction on r. We assume that (26) has already been proven for all $r < p$. In the Sylvester identity

$$Z\begin{pmatrix} n+1, n+2, \ldots, n+p-2 \\ k_1, \ k_2, \ldots, k_{p-2} \end{pmatrix} \ Z\begin{pmatrix} n, n+1, \ldots, n+p-1 \\ k_1, k_2, \ldots, k_p \end{pmatrix}$$

$$= \begin{vmatrix} Z\begin{pmatrix} n, n+1, \ldots, n+p-3, n+p-2 \\ k_1, \ k_2, \ldots, k_{p-2}, k_{p-1} \end{pmatrix}, & Z\begin{pmatrix} n, n+1, \ldots, n+p-3, n+p-2 \\ k_1, \ k_2, \ldots, k_{p-2}, k_p \end{pmatrix} \\ Z\begin{pmatrix} n+1, n+2, \ldots, n+p-2, n+p-1 \\ k_1, \ k_2, \ldots, k_{p-2}, k_{p-1} \end{pmatrix}, & Z\begin{pmatrix} n+1, n+2, \ldots, n+p-2, n+p-1 \\ k_1, \ k_2, \ldots, k_{p-2}, k_p \end{pmatrix} \end{vmatrix}$$

we use the induction hypothesis to simplify the $(p-2)^{th}$ order determinant on the left and the four $(p-1)^{th}$ order determinants on the right. This leads to

$$Z\begin{pmatrix} n, n+1, \ldots, n+p-1 \\ k_1, \ k_2, \ldots, k_p \end{pmatrix}$$

$$= \frac{\prod\limits_{\alpha=1}^{p-1} Q\begin{pmatrix} n+\alpha, n+\alpha+1, \ldots, n+m+\alpha-1 \\ y_1, \ y_2, \ldots, \ y_m \end{pmatrix}}{Q\begin{pmatrix} 0, \ 1, \ldots, m-1 \\ y_1, y_2, \ldots, y_m \end{pmatrix}^p} \cdot \frac{\Delta}{Q\begin{pmatrix} k_1, \ldots, k_{p-2}, n+p-1, n+p, \ldots, n+m \\ y_1, \ldots \ldots \ldots \ldots \ldots \ldots \ldots, y_m \end{pmatrix}}$$

where

$$\Delta = \begin{vmatrix} A_{11} & A_{12} \\ A_{21} & A_{22} \end{vmatrix}$$

and

$$A_{11} = Q\begin{pmatrix} k_1, \ldots, k_{p-2}, k_{p-1}, n+p-1, \ldots, n+m-1 \\ y_1, \ldots\ldots\ldots\ldots\ldots\ldots\ldots, y_m \end{pmatrix}$$

$$A_{12} = Q\begin{pmatrix} k_1, \ldots, k_{p-2}, k_p, n+p-1, \ldots, n+m-1 \\ y_1, \ldots\ldots\ldots\ldots\ldots\ldots\ldots, y_m \end{pmatrix}$$

$$A_{21} = Q\begin{pmatrix} k_1, \ldots, k_{p-2}, k_{p-1}, n+p, \ldots, n+m \\ y_1, \ldots\ldots\ldots\ldots\ldots\ldots, y_m \end{pmatrix}$$

$$A_{22} = Q\begin{pmatrix} k_1, \ldots, k_{p-2}, k_p, n+p, \ldots, n+m \\ y_1, \ldots\ldots\ldots\ldots\ldots\ldots, y_m \end{pmatrix}.$$

To evaluate Δ consider the $2m$ - square determinant

$$\begin{vmatrix} Q_{k_1}(y_1) & , \ldots, & Q_{k_1}(y_m), & 0 & , \ldots, & 0 \\ \vdots & & \vdots & & \vdots & \vdots \\ Q_{k_{p-2}}(y_1) & , \ldots, & Q_{k_{p-2}}(y_m), 0 & & , \ldots, & 0 \\ Q_{n+p}(y_1) & , \ldots, & Q_{n+p}(y_m), & 0 & , \ldots, & 0 \\ \vdots & & \vdots & & & \vdots \\ Q_{n+m-1}(y_1), & \ldots, & Q_{n+m-1}(y_m), 0 & & , \ldots, & 0 \\ Q_{k_1}(y_1) & , \ldots, & Q_{k_1}(y_m), Q_{k_1}(y_1) & & , \ldots, & Q_{k_1}(y_m) \\ \vdots & & \vdots & \vdots & & \vdots \\ Q_{k_p}(y_1) & & Q_{k_p}(y_m), Q_{k_p}(y_1) & & , \ldots, & Q_{k_p}(y_m) \\ Q_{n+p-1}(y_1), & \ldots, & Q_{n+p-1}(y_m), Q_{n+p-1}(y_1), & \ldots, & Q_{n+p-1}(y_m) \\ \vdots & & \vdots & \vdots & & \vdots \\ Q_{n+m}(y_1) & , \ldots, & Q_{n+m}(y_m), Q_{n+m}(y_1) & & , \ldots, & Q_{n+m}(y_m) \end{vmatrix}$$

in which $0 \le k_1 < \ldots < k_p \le n+p-1$. This determinant is zero since its last $m+1$ rows are linearly dependent. We employ Laplace's expansion by minors of the first m columns. The cofactor of such a minor will be zero unless the minor contains the first $m-2$ rows and even in this case the minor itself will be zero except if perhaps its remaining two rows are chosen from the rows with $Q_{k_{p-1}}$, Q_{k_p}, Q_{n+p-1}, Q_{n+m}. Therefore the Laplace expansion has at most six non-zero terms. These terms are moreover equal in pairs and we obtain the identity

$$(28) \quad \Delta = Q\begin{pmatrix} k_1, \ldots, k_{p-2}, n+p-1, \ldots, n+m \\ y_1, \ldots\ldots\ldots\ldots\ldots, y_m \end{pmatrix} Q\begin{pmatrix} k_1, \ldots, k_p, n+p, \ldots, n+m-1 \\ y_1, \ldots\ldots\ldots\ldots\ldots, y_m \end{pmatrix}.$$

Now combining (27) and (28) we obtain (26) for $r = p$ and this completes the induction.

The identity (26) with $r = p$ can be used to simplify the right member of (24). We first perform a parallel simplification of the left member of (24), which can be written in terms of the kernel

$$D(n, x) = Q\begin{pmatrix} n, n+1, \ldots, n+m \\ x, y_1, \ldots, y_m \end{pmatrix}$$

as

$$(29) \quad R\begin{pmatrix} n, n+1, \ldots, n+p-1 \\ x_1, x_2, \ldots, x_p \end{pmatrix} = \left[\prod_{r=1}^{p} \frac{c_{n+r-1}}{Q\begin{pmatrix} 0, 1, \ldots, m \\ x_r, y_1, \ldots, y_m \end{pmatrix}} \right] D\begin{pmatrix} n, n+1, \ldots, n+p-1 \\ x_1, x_2, \ldots, x_p \end{pmatrix}$$

where

$$(30) \quad c_n = \frac{\pi_n}{\pi_0} \prod_{r=1}^{m} \frac{\lambda_{n+r-1}}{\lambda_{r-1}}.$$

We will prove that for $2 \leq p \leq m$

541

$$(31) \quad D\begin{pmatrix} n, n+1, \ldots, n+p-1 \\ x_1, x_2, \ldots, x_p \end{pmatrix} = Q\begin{pmatrix} n, n+1, \ldots, n+p-1, n+p, \ldots, n+p+m-1 \\ x_1, x_2, \ldots, x_p, y_1, \ldots, y_m \end{pmatrix}$$

$$\prod_{r=1}^{p-1} Q\begin{pmatrix} n+r, n+r+1, \ldots, n+r+m-1 \\ y_1, y_2, \ldots, y_m \end{pmatrix}$$

The case $p = 2$ results directly from the Sylvester identity for determinant $Q\begin{pmatrix} n, n+1, n+2, \ldots, n+m+1 \\ x_1, x_2, y_1, \ldots, y_m \end{pmatrix}$ with the "pivot minor $Q\begin{pmatrix} n+1, \ldots, n+m \\ y_1, \ldots, y_m \end{pmatrix}$".

For $p > 2$ we proceed by induction on p. In the Sylvester identity

$$D\begin{pmatrix} n+1, \ldots, n+p-2 \\ x_1, \ldots, x_{p-2} \end{pmatrix} D\begin{pmatrix} n, n+1, \ldots, n+p-1 \\ x_1, x_2, \ldots, x_p \end{pmatrix}$$

$$= \begin{vmatrix} D\begin{pmatrix} n, \ldots, n+p-3, n+p-2 \\ x_1, \ldots, x_{p-2}, x_{p-1} \end{pmatrix}, & D\begin{pmatrix} n, \ldots, n+p-3, n+p-2 \\ x_1, \ldots, x_{p-2}, x_p \end{pmatrix} \\ D\begin{pmatrix} n+1, \ldots, n+p-2, n+p-1 \\ x_1, \ldots, x_{p-2}, x_{p-1} \end{pmatrix}, & D\begin{pmatrix} n+1, \ldots, n+p-2, n+p-1 \\ x_1, \ldots, x_{p-2}, x_p \end{pmatrix} \end{vmatrix}$$

we use the induction assumption to simplify the $(p-2)^{th}$ order determinant on the left and the four $(p-1)^{th}$ order determinants on the right. This leads to

$$D\begin{pmatrix} n, n+1, \ldots, n+p-1 \\ x_1, x_2, \ldots, x_p \end{pmatrix} = \frac{\left[\prod_{r=1}^{p-1} Q\begin{pmatrix} n+r, \ldots, n+r+m-1 \\ y_1, \ldots, y_m \end{pmatrix} \right] \Delta'}{Q\begin{pmatrix} n+1, \ldots, n+p-2, n+p-1, \ldots, n+p+m-2 \\ x_1, \ldots, x_{p-2}, y_1, \ldots, y_m \end{pmatrix}}$$

where

$$\Delta' = \begin{vmatrix} A'_{11} & A'_{12} \\ A'_{21} & A'_{22} \end{vmatrix}$$

and

$$A'_{11} = Q\begin{pmatrix} n, \ldots, n+p-3, n+p-2, \ldots, n+p+m-2 \\ x_1, \ldots, x_{p-2}, x_{p-1}, y_1, \ldots, \quad y_m \end{pmatrix}$$

$$A'_{12} = Q\begin{pmatrix} n, \ldots, n+p-3, n+p-2, n+p-1, \ldots, n+p+m-2 \\ x_1, \ldots, x_{p-2}, \quad x_p, \quad y_1, \quad \ldots, \quad y_m \end{pmatrix}$$

$$A'_{21} = Q\begin{pmatrix} n+1, \ldots, n+p-2, n+p-1, n+p, \ldots, n+p+m-1 \\ x_1, \ldots, x_{p-2}, x_{p-1}, \quad y_1, \ldots, \quad y_m \end{pmatrix}$$

$$A'_{22} = Q\begin{pmatrix} n+1, \ldots, n+p-2, n+p-1, n+p, \ldots, n+p+m-1 \\ x_1, \ldots, x_{p-2}, \quad x_p, \quad y_1, \ldots, \quad y_m \end{pmatrix} .$$

Then

$$\Delta' = Q\begin{pmatrix} n+1, \ldots, n+p-2, n+p-1, \ldots, n+p+m-2 \\ x_1, \ldots, x_{p-2}, \quad y_1, \quad \ldots, \quad y_m \end{pmatrix} .$$

$$\times Q\begin{pmatrix} n, n+1, \ldots, n+p-1, n+p, \ldots, n+p+m-1 \\ x_1, \ldots, \quad x_p, \quad y_1, \ldots, \quad y_m \end{pmatrix} .$$

From this we obtain (31) and the induction is complete.

Using (26), (29), (31) we can now write (24) in the form

$$(32) \qquad \left[\prod_{r=1}^{p} \frac{c_{n+r-1}}{Q\begin{pmatrix} 0, 1, \ldots, m \\ x_r, y_1, \ldots, y_m \end{pmatrix}} \right] Q\begin{pmatrix} n, \ldots, n+p-1, n+p, \ldots, n+p+m-1 \\ x_1, \ldots, \quad x_p, \quad y_1, \ldots, \quad y_m \end{pmatrix}$$

$$= \sum_{0 \le k_1 < \ldots < k_p \le n+p-1} Q\begin{pmatrix} k_1, \ldots, k_p \\ x_1, \ldots, x_p \end{pmatrix}$$

$$\frac{Q\begin{pmatrix} k_1, \ldots, k_p, n+p, n+p+1, \ldots, n+m-1 \\ y_1, \ldots\ldots\ldots\ldots\ldots\ldots\ldots, y_m \end{pmatrix}}{\left[Q\begin{pmatrix} 0, 1, \ldots\ldots\ldots\ldots\ldots, m-1 \\ y_1, y_2, \ldots\ldots\ldots\ldots\ldots, y_m \end{pmatrix} \right]^p} \; \pi_{k_1} \pi_{k_2} \cdots \pi_{k_p}$$

Setting $n = 0$ in (32) gives

$$(33) \qquad \frac{\left[Q\begin{pmatrix} 0, 1, \ldots, m-1 \\ y_1, y_2, \ldots, y_m \end{pmatrix} \right]^{p-1}}{Q\begin{pmatrix} 0, 1, \ldots, p-1 \\ x_1, x_2, \ldots, x_p \end{pmatrix}} \quad \prod_{r=1}^{p} \quad \frac{c_{r-1}}{Q\begin{pmatrix} 0, 1, \ldots, m \\ x_r, y_1, \ldots, y_m \end{pmatrix}}$$

$$= \frac{1}{Q\begin{pmatrix} 0, 1, \ldots, p-1, p, p+1, \ldots, p+m-1 \\ x_1, x_2, \ldots, x_p, y_1, y_2, \ldots, y_m \end{pmatrix}} \quad .$$

Finally combining (32) and (33) we obtain

$$(34) \qquad \left(\prod_{r=1}^{p} \frac{c_{n+r-1}}{c_{r-1}} \right) \frac{Q\begin{pmatrix} n, \ldots, n+p-1, n+p, \ldots, n+p+m-1 \\ x_1, \ldots, x_p, \quad y_1, \ldots, \quad y_m \end{pmatrix}}{Q\begin{pmatrix} 0, \ldots, p-1, p, \ldots, p+m-1 \\ x_1, \ldots, x_p, y_1, \ldots, \quad y_m \end{pmatrix}}$$

$$= \sum_{0 \le k_1 < \ldots < k_p \le n+p-1} \frac{Q\begin{pmatrix} k_1, \ldots, k_p \\ x_1, \ldots, x_p \end{pmatrix}}{Q\begin{pmatrix} 0, \ldots, p-1 \\ x_1, \ldots, x_p \end{pmatrix}}$$

$$\times \frac{Q\begin{pmatrix} k_1, \ldots, k_p, n+p, \ldots, n+m-1 \\ y_1, \ldots\ldots\ldots\ldots\ldots, y_m \end{pmatrix}}{Q\begin{pmatrix} 0, \ldots, m-1 \\ y_1, \ldots, y_m \end{pmatrix}} \; \pi_{k_1} \cdots \pi_{k_p}$$

and this is precisely (13).

The multi determinant Wronskian identity: For $0 \le i_1 < i_2 < \ldots < i_r$

$< i_{r+1} < \ldots < i_n$ define

$$Q\begin{pmatrix} i_1, i_2, \ldots, i_r & i_{r+1}, i_{r+2}, \ldots, i_n \\ x_1, x_2, \ldots, x_r & x_{r+1}, x_{r+2}, \ldots, x_n \end{pmatrix}$$

$$(35) = \begin{vmatrix} Q_{i_1}(x_1), Q_{i_1}(x_2), \ldots, Q_{i_1}(x_r), Q_{i_1}^{(0)}(x_{r+1}), \ldots, Q_{i_1}^{(0)}(x_n) \\ Q_{i_2}(x_1), Q_{i_2}(x_2), \ldots, Q_{i_2}(x_r), Q_{i_2}^{(0)}(x_{r+1}), \ldots, Q_{i_2}^{(0)}(x_n) \\ \vdots \\ Q_{i_n}(x_1), Q_{i_n}(x_2), \ldots, Q_{i_n}(x_r), Q_{i_n}^{(0)}(x_{r+1}), \ldots, Q_{i_n}^{(0)}(x_n) \end{vmatrix}$$

The formulas

$$Q\begin{pmatrix} n, n+1, \ldots, n+r & n+r+1, n+r+2, \ldots, n+r+k+1 \\ x, y_1, \ldots, y_r & x, x_1, \ldots, x_k \end{pmatrix}$$

$$(36) = \frac{(-1)^{r+1}(\lambda_0 \pi_0)^{r+k}}{\pi_n \lambda_n \lambda_{n+1} \cdots \lambda_{n+r+k}} \prod_{i=1}^{r} Q\begin{pmatrix} 0, 1 \\ x, y_i \end{pmatrix} \prod_{j=1}^{k} Q\begin{pmatrix} 0, 1 \\ x, x_j \end{pmatrix}$$

$$Q\begin{pmatrix} n+1, \ldots, n+r & n+r+1, \ldots, n+r+k \\ y_1, \ldots, y_r & x_1, \ldots, x_k \end{pmatrix}$$

and

545

$$Q\begin{pmatrix} n, n+1, \ldots, n+r-1 & n+r, \ldots, n+2r-1 \\ x_1, x_2, \ldots, x_r & x_1, \ldots, x_r \end{pmatrix}$$

(37)

$$= \frac{(-1)^{r(r+1)/2} (\lambda_0 \pi_0)^{r(r-1)} \prod_{1 \le i < j \le r} \left[Q\begin{pmatrix} 0, 1 \\ x_i, x_j \end{pmatrix} \right]^2}{\pi_n \pi_{n+1} \cdots \pi_{n+r-1} \lambda_n \lambda_n^2 \lambda_{n+2}^3 \cdots \lambda_{n+r-1}^r \lambda_{n+r}^{r+1} \lambda_{n+r+1}^r \cdots \lambda_{n+2r-2}}$$

are generalizations of the Wronskian identity for the polynomials cited before. The proofs of (36) and (37) run similarly to that of (13). If we denote the denominator on the right in (37) by $c(n, r)$. Then we find that

$$c(n, r) \, Q\begin{pmatrix} n, n+1, \ldots, n+r-1 & n+r, \ldots, n+2r-1 \\ x_1, x_2, \ldots, x_r & x_1, \ldots, x_r \end{pmatrix} \text{ is independent of } n.$$

3. Further Properties

Positivity

The normalization $Q_n(0) = 1$ entails that for $x \le 0$, $Q_n(x) > 0$. The analogous inequality for the determinants (1) is

$$(38) \quad (-1)^{n(n-1)/2} \, Q\begin{pmatrix} i_1, i_2, \ldots, i_n \\ x_1, x_2, \ldots, x_n \end{pmatrix} > 0 \text{ if } x_1 < x_2 < \ldots < x_n \le 0$$

and this together with a number of related and sharper inequalities has been discussed in detail in [4].

Completeness

We next show that if $\{Q_i(x)\}$ is a complete system in $L_2(\psi)$ then the system (1) is a complete system with respect to all skew symmetric functions $f(\underline{x}) = f(x_1, \ldots, x_n)$ in L_2 of the product measure $d\psi^{(n)}(x) = d\psi(x_1) d\psi(x_2) \ldots d\psi(x_n)$. The proof is elementary. Since $\{Q_i\}$ is by assumption complete, the expansion

546

(39) $$f(\underline{x}) \sim \sum a_{i_1, \ldots, i_n} Q_{i_1}(x_1) Q_{i_2}(x_2), \ldots, Q_{i_n}(x_n)$$

is valid in $L_2(\psi^{(n)})$ and

(40)
$$a_{\underline{i}} = a_{i_1, \ldots, i_n} = (\pi_{i_1} \pi_{i_2} \cdots \pi_{i_n}) \int f Q_{i_1} Q_{i_2} \cdots Q_{i_n} d\psi^{(n)}$$

$$= (\pi_{i_1} \pi_{i_2} \cdots \pi_{i_n}) \int_\Delta f(\underline{x}) \; Q\begin{pmatrix} i_1, \ldots, i_n \\ x_1, \ldots, x_n \end{pmatrix} d\psi^{(n)}$$

where the last identity ensues because of the stipulation that f is skew symmetric and Δ is the simplex $0 \le x_1 < x_2 < \ldots < x_n$.

Let τ be a permutation of the non-negative integers into itself. We confirm on the basis of (40) that

(41)
$$a_{\tau\underline{i}} = (-1)^{|\tau|} a_{\underline{i}} \; ,$$

where $|\tau| = 1(-1)$ if τ involves an even(odd) number of transpositions.

Combining the terms of (39) with the help of (41) we find that

$$f(\underline{x}) \sim \sum_{\underline{i}} |a_{\underline{i}}| \; Q\begin{pmatrix} i_1, \ldots, i_n \\ x_1, \ldots, x_n \end{pmatrix}$$

and thereby the determinantal polynomial system is complete.

Nodal curves of $Q\begin{pmatrix} n, & n+1 \\ x, & y \end{pmatrix}$

It can be easily proved that the nodal curves of $Q\begin{pmatrix} n, & n+1 \\ x, & y \end{pmatrix}$ divides the sector $0 < x < y$ into $n+1$ regions. In general, we can say that the zero set of $Q\begin{pmatrix} i, & n+1 \\ x, & y \end{pmatrix}$ involves at most $n+1$ curves but the precise number is unclear.

547

Continuous analogues

The function $\phi(x, \lambda) = \cos \lambda^{\frac{1}{2}}x$ satisfies

$$\frac{d^2}{dx^2} \phi = -\lambda \phi, \qquad 0 < x < \infty ,$$

(42)

$$\phi(0, \lambda) = 1, \quad \phi_x(0, \lambda) = 0 .$$

For fixed $x \geq 0$, ϕ is an entire function of λ and these entire functions are analogous to the polynomials. We form the determinants

(43)
$$\phi \begin{pmatrix} x_1, x_2, \ldots, x_n \\ \lambda_1, \lambda_2, \ldots, \lambda_n \end{pmatrix} = \det\|\phi(x_i, \lambda_j)\|$$

where $0 \leq x_1 < x_2 < \ldots < x_n$. For each of the above described properties of the determinants (1) there is a similar property of the determinants (43). For example in place of the system of n recurrence relations we have a system of n partial differential equations

(44)
$$(-1)^k E_k(\lambda_1, \ldots, \lambda_n)u = E_k\left(\frac{\partial^2}{\partial x_1^2}, \ldots, \frac{\partial^2}{\partial x_n^2}\right) u \qquad k = 1, 2, \ldots, n ,$$

where u is the determinant (43). The following uniqueness assertion holds. If $u = u(x_1, \ldots, x_n; \lambda_1, \ldots, \lambda_n)$ has continuous partial derivatives up to order $2n$ and satisfies all n equations (44) on the closed simplex $0 \leq x_1 < x_2 < \ldots < \infty$, and the boundary conditions

$$u_{x_1}(0, x_2, x_3, \ldots, x_n; \lambda_1, \ldots, \lambda_n) = 0 ,$$

$$u(x_1, x_2, \ldots, x_n; \lambda_1, \ldots, \lambda_n) = 0 \text{ if } x_i = x_{i+1}, \quad i = 1, \ldots, n-1 ,$$

then

548

$$(45) \qquad u = f(\lambda_1, \ldots, \lambda_n) \ \phi \begin{pmatrix} x_1, \ldots, x_n \\ \lambda_1, \ldots, \lambda_n \end{pmatrix} .$$

As a concrete illustration of this result let $w(x, y, z)$ on $0 \le x \le y \le z$ solve

$$w_{xx} + w_{yy} + w_{zz} = -(\lambda + \mu + \nu)w$$

$$w_{xxyy} + w_{yyzz} + w_{zzxx} = (\lambda\mu + \mu\nu + \lambda\nu)w$$

$$w_{xxyyzz} = -\lambda\mu\nu \ w$$

subject to the boundary conditions

$$w(x, x, z) = 0, \quad w(x, y, y) = 0, \quad w_x(0, y, z) = 0$$

then w has the form (45) where the component functions solve (42).

Similar results to (45) may be obtained when $\phi(x, \lambda)$ is replaced by the system of solutions of more general Sturm-Liouville problems.

REFERENCES

1. R. V. Atkinson, Multivariate Spectral Theory, MRC, Technical Report 431, (1964).

2. S. Karlin, Total Positivity, Stanford Univ. Press (1968).

3. S. Karlin and J. McGregor, Coincidence probabilities, Pacific J. Math. 9 (1959), 1141-1164.

4. S. Karlin and J. McGregor, Coincidence properties of birth and death processes, Pacific J. Math. 9 (1959), 1109-1140.

5. S. Karlin and J. McGregor, A characterization of birth and death processes, Proc. Nat. Acad. Sci. 45 (1959), 375-379.

6. S. Karlin and J. McGregor, Determinants of Orthogonal Polynomials, 68 (1962), 204-209.

Samuel Karlin, Department of Mathematics, The Weizmann
Institute of Science, Rehovot, Israel, and Stanford University,
Stanford, California

James McGregor, Department of Mathematics, Stanford
University, Stanford, California

This work was supported in part by N. S. F. grant No. MPS71-
02905 A03.

Index

551

A 5
B 6
C 7
D 8
E 9
F 0
G 1
H 2
I 3
J 4